148
Advances in Biochemical Engineering/Biotechnology

Aims and Scope

This book series reviews current trends in modern biotechnology and biochemical engineering. Its aim is to cover all aspects of these interdisciplinary disciplines, where knowledge, methods and expertise are required from chemistry, biochemistry, microbiology, molecular biology, chemical engineering and computer science.

Volumes are organized topically and provide a comprehensive discussion of developments in the field over the past 3–5 years. The series also discusses new discoveries and applications. Special volumes are dedicated to selected topics which focus on new biotechnological products and new processes for their synthesis and purification.

In general, volumes are edited by well-known guest editors. The series editor and publisher will, however, always be pleased to receive suggestions and supplementary information. Manuscripts are accepted in English.

In references, Advances in Biochemical Engineering/Biotechnology is abbreviated as *Adv. Biochem. Engin./Biotechnol.* and cited as a journal.

More information about this series at http://www.springer.com/series/10

Jens Schrader · Jörg Bohlmann
Editors

Biotechnology of Isoprenoids

With contributions by

Jorge Alonso-Gutierrez · Stephen Bell · Ralf G. Berger
Rita Bernhardt · Jörg Bohlmann · Albert Boronat
Joe Chappell · Yun Chen · Kevin W. George
Björn Hamberger · Allison Maree Heskes · Simon Janocha
Zuodong Jiang · Jay D. Keasling · Chase Kempinski
Bernd Markus Lange · Taek Soon Lee
Robin-Hagen Leonhardt · Yi Li · Marco Antonio Mirata
Jens Nielsen · Irini Pateraki · Blaine A. Pfeifer
Manuel Rodríguez-Concepción · Suriana Sabri
Gerhard Sandmann · Hendrik Schewe
Claudia Schmidt-Dannert · Daniela Schmitz · Jens Schrader
Verena Siewers · Dorothea Tholl · Claudia E. Vickers
Matthias Wüst · Philipp Zerbe · Guojian Zhang
Yongjin J. Zhou

Springer

Editors
Jens Schrader
Bioverfahrenstechnik
DECHEMA-Forschungsinstitut
Frankfurt am Main
Germany

Jörg Bohlmann
Michael Smith Laboratories
University of British Columbia
Vancouver, BC
Canada

ISSN 0724-6145 ISSN 1616-8542 (electronic)
Advances in Biochemical Engineering/Biotechnology
ISBN 978-3-319-20106-1 ISBN 978-3-319-20107-8 (eBook)
DOI 10.1007/978-3-319-20107-8

Library of Congress Control Number: 2015940727

Springer Cham Heidelberg New York Dordrecht London

Printed on acid-free paper

Springer International Publishing AG Switzerland is part of Springer Science+Business Media (www.springer.com)

Preface

Isoprenoids, also known as terpenoids, represent the largest class of natural products, comprising more than 40,000 different structures found in all kingdoms of life. Plants, in particular, are well known to harbor an impressive diversity of isoprenoids with a wide range of different ecological, physiological and structural functions, including light-harvesting pigments, hormones, phytoalexins and semiochemicals, among others. The functions of many complex plant isoprenoids, often referred to as secondary or specialized metabolites, in the natural world remain to be discovered. Likewise, the microbial realm is another rich source for isoprenoid molecules. Since ancient times, isoprenoids have been omnipresent in people's every day lifes and utilized for their many different properties, for example as medicines, flavors, and fragrances. Isoprenoids define many of the flavor impressions of foods and beverages and may help preserve them during storage. They are key fragrance compounds in perfumes, and beyond their fragrant properties the biological activities of terpenoid-rich essential oils have been used in many traditional medicines. Modern industrial exploitation of isoprenoids range from some high-value pharmaceuticals such as the anticancer drug paclitaxel (Taxol ®) to common components of personal hygiene and cosmetic products and inexpensive antimicrobial and organic solvent-like ingredients of household cleaners to commodity materials such as natural rubber. More recently, properties of terpenoids as fuels have been rediscovered and are being explored for the development of advanced biofuels.

Modern biotechnological approaches, including genomics, systems metabolic engineering and synthetic biology, along with biochemistry and chemistry, are bringing along a deeper understanding of isoprenoid biosynthesis in nature and are guiding strategies to harness this knowledge for sustainable production of isoprenoids based on renewable resources.

With the compilation of chapters published in "Biotechnology of Isoprenoids" we set out to cover a rather broad spectrum of research in this field to illustrate both the fascinating diversity and the great industrial potential of these natural products. Given the diversity and manifold applications of isoprenoids, this volume could only cover selected topics while inevitably missing others. Nevertheless, we hope this volume will serve as an introduction for those readers who are new to the field of isoprenoids and may also give the expert reader an up-to-date overview of recent advances in the field. As the editors, we are grateful to so many of our highly esteemed colleagues who accepted our invitation and generously contributed to their time and expertise to this compilation of sixteen chapters.

The first part of the book focuses on some of the fundamentals of isoprenoid biosynthesis in bacteria, fungi and plants and also offers a deeper insight in one of the key enzyme classes involved in isoprenoid biosynthesis, plant cytochrome P450 monooxygenases. Other important classes of enzymes such as prenyl transferases and terpenoid synthases are covered with different subchapters in several chapters across the book. The second part addresses fundamental techniques for metabolic engineering of microbes, algae and higher plants and also covers the potential of microbial P450s for isoprenoid hydroxylation. It also includes bioprocess engineering of microbial isoprenoid production and analytical methods for volatile isoprenoids. The third part covers a set of examples for the fascinating arena of isoprenoid biotechnology with emphasis on industrially important products. With ascending number of isoprenoid C atoms in mind, the seven chapters of this part are each devoted to a particular isoprenoid product or product class: the hemiterpene isoprene, p-menthane monoterpenoids, sesquiterpenoids such as artemisinic acid, farnesene and nootkatone, diterpenoids such as paclitaxel and ambroxide-related compounds, and carotenoids. The reader will also find a variety of other highly interesting isoprenoid products in the course of the chapters in Parts I and II.

We hope this book will provide all readers with new inspiration for their learning, teaching and research in the field of biotechnology of isoprenoids.

May 2015

Jens Schrader
Jörg Bohlmann

Contents

Part I
Biosynthesis and Function of Isoprenoids

Adv Biochem Eng Biotechnol (2015) 148: 3–18
DOI 10.1007/10_2014_285

Published Online: 19 December 2014

Terpenoid Biosynthesis in Prokaryotes

Albert Boronat and Manuel Rodríguez-Concepción

Abstract Prokaryotic organisms (archaea and eubacteria) are found in all habitats where life exists on our planet. This would not be possible without the astounding biochemical plasticity developed by such organisms. Part of the metabolic diversity of prokaryotes was transferred to eukaryotic cells when endosymbiotic prokaryotes became mitochondria and plastids but also in a large number of horizontal gene transfer episodes. A group of metabolites produced by all free-living organisms is terpenoids (also known as isoprenoids). In prokaryotes, terpenoids play an indispensable role in cell-wall and membrane biosynthesis (bactoprenol, hopanoids), electron transport (ubiquinone, menaquinone), or conversion of light into chemical energy (chlorophylls, bacteriochlorophylls, rhodopsins, carotenoids), among other processes. But despite their remarkable structural and functional diversity, they all derive from the same metabolic precursors. Here we describe the metabolic pathways producing these universal terpenoid units and provide a complete picture of the main terpenoid compounds found in prokaryotic organisms.

Keywords Terpenoid · MEP pathway · MVA pathway · Bacteriochlorophylls · Carotenoids · Rhodopsins · Isopentenyl tRNA · Heme · Ubiquinone · Menaquinone · Bactoprenol · Hopanoids · Ether-type lipids

---

A. Boronat · M. Rodríguez-Concepción (✉)
Centre for Research in Agricultural Genomics (CRAG) CSIC-IRTA-UAB-UB,
Campus UAB Bellaterra, 08193 Barcelona, Spain
e-mail: manuel.rodriguez@cragenomica.es

A. Boronat
Department of Biochemistry and Molecular Biology, University of Barcelona,
08028 Barcelona, Spain

Contents

1 Pathways Producing Universal Terpenoid Precursors in Prokaryotic Organisms

Despite the remarkable diversity of structures and functions of terpenoids, they all derive from the basic five-carbon (C5) precursor units isopentenyl diphosphate (IPP) and its double-bond isomer dimethylallyl diphosphate (DMAPP). Two major pathways produce these universal precursors of all terpenoids: the mevalonate (MVA) pathway and the 2-*C*-methyl-D-erythritol 4-phosphate (MEP) pathway. For many years it was accepted that the MVA pathway was the only one present in all organisms, including prokaryotes. In the last decade of the last century, however, a second pathway (currently known as the MEP pathway) was identified in bacteria for the biosynthesis of IPP and DMAPP [48]. Strikingly, later work has shown that the plasticity displayed by prokaryotic organisms to produce their terpenoid precursors is much higher than originally expected.

1.1 MVA Pathway

The MVA pathway (Fig. 1) is the only one producing terpenoid precursors in animals (including humans) and fungi. In plant cells, it produces cytosolic IPP, whereas plastidial terpenoid precursors are made by the MEP pathway. The MVA pathway in these eukaryotic organisms is well established. In the first steps of the pathway, 3-hydroxy-3-methylglutaryl-CoA (HMG-CoA) is produced from the sequential condensation of three molecules of acetyl-CoA catalyzed by the enzymes acetoacetyl-CoA thiolase (AACT) and HMG-CoA synthase (HMGS). In the first committed step of the pathway, HMG-CoA is then converted to MVA in a functionally irreversible reaction catalyzed by HMG-CoA reductase (HMGR).

Then, MVA is sequentially phosphorylated to 5-phosphomevalonate and 5-diphosphomevalonate and decarboxylated to generate IPP by the enzymes mevalonate kinase (MVK), 5-phosphomevalonate kinase (PMVK), and 5-diphosphomevalonate decarboxylase (DPMD). The activity of an IPP/DMAPP isomerase enzyme (IDI) is required to form DMAPP from IPP (Fig. 1).

In prokaryotic cells, the MVA pathway is found in archaea and a few eubacterial strains, including the spirochete *Borrelia burgdorferi* and the gram-positive cocci *Staphylococcus aureus* and *Streptococcus pneumoniae*. Among the eubacterial

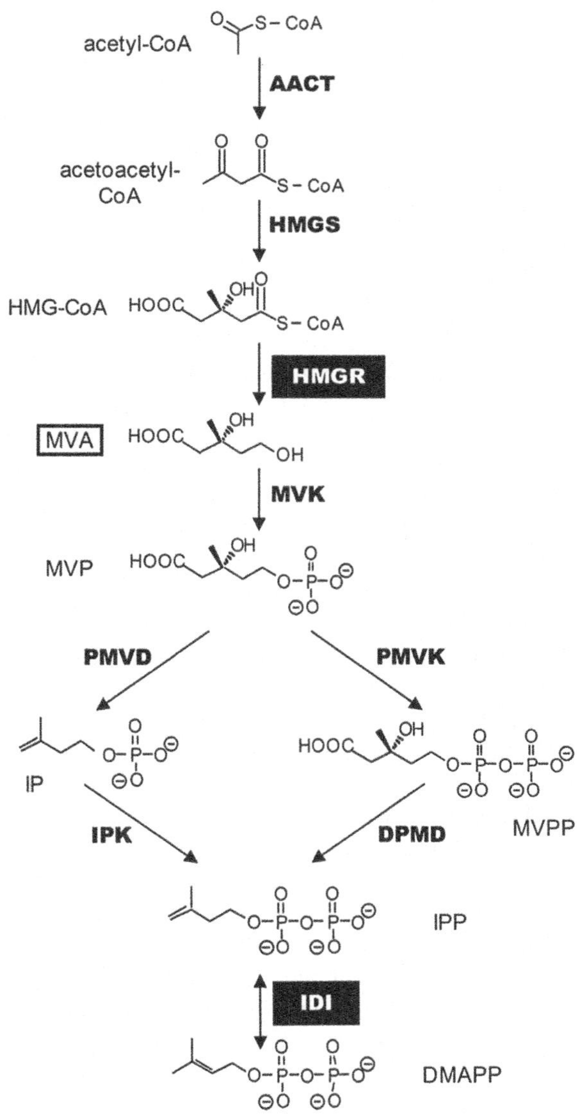

Fig. 1 The MVA pathway. Steps leading to the formation of IPP and DMAPP from acetyl CoA are represented. *HMG-CoA* 3-hydroxy-3-methylglutaryl-CoA; *MVA* mevalonic acid; *MVP* 5-phosphomevalonate; *MVPP* 5-diphosphomevalonate; *IP* isopentenyl phosphate. Enzyme acronyms (in *bold*) are described in the text. Boxed enzymes correspond to those with more than one type

species containing the MVA pathway some of them (including *Listeria monocytogenes* and some *Streptomyces* strains) also have the MEP pathway. Although all *Streptomyces* use the MEP pathway for the synthesis of essential terpenoids, some species can additionally use the MVA pathway for the production of secondary metabolites such as antibiotics [29].

The archaeal genomes sequenced to date have exclusively revealed the presence of genes encoding MVA pathway enzymes [4, 33, 38, 57]. However, the genomic analyses failed to identify the full complement of eukaryotic-like MVA pathway genes. In particular, the genes encoding PMVK and DPMD (Fig. 1) are absent in the genome of almost all archaea. Exceptions include halophilic archaea and *Thermoplasma* relatives (that possess a DPMD-like gene) and some thermoacidophilic archaea such as those in the genera *Sulfolobus*, *Acidianus*, and *Metallosphaera* (that have both PMVK and DPMD orthologues). Recent experiments have demonstrated that the conversion of 5-phosphomevalonate to IPP in most archaea is achieved by an alternate route involving the formation of isopentenyl phosphate (IP) from 5-phosphomevalonate by 5-phosphomevalonate decarboxylase (PMVD) and further conversion to IPP by IP kinase (IPK; Fig. 1). IPK was identified and characterized first [10, 21, 35], whereas PMVD-encoding sequences with homology to DPMD enzymes were isolated more recently [14, 61]. The presence of IPK homologues in prokaryotic and even eukaryotic cells that also show the so-called classical MVA pathway [14, 33, 38, 61] suggests that some organisms might have two functional branches of the MVA pathway (i.e., classical PMVK + DPMD—and alternative PMVD + IPK). It is possible that both branches might be functional, and their presence might represent an adaptive advantage.

Additional levels of plasticity in the prokaryotic MVA pathway have been described for two key enzymatic steps: the production of MVA by HMGR, and the isomerization of IPP and DMAPP by IDI. Two different classes of HMGR can be distinguished based on sequence alignments and phylogenetic analyses [2]. HMGR class I enzymes (HMGR-I) are found predominantly in archaea and eukaryotes, whereas HMGR class II enzymes (HMGR-II) are found in the few eubacteria that use the MVA pathway. Differences between these two classes, proposed to have arisen by divergent evolution from a common ancestor, are most apparent around the active site and the preceding region. As a consequence, HMGR types show a differential sensitivity to statins, a family of HMGR inhibitors originally isolated from *Penicillium* and *Aspergillus* fungi that are widely used to lower cholesterol levels in humans [1]. Statins are excellent inhibitors of the HMGR-I type but relatively poor inhibitors of HMGR-II enzymes, which might explain the progressive substitution of the original HMGR-I enzymes by HMGR-II proteins in eubacteria [3, 5]. Drugs targeting only HMGR-II could be potentially useful antibiotics against eubacterial pathogens harboring HMGR-I enzyme isoforms. However, no good inhibitors against HMGR-II have been reported yet.

Regarding IDI, two types of structurally unrelated enzymes with no sequence similarity and different reaction mechanisms and cofactor requirements have been reported [24, 50, 54, 59]. Type I IDI enzymes (IDI-I) are extensively studied metalloproteins found in eukaryotic cells and many eubacterial organisms [15, 64]. By contrast, the enzymatic mechanism of class II IDI enzymes (IDI-II), which require reduced flavin mononucleotide as a cofactor, is less known. IDI-II enzymes were discovered in *Streptomyces* and shown to be present in archaea and some bacteria but not in eukaryotic organisms [24, 29, 31]. Interestingly, there are bacteria that possess either IDI-I or IDI-II enzymes, others have both, and a large proportion of bacteria containing the MEP pathway do not contain IDI proteins. The latter is not surprising because IDI activity is essential to produce DMAPP in organisms that only contain the MVA pathway but it is not required in those harboring the MEP pathway, which simultaneously produces both IPP and DMAPP (Fig. 2).

1.2 MEP Pathway

The MEP pathway (Fig. 2) has been best characterized in *Escherichia coli*, a model bacterium that, like most eubacteria, lacks the MVA pathway [47, 49]. The first reaction of the MEP pathway is the condensation of (hydroxyethyl)thiamine derived from pyruvate with the C1 aldehyde group of D-glyceraldehyde 3-phosphate to produce 1-deoxy-D-xylulose 5-phosphate (DXP). This reaction is catalyzed by the enzyme DXP synthase (DXS). DXP can be used as a precursor for the production of cofactors such as pyridoxal phosphate (vitamin B_6) and thiamine (vitamin B_1) diphosphate [32, 41]. When used for the production of terpenoids, an intramolecular rearrangement and reduction of DXP produces MEP. The reaction, considered as the first truly committed step of the MEP pathway, is catalyzed by the enzyme DXP reductoisomerase (DXR). Then, the enzymes MEP cytidylyltransferase (MCT), 4-(cytidine 5′-diphospho)-2-*C*-methyl-D-erythritol kinase (CMK), and 2-*C*-methyl-D-erythritol 2,4-cyclodiphosphate (MEcPP) synthase (MDS) transform MEP into MEcPP, a compound recently found to act as a retrograde signal in plant cells [65]. An opening reduction of the MEcPP ring catalyzed by the enzyme 4-hydroxy-3-methylbut-2-enyl diphosphate (HMBPP) synthase (HDS) forms the next intermediate of the pathway, HMBPP. It has been demonstrated that in addition to being an intermediate in the MEP pathway, HMBPP is a powerful phosphoantigen that activates a specific group of human T cells [16]. In the last step of the MEP pathway, the enzyme HMBPP reductase (HDR) catalyzes the simultaneous formation of IPP and DMAPP in an approximate 5–1 proportion (Fig. 2).

The MEP pathway is absent from archaea, fungi, and animals (with the exception of apicomplexa protozoa), but it is found in most eubacteria and in plant plastids. However, reactions and enzymes that are alternative to those of the so-called canonical MEP pathway have been reported in some groups of organisms. For example, a route alternative to the MEP pathway appears to operate in *Synechocystis* [18, 46]. Although the genome of this cyanobacterium contains homologues of all MEP pathway genes [25], feeding with known pathway substrates or a pharmacological block of the pathway in cells grown photoautotrophically had no

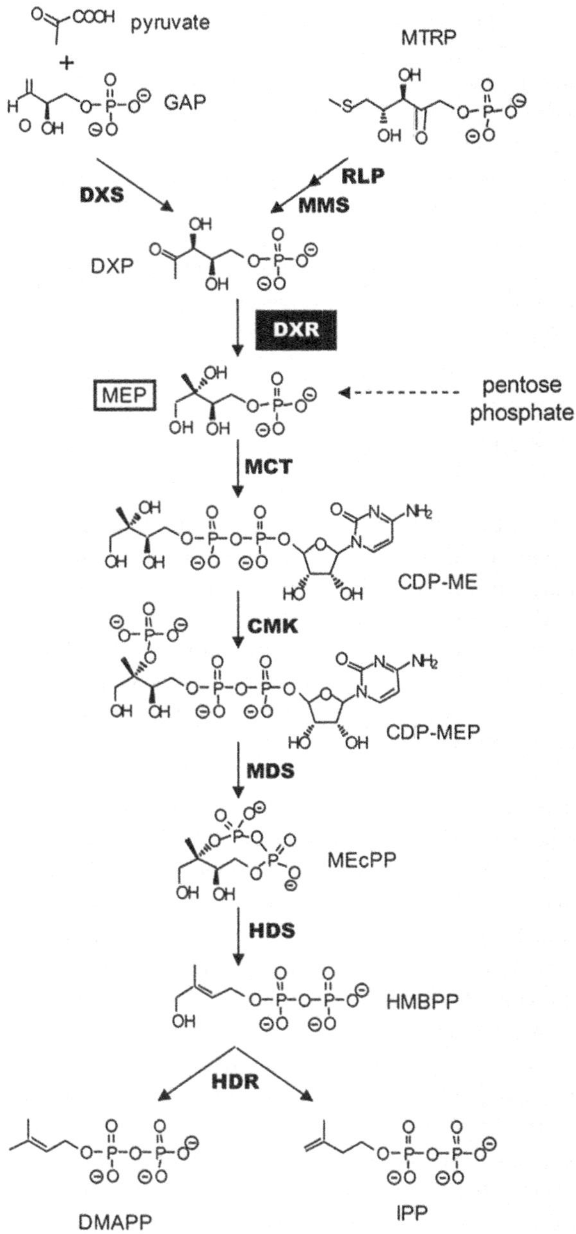

Fig. 2 The MEP pathway. Steps leading to the formation of IPP and DMAPP from pyruvate and D-glyceraldehyde 3-phosphate are shown. *MTRP* 5-methylthio-D-ribulose 1-phosphate; *GAP* D-glyceraldehyde 3-phosphate; *DXP* 1-deoxy-D-xylulose 5-phosphate; *MEP* 2-*C*-methyl-D-erythritol 4-phosphate; *CDP-ME* 4-(cytidine 5′-diphospho)-2-*C*-methyl-D-erythritol; *CDP-MEP* CDP-ME 2-phosphate; *MEcPP* 2-*C*-methyl-D-erythritol 2,4-cyclodiphosphate; *HMBPP* 4-hydroxy-3-methylbut-2-enyl diphosphate. Enzyme acronyms (in **bold**) are described in the text and correspond to those suggested by Phillips et al. [45]. Boxed enzymes correspond to those with more than one type

effect on terpenoid biosynthesis. By contrast, phosphorylated sugars of the pentose phosphate cycle stimulated the incorporation of labeled IPP into terpenoids in a cell-free system [18]. Together, the results suggested that photosynthesis-derived products of the pentose phosphate cycle could serve as substrates for IPP and

DMAPP synthesis in *Synechocystis*, probably entering the pathway downstream of MEP (Fig. 2).

Recent results have established the existence of a second alternative metabolic shunt in another photosynthetic bacterium, *Rhodospirillum rubrum* [17, 63]. In this case, DXP was found to be produced from 5-methylthio-D-ribulose 1-phosphate (MTRP), an intermediate of the methionine salvage pathway [17]. In this metabolic shunt to produce DXP, an atypical RubisCO-like protein (RLP) isomerizes MTRP to 1-methylthio-D-xylulose 5-phosphate (MTXP) and then the enzyme MTXP methylsulfurylase (MMS) converts MTXP into DXP (Fig. 2; [17, 63]). Because genes encoding RLP and MMS (as well as other enzymes required to produce MTRP) are found in several groups of bacteria, it is likely that this metabolic shunt can be functional in a number of other bacteria. Interestingly, these bacteria also have genes encoding DXS. It is therefore possible that bacteria harboring the canonical MEP pathway and alternative shunts such as those described in *Synechocystis* and *Rhodospirillum* potentially use the two pathways to synthesize their terpenoids, similar to that proposed above for archaea displaying the classical and alternative MVA pathways.

As described for the MVA pathway, different enzyme types can catalyze the same reaction on the MEP pathway in different bacteria (Fig. 2). In particular, two completely different enzymes have been found to catalyze the production of MEP from DXP in the first committed step of the MEP pathway [42, 52]. The vast majority of bacteria have a canonical (type I) DXR enzyme to catalyze this reaction (DXR-I). However, the absence of sequences encoding DXR-I from the genomes of some bacteria displaying the rest of the MEP pathway genes led to the discovery of a different (type II) DXR-like enzyme (DRL or DXR-II) catalyzing the same biochemical reaction [52]. DXR-II belongs to a family of previously uncharacterized proteins with oxidoreductase sequence features that only show some sequence similarity to DXR-I at the level of the NADPH-binding domain [9, 52]. Phylogenetic analyses recently supported a single origin of the DXR-II family through functional divergence, which constitutes an exceptional model of acquisition and maintenance of redundant gene functions between nonhomologous genes as a result of convergent evolution [9]. Although episodic events of horizontal gene transfer could not be excluded, the spotty distribution of DXR-II in specific pathogenic eubacteria (including animal and human pathogens such as *Brucella* and *Bartonella*) is best explained by a prevalent role of gene loss. As a consequence, there are bacteria with DXR-I but not DXR-II (the majority), with DXR-II but not DXR-I, and with both DXR-I and DXR-II enzymes. These observations highlight again the astonishing plasticity of bacterial terpenoid biosynthesis and open new avenues for research. For example, DXR-I and DXR-II enzymes show a different arrangement of their active sites [42], which opens the door to the design of highly specific antibiotics against only one of these two types of DXR enzymes. In particular, the design of antibiotics that selectively target pathogens using DXR-II without affecting beneficial or innocuous bacteria harboring DXR-I enzymes would be most useful.

Although the MEP pathway can be a good target for the design of new antibiotics to fight bacterial pathogens, attention must be paid to the remarkable diversity observed in bacteria for terpenoid biosynthesis. To date, the only drug targeting the MEP pathway that is being tested in clinical trials is fosmidomycin, an inhibitor of DXR [30]. However, a number of mechanisms allow bacteria to bypass a pharmacological blockage of terpenoid biosynthesis with this inhibitor, including a reduced uptake or an activated efflux of the drug [7, 20, 51, 52]. Antibiotic resistance can also result from the use of alternative pathways or enzymes not affected by the inhibitor. For example, spontaneous mutations in *E. coli* have been found to rescue the loss of genes encoding DXS or DXR [43, 53], whereas shunt pathways or alternative enzymes (such as DXR-II) have been found to be active in different organisms. To facilitate an effective drug design, genetic, biochemical, and crystallographic approaches should identify the most appropriate pathway enzymes to inhibit and the residues that play a relevant structural or catalytic role. On the other hand, the rich collection of genes and proteins developed by bacteria to synthesize their terpenoids represents a source of potential biotechnological tools.

2 Terpenoids Present in Prokaryotic Organisms and Their Biological Role

Addition of IPP units to a DMAPP core molecule results in the production of prenyl diphosphates of increasing length, including C15 farnesyl diphosphate (FPP) and C20 geranylgeranyl diphosphate (GGPP). Prenyl diphosphates such as IPP, FPP, and GGPP are the starting points of downstream pathways leading to the synthesis of the different terpenoid end-products (Fig. 3). Many of these products have a key role in biological processes that are essential for the growth and survival of prokaryotic organisms. They include, among others, cell-wall biosynthesis, membrane function, electron transport, and conversion of light into chemical energy. Some prokaryotes also produce secondary metabolites of terpenoid origin, some of them of biotechnological relevance. Here we describe the main terpenoids found in prokaryotic organisms, with a particular emphasis on those present in the model bacterium *E. coli*.

2.1 Isopentenyl tRNA

The adenosine at position 37 (adjacent to the 3′ position of the anticodon) of most tRNAs for codons starting with uracil is usually modified to *N*(6)-(Δ2)-isopentenyl adenosine (Fig. 3). Isopentenyl adenosine increases the efficiency of translation of the modified tRNAs and makes them less sensitive to codon context. The transfer of the isopentenyl group from IPP to the tRNAs is catalyzed by the enzyme tRNA

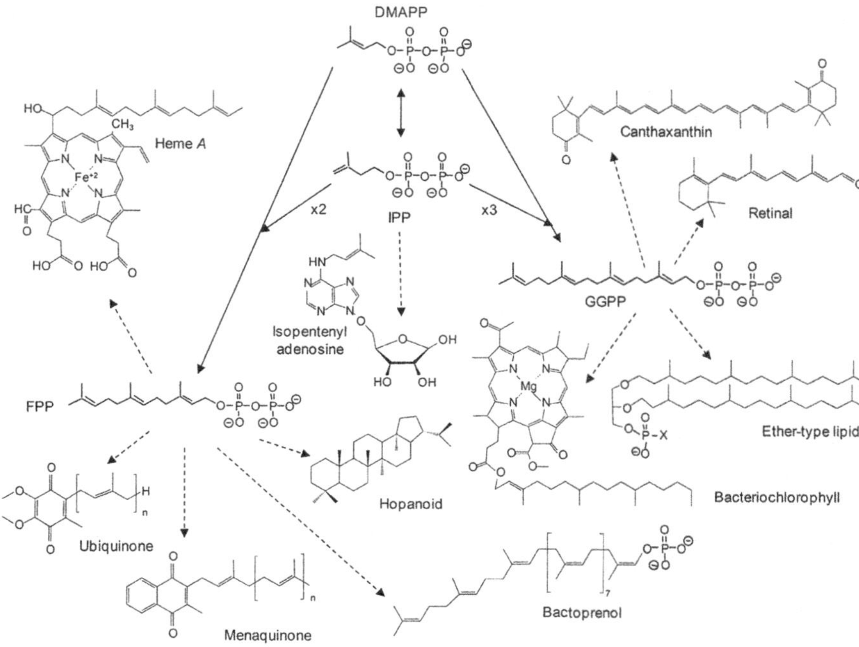

Fig. 3 Terpenoid metabolites found in prokaryotic organisms. Examples of terpenoids derived from IPP, FPP, and GGPP are shown. *Dashed arrows* indicate multiple steps

isopentenyltransferase, also known as tRNA Δ2-IPP transferase. This enzyme is encoded by the *miaA* gene in *E. coli*. Homologues of the *miaA* gene have been detected in other microorganisms [44].

2.2 Heme

Some proteins involved in electron transport chains contain heme as a prosthetic group. Heme groups are composed of a heterocyclic porphyrin ring and a metal ion (usually iron) in a central position. Heme A, which is found in cytochrome *c* and cytochrome *c* oxidase, is characterized by containing a hydroxyethylfarnesyl side-chain at position 8 of the tetrapyrrole ring (Fig. 3). *E. coli* has no cytochrome *c* and no equivalent to cytochrome *c* oxidase. Instead, *E. coli* contains cytochromes *bo* and *d*, able to oxidize ubiquinol and directly reduce molecular oxygen to water. Cytochrome *bo* contains heme O, which only differs from heme A by having a methyl group instead of a formyl group [40]. Heme O is formed from heme B (protoheme IX) by the transfer of the farnesyl group from FPP catalyzed by a farnesyltransferase, which is encoded by the *cyoE* gene in *E. coli* [40]. Heme A is derived from heme O by the action of heme A synthase, an enzyme activity absent in *E. coli* [23].

2.3 Ubiquinone and Menaquinone

Ubiquinone and menaquinone are essential components of the respiratory electron transport chain. Both compounds contain a quinone moiety linked to a terpenoid side-chain (Fig. 3). The hydrophobic terpenoid side-chain anchors these compounds to cell membranes whereas the quinone head group enables interaction with proteins. Ubiquinone (also known as coenzyme Q) has a benzoquinone group linked to a terpenoid chain of different length (6–10 isoprene units) depending on the organism. In *E. coli* the terpenoid side-chain of ubiquinone contains 8 isoprene units. Menaquinone has a naphthoquinone ring linked to a terpenoid chain that also contains 8 isoprene units in *E. coli*. Facultative anaerobes, including *E. coli*, use ubiquinone when growing under aerobic conditions and menaquinone under anaerobic conditions. By contrast, many gram-positive aerobes contain only menaquinone. The head group precursor and the terpenoid side-chain are synthesized separately. In *E. coli*, the terpenoid side-chain of ubiquinone and menaquinone is synthesized by the enzyme octaprenyl diphosphate synthase, which catalyzes the condensation of FPP with five molecules of IPP to generate octaprenyl diphosphate (C40). Octaprenyl diphosphate synthase is encoded by the essential gene *ispB* in *E. coli* [27, 39].

2.4 Bactoprenol (Undecaprenyl Phosphate)

The bacterial cell wall plays an essential role in maintaining cell shape and preventing the deleterious effect derived from the high internal osmotic pressure. It is a rigid structure that surrounds the cytoplasmic membrane and serves as a scaffold for anchoring other components such as proteins and polysaccharides (i.e., teichoic acids). Peptidoglycan is the major component of bacterial cell walls and consists of long glycan chains made up of alternating residues of *N*-acetylglucosamine (GlcNAc) and *N*-acetylmuramic acid (MurNAc) crosslinked by short chains of amino acids and amino acid derivatives [62]. Peptidoglycan biosynthesis is a complex process that takes place in the cytoplasm and the inner and outer sides of the cytoplasmic membrane. A key component in the synthesis of peptidoglycan is undecaprenyl phosphate (UP), a compound commonly referred to as bactoprenol (Fig. 3). UP is involved in the synthesis and transport of hydrophilic GlcNAc-MurNAc-peptide monomeric units outside the cytoplasmic membrane, the site for peptidoglycan polymerization [34]. UP derives from undecaprenyl diphosphate, a prenyl diphosphate synthesized from FPP. The enzymes involved in UP biosynthesis are undecaprenyl diphosphate synthase, which adds eight molecules of IPP (in *cis* form) onto FPP, and undecaprenyl diphosphate phosphatase, which removes a phosphate group. In *E. coli*, undecaprenyl diphosphate synthase and undecaprenyl diphosphate phosphatase are encoded by the genes *ispU* and *bacA*, respectively. In contrast to *E. coli*, some gram-positive bacteria contain significant amounts of free undecaprenol, which can be converted to UP by the action of the enzyme undecaprenol kinase [60]. The biosynthesis of UP and its role in peptidoglycan biosynthesis have recently been reviewed elsewhere [6].

2.5 Hopanoids

Sterols are known to play a key role in the regulation of membrane fluidity and permeability in eukaryotic cells. Because most prokaryotes lack sterols, the search for molecules playing an equivalent role in prokaryotes has been an ongoing issue during the last decades. It is now well established that many bacteria contain hopanoids, pentacyclic triterpenoids with a structure similar to that of sterols (Fig. 3). Hopanoids have been detected in a wide range of gram-negative and gram-positive bacteria. Although hopanoids predominantly occur in aerobic bacteria they have also been found in some facultative anaerobic bacteria [26]. Hopanoids have not been detected in archaea, which produce a particular membrane lipid of terpenoid nature (see below). The hopanoid skeleton is formed from squalene by the action of squalene–hopane cyclases. Structural and functional aspects related to squalene–hopane cyclases have recently been reviewed elsewhere [56].

2.6 Ether-Type Membrane Lipids

Archaea are often found inhabiting extreme environments. They are similar to eubacteria in being prokaryotes (and thus lacking a nucleus) but differ from them in ribosomal structure, the presence of introns (in some species), and in membrane structure and composition. One of the most remarkable features of archaea is the presence in their membranes of ether-type lipids containing hydrocarbon chains of terpenoid nature (usually C20 and C25 prenyl groups) linked to *sn*-glycerol 1-phosphate (Fig. 3). A unique type of lipid found in archaeal membranes is diphytanyldiglycerol tetraether, which consists of two hydrophilic head groups linked by C40 terpenoid chains. The particular lipid composition of archaeal membranes has been related to their ability to survive in extreme environments. The C20 phytanyl chains derive from GGPP. C25 farnesylgeranyl groups are synthesized by farnesylgeranyl diphosphate synthase either by the consecutive condensation of IPP molecules to DMAPP or to prenyl diphosphates such as GGPP [4, 38]. Archaeal polyprenyl diphosphate synthases can also synthesize products of shorter chain lengths. The biosynthesis of ether-type lipids in archaebacteria has been reviewed elsewhere [28, 38].

2.7 Bacteriochlorophylls, Carotenoids, and Rhodopsins

Bacteriochlorophylls are light-harvesting pigments present in phototropic bacteria. Similar to chlorophylls present in plants and photosynthetic bacteria, bacteriochlorophylls (Fig. 3) also contain a long terpenoid chain (usually phytol) that contributes to their localization in the photosynthetic membranes. It has been proposed that phytol is formed after the addition of a GGPP-derived geranylgeraniol group which is later sequentially saturated by geranylgeranyl reductase. However, some evidence suggests that geranylgeranyl reductase can saturate GGPP prior to the transfer of the phytol tail [11]. Although the phytol group is present in most bacterial chlorophylls, other terpenoids (such as farnesyl, geranylgeranyl, and 2,6-phytadienyl groups) have also been reported [11].

Carotenoids are found in photosynthetic and some nonphotosynthetic bacteria. As in plants, bacterial carotenoids are synthesized from GGPP (Fig. 3). They function primarily as photoprotective pigments but can also participate in the light-harvesting process. Carotenoids also function as natural pigments, some of them of very high commercial value, including β-carotene, astaxanthin, and canthaxanthin (Fig. 3). A number of novel and rare carotenoids isolated from marine bacteria have recently been shown to display a potent antioxidant activity [55]. The immense structural diversity of bacterial carotenoids and the biochemical aspects related to their biosynthesis are reviewed elsewhere [37].

Bacteriorhodopsins and proteorhodopsins are retinal-binding membrane proteins acting as light-dependent ion pumps or photoreceptors in aquatic bacteria [19, 22]. In archaea and most bacteria, the retinal chromophore (Fig. 3) is synthesized by the cleavage of β-carotene by the action of carotenoid cleavage dioxygenases. However, recent reports on the characterization of bacterial apocarotenoid oxygenases have shown that the retinal chromophore can also be produced by cleavage of some apocarotenoids [36, 58]. At present, the substrate(s) used for retinal production in some bacteria remains an open question [37].

2.8 Secondary Metabolites

In contrast with the high number of terpenoid secondary metabolites identified in plants, liverworts, and fungi, only a relatively small number of such terpenes have been isolated from prokaryotic organisms. Prokaryotic secondary terpenoids mainly act as odor constituents, pigments, and antibiotics. Geosmin, a sesquiterpene derivative responsible for the characteristic odor of moist soil, is a typical volatile produced by Actinomycetes and was identified more than 40 years ago. Since then, a variety of other terpenoid volatiles (hydrocarbons and alcohol derivatives) have been reported in *Streptomyces* and other Actinomycetes. As indicated above, many nonphotosynthetic microorganisms can synthesize carotenoids as pigments. Another example of terpenoid secondary metabolites is given by some antibiotics produced by many Actinomycete strains [12].

In recent years, the search of bacterial genome databases has provided new insights into the potential of prokaryotic organisms concerning terpenoid biosynthesis. In this regard, a high number of novel bacterial terpene synthases have currently been identified [8]. This is allowing the identification of novel terpenoid secondary metabolites produced by microorganisms. The identification of gene clusters containing terpenoid biosynthetic genes has also revealed the presence of oxidative enzymes (such as cytochrome P450s) involved in the modification of microbial terpenes [8, 13, 66].

Acknowledgments Work from our groups was mainly supported by grants from the Spanish Dirección General de Investigación (BIO2009-09523, BIO2011-23680, and PIM2010IPO-00660), Generalitat de Catalunya (2009SGR-26), Programa Iberoamericano de Ciencia y Tecnologia para el Desarrollo (IBERCAROT), and European Union FP7 (TiMet, contract 245143).

References

1. Barrios-Gonzalez J, Miranda RU (2010) Biotechnological production and applications of statins. Appl Microbiol Biotechnol 85:869–883
2. Bochar DA, Stauffacher CV, Rodwell VW (1999) Sequence comparisons reveal two classes of 3-hydroxy-3-methylglutaryl coenzyme A reductase. Mol Genet Metab 66:122–127
3. Boucher Y, Doolittle WF (2000) The role of lateral gene transfer in the evolution of isoprenoid biosynthesis pathways. Mol Microbiol 37:703–716
4. Boucher Y, Kamekura M, Doolittle WF (2004) Origins and evolution of isoprenoid lipid biosynthesis in archaea. Mol Microbiol 52:515–527
5. Boucher Y, Huber H, L'Haridon S, Stetter KO, Doolittle WF (2001) Bacterial origin for the isoprenoid biosynthesis enzyme hmg-coa reductase of the archaeal orders thermoplasmatales and archaeoglobales. Mol Biol Evol 18:1378–1388
6. Bouhss A, Trunkfield AE, Bugg TD, Mengin-Lecreulx D (2008) The biosynthesis of peptidoglycan lipid-linked intermediates. FEMS Microbiol Rev 32:208–233
7. Brown AC, Parish T (2008) Dxr is essential in Mycobacterium tuberculosis and fosmidomycin resistance is due to a lack of uptake. BMC Microbiol 8:78
8. Cane DE, Ikeda H (2012) Exploration and mining of the bacterial terpenome. Acc Chem Res 45:463–472
9. Carretero-Paulet L, Lipska A, Perez-Gil J, Sangari FJ, Albert VA, Rodriguez-Concepcion M (2013) Evolutionary diversification and characterization of the eubacterial gene family encoding DXR type II, an alternative isoprenoid biosynthetic enzyme. BMC Evol Biol 13:180
10. Chen M, Poulter CD (2010) Characterization of thermophilic archaeal isopentenyl phosphate kinases. Biochemistry 49:207–217
11. Chew AG, Bryant DA (2007) Chlorophyll biosynthesis in bacteria: the origins of structural and functional diversity. Annu Rev Microbiol 61:113–129
12. Dairi T (2005) Studies on biosynthetic genes and enzymes of isoprenoids produced by actinomycetes. J Antibiot 58:227–243
13. Dairi T (2013) Biosynthetic genes and enzymes of isoprenoids produced by Actinomycetes. In: Bach T, Rohmer M (eds) Isoprenoid synthesis in plants and microorganisms. Springer, New York, pp 29–49
14. Dellas N, Thomas ST, Manning G, Noel JP (2013) Discovery of a metabolic alternative to the classical mevalonate pathway. eLife 2:e00672

15. Durbecq V, Sainz G, Oudjama Y, Clantin B, Bompard-Gilles C, Tricot C, Caillet J, Stalon V, Droogmans L, Villeret V (2001) Crystal structure of isopentenyl diphosphate:dimethylallyl diphosphate isomerase. Embo J 20:1530–1537
16. Eberl M, Jomaa H (2003) A genetic basis for human gammadelta T-cell reactivity towards microbial pathogens. Trends Immunol 24:407–409
17. Erb TJ, Evans BS, Cho K, Warlick BP, Sriram J, Wood BM, Imker HJ, Sweedler JV, Tabita FR, Gerlt JA (2012) A RubisCO-like protein links SAM metabolism with isoprenoid biosynthesis. Nat Chem Biol 8:926–932
18. Ershov YV, Gantt RR, Cunningham FX Jr, Gantt E (2002) Isoprenoid biosynthesis in Synechocystis sp. strain PCC6803 is stimulated by compounds of the pentose phosphate cycle but not by pyruvate or deoxyxylulose-5-phosphate. J Bacteriol 184:5045–5051
19. Fuhrman JA, Schwalbach MS, Stingl U (2008) Proteorhodopsins: an array of physiological roles? Nature reviews. Microbiology 6:488–494
20. Fujisaki S, Ohnuma S, Horiuchi T, Takahashi I, Tsukui S, Nishimura Y, Nishino T, Kitabatake M, Inokuchi H (1996) Cloning of a gene from Escherichia coli that confers resistance to fosmidomycin as a consequence of amplification. Gene 175:83–87
21. Grochowski LL, Xu H, White RH (2006) Methanocaldococcus jannaschii uses a modified mevalonate pathway for biosynthesis of isopentenyl diphosphate. J Bacteriol 188:3192–3198
22. Grote M, O'Malley MA (2011) Enlightening the life sciences: the history of halobacterial and microbial rhodopsin research. FEMS Microbiol Rev 35:1082–1099
23. Hederstedt L (2012) Heme A biosynthesis. Biochimica et biophysica acta 1817:920–927
24. Kaneda K, Kuzuyama T, Takagi M, Hayakawa Y, Seto H (2001) An unusual isopentenyl diphosphate isomerase found in the mevalonate pathway gene cluster from Streptomyces sp. strain CL190. Proc Natl Acad Sci USA 98:932–937
25. Kaneko T, Sato S, Kotani H, Tanaka A, Asamizu E, Nakamura Y, Miyajima N, Hirosawa M, Sugiura M, Sasamoto S, Kimura T, Hosouchi T, Matsuno A, Muraki A, Nakazaki N, Naruo K, Okumura S, Shimpo S, Takeuchi C, Wada T, Watanabe A, Yamada M, Yasuda M, Tabata S (1996) Sequence analysis of the genome of the unicellular cyanobacterium Synechocystis sp. strain PCC6803. II. Sequence determination of the entire genome and assignment of potential protein-coding regions. DNA Res 3:109–136
26. Kannenberg EL, Poralla K (1999) Hopanoid biosynthesis and function in bacteria. Naturwissenschaften 86:168–176
27. Kawamukai M (2002) Biosynthesis, bioproduction and novel roles of ubiquinone. J Biosci Bioeng 94:511–517
28. Koga Y, Morii H (2007) Biosynthesis of ether-type polar lipids in archaea and evolutionary considerations. Microbiol Mol Biol Rev MMBR 71:97–120
29. Kuzuyama T, Seto H (2003) Diversity of the biosynthesis of the isoprene units. Nat Prod Rep 20:171–183
30. Kuzuyama T, Shimizu T, Takahashi S, Seto H (1998) Fosmidomycin, a specific inhibitor of 1-Deoxy-d-Xylulose 5-phosphate reductoisomerase in the nonmevalonate pathway for terpenoid biosynthesis. Tetrahedron Lett 39:7913–7916
31. Laupitz R, Hecht S, Amslinger S, Zepeck F, Kaiser J, Richter G, Schramek N, Steinbacher S, Huber R, Arigoni D, Bacher A, Eisenreich W, Rohdich F (2004) Biochemical characterization of Bacillus subtilis type II isopentenyl diphosphate isomerase, and phylogenetic distribution of isoprenoid biosynthesis pathways. Eur J Biochem 271:2658–2669
32. Leonardi R, Roach PL (2004) Thiamine biosynthesis in *Escherichia coli*: in vitro reconstitution of the thiazole synthase activity. J Biol Chem 279:17054–17062
33. Lombard J, Moreira D (2011) Origins and early evolution of the mevalonate pathway of isoprenoid biosynthesis in the three domains of life. Mol Biol Evol 28:87–99
34. Lovering AL, Safadi SS, Strynadka NC (2012) Structural perspective of peptidoglycan biosynthesis and assembly. Annu Rev Biochem 81:451–478
35. Mabanglo MF, Schubert HL, Chen M, Hill CP, Poulter CD (2010) X-ray structures of isopentenyl phosphate kinase. ACS Chem Biol 5:517–527

36. Maeda I, Inaba A, Koike H, Yoneyama K, Ueda S, Yoshida K (2013) Acyclic carotenoid and cyclic apocarotenoid cleavage by an orthologue of lignostilbene-alpha, beta-dioxygenase in Rhodopseudomonas palustris. J Biochem 154:449–454
37. Maresca JA, Graham JE, Bryant DA (2008) The biochemical basis for structural diversity in the carotenoids of chlorophototrophic bacteria. Photosynth Res 97:121–140
38. Matsumi R, Atomi H, Driessen AJ, van der Oost J (2011) Isoprenoid biosynthesis in Archaea–biochemical and evolutionary implications. Res Microbiol 162:39–52
39. Meganathan R (2001) Ubiquinone biosynthesis in microorganisms. FEMS Microbiol Lett 203:131–139
40. Mogi T, Saiki K, Anraku Y (1994) Biosynthesis and functional role of haem O and haem A. Mol Microbiol 14:391–398
41. Mukherjee T, Hanes J, Tews I, Ealick SE, Begley TP (2011) Pyridoxal phosphate: biosynthesis and catabolism. Biochim Biophys Acta 1814:1585–1596
42. Perez-Gil J, Calisto BM, Behrendt C, Kurz T, Fita I, Rodriguez-Concepcion M (2012) Crystal structure of Brucella abortus deoxyxylulose-5-phosphate reductoisomerase-like (DRL) enzyme involved in isoprenoid biosynthesis. J Biol Chem 287:15803–15809
43. Perez-Gil J, Uros EM, Sauret-Gueto S, Lois LM, Kirby J, Nishimoto M, Baidoo EE, Keasling JD, Boronat A, Rodriguez-Concepcion M (2012) Mutations in Escherichia coli aceE and ribB genes allow survival of strains defective in the first step of the isoprenoid biosynthesis pathway. PloS one 7:e43775
44. Persson BC, Esberg B, Olafsson O, Bjork GR (1994) Synthesis and function of isopentenyl adenosine derivatives in tRNA. Biochimie 76:1152–1160
45. Phillips MA, Leon P, Boronat A, Rodriguez-Concepcion M (2008) The plastidial MEP pathway: unified nomenclature and resources. Trends Plant Sci 13:619–623
46. Poliquin K, Ershov YV, Cunningham FX Jr, Woreta TT, Gantt RR, Gantt E (2004) Inactivation of sll1556 in Synechocystis strain PCC 6803 impairs isoprenoid biosynthesis from pentose phosphate cycle substrates in vitro. J Bacteriol 186:4685–4693
47. Rodriguez-Concepcion M, Boronat A (2013) Isoprenoid biosynthesis in prokaryotic organisms. In: Bach T, Rohmer M (eds) Isoprenoid synthesis in plants and microorganisms. Springer, New York, pp 1–16
48. Rodríguez-Concepción M, Boronat A (2002) Elucidation of the methylerythritol phosphate pathway for isoprenoid biosynthesis in bacteria and plastids. A metabolic milestone achieved through genomics. Plant Physiol 130:1079–1089
49. Rohmer M (2008) From molecular fossils of bacterial hopanoids to the formation of isoprene units: discovery and elucidation of the methylerythritol phosphate pathway. Lipids 43:1095–1107
50. Rothman SC, Johnston JB, Lee S, Walker JR, Poulter CD (2008) Type II isopentenyl diphosphate isomerase: irreversible inactivation by covalent modification of flavin. J Am Chem Soc 130:4906–4913
51. Sakamoto Y, Furukawa S, Ogihara H, Yamasaki M (2003) Fosmidomycin resistance in adenylate cyclase deficient (cya) mutants of *Escherichia coli*. Biosci Biotechnol Biochem 67:2030–2033
52. Sangari FJ, Perez-Gil J, Carretero-Paulet L, Garcia-Lobo JM, Rodriguez-Concepcion M (2010) A new family of enzymes catalyzing the first committed step of the methylerythritol 4-phosphate (MEP) pathway for isoprenoid biosynthesis in bacteria. Proc Natl Acad Sci USA 107:14081–14086
53. Sauret-Güeto S, Uros EM, Ibanez E, Boronat A, Rodriguez-Concepcion M (2006) A mutant pyruvate dehydrogenase E1 subunit allows survival of Escherichia coli strains defective in 1-deoxy-D-xylulose 5-phosphate synthase. FEBS Lett 580:736–740
54. Sharma NK, Pan JJ, Poulter CD (2010) Type II isopentenyl diphosphate isomerase: probing the mechanism with alkyne/allene diphosphate substrate analogues. Biochemistry 49:6228–6233
55. Shindo K, Misawa N (2014) New and rare carotenoids isolated from marine bacteria and their antioxidant activities. Mar Drugs 12:1690–1698

56. Siedenburg G, Jendrossek D (2011) Squalene-hopene cyclases. Appl Environ Microbiol 77:3905–3915
57. Smit A, Mushegian A (2000) Biosynthesis of isoprenoids via mevalonate in Archaea: the lost pathway. Genome Res 10:1468–1484
58. Sui X, Kiser PD, Lintig J, Palczewski K (2013) Structural basis of carotenoid cleavage: from bacteria to mammals. Arch Biochem Biophys 539:203–213
59. Unno H, Yamashita S, Ikeda Y, Sekiguchi SY, Yoshida N, Yoshimura T, Kusunoki M, Nakayama T, Nishino T, Hemmi H (2009) New role of flavin as a general acid-base catalyst with no redox function in type 2 isopentenyl-diphosphate isomerase. J Biol Chem 284:9160–9167
60. Van Horn WD, Sanders CR (2012) Prokaryotic diacylglycerol kinase and undecaprenol kinase. Annu Rev Biophys 41:81–101
61. Vannice JC, Skaff DA, Keightley A, Addo JK, Wyckoff GJ, Miziorko HM (2014) Identification in Haloferax volcanii of phosphomevalonate decarboxylase and isopentenyl phosphate kinase as catalysts of the terminal enzyme reactions in an archaeal alternate mevalonate pathway. J Bacteriol 196:1055–1063
62. Vollmer W, Blanot D, de Pedro MA (2008) Peptidoglycan structure and architecture. FEMS Microbiol Rev 32:149–167
63. Warlick BP, Evans BS, Erb TJ, Ramagopal UA, Sriram J, Imker HJ, Sauder JM, Bonanno JB, Burley SK, Tabita FR, Almo SC, Sweedler JS, Gerlt JA (2012) 1-methylthio-D-xylulose 5-phosphate methylsulfurylase: a novel route to 1-deoxy-D-xylulose 5-phosphate in Rhodospirillum rubrum. Biochemistry 51:8324–8326
64. Wouters J, Oudjama Y, Ghosh S, Stalon V, Droogmans L, Oldfield E (2003) Structure and mechanism of action of isopentenylpyrophosphate-dimethylallylpyrophosphate isomerase. J Am Chem Soc 125:3198–3199
65. Xiao Y, Savchenko T, Baidoo EE, Chehab WE, Hayden DM, Tolstikov V, Corwin JA, Kliebenstein DJ, Keasling JD, Dehesh K (2012) Retrograde signaling by the plastidial metabolite MEcPP regulates expression of nuclear stress-response genes. Cell 149:1525–1535
66. Yamada Y, Cane DE, Ikeda H (2012) Diversity and analysis of bacterial terpene synthases. Methods Enzymol 515:123–162

Adv Biochem Eng Biotechnol (2015) 148: 19–62
DOI 10.1007/10_2014_283

Published Online: 21 November 2014

Biosynthesis of Terpenoid Natural Products in Fungi

Claudia Schmidt-Dannert

Abstract Tens of thousands of terpenoid natural products have been isolated from plants and microbial sources. Higher fungi (Ascomycota and Basidiomycota) are known to produce an array of well-known terpenoid natural products, including mycotoxins, antibiotics, antitumor compounds, and phytohormones. Except for a few well-studied fungal biosynthetic pathways, the majority of genes and biosynthetic pathways responsible for the biosynthesis of a small number of these secondary metabolites have only been discovered and characterized in the past 5–10 years. This chapter provides a comprehensive overview of the current knowledge on fungal terpenoid biosynthesis from biochemical, genetic, and genomic viewpoints. Enzymes involved in synthesizing, transferring, and cyclizing the prenyl chains that form the hydrocarbon scaffolds of fungal terpenoid natural products are systematically discussed. Genomic information and functional evidence suggest differences between the terpenome of the two major fungal phyla—the Ascomycota and Basidiomycota—which will be illustrated for each group of terpenoid natural products.

Keywords Ascomycota · Basidiomycota · Isoprenoid · Terpene synthase · Natural products · Terpenoids · Prenyl transferase · Terpene cyclase · Sesquiterpenoids · Diterpenoids · Biosynthesis · Pathways · Gene cluster

Abbreviations

TPS	Terpene synthase
TPC	Terpene cyclase
PT	Prenyltransferase
FPP	Farnesyl diphosphate
IPP	Isoptenyl diphosphate
IPPS	Isoptenyl diphosphate synthase

C. Schmidt-Dannert (✉)
Department of Biochemistry, Molecular Biology and Biophysics,
University of Minnesota, St. Paul, Minneapolis, MN 55108, USA
e-mail: schmi232@umn.edu

DMAPP	Dimethylallyl diphosphate
GPP	Geranyl diphosphate
GGPP	Geranylgeranyl diphosphate
CPP	Copalyl diphosphate
ABBA	beta/alpha-fold with antiparallel beta strands
Me	Methyl
Ac	Acetyl
NPs	Natural products

Contents

1 Introduction

The fungal kingdom includes an enormous diversity of organisms, of which only a fraction (~100,000 species) has been described [1]. Conservative calculations estimate that fungal diversity exceeds the number of described species by at least one order of magnitude (~1.5 million species), making the fungal kingdom the second largest kingdom after the bacteria [2–4]. The phylum Ascomycota (filamentous fungi) accounts for ~60 % of the described species. This group includes many serious human, animal, and plant pathogens [5, 6], as well as fungi (e.g. *Aspergillus*, *Fusarium*, *Trichoderma*) with a long tradition in industrial biotechnology [7–10] and strains that are investigated for their secondary metabolism [11–13]. Basidiomycota (including mushroom-forming fungi) make up ~30 % of the known fungal species. Very few species of this phylum have been characterized, despite playing important roles in the decomposition of plant material (e.g. wood-rotting or saprophytic fungi) or as plant symbionts (e.g. mycorrhizal fungi) [14, 15]. Considering the enormous diversity of basidiomyceteous fungal species, very few of these fungi have yet been characterized, and biosynthesis of natural products has been investigated for an even smaller subset of taxa [16–27].

Studies aimed at deciphering secondary metabolic pathways on molecular and biochemical levels have been almost exclusively focused on Ascomycota (e.g. *Aspergillus*, *Penicillium*, *Fusarium*), which can be grown readily in the laboratory and are genetically tractable [9, 13, 28, 29]. The fungi remain a largely unchartered territory for the discovery of new natural products and their biosynthetic pathways, including isoprenoid-derived secondary metabolites. This is particularly true for the Basidiomycota; these fungi are typically difficult to grow or cannot be grown at all under laboratory conditions; except for a few species, they are not amenable to genetic manipulation. Yet, mushrooms have been used for millenia in traditional medicine and are known to make a range of bioactive compounds, including a plethora of antimicrobial, cytoxic, and anticancer compounds [30–37].

The slow progress made over the past decades in fungal natural products pathway identification and characterization will certainly accelerate drastically during the coming decade. Rapid progress in fungal genome sequencing and the development of synthetic biology approaches for heterologous refactoring of complex biosynthetic pathways with synthetic genes will drive the discovery process. Together with advances in bioinformatics and metabolomics, these approaches will eventually facilitate high-throughput discovery of natural products pathways from genomic sequence information alone independent of fungal cultivation. Identification of many fungal natural products pathways is also greatly aided by the fact that their genes are physically co-localized as clusters that facilitate efficient co-regulation of their expression by the fungus [29, 38–43].

This chapter reviews the current progress in the biosynthesis of isoprenoid natural products in fungi. After providing an overview of the different pathways and enzymes involved in generating isoprenoid precursor molecules that are then converted into complex biomolecules, the current knowledge on fungal biosynthesis of sesquiterpenoids, diterpenoids, and triterpenoids is discussed. Natural products of mixed biosynthetic origin, such as meroterpenoids and indole-diterpenoids, that contain a terpenoid derived moiety are discussed elsewhere in this series; their biosynthesis is only discussed here as it relates to the enzymes that install the terpenoid moieties in these compounds. Fungal carotenoid biosynthesis by basidiomycetous yeasts, such as the well-known carotenoid producers *Rhodotorula* and *Xanthophyllomyces* (*Phaffia*), has been described extensively in the literature [44, 45] and will therefore not be reviewed in this chapter.

Special emphasis is given to the differences observed between the isoprenoid natural products repertoire of the two major fungal phyla Asco- and Basidiomycota. This review concludes by discussing current bottlenecks in the identification and characterization isoprenoid natural products' biosynthetic pathways, stressing the significance of genomic-and bioinformatics-driven pathway discovery as well as opportunities offered through synthetic biology approaches for heterologous refactoring of natural products pathways.

2 Fungal Terpenoid Natural Product Biosynthesis

2.1 Types of Isoprenoid Biosynthetic Enzymes

2.1.1 Overview

All fungal terpenoid natural products are derived from the common five-carbon isoprenyl diphosphate intermediates isopentenyl diphosphate (IPP) and dimethylallyl diphosphate (DMAPP), which are synthesized from acetyl-CoA through the mevalonate pathway shown in Fig. 1. Successive head-to-tail 1'-4 condensation of one to three IPP extender units to DMAPP catalyzed by *all-trans* isoprenyl

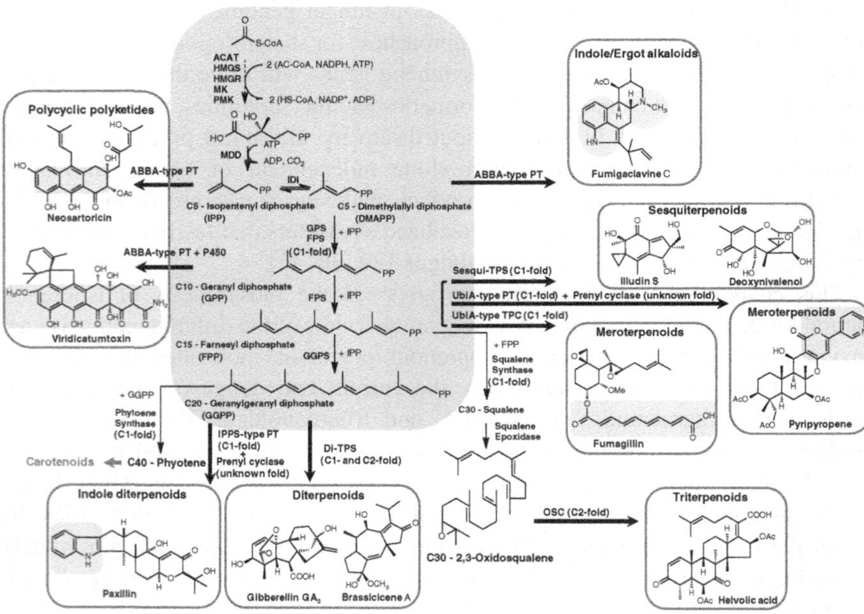

Fig. 1 Overview of isoprenoid derived natural products biosynthesis in fungi. The mevalonate pathway generates the isomeric C-5 isoprenoid chain precursors isopentenyl and dimethylallyl diphosphate (IPP and DMAPP). Head-to-tail condensation of these C-5 units yields prenyl-diphosphate chains of different lengths. Longer prenyl chains are formed by head-to-head condensation of two prenyl diphosphate chains. Linear prenyl chains are the substrates of prenyl transferases and cyclases as well as terpene synthases, which form the terpenoid scaffolds of different natural products classes, for which example compounds are shown. Prenyl chain modifying enzymes can be divided by structural folds (C1- and C2-fold, ABBA-type; see text) and by catalytic function. Abbreviations: ACAT: acetoacetyl-CoA thiolase, HMGS: 3-hydroxy 3-methylglutaryl CoA synthase, HMGR: 3-hydroxy 3-methylglutaryl CoA reductase, MK: mevalonate kinase, PMK: phosphomevalonate kinase, MDD: mevalonate diphosphate decarboxylase, IDI: isopentenyl diphosphate isomerase, GPS, FPS or GGPS: geranyl-, farnesyl- or geranylgeranyl diphosphate synthase, PT: prenyl transferase, TPS: terpene synthase, TPC: terpene cyclase, OSC: oxidosqualene synthase

diphosphate synthases (IPPS) gives rise to isoprenyl diphosphates with ten (C-10, geranyl, GPP), fifteen (C-15, farnesyl, FPP), or twenty (C-20, geranylgeranyl, GGPP) carbons. Longer chains (C-30 and C-40) are formed by a 1'-1 head-to-head condensation of two FPP (squalene synthase) or GGPP (phytoene synthase) molecules catalyzed by another group of prenyl chain synthases.

These linear chains are the substrates of different enzymes that either transfer a prenyl chain to another molecule, typically an aromatic compound, or trigger prenyl chain cyclization, thereby generating tens of thousands different isoprenoid-derived natural products [46]. Different types of prenyl transferases and cyclases have evolved in fungi and seem to be associated with specific natural products classes. Some biosynthetic pathways, such as the paxilline indole-diterpene pathway [47], involve two types of prenyl transferases that catalyze different prenylation reactions.

The different classes of characterized terpenoid natural products can be distinguished based on whether their scaffolds are derived solely from isoprenyl units or are of mixed biosynthetic origin. The first group is divided into mono-, sesqui-, di-, or triterpenoids, which contain two to six C-5 isoprene units. This group also includes the carotenoids and rare sesterterpenoids (C-25). The second group includes the meroterpenoids, the indole diterpenoids, and the structurally and biosynthetically diverse group of prenylated aromatic natural products.

As discussed in more detail elsewhere in this book series, the majority of characterized prenyl chain synthases, transferases (PTs), and cyclases share one of two structural folds (C1-and C2-fold) and associated reaction mechanisms (shown in Fig. 1) [48–50]. It should be noted, however, that despite clear structural homology and inferred evolutionary relatedness, sequence homology between these different enzymes is often not obvious.

A large group of aromatic prenyltransferases has a different fold (ABBA-fold) and uses a different reaction mechanism [51–53]. No structural or detailed mechanistic information is available for prenyl cyclases involved in the biosynthesis of indole diterpenoids [47, 54–56] and of some meroterpenoids [57–59].

The first major class of isoprenoid biosynthesis enzymes (class I) is characterized by an α-helical bundle fold (C1-fold, referred to as α-domain) that forms a hydrophobic active site cavity. Two aspartate-rich motifs (D(D/E)XX(D/E) and NSE/DTE) are situated at the active site entrance and coordinate binding of the substrate diphosphate group via Mg^{2+}. Catalysis is initiated by heterolytic cleavage of the diphosphate group, generating a reactive allylic carbocation for electrophilic addition to an electron-rich nucleophile. In the case of terpene synthases, the carbocation reacts with a C–C double bond, triggering in the active site a cascade of cyclization and rearrangement reactions of the prenyl chain until final quenching of the carbocation by proton abstraction or, in some cases, an electrophilic reaction with water. In the case of the *trans*-IPPS enzymes, a new C–C bond is formed between the carbocation of DMAP (or of GPP and FPP in subsequent chain extension) (head) and the terminal double-bond of IPP (tail). The head-to-head condensation of two FPP or GGPP molecules catalyzed by squalene or phytoene synthase, respectively, proceeds via a cyclopropyl diphosphate intermediate that is subsequently cleaved. Prenyl transferases (C1-fold and ABBA-type, see below) catalyze electrophilic alkylation between a prenyl diphosphate chain and an electron-rich indole or polyketide moiety.

The second major class of isoprenoid biosynthetic enzymes (class II) share a double α-helical barrel fold (C2-fold) with a hydrophobic active site cavity located between the two alpha-barrel domains (commonly referred to as γβ-domains). Catalysis is initiated by the addition of a proton from a conserved aspartate residue to a C–C double bond (e.g. copalyl diphosphate synthase domain of bifunctional diterpene synthases) or to an epoxide-ring (e.g. oxidosqualene lanosterol synthase). Formation of the ensuing carbocation triggers a cyclization cascade along the prenyl chain. If the substrate is a prenyl diphosphate (e.g. GGPP), the carbocation is generated at the distal double bond located at the tail end of the prenyl-chain, which leaves the diphosphate group attached. The conserved aspartate residue (bold) is typically located in a DX**D**D motif in eukaryotic class II enzymes that protonate a C–C double bond, whereas epoxide ring protonating enzymes contain a catalytic aspartate in a different motif (e.g. **D**CTAE in oxidosqualene synthases) [60, 61].

2.1.2 Aromatic Prenyltransferases

Known prenyl transferases in fungi fall into three general groups: the ABBA-, IPPS- or UbiA-type prenyltransferases (PTs), as shown in Fig. 1. The latter two share a common C1-fold, but while the IPPS-type synthases appear to be soluble and contain a canonical D(D/E)XX(D/E) motif, the UbiA-type prenyl transferases are integral membrane proteins.

The C-5 isoprenyl diphosphate DMAPP is the substrate of a large group of aromatic ABBA-type PTs identified in Ascomycota, which are mainly involved in the biosynthesis of diverse bioactive indole alkaloids, including many important pharmaceuticals and toxins made by strains of *Aspergillus*, *Penicillium*, *Claviceps,* and *Neosartorya* (reviewed in [62]). Most of these enzymes catalyze the regio- and stereoselective transfer of DMAPP to any position of a tryptophan-derived indole ring via condensation to either the C-1 or C-3 position of the prenyl chain [63–65]. Some enzymes catalyze DMAPP and GPP (C-10) prenyl transfer to other aromatic substrates, such as polycyclic polyketides and phenols (e.g. xanthones) [66, 67] or to O- and N-functional groups of aromatic moieties [68, 69]. More recently, an enzyme was identified that catalyzes O-prenylation of a glucose moiety of the diterpene fusicoccin A [70]. Mechanistic and structural studies revealed that these proteins adopt a new β-barrel fold containing repeating αββα-secondary structure elements, which gives this enzyme family their name: ABBA-type. This fold is conserved between bacterial and fungal aromatic PTs [51, 71]. These enzymes are soluble and do not contain aspartate-rich motifs that coordinate binding of the prenyl diphosphate group through Mg^{2+}.

IPPS-type PTs (e.g. LtmC, PaxC and AtmC) catalyze the transfer of GGPP to an indole group to create the core scaffold of the structurally diverse indole-diterpene natural products (NPs) isolated from Ascomycota, which include many potent mammalian mycotoxins [47, 54–56, 72–74]. In vitro studies have been only very recently carried out with a purified prenyl transferase (PaxC) from the paxilline pathway and confirm transfer of GGPP to the C3 position of tryptophan [47]. Surprisingly, the tryptophan precursor indole-3-glycerophosphate was the preferred substrate and C3 prenylation resulted in simultaneous elimination of glyceraldehyde-3-phosphate. Concurrent prenylation and elimination has only been known for a different type of prenyl transferase, MenA, which catalyzes napthol octaprenylation with simultaneous decarboxylation.

A third type of prenyl transferases (UbiA-type PT in Fig. 1) catalyzes the transfer of FPP to the polyketide-derived aromatic core in the biosynthesis of numerous bioactive meroterpenoids. Prenyl transferases (Trt2, Pyr6, AusN, and MpaA) have been identified in biosynthetic gene clusters for terretonin, pyripyropene, austinol, and mycophenolic acid synthesis [57–59, 75–77]. These enzymes are integral membrane proteins and belong to the UbiA/Coq2 quinone prenyl transferase group, which also includes enzymes involved in the secondary metabolite biosynthetic pathways of various plants [78]. Although these enzymes are integral membrane proteins, a homology model was built for UbiA based on a class I terpene synthase structure. Aspartate-rich motifs located at the entrance of the active site of UbiA and in a plant prenyl transferase homolog have subsequently been identified to bind Mg^{2+} and coordinate diphosphate binding of the prenyl substrate [79, 80]. Very recently, the structure of UbiA from *Aeropyrum pernix* was solved and shown to indeed possess an alpha-helical fold that is made up of nine transmembrane helices and three helices that cap a large active site cavity [81]. Aspartate residues located in two of the helices are involved in Mg^{2+} and diphosphate binding. The active site opens laterally into the membrane, thereby allowing binding of long-chain prenyl substrates and release of hydrophobic prenylated products directly into the membrane. The UbiA structure suggests that this enzyme may indeed be evolutionarily related to C-1 fold, IPPS-type PTs.

2.1.3 Prenyl Chain Cyclases

A large number of class I and class II terpene synthases have been characterized from plants and, to a lesser extent, from microorganisms, including fungi. These enzymes create the cyclic hydrocarbon scaffolds of many bioactive terpenoid antibiotics, toxins, and pheromones [82]. Enzymes that cyclize FPP or GGPP into structurally diverse sesqui- or diterpenoids have been characterized from several fungi, and structures are available for two fungal sesquiterpene synthases [83, 84]. As discussed later in more detail, the majority of biochemically characterized fungal terpene synthases are sequiterpene synthases. Relatively few diterpene synthases have been characterized from fungi, and fungal triterpenoid natural products biosynthesis is even less studied. So far, no *bona fide* monoterpene synthases have been identified in fungi, although in vitro biochemical studies have shown that fungal sesquiterpene synthases are able to cyclize GPP in addition to their native substrate FPP [22]. This is in stark contrast to the large number of mono-, sesqui-, di-, and triterpene synthases characterized from plants [85–87].

Until recently, it was thought that the enzymes that cyclize isoprenoid diphosphate molecules belong to one of the prototypical class I or II terpene synthase families [48]. Tang and co-workers, however, discovered a new, noncanonical type of sesquiterpene synthase in *Aspergillus fumigatus* that cyclizes FPP into the monocyclic β-trans-bergamotene, which is the biosynthetic terpenoid precursor of the meroterpenoid fumagillin [88] (Fig. 1). This enzyme is membrane-bound and shares a low, but recognizable, sequence similarity with UbiA-type prenyltransferases. Additional examples and biochemical studies are needed to gain insights into its role in the biosynthesis of fungal natural products, as well as its evolutionary relationship to the canonical class I terpene synthases. UbiA-type terpene cyclases could be the result of convergent evolution, or they could represent the common ancestor of all canonical class I terpene synthases.

Yet another group of isoprenyl-chain cyclizing enzymes acts on the farnesyl- and geranylgeranyl-chains attached to the polyketide and indole moieties during meroterpenoid [57–59, 75, 76] and indole-diterpene [47, 54, 56, 72, 74, 89–91] natural products biosynthesis (see also Sect. 2.5). Heterologous reconstitution of the pyripyropene (Fig. 1) meroterpenoid biosynthetic pathway from *A. fumigatus* in *A. oryzae* [59] identified a protein (Pyr4) predicted to be an integral membrane protein as farnesyl cyclase. This protein is unusually small (242 aa) and does not share any homology with known proteins in the National Center for Biotechnology Information (NCBI) database. It also lacks the aspartate-rich motifs commonly found in prenyl transferases and cyclases. Homologs are present in other meroterpenoid (e.g. Trt1, AusL, Pyr4) and indole-diterpenoid (e.g. PaxB, AtmB, LtmB) biosynthetic gene clusters. Each of these clusters also encodes an FAD-dependent monooxygenase. This enzyme was shown to catalyze epoxidation of the terminal (C10-C11) double-bond of the farnesyl moiety to allow for protonation initiated prenyl chain cyclization by the transmembrane cyclase, similar to the reaction catalyzed by oxidosqualene synthase [59]. Site-directed mutagenesis identified two residues (E62 and D218) conserved in all transmembrane cyclase homologs that when mutated abolished in vitro activity of Pyr4 [59]. These acidic residues are proposed to catalyze epoxide protonation in a similar fashion to oxidosqualene synthase [59]. Heterologous reconstitution of the indole-diterpene paxilline (Fig. 1) pathway from *Penicillium paxillii* in *A. oryzae* has confirmed epoxidation-dependent prenyl chain cyclization. Intriguingly, formation of the tricyclic terpenoid moiety of paxilline was shown to involve stepwise epoxidation by PaxM of the terminal and the C10-11 double bond of the prenyl moiety; each oxidation was followed by protonation-dependent cyclization of parts of the prenyl chain by PaxB [47].

The viridicatumtoxin biosynthetic gene cluster appeared to have no terpene cyclase that could cyclize the geranyl chain attached to the polyketide-derived naphthacene scaffold [67, 92, 93] (Fig. 1). Instead, a cytochrome P450–encoded by *vrt*K was identified through gene deletion in the native producer *P. aethiopicum* and heterologous expression in yeast as the enzyme responsible for prenyl cyclization [93]. Vrtk is proposed to initiate carbocation formation by oxidation of the allylic C4 of the geranyl chain. This leads to the formation of an allylic cation that triggers subsequent prenyl chain cyclization via a proposed tertiary carbocation intermediate [93].

The identification of novel cyclases that have little in common with canonical class I and II terpene synthases, except that they all take advantage of the reactivity of the allylic prenyl chain to create a carbocation intermediate, shows the diversity of potential enzyme active sites and protein scaffolds available in nature for catalysis of reactions with similar outcomes. Fungal natural product gene clusters tend to have multiple P450 enzymes as well as predicted genes with no conserved domains. These unknown and uncharacterized genes provide a rich source for the discovery of novel types of enzymes, as discussed above for the recently discovered prenyl cyclases.

2.1.4 Isoprenoid Biosynthetic Genes in Ascomycota and Basidiomycota

The majority of studies on the enzymes discussed in the above sections have been done with enzymes and pathways from filamentous fungi (Ascomycota). Terpenoid-derived natural products of mixed biosynthetic origin (prenylated indole-alkaloids, indole-diterpenoids, meroterpenoids) have so far received the most attention. Relatively few studies describe the identification and characterization of enzymes and pathways that give rise to sesqui-, di- and triterpene natural products in filamentous fungi. Based on literature reports and a survey in SciFinder, polyketides (PK) and non-ribosomal peptides (NRP) synthesized by thiotemplate mechanism synthetases (PKSs and NRPSs) clearly appear to be the major classes of natural products made by Ascomycota [94–96]. A similar survey for natural products isolated from Basidiomycota, however, returned less than 100 potentially PKS-or NRPS-derived compounds, whereas close to 500 reports ($\sim$1,000 compounds) described the isolation of terpenoids. Furthermore, although filamentous fungi tend to have between 20 and 50 NRPS and PKS biosynthetic gene clusters in their genomes [38], Basidiomycota typically have fewer than 10 PKS or NRPS genes, of which only a few have been expressed and characterized [16, 24, 25, 97–99].

These surveys indicate that Ascomycota and Basidiomycota may have evolved different arsenals of natural products. Ascomycota seem to rely on PK, NRP, and indole-derived compounds as their predominant natural product classes, while terpenoids are the major class made by Basidiomycota. To support this conclusion further, we performed a BLAST survey of all Ascomycota and Basidiomycota genomes available at the Joint Genome Institutes (JGI) Fungal Genome database at the time of writing, with known sequences of the different prenyl transferases, prenyl cyclases, and terpene synthases discussed in this chapter. As shown in Fig. 2, sesquiterpene synthases constitute the most abundant group of enzymes in Basidiomycota, with four times as many sequences identified per genome than in Ascomycota genomes. In contrast, Basidiomycota have comparatively few ABBA-type prenyltransferases involved in the biosynthesis of prenylated polyketides and indole alkaloids. It also appears that homologs of UbiA-type prenyl transferase and

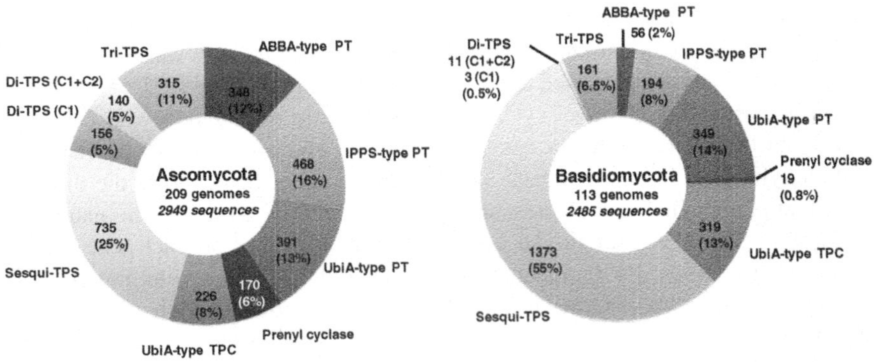

Fig. 2 Abundance of prenyl-chain modifying enzyme homologs identified in Ascomycota and Basidiomycota genomes. Shown are numbers (and rounded percentages) of homologs of enzymes identified in fungal genomes available at JGI using BLAST analysis with sequences of functionally characterized prenyl transferases, prenyl cyclases, and terpene synthases and cyclases discussed in this review. BLAST hits from different searches were cross-analyzed for duplicates and manually inspected for alignment coverage and scores. Enzyme abbreviations correspond to those in Fig. 1

cyclases are relatively more abundant in Basidiomycota compared to Ascomycota. Because none of these putative enzymes have been characterized, we do not know whether they catalyze similar reactions to the enzyme characterized from Ascomycota or are involved in the biosynthesis of yet-to-be-identified natural products. Notable is the scarcity of diterpene synthase homologs (both mono- and bifunctional enzymes) in genomes from both fungal phyla. For example, only 11 sequences were identified in six Basidiomycota genomes out of over 100 searched. Homologs of prenyl cyclases involved in meroterpenoid and indole-diterpene biosynthesis in filamentous fungi are likewise rare in Basidiomycota, suggesting that these compounds either represent a minor group of natural products in Basidiomycota or their synthesis involves a different type of cyclizing enzyme.

2.2 Sesquiterpenoids

2.2.1 Overview

The C-15-hydrocarbon scaffolds of the thousands of structurally diverse sesquiterpenoid natural products isolated from plants, bacteria, and fungi are synthesized from FPP by sesquiterpene synthases [46]. These enzymes bind the pyrophosphate group (PP) of FPP at the entrance of the active site via a Mg^{2+} cluster, which is coordinated by two conserved aspartate-rich motifs, DDXXD/E and NSE/DTE. Upon binding of the PP group, the prenyl chain becomes oriented in the hydrophobic active site cavity of the enzyme [100] and a conformational change is triggered, which results in the closure of the active site and concurrent PP cleavage to generate

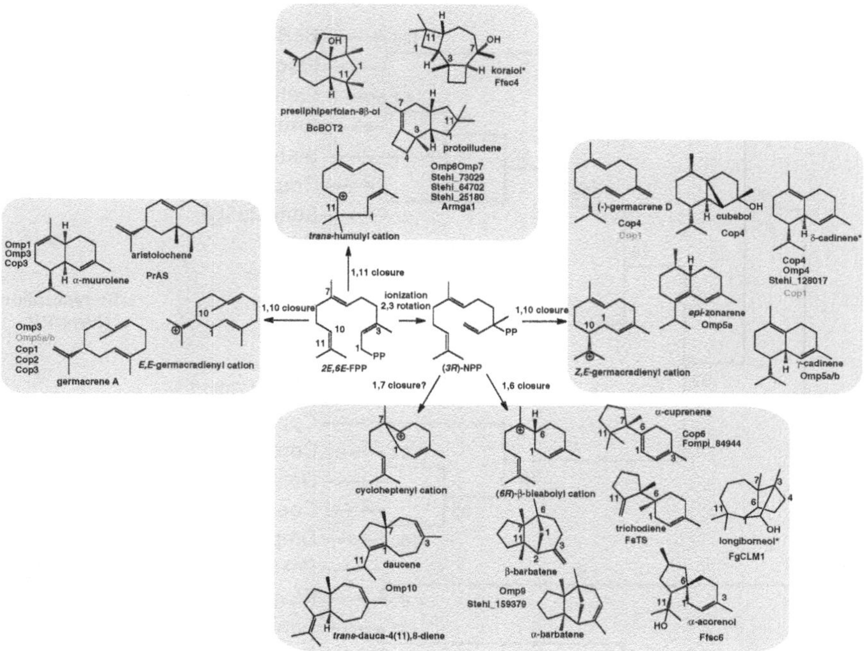

Fig. 3 Cyclization of FPP by characterized fungal sesquiterpene synthases. Enzymes catalyze different initial cyclization reactions en route to the final major sesquiterpenoid product(s) shown for each enzyme. Omp5 and Cop1 each appear to catalyze two different initial cyclization reactions leading to two major products. Asterisks denote that alternative cyclization pathways are possible. Key to characterized sesquiterpene synthases from fungal source, references, and NCBI sequence accession numbers: Cop: *Coprinus cinereus* [21–23]; Omp: *Omphalotus olearius* [20]; Stehi: *S. hirsutum* [19]; Fompi_84944: *Fomitopsis pinicola* [20]; Armga1: *Armillaria gallica* [105], Bc_BOT2: *Botrytis cinerea* (#AAQ16575) [106, 107], Pr_AS: *Penicillium roqueforti* (#Q03471) [84, 101, 108], Fs_TS: *Fusarium sporotrichoides* (#AAN05035) [83, 109, 110]; Fg_CLM1: *F. graminearum* (#ACY69978) [111]; Ff_sc4 & Ff_sc6: *F. fujikuroi* (#CCP20071 & #CCP20072 —product profiles were derived from strains containing corresponding gene knockouts) [112, 113]

an initial *transoid* allylic carbocation [101–103], shown in Fig. 3. This carbocation is then transferred along the isoprenyl chain and eventually quenched either by a water molecule or through proton abstraction. The binding pocket determines folding of the isoprenyl chain and chaperones the reactive carbocation intermediates until the final quenching step [104], thereby defining the product profile of a particular sesquiterpene synthase.

Sesquiterpene synthases catalyze different initial cyclization reactions generating secondary or tertiary cyclic carbocation intermediates, as shown in Fig. 3. For example, the C–C bond formation between C1 and C11 of the primary farnesyl carbocation yields a *trans*-humulyl-carbocation, which is a 1,11-cyclization product. This secondary carbocation can then undergo additional cyclizations and rearrangements until carbocation quenching in the active site and subsequent release of

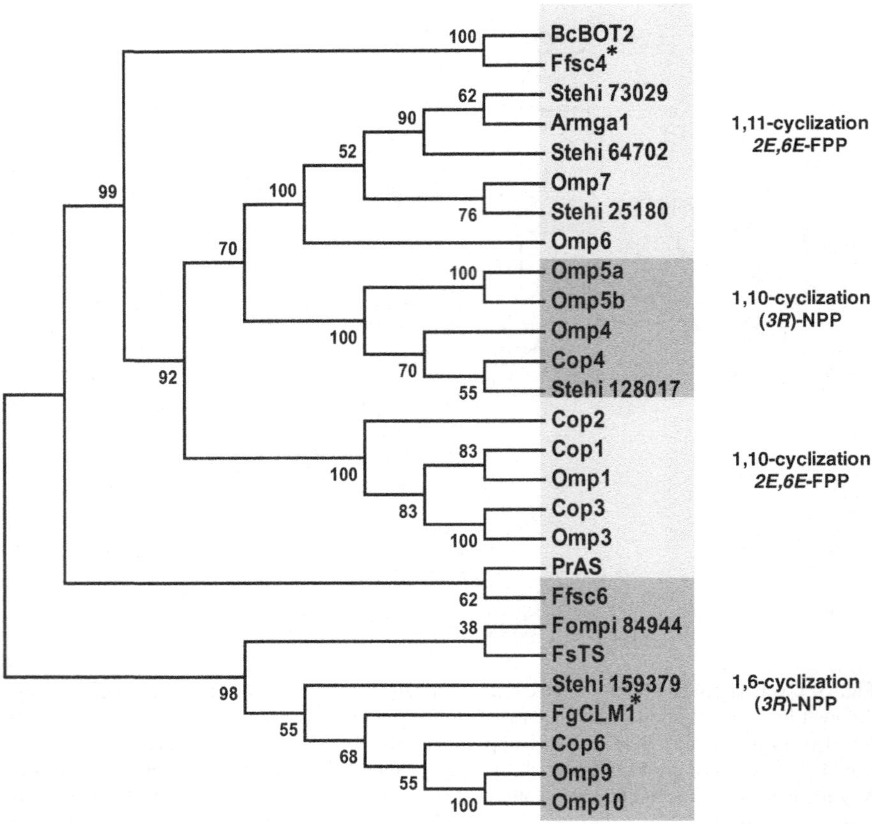

Fig. 4 Phylogenetic analysis of characterized fungal sesquiterpene synthases. Sesquiterpene synthases form different clades based on their cyclization mechanisms. The unrooted neighbor-joining phylogram was built in MEGA6 [114] with the sequences listed in Fig. 3. Branches are labeled with their bootstrap values

the final sesquiterpenoid scaffold by the enzyme. Some sesquiterpene synthases catalyze first *trans-cis* isomerization of the 2,3-double-bond of (*2E,6E*)-FPP, which yields a *cisoid,* allylic nerolidyl-carbocation after PP_i cleavage. The *cis*-configuration of the 2,3-double bond now allows initial cyclizations between the C1 and C6 or C7 as well as C10. Figure 3 shows the major cyclization products of all currently cloned and characterized fungal sesquiterpene synthases. Phylogenetic analysis indicates that the majority of these enzymes form distinct clades related to their cyclization mechanisms (Fig. 4).

Crystal structures have been solved for several microbial and plant sesquiterpene synthases [83, 84, 115–121]. Aristolochene synthase from *Penicillium roqueforti* and *Aspergillus terreus* [84, 117, 122, 123] and trichodiene synthase from *Fusarium sporotrichioides* [83, 103, 124, 125] are the only two fungal enzymes for which crystal structures are known; structures are available in both the open (ligand-free) and closed (substrate/substrate analog-complexed) conformations.

The active site of all sesquiterpene synthases is located in a α-helical bundle (α-domain), which is characteristic for ionization-dependent class I terpene synthases as discussed above. Plant sesquiterpene synthases have catalytically inactive domains that are believed to be remnants from an ancient fusion event between the α-bundle domain of class I terpene synthase and the double α-barrel domains (γβ-domains) of a protonation-dependent class II terpene synthase [50]. As described below, all three domains are present in diterpene synthases and both are functional in bifunctional enzymes. Plant sesquiterpene synthases have retained either one (β-domain) [118] or both α-barrel domains (γβ-domains) [120], in addition to the catalytically active α-domain.

2.2.2 Sesquiterpenoid Biosynthetic Pathways in Ascomycota and Basidiomycota

Filamentous fungi, such as *Fusarium*, *Botrytis*, *Aspergillus,* and *Penicillium,* produce potent mycotoxins that can pose major health risks for humans and animals if these fungi infect plants or grow on feeds or foods (reviewed in [12]). Many well-known mycotoxins are sesquiterpenoids that are produced by plant pathogenic fungi such as *Fusarium* and *Botrytis* strains, and by fungi such as *P. roqueforti* or *Aspergillus oryzae* used in food processing. These toxins often play an important role in fungal virulence.

Fusarium strains infect cereals, which causes head blight or crown and root rot; it also leads to contamination of cereal crops with trichothecene mycotoxins, as shown in Fig. 5. Not surprisingly, trichothecene biosynthesis has been intensely studied. Trichodiene synthase, which creates the tricyclic sesquiterpenoid scaffold for these compounds, was among the first fungal sesquiterpenoid synthases to be purified, characterized, and subsequently crystallized (trichothecene biosynthesis is reviewed in [126, 127]). This enzyme catalyzes the 1,6-cyclization of a *cisoid,* allylic nerolidyl-carbocation, followed by additional cyclization and rearrangement reactions that yield the 5-,6-membered bicyclic trichodiene structure shown in Fig. 5. *Fusarium* strains modify this scaffold into toxic trichothecenes, which include deoxynivalenol and nivalenol made by *F. graminearum* and the T-2 toxin produced by *F. sporotrichoides* (Fig. 5) [126]. Most of the biosynthetic genes are found in one gene cluster locus, but in several strains three additional genes are located in two loci outside this core cluster.

Fusarium strains also produce tricyclic diols, such as the antifungal culmorin, that are proposed to follow a similar 1,6-cyclization pathway than trichodiene. A sesquiterpene synthase that makes the mono-hydroxylated culmorin precursor longiborneol has been identified in *F. graminearum* [111] (Fig. 3). However, the surrounding genomic region of the gene that encodes this protein (ACY69978, FG_10397, Broad Institute *F. graminearum* genome sequence) does not appear to include predicted genes that encode for a P450 or other oxidoreductase that could

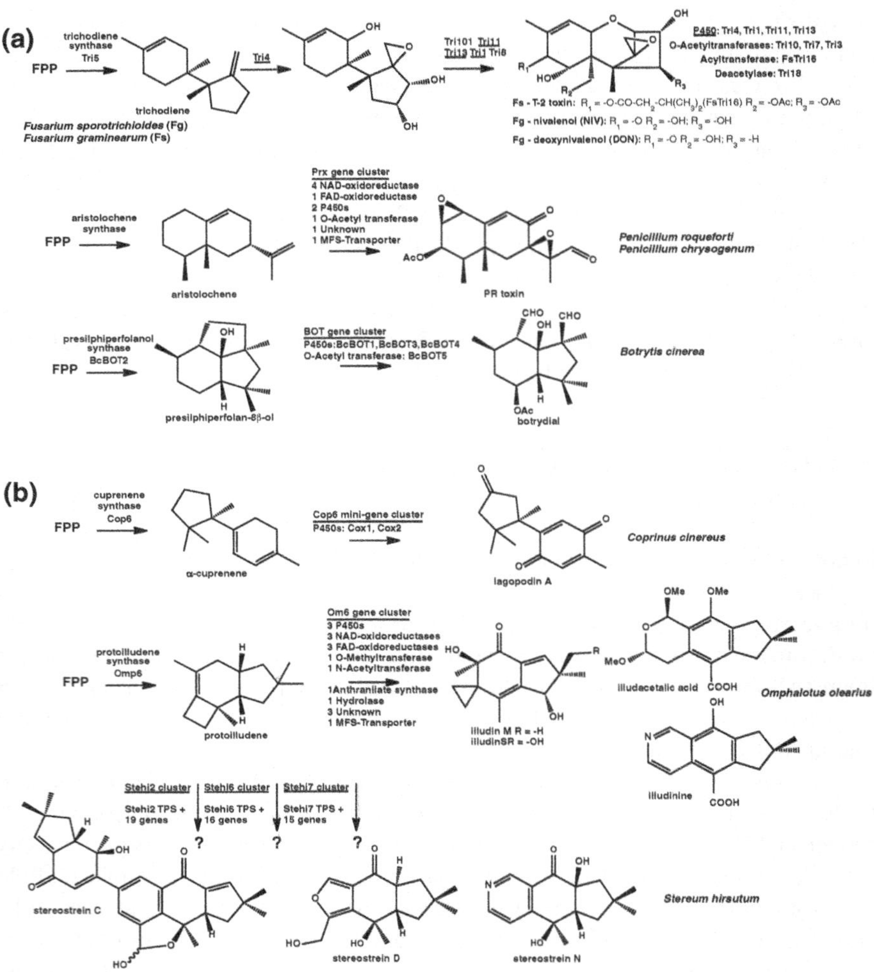

Fig. 5 Identified sesquiterpenoid biosynthetic gene clusters in Ascomycota (**a**) and Basidiomycota (**b**). Terpene synthases cyclize FPP into sesquiterpene hydrocarbon scaffolds that are modified by tailoring enzymes into the final products shown. Function of the majority of biosynthetic genes shown (except for the trichodiene and cuprenene gene clusters) has been inferred based on sequence homology only. The gene clusters identified in *Stereum hirsutum* and *Omphalotus olearius* are expected to make a range of structurally similar compounds that have been isolated from these fungi. See text for details

yield the dihydroxylated culmorin. The closest homolog among the nine sesquiterpene synthases found in the genome of this fungus, FG_06444, appears to be the only other putative sesquiterpene synthase in addition to the characterized trichodiene synthases that may be part of a gene cluster that contains P450s.

The rice pathogen *Fusarium fujikuroi* is well known for its production of diterpenoid phytohormones, as discussed in the next section [112]. It also produces a number of volatile sesquiterpenoids, and its genome contains putative sesquiterpene synthases [113]. Gene deletion studies have identified two terpene synthases responsible for the production of α-acorenol and koraiol [113]. Although α-acorenol is derived from an initial 1,6-cyclization intermediate, koraiol synthesis likely proceeds through an initial 1,11-cyclization reaction to generate a humulyl cation (Fig. 3).

Synthesis of the sesquiterpenoid alcohols longiborneol, koraiol, and α-acorenol by their respective fungal terpene synthases involves quenching of the final carbocation intermediate by water. Formation of such a sesquiterpenoid alcohol by a fungal sesquiterpenoid synthase was first characterized for an enzyme identified in the biosynthetic gene cluster responsible for the production of the phytotoxin botrydial by the gray mold *Botrytis cinerea* [106, 107] (Fig. 5). This enzyme catalyzes a 1,11-cyclization reaction, which is followed by additional cyclizations to yield the tricyclic alcohol presilphiperfola-8β-ol [107] (Fig. 3).

Other filamentous fungi, such as *Aspergillus* and *Penicillium* strains, produce sesquiterpenoid toxins, including the PR-toxin, sporogen-AO1, and phomenone, which are derived from the aristolochene scaffold [128] (Fig. 3). The corresponding 1,10-cyclizing sesquiterpene synthases were cloned more than a decade ago [101, 129], but only recently have studies begun to elucidate the biosynthesis of these toxins [130, 131]. A gene cluster encoding the PR-toxin biosynthetic pathway has been identified in the blue cheese mold *P. roqueforti* by screening a genomic phage library and subsequent comparison with orthologous genes identified in the sequenced genome of a related strain, *P. chysogenum* [131] (Fig. 5).

2.2.3 Sesquiterpenoid Biosynthetic Pathways in Basidiomycota

Until recently, no sesquiterpene synthase had been cloned and characterized from Basidiomycota. This lack of biosynthetic information is surprising considering that this fungal phylum is known to produce a plethora of structurally unique and bioactive sesquiterpenoids that likely are responsible of the medicinal properties of many traditionally used mushrooms [33–35, 107, 132–135]. Many of the isolated sesquiterpenoids have unique skeletons derived from the humulyl pathway that are not made by other organisms [35, 132, 136–139]. For example, protoilludenes have only been identified in one Ascomycete [140], a fern [141], and an octacorallia [142].

More than 60 genera of Basidiomycota (e.g. *Boletales*, *Cantharellales*, *Lactarius*, *Laccaria*, *Amanita*) are ectomycorrhizal fungi that associate with tree roots and form a symbiotic partnership with their hosts, which is crucial to plant nutrition in terrestrial ecosystems [15]. *Lactarius* species are known to produce modified lactarane- and protoilludane-derived sesquiterpenoids that have plant growth promoting activities [143–145]. Although the mechanism is not known, they may play similar roles as the diterpenoid gibberellic acids produced by filamentous fungi discussed later.

As shown in Fig. 3, different sesquiterpene synthases catalyze the 1,11-cyclization of FPP into various tricyclic sesquiterpene scaffolds. Strained rings may then undergo secondary rearrangements triggered by other enzymes, such as P450s.

Recognizing Basidiomycota as virtually unexplored territory for the discovery of novel terpenoid biosynthetic pathways, we chose to investigate sesquiterpenoid biosynthesis in the model basidiomycete *Coprinus cinerea*, which was the first mushroom-forming basidiomycete for which a genome sequence had been released [2]. We cloned and characterized six sesquiterpene synthases (Cop1-6) [23]. Three of the enzymes—Cop3, Cop4, and Cop6—had new activities shown in Fig. 3. Biochemical studies as well as structural modeling and mutagenesis provided new insights into their cyclization mechanism as well as their product specificities [21, 22]. Cuprenene-synthase Cop6 proved to be a highly product-specific enzyme, while all the other Cop enzymes were less selective; in the case of Cop4, product selectivity could be influenced by reaction conditions as well as site-directed mutagenesis [21, 22]. Cop6 was also the only enzyme located in an apparent mini-cluster that includes to P450s. Co-expression of Cop6 with its two P450s in *S. cerevisiae* [23] led to the discovery of the pathway for the biosynthesis of the antimicrobial NP lagopodin [146] (Fig. 5).

Although we detected the production of the 1,11-cyclization product pentalenene by *C. cinerea*, none of the functionally cloned enzymes produced sesquiterpenoids that could be derived from the humulyl-cation intermediate (Fig. 3). One enzyme, Cop5, was nonfunctional and is hypothesized to be responsible for this cyclization activity. Because humulyl-cation derived sesquiterpenoids are the major bioactive sesquiterpenoid natural products isolated from Basidiomycota (Fig. 6), we continued

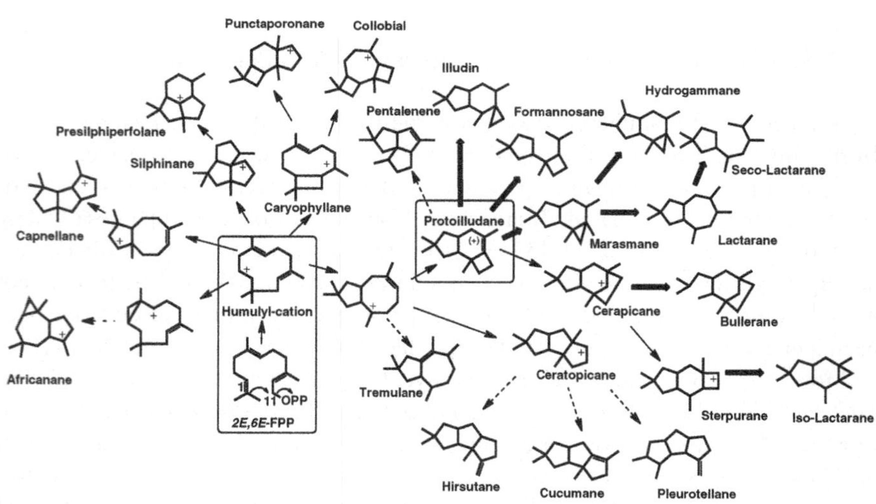

Fig. 6 Structurally diverse tricyclic sesquiterpenoid scaffolds known to be made by Basidiomycota. Possible cyclization routes and cyclic intermediates are shown. Fungal enzymes catalyzing cyclization of FPP to presilphiperfolane, protoilludane, and pentalenene have been biochemically characterized (see Fig. 3). *Bold arrows* indicate secondary ring-opening and ring-contraction reactions that are proposed to be catalyzed by terpenoid scaffold-modifying enzymes, such as cytochrome P450s. In the absence of any characterized enzymes that catalyze these reactions, only the undecorated, rearranged terpenoid hydrocarbon scaffolds are shown

our quest to identify 1,11-cyclizing sesquiterpene synthases in other Basidiomycota, which would facilitate the discovery of the corresponding biosynthetic gene clusters for these compounds.

Illudane-type sesquiterpenoids are among the best-known bioactive compounds made by Basidiomycota. They exhibit potent antitumor, antiviral, and antibacterial activity [147–152] and have been isolated from several mushrooms [153–166]. *Omphalotus olearius* was identified decades ago as a prolific producer of anticancer illudin M and S [152, 167–169]. Illudins are derived from the protoilludane scaffold (Fig. 6). We aimed to identify the corresponding sesquiterpene synthase and expected surrounding genes encoding the illudin biosynthetic pathway.

We sequenced the genome of *O. olearius* (deposited at JGI) and discovered a surprisingly large sesquiterpene synthase family (Omp1-10) [20]. The recombinant enzymes catalyze all but one possible initial cyclization reaction of FPP, shown in Fig. 3. Several sesquiterpene synthases catalyze new cyclization reactions, including synthesis of Δ6-protoilludene by Omp6 and 7, and of barbatene and daucene by Omp9 and 10, respectively. Unlike many characterized sesquiterpene synthases, protoilludene synthases Omp6 and 7are highly product specific. Omp6 is part of a large biosynthetic cluster comprised of 18 genes that are proposed to synthesize the different illudin compounds isolated from *O. olearius* (Fig. 5) [20].

The sesquiterpene synthase families characterized from *C. cinereus* (Cop1-6) and *O. olearius* (Omp1-10) provided a diverse set of sequences for bioinformatics analysis of Basidiomycota genomes, with the goal of developing a framework for the prediction of sesquiterpenoid biosynthesis in these fungi. BLAST analysis of the 40 Basidiomycota genomes available at JGI's Fungal Genomes database [170] in 2011–2012 led to the identification of 500 putative STS (this number has since more than doubled, with currently more than 100 genomes sequenced; see Fig. 2), which could be grouped into five distinct clades based on sequence homology and cyclization mechanism [20]. Protoilludene synthases Omp6 and 7 are located in one clade (clade III), suggesting that all of its members carry out an initial 1,11-cyclization of FPP (Fig. 3).

Considering that the vast majority of bioactive sesquiterpenoids in Basidiomycota are derived from the 1,11-cyclization pathway, sesquiterpene synthases in clade III should therefore be considered as prime candidates for the identification of gene clusters associated with the biosynthesis of bioactive sesquiterpenoid natural products. To test our prediction, we recently set out to identify terpenoid pathways in the published genome sequence of *Stereum hirsutum* [171]. This wood-rotting mushroom was primarily sequenced to elucidate its lignin decomposition biochemistry. However, it is also among the few Basidiomycota with a sequenced genome for which a number of bioactive natural products have been isolated, including many humulyl-pathway derived sesquiterpenoids [138, 172–178].

We identified 16 putative sesquiterpene synthase sequences in *S. hirsutum* and predicted their cyclization mechanisms [19]. Seven enzymes were predicted to catalyze 1,11-cyclization of FPP, while the remaining enzymes were predicted to catalyze other cyclizations of FPP. Representative sequences, including three sequences predicted to encode 1,11-cyclizing sesquiterpene synthases (Stehi2, 6, and

7), were cloned and the recombinant enzymes characterized [19]. Stehi2, 6, and 7 proved to be highly specific protoilludane synthases that are each located in large gene clusters, which are presumed to be responsible for the synthesis of protoilludane-derived terpenoids previously isolated from *Stereum* (Fig. 6) [139, 174]. Other representative enzymes cloned from *Stereum* catalyzed as predicted 1,6- and 1,10-cyclizations of FPP [19].

BLAST searches of Ascomycota and Basidiomycota genome sequences available at JGI for putative sesquiterpene synthase sequences uncovered major differences between these two fungal phyla. Not only do Basidiomycota genomes have an average of 2–3 times larger families of sesquiterpene synthases (10–20 sequences compared to 4–10 in Ascomycota), but their genomes typically encode several predicted protoilludene synthase homologs, of which one or more are located in biosynthetic gene clusters. Only about a dozen sequence homologs of 1,11-cyclizing enzymes (BLAST searches with Omp6, 7, and Stehi2, 6, 7 sequences; E-value cutoff: $<10^{-50}$) are found scattered in different Ascomycota genomes, compared to about one-fifth of all sesquiterpene sequences identified in Basidiomycota (Fig. 2). The sequenced Ascomycota genomes seem to have a relatively large complement of 1,6-cyclizing enzymes; representatives include the characterized trichodiene, longiborneol and α-acorenol synthases from *Fusarium* strains. However, only trichodiene synthase appears to be located in a biosynthetic gene cluster.

Mining of fungal genomes for sesquiterpene synthase homologs revealed that only a few of the homologs in each fungal genome are part of identifiable biosynthetic gene clusters that could synthesize modified and probably nonvolatile products. In Basidiomycota, 1,11-cyclizing enzymes are frequently located in predicted biosynthetic gene clusters. A random survey of different Ascomycota genomes did not reveal a similar preference for one particular clade of sesquiterpene synthase homologs. Enzymes that are not part of gene clusters in a fungus are probably responsible for the production of an array of volatile terpene products; their biological significance is currently unknown, but they could potentially play a role in signaling, such as for interactions with insects [179].

2.3 *Diterpenoids*

2.3.1 Overview

The C-20 scaffolds of tens of thousands of known diterpenoid compounds [82] are generated from geranylgeranyl diphosphate via two different cyclization routes involving either one or two separate cyclase activities (Fig. 7).

Diterpenoids cyclized by the first, one-step route involve a monofunctional class I diterpene synthase that catalyzes ionization-dependent diphosphate cleavage and subsequent carbocation migration and quenching using a mechanism similar to sesquiterpene synthases, except the prenyl chain is now longer by one isoprene unit.

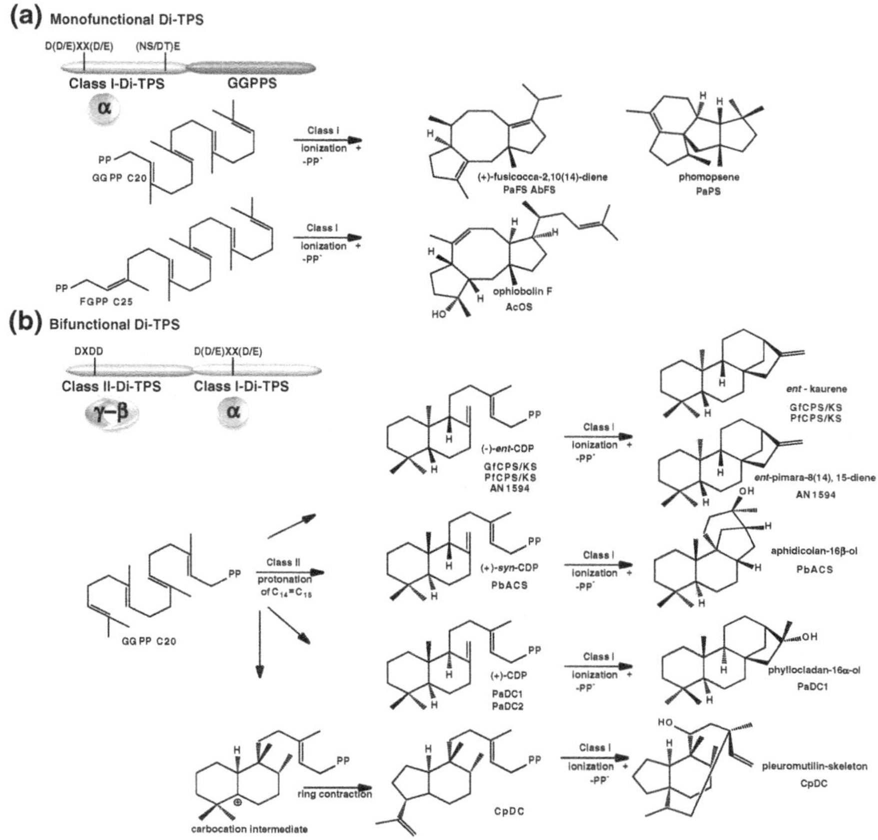

Fig. 7 Cyclization of GGPS (FGPP) by characterized fungal diterpene synthases (Di-TPS). **a** Monofunctional class I Di-TPS fused to GGPP synthase catalyzes a one-step ionization dependent cyclization of GGPP or FGPP to generate tricyclic (5-8-5 ring system) or tetracyclic (5-6-5-5) hydrocarbon scaffolds. **b** Bifunctional Di-TPS catalyzes a two-step cyclization that is performed by two separate enzyme functions involving a bicyclic diphosphate intermediate that leads to the labdane-related diterpenoids. The N-terminal class II domain catalyzes a protonation dependent cyclization that yields a bicyclic 6-6 copalyl- or 5-6 diphosphate. The C-terminal class I domain subsequently catalyzes ionization-dependent cyclization to yield the final cyclic products. Note that different copalyl diphosphate stereoisomers are generated. Cartoons illustrate domain organizations (class I: C1-fold or α-bundle fold, referred to as α-domain; class II: C2-fold or α-barrel fold, referred to as βγ-domains) and conserved catalytic motifs. Enzyme names refer to fungal source organisms and cyclization activities described in the text. CDP: copalyl diphosphate

Conserved aspartate-rich motifs in these enzymes likewise facilitate Mg-ion mediated binding of the diphosphate group.

Diterpenoids generated by the second, two-step route involve two separate enzyme activities. First, a class II-type, protonation-dependent mechanism generates a carbocation at the terminal C14-C15 double bond of the prenyl diphosphate chain that is cyclized into a bicyclic diphosphate characteristic of labdane-related diterpenoids [180]. Second, a class I ionization-dependent cleavage of the diphosphate group is followed by carbocation-triggered cyclization to yield the final cyclic scaffold. In fungi, these two class I and class II activities are present in one large bifunctional enzyme. The N-terminal region contains the α-domain of class I terpenoid synthases, whereas the C-terminal domain contains the α-barrel (or γβ-) domains of class II terpenoid synthases (Fig. 7).

It is believed that these three domain (γβα) proteins found in fungi and plants are the result of an ancient fusion event of bacterial class I (α) and class II (γβ) terpenoid synthases. In fact, typical bacterial diterpene synthases are not fused, and cyclization involves class II (γβ) and class I (α) terpenoid synthases for this two-step cyclization mechanism [181]. In plants, however, many three-domain (γβα) diterpene synthases are monofunctional and catalyze either a protonation-or an ionization-dependent cyclization reaction (on a bicyclic diphosphate or GGPP) because the other domain is not functional [180]. Structures have been solved for the monofunctional taxadiene synthase [182], which converts GGPP directly into taxadiene; bifunctional abietadiene synthase [183], which has two functional domains like the fungal enzymes; and a monofunctional copalyl diphosphate synthase [184]. Although the characterized fungal bifunctional diterpene synthases share little sequence homology with plant enzymes except for some conserved motifs (Fig. 7), structural modeling of the fungal proteins in Phyre2 [185] reveals that they have the same three-domain structure as abietadiene and taxadiene synthase.

2.3.2 Fusicoccanes and Other Diterpenoids Made by Monofunctional Enzymes

Fusicoccanes are potent phytotoxins known to be synthesized by a few fungal species [186]. *Phomopsis amygdali* causes wilting disease in trees and produces the diterpenoid fusicoccin A that binds and permanently activates plasma membrane H^+-ATPase [186], which causes severe physiological effects in plants. A related compound, cotylenin A, made by the fungus *Cladosporium* sp. was shown to have the same mode of action (structures shown in Fig. 8). Related fusicoccane phytotoxins, brassicene A-F, were subsequently isolated from the phytopathogenic fungus *Alternaria brasscicola* that causes leaf spot in Brassica plants [186]. Fusicoccanes bind to a highly conserved family of 14-3-3 proteins in eukaryotes, which regulate a wide range of cellular functions. Interaction of cotylenin A with 14-3-3 proteins has been shown to induce differentiation of leukemia cells and apoptosis of cancer cells [186].

Identification of the gene clusters encoding fusicoccane biosynthetic pathways followed a strategy that was previously successfully applied to the cloning of bifunctional diterpene synthases described below. All GGPP using biosynthetic pathways (including carotenoid, indole-diterpene, and diterpene pathways) require a dedicated GGPP synthase that adds one isoprene unit to FPP, which is a

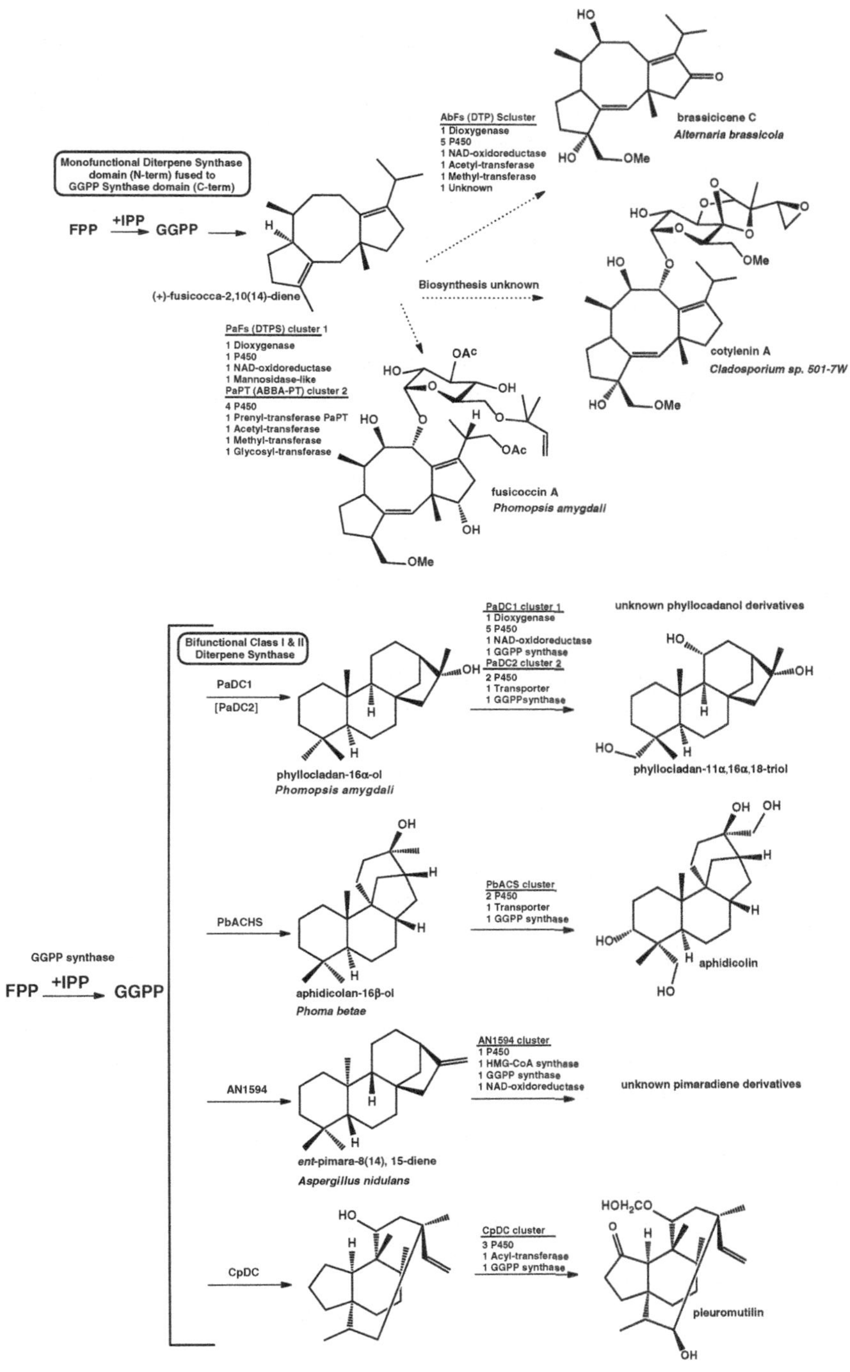

Fig. 8 Biosynthesis of fusicoccane- and labdane-related diterpenoids. **a** Fusioccoa-2,10(14)-diene is modified into different fusicoccane compounds. **b** Bifunctional diterpene synthases make different labdane-related scaffolds that are modified into bioactive compounds. Biosynthetic gene clusters, if known, are shown and named based on their corresponding diterpene synthases (or prenyl-transferase in one case). See text and Fig. 7 for details

prenyldiphosphate made by all eukaryotic cells. In fungi (and bacteria), GGPP synthase genes are typically located in gene clusters (operons in bacteria) of GGPP requiring pathways. Reverse-transcriptase polymerase chain reaction with degenerate primers to amplify GGPP synthase cDNA sequences and subsequent genome walking to identify flanking regions therefore has been a successful strategy to clone several monofunctional diterpene synthases.

Fusicocca-2,10(14)diene synthase (PaFS) from *P. amygdali* was the first monofunctional diterpene synthase cloned and characterized in *E. coli* [187] (Fig. 7). This enzyme, like all other fungal monofunctional diterpene synthases characterized so far, is a chimeric protein where the GGPP synthase is fused to the C-terminus of the terpene synthase domain. Genome walking led to the identification of four additional biosynthetic genes clustered with the chimeric terpene synthase gene [187]. Enzymes responsible for oxidative modification of the hydrocarbon scaffold were subsequently biochemically characterized [188]. Later, a second gene cluster was identified that encodes nine additional genes needed for the biosynthesis of fusicoccin A [189] (Fig. 8). This cluster was identified in the draft genome sequence of *P. amygdali* based on a previously identified ABBA-type prenyltransferases that catalyzed in vitro the reverse O-prenylation of glucose with DMAPP [70], which is a prenylation that is also present in fusiccocin A. A combination of in vitro assays with recombinant enzymes and gene disruption allowed Dairi's group to then establish the complete fusicoccin A pathway [189]. His group also identified and functionally characterized the brassicicene C biosynthetic pathway from *A. brassicola*, which includes, in addition to a fusicoccadiene synthase (AbFs), a dioxygenase and oxidoreductase as well as five P450s [188, 190, 191].

Two additional fusicoccadiene synthase homologs (PaPS, AcOS) with new cyclization activities were identified and biochemically characterized (Fig. 7). A second chimeric diterpene synthase (PaPS) was cloned from *P. amygdali* that cyclizes GGPP into the tetracyclic (5/6/5/5) phomopsene. This tetracyclic hydrocarbon is the precursor for methylphomopsenonate made by *P. amygdali* and for related spirocyclic diterpenoids isolated from other fungi [192]. Mining of the genome of *Aspergillus clavatus* recently led to the identification of a chimeric sesterterpene (C-25) synthase (AcOS) [193]. The prenyldiphosphate synthase domain of AcOS converts GPP and FPP (not GGPP) into geranylfarnesyl diphosphate (GFPP) to provide the substrate for subsequent cyclization by terpene synthase domain to ophiobolin F, which is a fusicoccadienol tricycle appended with a 5-carbon extension.

The few monofunctional diterpene synthases and associated biosynthetic genes characterized so far are all found in Ascomycota. Except for pleuromutulin biosynthesis (see below), no genes encoding diterpenoid biosynthetic enzymes from Basidiomycota have been identified. This may not be surprising considering that only a handful of homologs to known diterpene synthases (both monofunctional and bifunctional) can be found in currently available Basidiomycota genomes (Fig. 2).

The scarcity of diterpene synthase homologs in our genome survey seems to be supported by the relatively small number of diterpenoid compounds characterized from Basidiomycota [194] compared to the large number of sesquiterpenoid natural products isolated from these fungi. However, the assumption that diterpenoids do not play a major role in the natural products portfolio of Basidiomycota may not be correct. So far, only relatively few genome sequences are available for Basidiomycota; the vast majority have been sequenced for reasons other than natural product discovery. Most of the well-known medicinal mushrooms [37] have not been genome sequenced, but a large number of the known diterpenoids from Basidiomycota have been isolated from these mushrooms (e.g. *Cyathus helence* [bird's nest fungus], *Hericium erinaceum* [lion's mane fungus], *Sarcodon scabrosus* [bitter tooth fungus]) [194]. Furthermore, our genome searches were carried out with known fungal diterpene synthase sequences, including the full chimeric sequences of fungal monofunctional enzymes, such as PaFS. Basidiomycota may not use chimeric diterpene synthases and may instead use enzymes that are currently grouped with other terpene/prenyl synthases/cyclase sequences. The most likely class of such enzymes would be the sesquiterpene synthases. In fact, except for a few examples for labdane-related diterpenoids, most of the identified diterpenoids from Basidiomycota have tricyclic structures, which suggest an ionization-dependent cyclization mechanism (Fig. 9).

2.3.3 Labdane-Related Diterpenoids Made by Bifunctional Enzymes

As mentioned above, biosynthesis of labdane-type diterpenoids requires a two-step cyclization pathway involving first a protonation dependent cyclization of GGPP to form the characteristic labdane bicycle and, in a second step, ionization-dependent cyclization at a separate active site to generate the final cyclic product (Fig. 7). Cyclization of GGPP to *ent*-CDP and then to the tetracyclic *ent*-kaurene generates the precursor for gibberellin (gibberellic acids, GA) phytohormones that are major regulators of plant growth and development. It is believed that because of its essential role in plants, *ent*-kaurene represents the ancestral diterpenoid cyclization pathway from which alternative cyclization routes evolved to generate the large diversity of labdane-type compounds known today [180]. In fact, it has been shown that single amino acid changes are sufficient to alter the product profile of the class-I *ent*-kaurene synthase to form new cyclic scaffolds [195, 196].

Despite their critical function in plants, GAs were first isolated in the early 20th century from a fungal phytopathogen, the rice pathogen *F. fujikuroi* (teleomorph: *Gibberella*). Following the realization of the importance of GAs for plant growth, biotechnological production methods were established. In the early to mid-1990s, GA biosynthetic genes, including dedicated *ent*-CDP and *ent*-kaurene synthases, were identified in plants (reviewed in [197–199]). Structural studies showed that both synthases are tri-domain proteins that share the same ancestor, but each enzyme has only one functional active site (see above) [182–184].

Identification of fungal GA biosynthetic genes followed later and uncovered fundamental differences between the fungal and plant pathways, pointing to a largely divergent evolution except for presumably a shared ancestral *ent*-CDP,

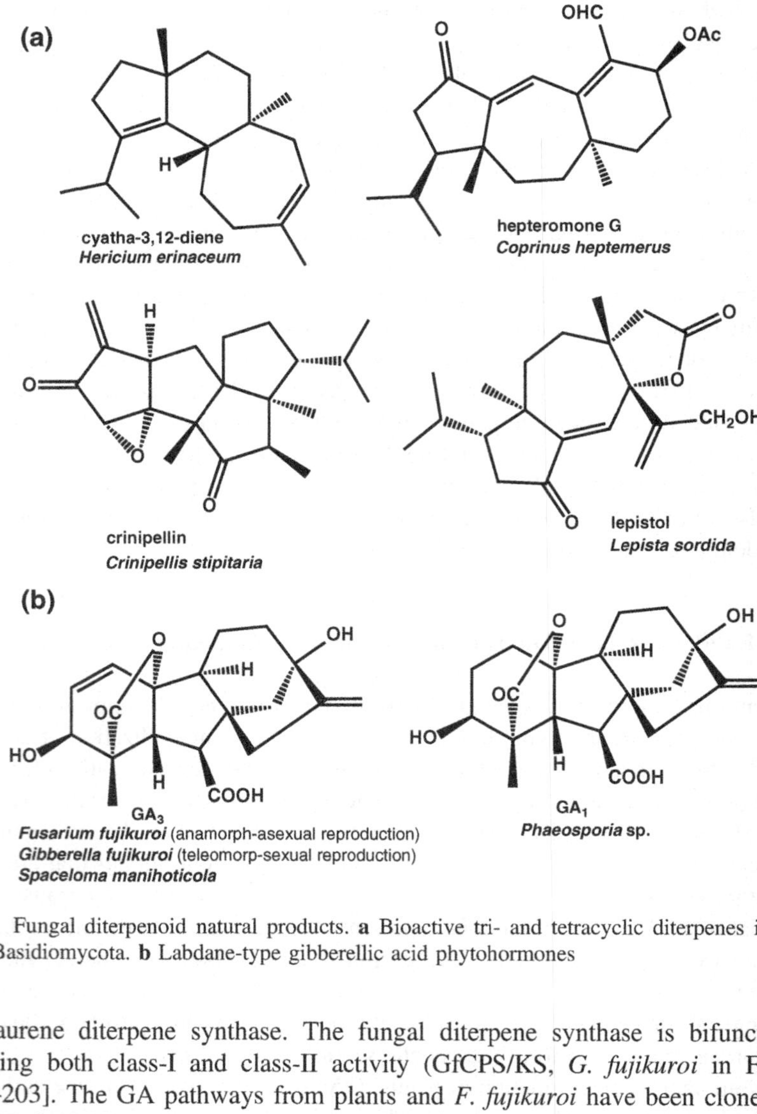

Fig. 9 Fungal diterpenoid natural products. **a** Bioactive tri- and tetracyclic diterpenes isolated from Basidiomycota. **b** Labdane-type gibberellic acid phytohormones

ent-kaurene diterpene synthase. The fungal diterpene synthase is bifunctional, retaining both class-I and class-II activity (GfCPS/KS, *G. fujikuroi* in Fig. 7) [200–203]. The GA pathways from plants and *F. fujikuroi* have been cloned and characterized; they are reviewed in detail in [197]. Notable are the many oxidation steps involved in the production of GAs. (Note that GAs are numbered according to their order of discovery; only few GA structures are bioactive). In GA_3 biosynthesis by *F. fujikuroi*, oxidations are carried out by four P450s and one desaturase; each P450 catalyzes multiple scaffold modifications. Functional GA pathways were also identified in *Phaeosphaeria sp.* (PfCPS/KS in Fig. 7), *Sphaceloma manihoticola.* These fungi produce less oxidized GAs (Fig. 9) because their biosynthetic gene clusters lack a desaturase or a desaturase and P450, respectively, which are present in the *F. fujikuroi* GA cluster [197].

Although only distantly related, these GA-producing fungi share conserved biosynthetic genes that are rearranged in their respective clusters. Horizontal gene transfer is presumed to be responsible for the spread of GA clusters among fungi. GA clusters (complete or remnants thereof) were also found in other fungi closely related to *F. fujikuroi*. However, GA production could only be established in a few strains; in some clusters, mutations were shown to render pathways nonfunctional (see detailed overview in [197]). This serves as a reminder that some biosynthetic gene clusters identified by in silico genome mining may be evolutionary junk. Careful sequence analysis and re-annotation should precede efforts toward functional characterization via heterologous means.

As hypothesized for labdane-diterpenoid biosynthesis in plants, ancestral GA-type pathways in fungi may have diversified to produce other bioactive compounds. New modifying enzymes may have been recruited into the cluster and activities of GA-biosynthetic cluster genes evolved to act on new labdane scaffolds. Although our genome survey suggests that more than 100 such putative clusters may be found in a quarter of the sequenced Ascomycota genomes (Fig. 2), only three such clusters have so far been characterized from these fungi; this is in contrast to the many labdane-type diterpenoid biosynthetic genes characterized from plants [204]. The first bifunctional diterpene synthase leading to alternative labdane skeletons cloned and characterized was the aphdicolan-16β-ol synthase (PbACS in Fig. 7) [205] from *Phoma betae*. This enzyme generates the scaffold for the DNA polymerase inhibitor aphidicolin. Genome walking was used to identify and clone the clustered aphidicolin biosynthetic genes [206] (Fig. 8), and then heterologously express this pathway in *A. oryzae* [207].

A second fungal labdane-diterpenoid pathway was identified in the fusiccocin producing fungus *P. amygdali*. Several GGPP synthase sequences were amplified; genome walking led to the identification of three diterpene synthase sequences, including the monofunctional fusicoccadiene synthase (PaFS) described above and two bifunctional CDP/KS synthase homologs, PaDC1 and PaDC2 [208]. PaDC1 and PaDC2 (in Fig. 7) are each located in gene clusters; however, although only the CDP-synthase domain is functional in PaDC2, PaDC1 converts GGDP into a novel diterpenoid phyllocladan-16α-ol (Fig. 8). The two gene clusters together are proposed to be responsible for the biosynthesis of phyllocladan-11α,16α,18-triol [208]. Most recently, a diterpene cluster has been identified through genome mining in *A. niger*. In this case, overexpression of a transcription factor located within the cluster resulted in the production of *ent*-pimara-8(14),15 diene, suggesting that the bifunctional diterpene synthase located in this cluster is a pimaradiene synthase [209].

Finally, only one labdane-type diterpene synthase gene has been cloned from Basidiomycota. A patent application described the identification of the biosynthetic pathway for pleuromutilin (Fig. 8) from *Clitopilus passeckeranus* [210]. This diterpenoid has been isolated from *Clitopilus* and related fungi [211]. Derivatives of pleuromutilin are commercially important antibiotics and are used in veterinary medicine (tiamulin, valnemurin) and for human treatments (retapamulin) [212, 213]. Transfer of the biosynthetic gene cluster into a heterologous production host is expected to yield increased production titers and allow modification of the pleuromutilin scaffold to afford production of derivatives. A cyclization pathway forming the tricyclic pleuromutilin scaffold has been proposed in [180] and is thought to involve the generation of a rearranged bicycle (5-6) diphosphate, which is subsequently cyclized upon diphosphate cleavage into a tricyclic skeleton.

Homology searches in Basidiomycota genomes for bifunctional diterpene synthases, however, result in only 11 putative homologs in the genome sequences of five strains (Fig. 2). Few labdane diterpenoids have been isolated from Basidiomycota [194], which could mean that this class of diterpenoid metabolites does not play a significant role in these fungi.

2.4 Triterpenoids

2.4.1 Overview

Triterpenoids constitute a large group of natural products that are particularly ubiquitous in plants, and more than 100 different skeletons have been described (reviewed in [214, 215]). Class II triterpene cyclases catalyze proton-initiated cyclization of either squalene (bacterial squalene hopene cyclases) or oxidosqualene (eukaryotic cyclases) into mostly penta- or tetracyclic ring systems, although other cyclizations are also known [214, 216]. Triterpene cyclases are considered to catalyze among the most complex enzyme reactions by guiding in a single-step carbocation-driven multicycle formation with frequently exquisite product specificity.

The most basic function of triterpenoids is their essential role as membrane sterols in eukaryotes (and in some bacteria). The three eukaryotic kingdoms (plants, fungi, animals) make different types of sterols: phytosterols (sitosterol, stigmasterol, campestrol) in plants, cholesterol in animals, and ergosterol in fungi [217]. All of these structures are derived from cyclization of oxidosqualene into either cycloarteneol in plants or lanosterol in animals and fungi (Fig. 10). Multiple oxidation steps subsequently generate the different sterols.

In addition to the housekeeping cycloartenol synthase, plants have additional cyclase homologs that make other triterpenoid skeletons. A large number of P450s have been identified that oxidize these skeletons into diverse secondary metabolites, which are frequently further glycosylated (reviewed in [218]). Lanosterol is the precursor to both ergosterol and triterpenoid metabolites in fungi. Some fungi cyclize oxidosqualene into the related protostadienol, which is modified into bioactive compounds (Fig. 10).

2.4.2 Fungal Triterpenoid Biosynthesis

A large body of publications describes triterpenoid biosynthesis and metabolic engineering in plants, while very little is known about the structural diversity,

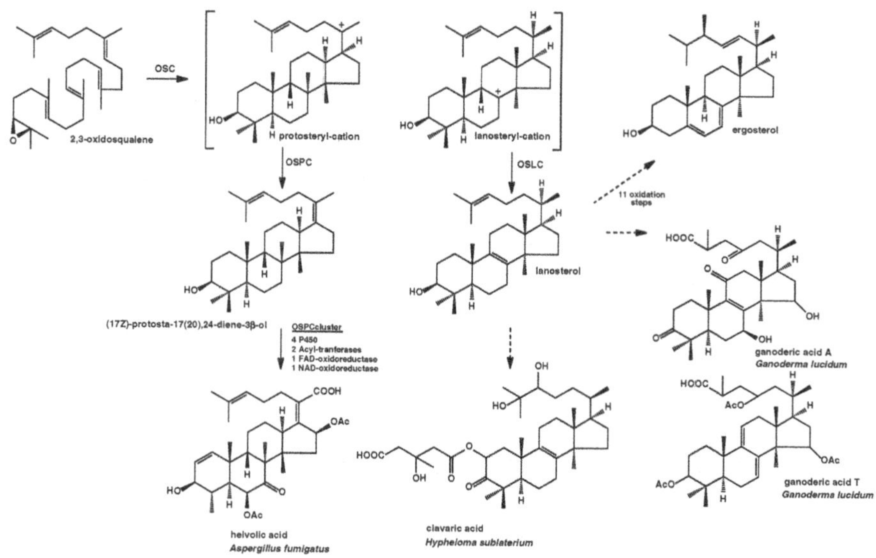

Fig. 10 Fungal triterpenoid biosynthesis. Oxidosqualene is cyclized by a cyclase (OSC) into tetracycles hydrocarbon scaffolds as biosynthetic precursors for the membrane sterol ergosterol and bioactive triterpenoids

biological significance, and biosynthesis of fungal triterpenoids. Bioactive triterpenoids, such as the fusidane antibiotics [219], have been identified from Ascomycota, but the vast majority of fungal triterpenoid natural products have been isolated from Basidiomycota [220]. Genera such as *Ganoderma*, *Innonotus*, (*Wolfi*) *Poria*, *Laetiporus*, *Antrodia*, and *Daedalea* are well-known producers of diverse lanosterane-type triterpenoids, with pharmacological properties that include anti-tumor, apoptotic, and antimalarial activities (reviewed in [220]). Ganoderic acids produced by the medicinal mushroom *Ganodera lucidum* are among the best known fungal triterpenoids [221], and considerable effort has been devoted to the development of fermentative production processes [32].

Two fungal triterpenoid cyclases involved in secondary metabolite biosynthesis have so far been characterized on a molecular level. *A. fumigatus* produces fusidane-type triterpenoids, including the anticancer helvolic acid derived from the protostadienol scaffold (Fig. 10). The corresponding triterpene synthase and associated fusidane biosynthetic gene cluster has been identified by two groups [222, 223] (Fig. 10). It was shown that the cyclization outcome of this enzyme (as for diterpene synthases, see above) can be readily altered; substitution of one amino acid residue was sufficient to direct cyclization either towards lanosterol or protostadienol [224]. Two genes, oxidosqualene synthase and squalene epoxidase, were identified in the antitumor clavaric acid producer *Hypholoma sublaterium* (a basidiomycete), and are expected to be required for the production of this secondary metabolite [225, 226].

In an effort to elucidate the biosynthesis of lanosterane-type triterpenoids in *Wolfiporia* and *Ganoderma*, genomic and transcriptomic studies were published by several groups [227–230]. These studies suggested that *Ganoderma* has two oxidosqualene-lanosterol cyclases, with one homolog dedicated to either ergosterol or triterpenoid natural product biosynthesis. Our genome survey (Fig. 2) also showed that Basidiomycota genomes commonly have one or two (sometimes three) triterpene cyclase homologs. Several Ascomycota genera, such as *Aspergillus* and *Neosartoryam*, however, have between 2 and 6 homologs, of which some are part of predicted biosynthetic clusters. In Basidiomycota, triterpene cyclases seem not to be located in closely arranged gene clusters, but tailoring enzymes are likely to be coregulated with a dedicated secondary metabolic triterpene cyclase. In fact, genome analysis of *Ganoderma* revealed a large complement of cytochrome P450s (CYPome), including 24 P450 clusters and 78 P450s that appear to be co-regulated with lanosterol cyclase [228].

In plants, P450s are considered to be drivers of chemical diversity, and they play a major role in the diversification of triterpenoid secondary metabolism [231]. The vast majority of triterpenoid scaffold modifications are catalyzed by P450s. Although genes are typically not clustered in plants, notable exceptions have been identified in *Arabidopsis,* where triterpene cylases and P450s form co-regulated gene clusters [232, 233]. Considering the massive expansion of the cytochrome P450 family in Basidiomycota, P450s likely play a similar role in triterpenoid secondary metabolism in these fungi. After plants, Basidiomycota have the largest and functionally most diverse CYPome, with 100–300 P450s per genome in the *Agaricomycotina* subdivison [234, 235], which includes ~70 % of the known Basidiomycota [4]. Only few Basidiomycota subdivisions, including pathogenic Basidiomycota such as *Puccinia*, *Ustilago*, and *Malessizia,* have a drastically reduced CYPome (less than 20), which is also reflected in a largely absent secondary metabolome in these fungi.

2.5 Terpenoids of Mixed Biogenetic Origin

Many natural products are hybrid compounds built from precursors derived from different biosynthetic pathways. Ascomycota are well known for their ability to produce structurally diverse meroterpenoids and indole-diterpenoids that are hybrid scaffolds containing terpenoid and polyketide or indole domains. Their biosynthetic gene clusters encode enzymes for the generation and tailoring of both domains. In this section, we focus on the installation of the terpenoid domains shown in Fig. 11.

Two different types of prenyl transferases (see Sect. 2.1.2) catalyze transfer of a prenyl-chain to either a polyketide or indole precursor. UbiA-type prenyl transferases catalyze FPP transfer in the biosynthesis of meroterpenoids such as terretonin, pyripyropene, austinol, and mycophenolic acid [57–59, 75–77]. IPPS-type prenyl transferases catalyze the transfer of GGPP in the biosynthesis of a large group of structurally diverse indole-diterpenoids, such as the lolitrems made by

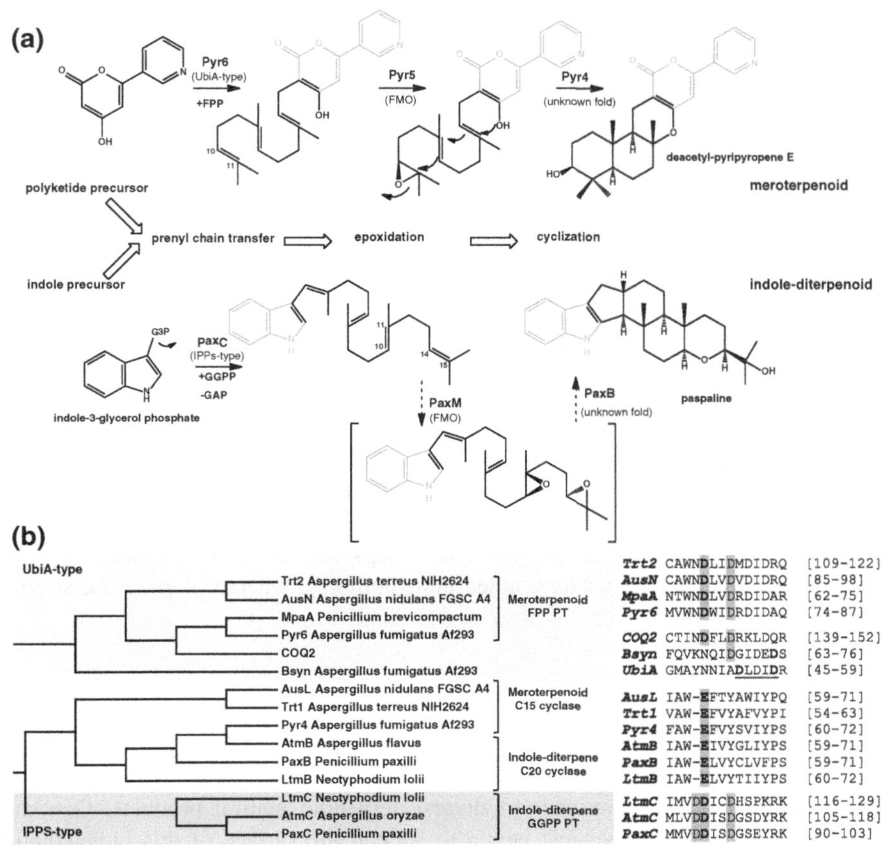

Fig. 11 Meroterpenoids and indole-diterpenoids are of mixed biogenetic origin. **a** Biosynthesis of most known hybrid compounds containing a terpenoid- and a polyketide or indole-derived moiety involves transfer, epoxidation, and cyclization of a prenyl chain (black) to a nonterpenoid precursor (gray). Shown are steps involved in pyripyropene and paxilline biosynthesis. **b** A dendogram showing relatedness of different prenyl cyclases and transferases involved in indole-diterpene and meroterpenoid biosynthesis. Sequences were aligned by ClustalW (Gonnet) and subsequently subjected to phylogenetic analysis using the minimum evolution method in MEGA6 [114]. Conserved acidic residues (D/E) likely involved in catalysis (underlined: catalytic aspartate residues in UbiA structure [81], boxed: residue numbers) of each sequence are shown. Cyclases and prenyl transferases separate into three groups, quinone prenyltransferases, *E. coli* UbiA, and *S. cerevisiae* CoQ2, as well as β-trans-bergamotene cylase Bsyn, are included as UbiA-type reference sequences [88]. NCBI sequence accession numbers: Trt2: XP_001209380, AusN: XP_682528, MpaA: ADY00128, Pyr6: XP_751272, COQ2: CAA96321, Bsyn: AGI05042, UbiA: AAC43134, AusL: XP_682526, Trt1: XP_001209379, Pyr4: XP_751270, AtmB: CAP53939, PaxB: ADO29934, LtmB: ABF20226, LtmC: ABF20225, AtmC: EIT82606, PaxC: AAK11529

grass endophytes and aflatreme, paxicilline mycotoxins produced by *Aspergillus* and *Penicillium*, respectively [47, 54–56, 72–74].

Cyclization of prenyl chains attached to either the polyketide ore indole moieties presumably involves a protonation-dependent cyclization reaction catalyzed by membrane-bound cyclases (see Sect. 2.1.3) that have no homology to any known proteins. In each of the characterized biosynthetic gene clusters, a flavin-mono-oxygenase-type FMO is present that is required to catalyze prenyl-chain epoxidation as a prerequisite for subsequent cyclization (see Sect. 2.1.3).

Meroterpenoid and indole-terpenoid biosynthesis has so far only been studied in Ascomycota, and it is not known if these natural products are synthesized by Basidiomycota. A SciFinder search for indole-diterpenoids or meroterpenoids from Basidiomycota yielded no hits for the first class and less than 10 results that described the isolation of compounds that contain an aromatic ring attached to an acyclic farnesyl chain (e.g., [236]). These two groups of natural products may therefore not play an important role in Basidiomycota. A search of fungal genomes in the JGI Fungal Genome database with sequences of prenyl-chain cyclases (AtmB, PaxB, LtB, Pyr4, Trt1, AusL; Sect. 2.1.3) yielded only 19 hits in 5 Basidiomycota strains, whereas ~200 homologs are found in Ascomycota genomes. Half of these genomes have at least one homolog, whereas *Aspergillus* strains have 2–5 homologs.

3 Conclusions and Outlook

Fungi synthesize a vast number of diverse terpenoid natural products. Genome surveys show that we have just begun to scratch the surface of this biosynthetic diversity. The majority of explorations into the fungal terpenome have so far been in Ascomycota, whereas we have only recently begun to investigate the natural product potential of Basidiomycota using molecular and biochemical approaches. From our studies [19–23] and the observations described in this chapter, it is clear that the two fungal phyla have evolved different portfolios of terpenoid products and associated biosynthetic enzymes. Application of our current knowledge of secondary metabolite biosynthesis to searches of fungal genomes suggests that the secondary metabolome of Ascomycota is dominated by thiotemplated biosynthetic machineries. Sesqui- and triterpenoid natural products, on the other hand, seem to be playing major roles in the secondary metabolome of Basidiomycota.

It should be pointed out that these observed differences are based on the biosynthetic pathways and enzymes that have been characterized to date, as well as on the small number of available fungal genomes that represent just a tiny fraction of the fungal diversity. Many genes annotated in fungal genomes, and in particular in Basidiomycota genomes, have no known homologs. These "unknown" predicted or putative genes may well encode entirely new biosynthetic activities, as was seen for terpene cyclases discovered in indole-diterpene and meroterpenoid biosynthesis. Furthermore, many other genes may have been wrongly annotated and may catalyze very different activities. Of particular note here is the discovery of a new UbiA-type sesquiterpene cyclase in the fumagillin biosynthetic pathway. Genome mining efforts with these new sequences will likely lead to new terpenoid natural products pathways.

To what extent we will be able to characterize the fungal terpenome is limited by our ability to biochemically characterize biosynthetic genes and their functions. Our current technologies for pathway identification and discovery are slow and inadequate to keep up with the massive influx of sequencing data. We need to develop strategies that will enable us to move rapidly from in silico biosynthetic pathway identification to high-throughput assembly and expression of predicted pathways, with concurrent analytical profiling of produced compounds. For this to happen, a number of bottlenecks and obstacles will need to be tackled. The development of such strategies will be crucial for reviving the natural products drug discovery pipeline.

The first step for the implementation of an in silico to metabolite integrated pipeline involves the accurate identification of natural products genes and pathways in genomes. A number of bioinformatics tools for genome mining (e.g. antiSMASH 2.0 [237]) have been developed for this purpose (for a recent review, see [238]), but these tools rely on algorithms trained with hidden Markov models derived from known biosynthetic genes—the majority of which come from Ascomycota and Actinomycetes. In our experience, when applied to terpenoid pathways from Basidiomycota, they tend to identify only a subset of the pathways and have difficulties identifying cluster boundaries.

Accurate structural annotation of biosynthetic genes will be crucial for any high-throughput synthetic biology workflow involving the assembly of pathways from synthetic genes based on in silico annotated gene structures. Our own experience has shown that gene predictions in the automated genome annotations of Basidiomycota are frequently incorrect, requiring manual reannotation and several attempts at amplifying spliced genes from cDNA. Basidiomycota have very intron-rich genomes [239], and we found many very small and unpredicted introns/exons (sometimes only 6–9 bps in size) in genes cloned from cDNA. Genes synthesized based on cDNA predictions are therefore often nonfunctional.

Deep RNA sequencing has been shown to significantly improve the accuracy of a large fraction of the gene models for *Laccaria bicolor* by finding all splice sites in genes through deep sequence coverage; one single transcriptome data set of 30 million reads provided 30 times more sequence coverage and hence much greater resolution for gene structure annotation than the EST libraries available for this fungus [240]. So far, only some of the more recently sequenced genomes have associated deep RNA sequence data [228, 230]. High-resolution transcriptomics analysis will be essential not only for direct translation of genomic information into synthetic gene assembly of pathway discovery, but also for improved functional annotation important for biosynthetic pathway identification.

Construction of gene co-expression networks built on the physical distance to natural biosynthesis product-associated seed genes represents a powerful tool for pathway discovery. The fact that natural product pathway genes are generally co-regulated through levels of shared transcriptional control elements [241] represents yet another approach for network analysis within and also across species. Significant advances have been made in understanding the regulatory control elements of NP pathways in filamentous fungi, including the velvet family of regulatory proteins that are conserved among Ascomycota and Basidiomycota [38, 41, 242]. Genome analysis of *Ganoderma* and *Schizophyllum* [228, 243, 244] suggests high conservation of regulatory networks among mushroom-forming fungi, which could be exploited for network building. Yet, gene co-expression network analysis so far has been largely applied for the discovery of natural product genes in plants [245, 246]. Analysis based on co-expression arrays was only recently applied to natural product gene cluster analysis in *A. nidulans* [28]. Comprehensive deep transcriptome analysis of fungi grown under conditions that activate a broad range of natural products pathways [29, 39, 247, 248] should enable accurate prediction of complete pathways, including satellite or super-clusters of split pathways known from fungi [57, 249–251].

Deletion of biosynthetic genes represents the classic approach used for functional characterization of pathways, and it has been used for many of the pathways described in this chapter. However, this is a slow process, requiring that the pathway is expressed under given laboratory growth conditions; and most importantly, the fungal producer strain can be genetically manipulated. Alternatively, gene clusters may be transferred into a heterologous host if a compatible host is available. These strategies do not work for the vast majority of fungal strains for which no genetic approaches are available, are difficult to cultivate, and where the conditions to induce natural products biosynthesis are unknown. To access the natural product potential of these fungi, the development of synthetic biology approaches (discussed elsewhere in this book series) will be crucial for the rapid assembly of pathways from synthetic genes in suitable designed hosts. Such hosts could be a filamentous fungus or a yeast strain equipped with mechanisms to facilitate the expression of a potentially large complement of microsomal P450s and deal with the potential toxic metabolites produced by the pathways. Finally, refactoring of natural products pathways in heterologous platforms designed for high-throughput assembly and screening would provide the means for combinatorial approaches to explore an even larger diversity of natural products.

Acknowledgment Research on terpenoid biosynthesis in the author's laboratory is supported by grant GM080299 from the National Institutes of Health.

References

1. Bass D, Richards TA (2011) Three reasons to re-evaluate fungal diversity on Earth and in the ocean. Fungal Biol Rev 25:159–164
2. Stajich JE et al (2010) Insights into evolution of multicellular fungi from the assembled chromosomes of the mushroom *Coprinopsis cinerea (Coprinus cinereus)*. Proc Natl Acad Sci U S A 107(26):11889–11894

3. Blackwell M (2011) The fungi: 1, 2, 3 … 5.1 million species? Am J Bot 98(3):426–438
4. Hibbett DS et al (2007) A higher-level phylogenetic classification of the fungi. Mycol Res 111(Pt 5):509–547
5. Ma LJ et al (2013) *Fusarium* pathogenomics. Annu Rev Microbiol 67:399–416
6. Gauthier GM, Keller NP (2013) Crossover fungal pathogens: the biology and pathogenesis of fungi capable of crossing kingdoms to infect plants and humans. Fungal Genet Biol 61:146–157
7. Punt PJ et al (2002) Filamentous fungi as cell factories for heterologous protein production. Trends Biotechnol 20(5):200–206
8. Dufosse L et al (2014) Filamentous fungi are large-scale producers of pigments and colorants for the food industry. Curr Opin Biotechnol 26C:56–61
9. Van Den Berg M et al (2010) Biosynthesis of active pharmaceuticals: beta-lactam biosynthesis in filamentous fungi. Biotechnol Genet Eng Rev 27:1–32
10. Archer DB (2000) Filamentous fungi as microbial cell factories for food use. Curr Opin Biotechnol 11(5):478–483
11. Evidente A et al (2014) Fungal metabolites with anticancer activity. Nat Prod Rep 31 (5):617–627
12. Brase S et al (2009) Chemistry and biology of mycotoxins and related fungal metabolites. Chem Rev 109(9): 3903–3990
13. Bladt TT et al (2013) Anticancer and antifungal compounds from Aspergillus, Penicillium and other filamentous fungi. Molecules 18(9):11338–11376
14. Morel M et al (2013) Xenomic networks variability and adaptation traits in wood decaying fungi. Microb Biotechnol 6(3):248–263
15. Plett JM, Martin F (2011) Blurred boundaries: lifestyle lessons from ectomycorrhizal fungal genomes. Trends Genet 27(1):14–22
16. Welzel K et al (2005) Characterization of the ferrichrome a biosynthetic gene cluster in the homobasidiomycete *Omphalotus olearius*. FEMS Microbiol Lett 249(1):157–163
17. Bushley, KE, Turgeon BG (2010) Phylogenomics reveals subfamilies of fungal nonribosomal peptide synthetases and their evolutionary relationships. BMC Evol Biol 10:26
18. Bushley KE, Ripoll DR, Turgeon BG (2008) Module evolution and substrate specificity of fungal nonribosomal peptide synthetases involved in siderophore biosynthesis. BMC Evol Biol 8:328
19. Quin MB et al (2013) Mushroom hunting using bioinformatics: application of a predictive framework facilitates the selective identification of sesquiterpene synthases in Basidiomycota. Chembiochem 14:2480–2491
20. Wawrzyn GT et al (2012) Draft genome of *Omphalotus olearius* provides a predictive framework for sesquiterpenoid natural product biosynthesis in basidiomycota. Chem Biol 19 (6):772–783
21. Lopez-Gallego F, Wawrzyn GT, Schmidt-Dannert C (2010) Selectivity of fungal sesquiterpene synthases: role of the active site's H-1 α loop in catalysis. Appl Environ Microbiol 76(23):7723–7733
22. Lopez-Gallego F et al (2010) Sesquiterpene synthases Cop4 and Cop6 from *Coprinus cinereus*: catalytic promiscuity and cyclization of farnesyl pyrophosphate geometric isomers. ChemBiochem 11(8):1093–1106
23. Agger S, Lopez-Gallego F, Schmidt-Dannert C (2009) Diversity of sesquiterpene synthases in the basidiomycete *Coprinus cinereus*. Mol Microbiol 72(5):1181–1195
24. Winterberg B et al (2010) Elucidation of the complete ferrichrome a biosynthetic pathway in *Ustilago maydis*. Mol Microbiol 75(5):1260–1271
25. Wackler B et al (2012) Characterization of the *Suillus grevillei* quinone synthetase GreA supports a nonribosomal code for aromatic alpha-keto acids. Chembiochem 13(12):1798–1804
26. Condon BJ et al (2013) Comparative genome structure, secondary metabolite, and effector coding capacity across *Cochliobolus pathogens*. PLoS Genet 9(1):e1003233

27. Quin MB, Wawrzyn G, Schmidt-Dannert C (2013) Purification, crystallization and preliminary X-ray diffraction analysis of Omp6, a protoilludene synthase from *Omphalotus olearius*. Acta Crystallogr Sect F Struct Biol Cryst Commun 69(Pt 5):574–577
28. Andersen MR et al (2013) Accurate prediction of secondary metabolite gene clusters in filamentous fungi. *Proc Natl Acadamy* Sci U S A 110(1):E99–E107
29. Lim FY et al (2012) Toward awakening cryptic secondary metabolite gene clusters in filamentous fungi. Methods Enzymol 517:303–324
30. Zjawiony JK (2004) Biologically active compounds from *Aphyllophorales* (polypore) fungi. J Nat Prod 67(2):300–310
31. Lindequist U, Niedermeyer THJ, Julich WD (2005) The pharmacological potential of mushrooms. Evid-Based Complement Altern Med 2(3):285–299
32. Xu JW, Zhao W, Zhong JJ (2010) Biotechnological production and application of ganoderic acids. Appl Microbiol Biotechnol 87(2):457–466
33. Zhong JJ, Xiao JH (2009) Secondary metabolites from higher fungi: discovery, bioactivity, and bioproduction. Adv Biochem Eng Biotechnol 113:79–150
34. Alves MJ et al (2012) A review on antimicrobial activity of mushroom (Basidiomycetes) extracts and isolated compounds. Planta Med 78(16):1707–1718
35. Abraham WR (2001) Bioactive sesquiterpenes produced by fungi: are they useful for humans as well? Curr Med Chem 8(6):583–606
36. Elisashvili V (2012) Submerged cultivation of medicinal mushrooms: bioprocesses and products (review). Int J Med Mushrooms 14(3):211–239
37. Wasser SP (2011) Current findings, future trends, and unsolved problems in studies of medicinal mushrooms. Appl Microbiol Biotechnol 89(5):1323–1332
38. Brakhage AA (2013) Regulation of fungal secondary metabolism. Nat Rev Microbiol 11 (1):21–32
39. Brakhage AA, Schroeckh V (2011) Fungal secondary metabolites—strategies to activate silent gene clusters. Fungal Genet Biol 48(1):15–22
40. Osbourn A (2010) Secondary metabolic gene clusters: evolutionary toolkits for chemical innovation. Trends Genet 26(10):449–457
41. Yin W, Keller NP (2011) Transcriptional regulatory elements in fungal secondary metabolism. J Microbiol 49(3):329–339
42. Keller NP, Turner G, Bennett JW (2005) Fungal secondary metabolism—from biochemistry to genomics. Nat Rev Microbiol 3(12):937–947
43. Keller NP, Hohn TM (1997) Metabolic pathway gene clusters in filamentous fungi. Fungal Gen Biol 21(1):17–29
44. Johnson EA (2013) Biotechnology of non-*Saccharomyces* yeasts-the basidiomycetes. Appl Microbiol Biotechnol 97(17):7563–7577
45. Johnson EA (2003) *Phaffia rhodozyma*: colorful odyssey. Int Microbiol 6(3):169–174
46. Christianson DW (2008) Unearthing the roots of the terpenome. Curr Opin Chem Biol 12 (2):141–150
47. Tagami K et al (2013) Reconstitution of biosynthetic machinery for indole-diterpene paxilline in *Aspergillus oryzae*. J Am Chem Soc 135(4):1260–1263
48. Gao Y, Honzatko RB, Peters RJ (2012) Terpenoid synthase structures: a so far incomplete view of complex catalysis. Nat Prod Rep 29(10):1153–1175
49. Oldfield E, Lin FY (2012) Terpene biosynthesis: modularity rules. Angew Chem Int Ed Engl 51(5):1124–1137
50. Cao R et al (2010) Diterpene cyclases and the nature of the isoprene fold. Proteins 78 (11):2417–2432
51. Bonitz T et al (2011) Evolutionary relationships of microbial aromatic prenyltransferases. PLoS One 6(11):e27336
52. Saleh O et al (2009) Prenyl transfer to aromatic substrates in the biosynthesis of aminocoumarins, meroterpenoids and phenazines: the ABBA prenyltransferase family. Phytochemistry 70(15–16):1728–1738

53. Heide L (2009) Prenyl transfer to aromatic substrates: genetics and enzymology. Curr Opin Chem Biol 13(2):171–179
54. Saikia S et al (2012) Functional analysis of an indole-diterpene gene cluster for lolitrem B biosynthesis in the grass endosymbiont *Epichloe festucae*. FEBS Lett 586(16):2563–2569
55. Saikia S et al (2008) The genetic basis for indole-diterpene chemical diversity in filamentous fungi. Mycol Res 112(Pt 2):184–199
56. Nicholson MJ et al (2009) Identification of two aflatrem biosynthesis gene loci in Aspergillus flavus and metabolic engineering of *Penicillium paxilli* to elucidate their function. Appl Environ Microbiol 75(23):7469–7481
57. Lo HC et al (2012) Two separate gene clusters encode the biosynthetic pathway for the meroterpenoids austinol and dehydroaustinol in *Aspergillus nidulans*. J Am Chem Soc 134 (10):4709–4720
58. Guo CJ et al (2012) Molecular genetic characterization of a cluster in *A. terreus* for biosynthesis of the meroterpenoid terretonin. Org Lett 14(22):5684–5687
59. Itoh T et al (2010) Reconstitution of a fungal meroterpenoid biosynthesis reveals the involvement of a novel family of terpene cyclases. Nat Chem 2(10):858–864
60. Thoma R et al (2004) Insight into steroid scaffold formation from the structure of human oxidosqualene cyclase. Nature 432(7013):118–122
61. Racolta S et al (2012) The triterpene cyclase protein family: a systematic analysis. Proteins 80(8):2009–2019
62. Li SM (2010) Prenylated indole derivatives from fungi: structure diversity, biological activities, biosynthesis and chemoenzymatic synthesis. Nat Prod Rep 27(1):57–78
63. Yu X, Li SM (2012) Prenyltransferases of the dimethylallyltryptophan synthase superfamily. Methods Enzymol 516:259–278
64. Mundt K, Li SM (2013) CdpC2PT, a reverse prenyltransferase from *Neosartorya fischeri* with a distinct substrate preference from known C2-prenyltransferases. Microbiology 159(Pt 10):2169–2179
65. Unsold IA, Li SM (2006) Reverse prenyltransferase in the biosynthesis of fumigaclavine C in *Aspergillus fumigatus*: gene expression, purification, and characterization of fumigaclavine C synthase FGAPT1. Chembiochem 7(1):158–164
66. Pockrandt D et al (2012) New insights into the biosynthesis of prenylated xanthones: Xptb from *Aspergillus nidulans* catalyses an O-prenylation of xanthones. Chembiochem 13 (18):2764–2771
67. Chooi YH et al (2012) Discovery and characterization of a group of fungal polycyclic polyketide prenyltransferases. J Am Chem Soc 134(22):9428–9437
68. Zou HX et al (2011) The tyrosine O-prenyltransferase SirD catalyzes O-, N-, and C-prenylations. Appl Microbiol Biotechnol 89(5):1443–1451
69. Kremer A, Li SM (2010) A tyrosine O-prenyltransferase catalyses the first pathway-specific step in the biosynthesis of sirodesmin PL. Microbiology 156(Pt 1):278–286
70. Noike M et al (2012) An enzyme catalyzing O-prenylation of the glucose moiety of fusicoccin A, a diterpene glucoside produced by the fungus *Phomopsis amygdali*. Chembiochem 13(4):566–573
71. Metzger U et al (2009) The structure of dimethylallyl tryptophan synthase reveals a common architecture of aromatic prenyltransferases in fungi and bacteria. Proc Natl Acad Sci USA 106(34):14309–14314
72. Young CA et al (2006) A complex gene cluster for indole-diterpene biosynthesis in the grass endophyte *Neotyphodium lolii*. Fungal Genet Biol 43(10):679–693
73. Zhang S et al (2004) Indole-diterpene gene cluster from *Aspergillus flavus*. Appl Environ Microbiol 70(11):6875–6883
74. Young C et al (2001) Molecular cloning and genetic analysis of an indole-diterpene gene cluster from *Penicillium paxilli*. Mol Microbiol 39(3):754–764
75. Matsuda Y et al (2012) Terretonin biosynthesis requires methylation as essential step for cyclization. Chembiochem 13(12):1738–1741

76. Itoh T et al (2012) Identification of a key prenyltransferase involved in biosynthesis of the most abundant fungal meroterpenoids derived from 3,5-dimethylorsellinic acid. Chembiochem 13(8):1132–1135
77. Regueira TB et al (2011) Molecular basis for mycophenolic acid biosynthesis in *Penicillium brevicompactum*. Appl Environ Microbiol 77(9):3035–3043
78. Yazaki K, Sasaki K, Tsurumaru Y (2009) Prenylation of aromatic compounds, a key diversification of plant secondary metabolites. Phytochemistry 70(15–16):1739–1745
79. Brauer L et al (2008) A structural model of the membrane-bound aromatic prenyltransferase UbiA from E. coli. Chembiochem 9(6):982–992
80. Ohara K et al (2009) Functional characterization of LePGT1, a membrane-bound prenyltransferase involved in the geranylation of p-hydroxybenzoic acid. Biochem J 421 (2):231–241
81. Cheng W, Li W (2014) Structural insights into ubiquinone biosynthesis in membranes. Science 343(6173):878–881
82. Buckingham J Dictionary of natural products (online version). CHEMnetBASE Chapman & Hall (2009) version 17.1: p. http://dnp.chemnetbase.co
83. Rynkiewicz MJ, Cane DE, Christianson DW (2001) Structure of trichodiene synthase from *Fusarium sporotrichioides* provides mechanistic inferences on the terpene cyclization cascade. Proc Natl Acad Sci USA 98(24):13543–13548
84. Caruthers JM et al (2000) Crystal structure determination of aristolochene synthase from the blue cheese mold *Penicillium roqueforti*. J Biol Chem 275(33):25533–25539
85. Chen F et al (2011) The family of terpene synthases in plants: a mid-size family of genes for specialized metabolism that is highly diversified throughout the kingdom. Plant J 66 (1):212–229
86. Degenhardt J, Kollner TG, Gershenzon J (2009) Monoterpene and sesquiterpene synthases and the origin of terpene skeletal diversity in plants. Phytochemistry 70(15–16):1621–1637
87. Sawai S, Saito K (2011) Triterpenoid biosynthesis and engineering in plants. Front Plant Sci 2:25
88. Lin HC et al (2013) The fumagillin biosynthetic gene cluster in *Aspergillus fumigatus* encodes a cryptic terpene cyclase involved in the formation of beta-trans-bergamotene. J Am Chem Soc 135(12):4616–4619
89. Liu X, Walsh CT (2009) Characterization of cyclo-acetoacetyl-L-tryptophan dimethylallyltransferase in cyclopiazonic acid biosynthesis: substrate promiscuity and site directed mutagenesis studies. Biochemistry 48(46):11032–11044
90. Saikia S et al (2006) Four gene products are required for the fungal synthesis of the indole-diterpene, paspaline. FEBS Lett 580(6):1625–1630
91. Young CA et al (2009) Indole-diterpene biosynthetic capability of epichloe endophytes as predicted by ltm gene analysis. Appl Environ Microbiol 75(7):2200–2211
92. Chooi YH, Cacho R, Tang Y (2010) Identification of the viridicatumtoxin and griseofulvin gene clusters from *Penicillium aethiopicum*. Chem Biol 17(5):483–494
93. Chooi YH et al (2013) A cytochrome P450 serves as an unexpected terpene cyclase during fungal meroterpenoid biosynthesis. J Am Chem Soc 135(45):16805–16808
94. Chooi YH, Tang Y (2012) Navigating the fungal polyketide chemical space: from genes to molecules. J Org Chem 77(22):9933–9953
95. Evans BS, Robinson SJ, Kelleher NL (2011) Surveys of non-ribosomal peptide and polyketide assembly lines in fungi and prospects for their analysis in vitro and in vivo. Fungal Genet Biol 48(1):49–61
96. Boettger D, Hertweck C (2013) Molecular diversity sculpted by fungal PKS-NRPS hybrids. Chembiochem 14(1):28–42
97. Schneider P, Bouhired S, Hoffmeister D (2008) Characterization of the atromentin biosynthesis genes and enzymes in the homobasidiomycete *Tapinella panuoides*. Fungal Genet Biol 45(11):1487–1496

98. Misiek M, Braesel J, Hoffmeister D (2011) Characterisation of the ArmA adenylation domain implies a more diverse secondary metabolism in the genus *Armillaria*. Fungal Biol 115 (8):775–781
99. Ishiuchi K et al (2012) Establishing a new methodology for genome mining and biosynthesis of polyketides and peptides through yeast molecular genetics. Chembiochem 13(6):846–854
100. Davis EM, Croteau R (2000) Cyclization enzymes in the biosynthesis of monoterpenes, sesquiterpenes, and diterpenes. In: Biosynthesis: Aromatic Polyketides, Isoprenoids, Alkaloids. Springer, Berlin, pp 53–95
101. Cane DE, Kang I (2000) Aristolochene synthase: purification, molecular cloning, high-level expression in *Escherichia coli*, and characterization of the *Aspergillus terreus* cyclase. Arch Biochem Biophys 376(2):354–364
102. Christianson DW (2006) Structural biology and chemistry of the terpenoid cyclases. Chem Rev 106(8):3412–3442
103. Vedula LS et al (2008) Structural and mechanistic analysis of trichodiene synthase using site-directed mutagenesis: probing the catalytic function of tyrosine-295 and the asparagine-225/serine-229/glutamate-233-Mg^{2+} B motif. Arch Biochem Biophys 469(2):184–194
104. Lesburg CA et al (1998) Managing and manipulating carbocations in biology: terpenoid cyclase structure and mechanism. Curr Opin Struct Biol 8(6):695–703
105. Engels B et al (2011) Cloning and characterization of an *Armillaria gallica* cDNA encoding protoilludene synthase, which catalyzes the first committed step in the synthesis of antimicrobial melleolides. J Biol Chem 286(9):6871–6878
106. Pinedo C et al (2008) Sesquiterpene synthase from the botrydial biosynthetic gene cluster of the phytopathogen *Botrytis cinerea*. ACS Chem Biol 3(12):791–801
107. Wang CM et al (2009) Biosynthesis of the sesquiterpene botrydial in *Botrytis cinerea*. Mechanism and stereochemistry of the enzymatic formation of presilphiperfolan-8beta-ol. J Am Chem Soc 131(24):8360–8361
108. Felicetti B, Cane DE (2004) Aristolochene synthase: mechanistic analysis of active site residues by site-directed mutagenesis. J Am Chem Soc 126(23):7212–7221
109. Cane DE et al Overproduction of soluble trichodiene synthase from *Fusarium sporotrichioides* in *Escherichia coli*. Arch Biochem Biophys 300(1):416–22
110. Cane DE, Xue Q (1996) Trichodiene synthase. Enzymatic formation of multiple sesquiterpenes by alteration of the cyclase active site. J Am Chem Soc 118(6):1563–1564
111. McCormick SP, Alexander NJ, Harris LJ (2010) CLM1 of *Fusarium graminearum* encodes a longiborneol synthase required for culmorin production. Appl Environ Microbiol 76 (1):136–141
112. Brock NL, Tudzynski B, Dickschat JS (2011) Biosynthesis of sesqui- and diterpenes by the gibberellin producer *Fusarium fujikuroi*. Chembiochem 12:2667–2676
113. Brock NL et al (2013) Genetic dissection of sesquiterpene biosynthesis by *Fusarium fujikuroi*. Chembiochem 14(3):311–315
114. Tamura K et al (2013) MEGA6: molecular evolutionary genetics analysis version 6.0. Mol Biol Evol 30(12):2725–2729
115. Lesburg CA et al (1997) Crystal structure of pentalenene synthase: mechanistic insights on terpenoid cyclization reactions in biology. Science 277(5333):1820–1824
116. Starks CM et al (1997) Structural basis for cyclic terpene biosynthesis by tobacco 5-epi-aristolochene synthase. Science 277(5333):1815–1820
117. Shishova EY et al (2007) X-ray crystal structure of aristolochene synthase from *Aspergillus terreus* and evolution of templates for the cyclization of farnesyl diphosphate. Biochemistry 46(7):1941–1951
118. Gennadios HA et al (2009) Crystal structure of (+)-delta-cadinene synthase from *Gossypium arboreum* and evolutionary divergence of metal binding motifs for catalysis. Biochemistry 48(26):6175–6183
119. Aaron JA et al (2010) Structure of epi-isozizaene synthase from *Streptomyces coelicolor* A3 (2), a platform for new terpenoid cyclization templates. Biochemistry 49(8):1787–1797

120. McAndrew RP et al (2011) Structure of a three-domain sesquiterpene synthase: a prospective target for advanced biofuels production. Structure 19(12):1876–1884
121. Baer P et al (2014) Hedycaryol synthase in complex with nerolidol reveals terpene cyclase mechanism. Chembiochem 15(2):213–216
122. Shishova EY et al (2008) X-ray crystallographic studies of substrate binding to aristolochene synthase suggest a metal ion binding sequence for catalysis. J Biol Chem 283(22):15431–15439
123. Faraldos JA, Gonzalez V, Allemann RK (2012) The role of aristolochene synthase in diphosphate activation. Chem Commun (Camb) 48(26):3230–3232
124. Rynkiewicz MJ, Cane DE, Christianson DW (2002) X-ray crystal structures of D100E trichodiene synthase and its pyrophosphate complex reveal the basis for terpene product diversity. Biochemistry 41(6):1732–1741
125. Vedula LS, Cane DE, Christianson DW (2005) Role of arginine-304 in the diphosphate-triggered active site closure mechanism of trichodiene synthase. Biochemistry 44 (38):12719–12727
126. McCormick SP et al (2011) Trichothecenes: from simple to complex mycotoxins. Toxins (Basel) 3(7):802–814
127. Kimura M et al (2007) Molecular and genetic studies of fusarium trichothecene biosynthesis: pathways, genes, and evolution. Biosci Biotechnol Biochem 71(9):2105–2123
128. Motohashi K et al (2009) New sesquiterpenes, JBIR-27 and -28, isolated from a tunicate-derived fungus, Penicillium sp. SS080624SCf1. J Antibiot (Tokyo) 62(5):247–250
129. Cane DE et al (1993) Overexpression in *Escherichia coli* of soluble aristolochene synthase from *Penicillium roqueforti*. Arch Biochem Biophys 304(2):415–419
130. Brock NL, Dickschat JS (2013) PR toxin biosynthesis in *Penicillium roqueforti*. Chembiochem 14(10):1189–1193
131. Hidalgo PI et al (2014) Molecular characterization of the PR-toxin gene cluster in *Penicillium roqueforti* and *Penicillium chrysogenum*: cross talk of secondary metabolite pathways. Fungal Genet Biol 62:11–24
132. Erkel G, Anke T (2008) Products from Basidiomycetes. In: Biotechnology set. Wiley-VCH Verlag GmbH, pp 489–533
133. Ito-Kobayashi M et al (2008) Sterenin A, B, C and D, novel 11beta-hydroxysteroid dehydrogenase type 1 inhibitors from Stereum sp. SANK 21205. J Antibiot (Tokyo) 61 (3):128–135
134. Nasini G et al (2012) Isolation and structure elucidation of aza-sesquiterpenoids of protoilludane origin formed by shaken cultures of the fungus *Clavicorona divaricata*. Phytochem Lett 5(1):224–227
135. Zaidman BZ et al (2005) Medicinal mushroom modulators of molecular targets as cancer therapeutics. Appl Environ Microbiol 67(4):453–468
136. Kramer R, Abraham W-R (2011) Volatile sesquiterpenes from fungi: what are they good for? Phytochem Rev 11(1):15–37
137. Schueffler A, Anke T (2009) Secondary metabolites of basidiomycetes. In: Anke T, Weber E (eds) The Mycota: physiology and genetics XV. Springer, Berlin, pp 209–231
138. Liermann JC et al (2010) Hirsutane-type sesquiterpenes with uncommon modifications from three basidiomycetes. J Org Chem 75(9):2955–2961
139. Opatz T, Kolshorn H, Anke H (2008) Sterelactones: new isolactarane type sesquiterpenoids with antifungal activity from *Stereum* sp. IBWF 01060. J Antibiot (Tokyo) 61(9):563–567
140. Hanssen HP, Sprecher E, Abraham WR (1986) 6-Protoilludene, the major volatile metabolite from *Ceratocystes piceae* liquid cultures. Phytochemistry 25:1979–1980
141. Castillo UF et al (1999) Pteridanoside, the first protoilludane sesquiterpene glucoside as a toxic component of the neotropical bracken fern *Pteridium aquilinum var. caudatum*. Tetrahedron 55(42):12295–12300
142. Cimino G et al (1989) New sterpurane sesquiterpenoid from the mediterrenean *Alcyonium acaule*: structure of 3-acetoxy-sterpurene. Tetrahedron 45:6479–6484
143. Kashiwabara M et al (2006) Repraesentins D, E and F, new plant growth promoters from *Lactarius repraesentaneus*. Biosci Biotechnol Biochem 70(6):1502–1505

144. Kamo T et al (2006) 1,2-Dehydrolactarolide A, a new plant growth regulatory lactarane sesquiterpene from *Lactarius vellereus*. Biosci Biotechnol Biochem 70(9):2307–2309
145. Hirota M et al (2003) New plant growth promoters, repraesentins A, B and C, from Lactarius repraesentaneus. Biosci Biotechnol Biochem 67(7):1597–1600
146. Bulock JD, Darbyshire J (1976) Lagopodin metabolites and artifacts in cultures of *Coprinus*. Phytochemistry 15(12):2004–2004
147. McMorris TC et al (2001) Structure-activity studies of antitumor agent irofulven (hydroxymethylacylfulvene) and analogues. J Organ Chem 66(18):6158–6163
148. Jaspers NGJ et al (2002) Anti-tumour compounds illudin S and Irofulven induce DNA lesions ignored by global repair and exclusively processed by transcription—and replication-coupled repair pathways. DNA Repair 1(12):1027–1038
149. Lehmann VKB et al (2003) Illudin S, the sole antiviral compound in mature fruiting bodies of *Omphalotus illudens*. J Nat Prod 66(9):1257–1258
150. McMorris TC, Cong Q, Kelner MJ (2003) Structure-activity relationship studies of illudins: analogues possessing a spiro-cyclobutane ring. J Organ Chem 68(25):9648–9653
151. Dick RA, Yu X, Kensler TW (2004) NADPH alkenal/one oxidoreductase activity determines sensitivity of cancer cells to the chemotherapeutic alkylating agent irofulven. Clin Canc Res 10(4):1492–1499
152. Wiltshire T et al (2007) BRCA1 contributes to cell cycle arrest and chemoresistance in response to the anticancer agent irofulven. Mol Pharmacol 71(4):1051–1060
153. Nair MS et al (1969) Metabolites of *Clitocybe illudens*: IV. Illudalic acid, a sesquiterpenoid, and illudinine, a sesquiterpenoid alkaloid. J Org Chem 34(1):240–243
154. Singh P et al (1971) Metabolites of *Clitocybe illudens*: VI. Isolation of dihydroilludin M from *Clitocybe illudens*. Phytochemistry 10(9):22229–22230
155. Nair MS, Anchel M (1972) Metabolic products of *Clitocybe illudens*: IX structure of Illudacetalic acid and its conversion to Illudinine. Tetrahedron Lett 27:2753–2754
156. Bradshaw A et al (1982) 4α-Hydroxydihydroilludin M: a new sesquiterpenoid metabolite of *Clitocybe illudens*. Phytochemistry 21(4):942–943
157. Lee IK et al (1996) Illudins C2 and C3, new illudin C derivatives from *Coprinus atramentarius* ASI20013. J Antibiot (Tokyo) 49(8):821–822
158. Dufresne C et al (1997) Illudinic acid, a novel illudane sesquiterpene antibiotic. J Nat Prod 60(2):188–190
159. Burgess ML, Barrow KD (1999) Biosynthesis of illudosin, a fomannosane-type sesquiterpene, by the Basidiomycete *Omphalotus nidiformis*. J Chem Soc-Perkin Trans 1 (17):2461–2466
160. Burgess ML, Zhang YL, Barrow KD (1999) Characterization of new illudanes, illudins F, G, and H from the basidiomycete *Omphalotus nidiformis*. J Nat Prod 62(11):1542–1544
161. Kirchmair M, Poder R, Huber CG (1999) Identification of illudins in Omphalotus nidiformis and *Omphalotus olivascens* var. *indigo* by column liquid chromatography atmospheric pressure chemical ionization tandem mass spectrometry. J Chromatogr A 832(1–2):247–252
162. Del Val AG et al (2003) Novel illudins from *Coprinopsis episcopalis* (syn. *Coprinus episcopalis*), and the distribution of illudin-like compounds among filamentous fungi. Mycol Res 107:1201–1209
163. McMorris TC et al (2002) Sesquiterpenes from *Omphalotus illudens*. Phytochemistry 61 (4):395–398
164. McMorris TC et al (2000) Sesquiterpenes from the basidiomycete *Omphalotus illudens*. J Nat Prod 63(11):1557–1559
165. Wang G et al (2011) Illudin T, a new sesquiterpenoid from basidiomycete *Agrocybe salicacola*. J Asian Nat Prod Res 13(5):430–433
166. Zhu YC, Wang G, Liu JK (2010) Two new sesquiterpenoids from basidiomycete *Agrocybe salicacola*. J Asian Nat Prod Res 12(6):464–469
167. McMorris TC (1999) Discovery and development of sesquiterpenoid derived hydroxymethylacylfulvene: A new anticancer drug. Bioorg Medic Chem 7(5):881–886

168. McMorris TC et al (2010) Structure-activity studies of urea, carbamate, and sulfonamide derivatives of acylfulvene. J Med Chem 53(3):1109–1116
169. Schobert R et al (2011) Anticancer active illudins: recent developments of a potent alkylating compound class. Curr Med Chem 18(6):790–807
170. Martin F et al (2011) Sequencing the fungal tree of life. New Phytol 190(4):818–821
171. Floudas D et al (2012) The Paleozoic origin of enzymatic lignin decomposition reconstructed from 31 fungal genomes. Science 336(6089):1715–1719
172. Ainsworth AM et al (1990) Production and properties of the sesquiterpene, (+)-torreyol, in degenerative mycelial interactions between strains of *stereum*. Mycol Res 94:799–809
173. Isaka M et al (2011) Sterostreins A-E, new terpenoids from cultures of the basidiomycete *Stereum ostrea* BCC 22955. Org Lett 13(18):4886–4889
174. Isaka M et al (2012) Sterostreins F-O, illudalanes and norilludalanes from cultures of the basidiomycete *Stereum ostrea* BCC 22955. Phytochemistry 79:116–120
175. Kim YH et al (2006) Methoxylaricinolic acid, a new sesquiterpene from the fruiting bodies of Stereum ostrea. J Antibiot (Tokyo) 59(7):432–434
176. Li G et al (2011) Stereumins H-J, stereumane-type sesquiterpenes from the fungus *Stereum* sp. J Nat Prod 74(2):296–299
177. Yoo NH et al (2006) Hirsutenols D, E and F, new sesquiterpenes from the culture broth of *Stereum hirsutum*. J Antibiot (Tokyo) 59(2):110–113
178. Yun BS et al (2002) Sterins A and B, new antioxidative compounds from *Stereum hirsutum*. J Antibiot (Tokyo) 55(2):208–210
179. Davis TS et al (2013) Microbial volatile emissions as insect semiochemicals. J Chem Ecol 39 (7):840–859
180. Peters RJ (2010) Two rings in them all: the labdane-related diterpenoids. Nat Prod Rep 27 (11):1521–1530
181. Smanski MJ et al (2012) Bacterial diterpene synthases: new opportunities for mechanistic enzymology and engineered biosynthesis. Curr Opin Chem Biol 16(1–2):132–141
182. Koksal M et al (2011) Taxadiene synthase structure and evolution of modular architecture in terpene biosynthesis. Nature 469(7328):116–120
183. Zhou K et al (2012) Insights into diterpene cyclization from structure of bifunctional abietadiene synthase from *Abies grandis*. J Biol Chem 287(9):6840–6850
184. Koksal M et al (2011) Structure and mechanism of the diterpene cyclase ent-copalyl diphosphate synthase. Nat Chem Biol 7(7):431–433
185. Kelley LA, Sternberg MJ (2009) Protein structure prediction on the web: a case study using the Phyre server. Nat Protoc 4(3):363–371
186. de Boer AH, de Vries-van Leeuwen IJ (2012) Fusicoccanes: diterpenes with surprising biological functions. Trends Plant Sci 17(6):360–368
187. Toyomasu T et al (2007) Fusicoccins are biosynthesized by an unusual chimera diterpene synthase in fungi. Proc Natl Acad Sci USA 104(9):3084–3088
188. Ono Y et al (2011) Dioxygenases, key enzymes to determine the aglycon structures of fusicoccin and brassicicene, diterpene compounds produced by fungi. J Am Chem Soc 133 (8):2548–2555
189. Noike M et al (2012) Molecular breeding of a fungus producing a precursor diterpene suitable for semi-synthesis by dissection of the biosynthetic machinery. PLoS One 7(8): e42090
190. Minami A et al (2009) Identification and functional analysis of brassicicene C biosynthetic gene cluster in *Alternaria brassicicola*. Bioorg Med Chem Lett 19(3):870–874
191. Hashimoto M et al (2009) Functional analyses of cytochrome P450 genes responsible for the early steps of brassicicene C biosynthesis. Bioorg Med Chem Lett 19(19):5640–5643
192. Toyomasu T et al (2009) Biosynthetic gene-based secondary metabolite screening: a new diterpene, methyl phomopsenonate, from the fungus *Phomopsis amygdali*. J Org Chem 74 (4):1541–1548
193. Chiba R et al (2013) Identification of ophiobolin F synthase by a genome mining approach: a sesterterpene synthase from *Aspergillus clavatus*. Org Lett 15(3):594–597

194. Shen JW, Ruan Y, Ma BJ (2009) Diterpenoids of macromycetes. J Basic Microbiol 49 (3):242–255
195. Wilderman PR, Peters RJ (2007) A single residue switch converts abietadiene synthase into a pimaradiene specific cyclase. J Am Chem Soc 129(51):15736–15737
196. Xu M, Wilderman PR, Peters RJ (2007) Following evolution's lead to a single residue switch for diterpene synthase product outcome. Proc Natl Acad Sci U S A 104(18):7397–7401
197. Bomke C, Tudzynski B (2009) Diversity, regulation, and evolution of the gibberellin biosynthetic pathway in fungi compared to plants and bacteria. Phytochemistry 70 (15–16):1876–1893
198. Kawaide H (2006) Biochemical and molecular analyses of gibberellin biosynthesis in fungi. Biosci, Biotechnol, Biochem 70(3):583–590
199. Toyomasu T (2008) Recent advances regarding diterpene cyclase genes in higher plants and fungi. Biosci Biotechnol Biochem 72(5):1168–1175
200. Kawaide H et al (1997) *ent*-kaurene synthase from the fungus *Phaeosphaeria* sp. L487—cDNA isolation, characterization, and bacterial expression of a bifunctional diterpene cyclase in fungal gibberellin biosynthesis. J Biol Chem 272(35):21706–21712
201. Kawaide H, Sassa T, Kamiya Y (2000) Functional analysis of the two interacting cyclase domains in ent-kaurene synthase from the fungus *Phaeosphaeria* sp. L487 and a comparison with cyclases from higher plants. J Biol Chem 275(4):2276–2280
202. Tudzynski B, Holter K (1998) Gibberellin biosynthetic pathway in *Gibberella fujikuroi*: evidence for a gene cluster. Fungal Genet Biol 25(3):157–170
203. Tudzynski B, Kawaide H, Kamiya Y (1998) Gibberellin biosynthesis in *Gibberella fujikuroi*: cloning and characterization of the copalyl diphosphate synthase gene. Curr Genet 34 (3):234–240
204. Zi J, Mafu S, Peters RJ (2014) To gibberellins and beyond! Surveying the evolution of (di) terpenoid metabolism. Annu Rev Plant Biol 65:259–286
205. Oikawa H et al (2001) Cloning and functional expression of cDNA encoding aphidicolan-16 beta-ol synthase: a key enzyme responsible for formation of an unusual diterpene skeleton in biosynthesis of aphidicolin. J Am Chem Soc 123(21):5154–5155
206. Toyomasu T et al (2004) Cloning of a gene cluster responsible for the biosynthesis of diterpene aphidicolin, a specific inhibitor of DNA polymerase alpha. Biosci, Biotechnol, Biochem 68(1):146–152
207. Fujii R et al (2011) Total biosynthesis of diterpene aphidicolin, a specific inhibitor of DNA polymerase alpha: heterologous expression of four biosynthetic genes in *Aspergillus oryzae*. Biosci Biotechnol Biochem 75(9):1813–1817
208. Toyomasu T et al (2008) Identification of diterpene biosynthetic gene clusters and functional analysis of labdane-related diterpene cyclases in *Phomopsis amygdali*. Biosci Biotechnol Biochem 72(4):1038–1047
209. Bromann K et al (2012) Identification and characterization of a novel diterpene gene cluster in *Aspergillus nidulans*. PLoS One 7(4):e35450
210. Mitterbauer R, Specht T (2011) Cloning and sequence of *Clitopilus passeckerianus* diterpene synthase and pleuromutilin biosynthesis gene cluster, and use for producing pleuromutilin. WO/2011/110610. EP2011/053571: p WO2011110610A1
211. Hartley AJ et al (2009) Investigating pleuromutilin-producing *Clitopilus* species and related basidiomycetes. FEMS Microbiol Lett 297(1):24–30
212. Novak R (2011) Are pleuromutilin antibiotics finally fit for human use? Ann NY Acad Sci 1241:71–81
213. Novak R, Shlaes DM (2010) The pleuromutilin antibiotics: a new class for human use. Curr Opin Investig Drugs 11(2):182–191
214. Xu R, Fazio GC, Matsuda SP (2004) On the origins of triterpenoid skeletal diversity. Phytochemistry 65(3):261–291
215. Phillips DR et al (2006) Biosynthetic diversity in plant triterpene cyclization. Curr Opin Plant Biol 9(3):305–314
216. Abe I (2007) Enzymatic synthesis of cyclic triterpenes. Nat Prod Rep 24(6):1311–1331

217. Dupont S et al (2012) Ergosterol biosynthesis: a fungal pathway for life on land? Evolution 66(9):2961–2968
218. Moses T et al (2013) Bioengineering of plant (tri)terpenoids: from metabolic engineering of plants to synthetic biology in vivo and in vitro. New Phytol 200(1):27–43
219. Zhao M et al (2013) Protostane and fusidane triterpenes: a mini-review. Molecules 18 (4):4054–4080
220. Rios JL et al (2012) Lanostanoids from fungi: a group of potential anticancer compounds. J Nat Prod 75(11):2016–2044
221. Boh B et al (2007) *Ganoderma lucidum* and its pharmaceutically active compounds. Biotechnol Annu Rev 13:265–301
222. Mitsuguchi H et al (2009) Biosynthesis of steroidal antibiotic fusidanes: functional analysis of oxidosqualene cyclase and subsequent tailoring enzymes from *Aspergillus fumigatus*. J Am Chem Soc 131(18):6402–6411
223. Lodeiro S et al (2009) Protostadienol biosynthesis and metabolism in the pathogenic fungus *Aspergillus fumigatus*. Org Lett 11(6):1241–1244
224. Kimura M et al (2010) Protostadienol synthase from *Aspergillus fumigatus*: functional conversion into lanosterol synthase. Biochem Biophys Res Commun 391(1):899–902
225. Godio RP, Martin JF (2009) Modified oxidosqualene cyclases in the formation of bioactive secondary metabolites: biosynthesis of the antitumor clavaric acid. Fungal Genet Biol 46 (3):232–242
226. Godio RP, Fouces R, Martin JF (2007) A squalene epoxidase is involved in biosynthesis of both the antitumor compound clavaric acid and sterols in the basidiomycete *H. sublateritium*. Chem Biol 14(12):1334–1346
227. Shu S et al (2013) De novo sequencing and transcriptome analysis of *Wolfiporia cocos* to reveal genes related to biosynthesis of triterpenoids. PLoS One 8(8):e71350
228. Chen S et al (2012) Genome sequence of the model medicinal mushroom *Ganoderma lucidum*. Nat Commun 3:913
229. Liu D et al (2012) The genome of *Ganoderma lucidum* provides insights into triterpenes biosynthesis and wood degradation (corrected). PLoS One 7(5):e36146
230. Yu GJ et al (2012) Deep insight into the *Ganoderma lucidum* by comprehensive analysis of its transcriptome. PLoS One 7(8):e44031
231. Hamberger B, Bak S (2013) Plant P450s as versatile drivers for evolution of species-specific chemical diversity. Philos Trans R Soc Lond B Biol Sci 368(1612):20120426
232. Field B et al (2011) Formation of plant metabolic gene clusters within dynamic chromosomal regions. Proc Natl Acad Sci U S A 108(38):16116–16121
233. Field B, Osbourn AE (2008) Metabolic diversification–independent assembly of operon-like gene clusters in different plants. Science 320(5875):543–547
234. Syed K et al (2014) Systematic identification and evolutionary analysis of catalytically versatile cytochrome p450 monooxygenase families enriched in model basidiomycete fungi. PLoS One 9(1):e86683
235. Moktali V et al (2012) Systematic and searchable classification of cytochrome P450 proteins encoded by fungal and oomycete genomes. BMC Genomics 13:525
236. Liu LY et al (2013) Meroterpenoid pigments from the basidiomycete *Albatrellus ovinus*. J Nat Prod 76(1):79–84
237. Blin K et al (2013) Antismash 2.0–a versatile platform for genome mining of secondary metabolite producers. Nucleic Acids Res. 41(Web Server issue): p W204–12
238. Geris R, Simpson TJ (2009) Meroterpenoids produced by fungi. Nat Prod Rep 26 (8):1063–1094
239. Da Lage JL et al (2013) Gene make-up: rapid and massive intron gains after horizontal transfer of a bacterial alpha-amylase gene to Basidiomycetes. BMC Evol Biol 13:40
240. Larsen PE et al (2010) Using deep RNA sequencing for the structural annotation of the *Laccaria bicolor* mycorrhizal transcriptome. PLoS One 5(7):e9780
241. Noble LM, Andrianopoulos A (2013) Fungal genes in context: genome architecture reflects regulatory complexity and function. Genome Biol Evol 5:1336–1352

242. Bayram O, Braus GH (2012) Coordination of secondary metabolism and development in fungi: the velvet family of regulatory proteins. FEMS Microbiol Rev 36(1):1–24
243. Ohm RA et al (2011) Transcription factor genes of *Schizophyllum commune* involved in regulation of mushroom formation. Mol Microbiol 81(6):1433–1445
244. Ohm RA et al (2010) Genome sequence of the model mushroom *Schizophyllum commune*. Nat Biotechnol 28(9):957–963
245. Hur M et al (2013) A global approach to analysis and interpretation of metabolic data for plant natural product discovery. Nat Prod Rep 30(4):565–583
246. Higashi Y, Saito K (2013) Network analysis for gene discovery in plant-specialized metabolism. Plant Cell Environ 36:1597–1606
247. Chiang YM et al (2009) Unlocking fungal cryptic natural products. Nat Prod Commun 4 (11):1505–1510
248. Nutzmann HW et al (2011) Bacteria-induced natural product formation in the fungus *Aspergillus nidulans* requires Saga/Ada-mediated histone acetylation. Proc Natl Acad Sci U S A 108(34):14282–14287
249. Hansen BG et al (2011) Versatile enzyme expression and characterization system for *Aspergillus nidulans*, with the *Penicillium brevicompactum* polyketide synthase gene from the mycophenolic acid gene cluster as a test case. Appl Environ Microbiol 77(9):3044–3051
250. Sanchez JF et al (2011) Genome-based deletion analysis reveals the prenyl xanthone biosynthesis pathway in *Aspergillus nidulans*. J Am Chem Soc 133(11):4010–4017
251. Desjardins AE (2009) From yellow rain to green wheat: 25 years of trichothecene biosynthesi research. J Agric Food Chem 57(11):4478–4484

Adv Biochem Eng Biotechnol (2015) 148: 63–106
DOI 10.1007/10_2014_295

Published Online: 13 January 2015

Biosynthesis and Biological Functions of Terpenoids in Plants

Dorothea Tholl

Abstract Terpenoids (isoprenoids) represent the largest and most diverse class of chemicals among the myriad compounds produced by plants. Plants employ terpenoid metabolites for a variety of basic functions in growth and development but use the majority of terpenoids for more specialized chemical interactions and protection in the abiotic and biotic environment. Traditionally, plant-based terpenoids have been used by humans in the food, pharmaceutical, and chemical industries, and more recently have been exploited in the development of biofuel products. Genomic resources and emerging tools in synthetic biology facilitate the metabolic engineering of high-value terpenoid products in plants and microbes. Moreover, the ecological importance of terpenoids has gained increased attention to develop strategies for sustainable pest control and abiotic stress protection. Together, these efforts require a continuous growth in knowledge of the complex metabolic and molecular regulatory networks in terpenoid biosynthesis. This chapter gives an overview and highlights recent advances in our understanding of the organization, regulation, and diversification of core and specialized terpenoid metabolic pathways, and addresses the most important functions of volatile and nonvolatile terpenoid specialized metabolites in plants.

Keywords MVA pathway · MEP pathway · Prenyltransferase · Terpene synthase · Volatiles · Plant defense

Contents

D. Tholl (✉)
Department of Biological Sciences, Virginia Tech, 409 Latham Hall,
24061 Blacksburg, VA, USA
e-mail: tholl@vt.edu

1 Introduction

Introductory chapters on terpenoid biosynthesis usually highlight the large number of terpenoid compounds found in nature. Indeed, the structural diversity associated with at least 40,000 compounds makes the class of terpenoids one of the most impressive examples in the divergent evolution of plant chemicals. The evolutionary success of this compound class is in part based on the simplicity of constructing different size molecules. According to the isoprene rule recognized by Wallach and Rutzicka in the late nineteenth and mid-twentieth centuries [1], all terpenoids are derived from the universal five-carbon building blocks, isopentenyl diphosphate (IPP) and its allylic isomer dimethylallyl diphosphate (DMAPP). The prenyl diphosphate intermediates built by condensation of these five-carbon units are used as precursors for the biosynthesis of terpenoids with fundamental functions in growth and development and for the formation of a large number of terpenoid compounds with more specialized roles in the interaction of plants with their environment. It is the latter group of terpenoids that is characterized by its tremendous structural diversity as a consequence of divergent biosynthetic gene evolution. Specialized terpenoids have a long history of being used as flavors, fragrances, pharmaceuticals, insecticides, and industrial compounds, several of which are addressed in this book. With the growing need for sustainable production platforms of plant-based drugs and the emerging use of terpenoids in the production of alternative fuels, substantial progress has been made in the engineering of terpenoid biosynthetic pathways in microbes and plants [2, 3]. Advanced functional genomics approaches provide unlimited access to the biosynthetic genes and molecular regulators of terpenoid-producing plants, and, at the same time, allow deeper insight to the complexity of plant terpenoid metabolism and regulation. In this chapter, I provide an overview of the organization of the early and core terpenoid metabolic pathways and give updates on the regulation and functional diversification of their genes and enzymes. Furthermore, I summarize the function of terpene synthases and describe aspects of their coordinated and tissue-specific regulation in specialized metabolism prior to addressing the diverse roles of terpenoids in plant–environment interactions.

2 Core Terpenoid Biosynthetic Pathways and Their Regulation

Successful engineering of terpenoid products in plants critically depends on the flux of precursors delivered by the core isoprenoid biosynthetic pathways and, consequently, on the dynamic regulation of these biosynthetic routes. Plants use two independent pathways to produce IPP and DMAPP: the primarily cytosolic mevalonic acid (MVA) pathway and the plastidial methylerythritol phosphate (MEP) pathway (Fig. 1). The MVA pathway predominantly provides the precursors for the cytosolic biosynthesis of sesquiterpenoids, polyprenols, phytosterols, brassinosteroids, and triterpenoids, and for terpenoid biosynthesis in mitochondria (e.g., ubiquinones, polyprenols), and the five-carbon units derived from the MEP pathway are preferably used for the biosynthesis of hemiterpenoids (e.g., isoprene), monoterpenoids, diterpenoids, carotenoids and their breakdown products, cytokinins, gibberellins, chlorophyll, tocopherols, and plastoquinones (Fig. 2). It has become evident that both pathways are heavily regulated at multiple levels as was discussed in two recent reviews by Hemmerlin and coworkers [4, 5]. In addition to the transcriptional regulation of MVA and MEP pathway genes and their different paralogues, isoprenoid-pathway fluxes are controlled at posttranscriptional/-translational levels and by feedback regulation. Recent studies have given a more global view of the dynamics and networks of the core isoprenoid pathways and the regulation of metabolic flux during plant development and in response to external stimuli (reviewed in [6, 7]). Therefore, this chapter primarily gives an overview of both pathways with some emphasis on those in *Arabidopsis* and provides updates on the different modes of regulation.

2.1 MVA and MEP Pathways—A Brief Summary of Their Biosynthetic Steps

The MVA pathway in plants (Fig. 1a) consists of six steps and starts with the Claisen-type condensation of two molecules acetyl-CoA to acetoacetyl-CoA (AcAc-CoA) catalyzed by acetoacetyl-CoA thiolase (AACT). In a subsequent aldol condensation reaction catalyzed by HMG-CoA synthase (HMGS), AcAc-CoA is combined with a third molecule of acetyl-CoA to form the C6-compound *S*-3-hydroxy-3-methylglutaryl-CoA (HMG-CoA). Response to different stresses, feedback regulation, and the role of HMGS in sterol metabolism (see below)

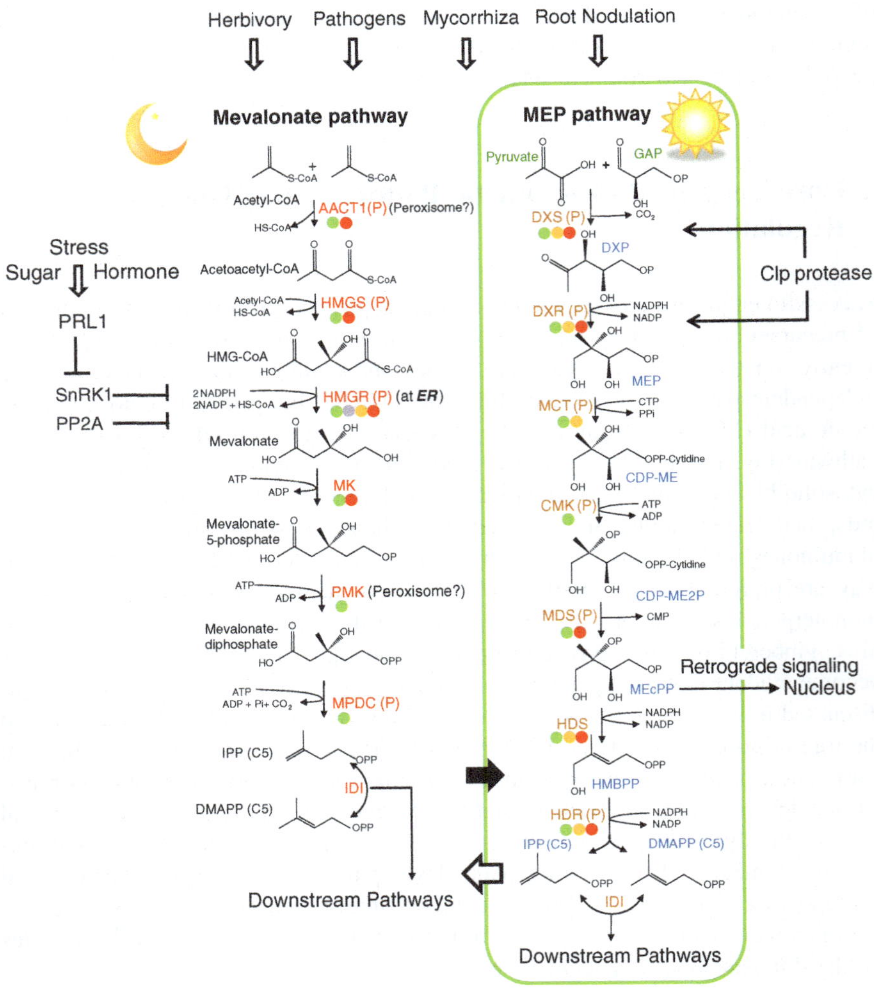

Fig. 1 Enzymatic steps of the MVA and MEP pathways and their regulation in isoprenoid precursor biosynthesis. *Colored dots* indicate the different levels of regulation for each enzyme according to the current status of knowledge [5]: *green*—transcriptional, *purple*—posttranscriptional, *yellow*—translational, *red*—posttranslational including feedback modulation. Other selected external and internal regulatory factors and posttranslational modifications of main regulatory enzymes are depicted as mentioned in the text. One or more gene paralogues as described from different plant species [4] are indicated by (*P*). *Arrows* indicate preferred trafficking of isoprenoid precursors between the cytosol and plastids in light (*white*) and dark (*black*) exposed tissues. Abbreviations for enzymes (*red*) and metabolites (*black*—MVA pathway; *blue*—MEP pathway) are as described in the text

support a key function of HMGS in the MVA pathway. In the following rate-limiting step, HMG-CoA reductase (HMGR) catalyzes the conversion of *S*-HMG-CoA to *R*-mevalonate in two NADPH-dependent reduction steps. All plant

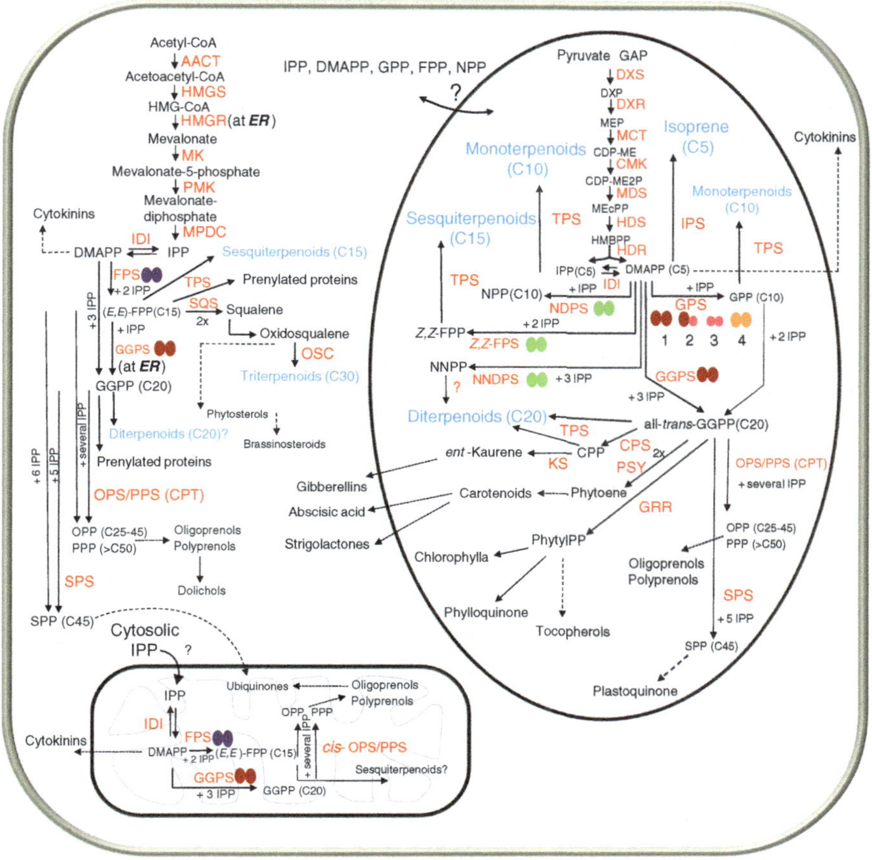

Fig. 2 Terpenoid biosynthetic pathways and their subcellular organization. Enzymes are marked in *red*; specialized terpenoids are marked in *blue*; all other intermediates and terpenoid end products are in *black*. *Solid* and *dashed arrows* indicate single and multiple enzymatic steps, respectively. *Colored ovals* indicate the homodimeric and heterodimeric composition of prenyltransferases involved in specialized terpenoid biosynthesis: GPS1—GGPS type, GPS2—heterodimer (heterotetramer) with large subunit (LSU) and small subunit (SSU I or SSU II), GPS3—homodimeric SSII type [136], GPS4—PPS type. Short-chain *cis*-prenyltranferases are marked in *green*. Abbreviations not mentioned in the text: *IPS* isoprene synthase; *GRR* geranylgeranyl reductase; *OPS* oligoprenyl diphosphate (*OPP*) synthase; *SPS* solanesyl diphosphate (*SPS*) synthase; *SQS* squalene synthase

HMGR proteins are membrane-bound with two membrane-spanning sequences and a highly conserved catalytic C-terminal domain. The presence of ER-specific retention motifs indicates a primary association of the membrane-spanning domain with the ER, whereas the N-terminal and C-terminal ends are positioned on the cytosolic side [8–13]. The association of HMGR to membranes seems to regulate its activity negatively, thereby limiting the accumulation of terpenoid end products such as sterols (e.g., [14, 15]). Many studies have reported on the critical regulatory role of HMGR in the biosynthesis of phytosterols, triterpenoids, and sesquiterpenoid phytoalexins, although flux control often involves additional downstream enzymes such as sesquiterpene synthases (e.g., [16–21]). MVA produced by HMGR is finally converted into IPP via three enzymatic steps: two ATP-dependent phosphorylation steps, catalyzed by mevalonate kinase (MK) and phosphomevalonate kinase (PMK), and an ATP-driven decarboxylative elimination catalyzed by mevalonate diphosphate decarboxylase (MVD or MPDC).

The MEP pathway (Fig. 1b), which occurs in all photosynthetic eukaryotes and in cyanobacteria, apicomplexan protozoa, and most eubacteria [22–25] consists of seven enzymatic steps. In the first reaction, 1-deoxy-D-xylulose 5-phosphate (DXP) is formed by DXP synthase (DXS) from (hydroxyethyl) thiamine diphosphate, which is derived from pyruvate, and glyceraldehyde-3-phosphate (GAP) in a transketolase-like condensation. Plant DXS enzymes carry a highly conserved thiamine phosphate binding domain and are divided in the class-I type enzymes with primary expression in photosynthetic and floral tissues and the class-II type enzymes with more distinct roles in specialized metabolism (see below). Numerous studies have confirmed that DXS functions as an important regulatory and rate-limiting enzyme in the biosynthesis of plastidial terpenes [26–31]. Consequently, *DXS* mutants such as those of the single functional *Arabidopsis* class-I type *DXS* gene (*DXS1*) exhibit albino phenotypes [32–34].

The enzyme 1-deoxy-D-xylulose 5-phosphate reductoisomerase (DXR) catalyzes the second step of the MEP pathway, in which DXP is converted into 2-*C*-methyl-D-erythritol 4-phosphate (MEP) by an intramolecular rearrangement of DXP into 2-*C*-methyl-D-erythrose 4-phosphate, followed by an NADPH-dependent reduction [35, 36]. The reaction can be specifically inhibited by fosmidomycin, a structure analogue of the DXR substrate [37–39] thereby blocking the biosynthesis of downstream plastidial terpene biosynthesis [40–42]. The reaction catalyzed by DXR is in some cases considered a rate-limiting step depending on the species, tissue, and developmental stage. In *Arabidopsis*, *DXR1* is expressed in different plant organs [36] and *dxr* mutants show, similar to those of *DXS1*, an albino phenotype and deficiencies in gibberellin and abscisic acid (ABA) biosynthesis [43].

MEP is further converted in a CTP-dependent reaction to 4-diphosphocytidyl-2-*C*-methyl-D-erythritol (CDP-ME) by the enzyme 4-diphosphocytidyl-2-*C*-methyl-D-erythritol synthase (MCT or IpsD) [44, 45]. Phosphorylation of CDP-ME by the enzyme 4-diphosphocytidyl-2-*C*-methyl-D-erythritol kinase (CMK, IspE) then leads to the formation of 4-diphosphocytidyl-2-*C*-methyl-D-erythritol 2-phosphate (CDP-ME2P) [46–48], which is subsequently cyclized by 2-*C*-methyl-D-erythritol 2,4-cyclodiphosphate synthase (MDS, IspF) into 2-*C*-methyl-D-erythritol 2,4-cyclodiphosphate (MEcPP) upon loss of CMP. In the last two steps of the MEP pathway, the enzyme 4-hydroxy-3-methylbut-2-enyl diphosphate synthase (HDS, IspG) first converts MEcPP in a two-electron reduction to 4-hydroxy-3-methylbut-2-enyl diphosphate (HMBPP). In a final branching step, HMBPP is converted by 4-hydroxy-3-methylbut-2-enyl diphosphate reductase (HDR, IspH) to a mixture of IPP and DMAPP with a ratio of 5 to 6:1 [49–51].

Mutants of *MCT*, *MDS*, and *CMK* exhibit similar albino phenotypes and downregulation of photosynthetic genes [52, 53]. Likewise, *hds* and *hdr-1* mutants have defects in chloroplast development [54, 55]. Interestingly, a partial loss-of-function mutant of *Arabidopsis* HDS, *hds-3* (*csb3*), was shown to be more resistant to biotrophic pathogens suggesting a link between the MEP pathway and plant defense responses [56].

2.2 Differential Expression of MVA and MEP Pathway Isozymes

Several enzymes of the MVA and MEP pathways, especially those with important regulatory roles, are encoded by small gene families, which allow for functional redundancy and divergence (summarized in [4]; Fig. 1). In the MVA pathway, paralogues have been identified for AACT, HMGS, HMGR, and MPDC, whereas the MEP pathway enzymes DXS, DXR, MCT, CMK, MDS, or HDR were found to be encoded by two or more isogenes [4]. The different roles of many of the MVA and MEP pathway isozymes depend on their expression in specific cellular tissues and are often divided into essential functions to provide terpenoid precursors in primary metabolism, growth, and development, and more specific functions in stress response and specialized metabolism. For example, in *Brassica juncea*, *HMGS* is represented by a four-member gene family. Two genes are highly expressed at early stages of floral development [57–59] and play a role in reproduction, as was also shown for the single *HMGS* gene in *Arabidopsis*, [60] whereas expression of the two other paralogues is restricted to leaves [58, 59].

Notably, paralogues of the *HMGR* family exhibit different developmental and tissue-specific expression patterns and can be distinguished by their response to endogenous molecules such as phytohormones and sterol metabolites as well as external stimuli that include light, wounding, elicitor treatment, and pest and pathogen attack (Fig. 1; reviewed in [4]). The differential expression of HMGR isozymes, as demonstrated by early studies of the *HMGR* gene families in Solanaceous plants (tomato, potato) [61, 62], is important for channeling and counterbalancing carbon flux to the different downstream pathways of stress response or development. This functional differentiation, however, does not seem to occur in all plants inasmuch as both *HMGR* genes in *Arabidopsis* do not respond to stress but are essential for the production of sterols for cell elongation, senescence, gametophyte development, and fertility [63, 64].

Functional divergence of MEP pathway genes has been primarily observed in the *DXS* gene family. Class II-type DXS genes respond to biotic interaction and are induced in the biosynthesis of apocarotenoids upon mycorrhizal colonization in legumes and other plant families [30, 65, 66]. Several studies also demonstrated that type II *DXS* genes are induced in response to pathogen and herbivore attack in association with the production of specialized metabolites (summarized in [4]).

2.3 Metabolic Regulation and Networks

There is clear evidence for the role of pathway intermediates and downstream metabolites in the regulation of the core terpenoid biosynthetic steps at transcriptional and posttranslational levels (Fig. 1). Feedback inhibition by free CoA has been demonstrated for AACT and HMGR and for the enzymatic products of HMGS [59, 67, 68]. Furthermore, plant MKs respond to feedback inhibition by the prenyl diphosphates, IPP, DMAPP, geranyl diphosphate (GPP), and farnesyl diphosphate (FPP) that modulate enzyme activity by acting as competitive inhibitors of ATP [69]. Similarly, in vitro feedback inhibition was found for a DXS protein from poplar by IPP and DMAPP and a structural analysis suggested possible binding of the prenyl diphosphates to the enzyme in competition with its thiamine pyrophosphate substrate [70]. This feedback inhibition has also been supported in vivo by recent metabolic flux studies in poplar [71].

The complexity of the regulatory network also becomes apparent when metabolic disturbances and changes in metabolic flux generated by overexpression or reduced expression of genes of the core isoprenoid pathways promote pathway feedback or feedforward signals that modify the expression of up- or downstream genes. For example, overexpression of *B. juncea* wild-type and mutated *HMGS1* in *Arabidopsis* caused an upregulation of *HMGR* and genes in sterol biosynthesis such as sterol methyltransferase 2, delta-24 sterol reductase, and C-22 sterol desaturase, which led to an elevated sterol content in leaves and seedlings and increased stress tolerance [57]. A similar response was observed for *HMGS* overexpression in tobacco resulting in improved sterol content, growth, pod size, and seed yield [72]. Conversely, knockdown of *AACT2* expression led to lower levels and altered profiles of sterols and caused reduced expression of downstream genes encoding FPP synthases and sterol methyltransferase [73]. HMGR activity also exhibits a positive feedback response to downstream metabolic changes such as reduced cycloartenol levels in transgenic tobacco expressing sterol methyltransferase type 1 (SMT1) and the depletion of endogenous sterols due to the inhibition of squalene synthase [74, 75].

The simultaneous response of several genes to pathway perturbations is further observed in mutants of the MEP pathway. For instance, silencing of *CMK* in *Arabidopsis* causes upregulation of *MCT*, *MDS*, and *HDS* expression [76]. Moreover, in rice, MEP pathway genes were found to be coexpressed with downstream genes in carotenoid and phytyl biosynthesis [77]. In line with these observations, detailed transcriptional coexpression network analyses in *Arabidopsis* demonstrated that gene modules in both MVA and MEP pathways are coregulated together with genes of downstream pathways and these findings have set the stage to identify regulatory elements of these gene modules [78–80]. Consequently, *cis* elements were mapped showing that the promoters of the *Arabidopsis* genes *DXS*, *DXR*, *CMK*, *HDR*, and phytoene synthase share a *cis*-regulatory element interacting with RAP2.2, a member of the ethylene response factor B-2 subfamily [6].

In conjunction with their regulation by light (see below), MEP and MVA pathways respond to regulators in sugar metabolism. *Arabidopsis* mutants of pleiotropic regulatory locus 1 (PRL1), a global regulator of sugar, stress, and hormone responses, accumulate MEP pathway-derived end products (Fig. 1a) [81]. The same mutants have reduced HMGR activity but no change in *HMGR* transcript or protein because of posttranslational modification. PRL1 inhibits the SNF1 (sucrose nonfermenting)-related protein kinase 1 (SnRK1), which negatively regulates HMGR1 by phosphorylation and inactivation of the catalytic domain (Fig. 1a) [82]. HMGR1 is also negatively regulated during normal development and in response to salt stress by protein phosphatase 2A (PP2A), which dephosphorylates the HMGR protein (most likely at a site different from the phosphorylation by SnRK1; Fig. 1a) [83]. Modulation of *HMGR* transcripts at the initiation of translation [84] and glycosylation of HMGR isoforms [11], respectively, have been discussed previously as other mechanisms of posttranscriptional or posttranslational regulation of stress-induced *HMGR* genes.

There are several possible connections of the isoprenoid pathway to other metabolic routes by delivery and competition for carbon precursors (e.g., amino acid degradation) [4], which will require further attention to gain a more comprehensive understanding of flux in terpenoid biosynthesis. A link of isoprenoid metabolism with lipid biosynthesis was described by Nieto et al. [85], who found that inhibition of sphingolipid biosynthesis in *Arabidopsis* caused posttranslational downregulation of HMGR activity decoupled from HMGR transcript and protein levels and a reduction in sterol content. Recently, an unexpected simultaneous downregulation of flavonoid and terpenoid metabolite levels was observed in trichomes of tomato mutants of the flavonoid biosynthetic enzyme chalcone isomerase (CHI) [86]. These results have led to several hypotheses about the regulatory connections between both pathways. It is possible that changes in the levels of flavonoids by accumulation (upstream of CHI) or depletion (downstream of CHI) modify terpenoid biosynthetic gene expression or directly inhibit biosynthetic and regulatory proteins [87, 88]. Based on previous findings, there is also the possibility that CHI itself might interact with proteins involved in terpenoid production or its regulation [89]. Furthermore, it will be important to examine regulatory factors that coordinate the metabolic flux through both pathways [90].

2.4 Regulation by Light and External Stimuli

Thanks to recent efforts to identify MVA and MEP pathway gene expression patterns by transcriptome and hierarchical cluster analyses it was shown that the genes of both pathways have opposite expression patterns during light or dark (Fig. 1) [6]. Whereas exposure to light leads to the downregulation of MVA pathway genes and reduced levels of sterols [79], it stimulates transcript accumulation of MEP pathway genes and genes in the carotenoid and chlorophyll biosynthetic pathways such as PSY (phytoene synthase) and HEMA1 (glutamyl-tRNA reductase), which are essential for chloroplast differentiation [79, 80, 91–94]. Light also upregulates tocopherol, and plastoquinone biosynthetic genes such as VTE3 (vitamin E defective 3) [79]. The results are supported by studies that observed an increased carbon flux through the MEP pathway under enhanced light conditions by measuring the accumulation of MEcDP when 2-C-methyl-D-erythritol 2,4-cyclodiphosphate reductase activity was inhibited [95]. In contrast to the upregulation by light, expression of MEP pathway genes with the exception of HDR [55] is reduced during light–dark transition [6]. Dark exposure can induce HMGR activity as was shown in ginseng where HMGRs play a regulatory role in triterpene ginsenoside biosynthesis [96]. The light-dependent response of *Arabidopsis* MEP and MVA pathway genes is controlled by phytochrome B (PHYB) because *phyB* mutants have enhanced transcript levels and enzyme activity of HMGR but reduced levels of MEP pathway products [92]. Consequently, phytochrome interacting factors (PIFs) of the basic helix–loop–helix (bHLH) transcription factor family were identified as regulators that are involved in the light control of MEP and carotenoid biosynthetic pathway genes [93, 97]. Turnover of the MEP pathway enzymes DXS and DXR was also found to be correlated with the activity of Clp, a major plastid stromal protease (Fig. 1b) [98].

Downregulation of MEP pathway enzymes in the dark provides a dilemma for the biosynthesis of carotenoids and gibberellins required for the development of etiolated seedlings. Supported by observations from treatments with the MEP pathway inhibitor fosmidomycin, Rodriguez-Concepcion and coworkers suggested that during seedling germination in the dark, prenyl diphosphates derived from the MVA pathway are transported into etioplasts for gibberellin and carotenoid synthesis prior to the induction of MEP pathway enzymes upon illumination [92]. Given the responses of MEP and MVA pathway genes in light and dark, it is not surprising that the expression of several genes is under circadian control [7]. Coexpression analyses in *Arabidopsis* photosynthetic tissue connect several MEP pathway genes with core circadian oscillators (*LHY*, *CCA1*, *PRR9*) whereas only *AACT2* of the MVA pathway follows the expression of circadian regulators peaking in the dark [6]. However, in roots, expression of several MVA pathway genes such as *HMGR1* is correlated with that of circadian regulators (*TOC1*, *TIC*) showing clear differences in the circadian control of early pathway genes in above- and belowground tissues. Interestingly, in triple mutants of the TOC1 related pseudo-response regulator (PRR) proteins PRR9, PRR7, and PRR5, genes and metabolites of carotenoid, chlorophyll, and tocopherol pathways are upregulated, which suggests a function of these proteins as negative regulators of the MEP pathway-dependent metabolic routes [99]. To what extent the oscillation of MVA and MEP pathway gene transcripts directly corresponds to changes in enzyme activity and downstream metabolites requires further attention. In snapdragon flowers, the rhythmic emission of volatile monoterpenes in plastids and sesquiterpenes in the cytosol depends on the MEP pathway that is controlled by the circadian clock [100].

In addition to their differential response to light, MVA and MEP pathways respond to multiple other external stimuli at gene transcript and posttranslational levels (Fig. 1; summarized in [4]). To support the production of terpenoids for protection against temperature stress, carbon flux through the MEP pathway increases under elevated temperatures [95]. In the MVA pathway, not only HMGR but other enzymes such as AACT show induced responses under abiotic stress and appear to be involved in MVA pathway-mediated abiotic stress adaptation [68]. Changes in redox state also directly affect MVA and MEP pathway enzymes. Both HDS and HDR, which function as iron–sulfur reductases, have been identified as targets of the redox protein thioredoxin [101, 102] and thioredoxin-dependent regulation has also been suggested for DXR [102]. Moreover, it has been shown that HDS can receive electrons directly through the photosynthetic electron-transport chain via ferredoxin without any reducing cofactor, which is different from the flavodoxin/flavodoxin reductase and NADPH-dependent reducing system of HDS in bacteria [103].

Biotic stress such as pathogen attack often upregulates individual genes of *HMGR* families to direct flux toward the production of sesquiterpene phytoalexins under simultaneous downregulation of squalene synthase and sterol biosynthesis [62, 104]. Studies in tobacco showed that the regulation of pathogen-activated expression of HMGR involves the MEK2-SIPK/WIPK MAP kinase cascade [105, 106]. Another example highlights the importance of HMGR in root nodule development. The HMGR1 protein of *Medicago truncatula* directly interacts with NORK, which is a receptor-like kinase required for Nod factor signaling. Reduced expression of *HMGR1* in transgenic plants causes a severe decrease of root nodulation [107].

2.5 Regulation and Metabolite Exchange Across Subcellular Compartments

The compartmentalization of MEP and MVA pathways and associated downstream pathways allows for the subcellular regulation and coordination of photosynthesis-dependent and independent terpenoid biosynthetic routes. Despite the general notion that the MVA pathway enzymes are located in the cytosol or associated with the ER, peroxisomes have been discussed as localization sites for AACT (particularly AACT1 in *Arabidopsis*), PMK, and MVD based on the prediction of peroxisomal PTS targeting peptides and transient protein peroxisome import studies in *Catharanthus roseus* cells [108–110]. For MVD1 in *Arabidopsis*, however, mass spectrometry analysis suggests a cytosolic localization and MVD2 is predicted to reside in the cytosol [6]. In the absence of additional evidence for a partial localization of the MVA pathway in peroxisomes and possible transporters of isoprenoid precursors between the compartments, our current view on the subcellular organization of the MVA pathway remains incomplete.

The exchange of intermediates between the cytosol and plastids is usually not sufficient to rescue *Arabidopsis* mutants of biosynthetic enzymes in the MVA or

MEP pathways [31, 43, 64]. However, studies on *dxs2* mutants in tomato suggested that both pathways can, to some extent, compensate each other [66]. Moreover, in numerous cases, some degree of exchange of isoprenoid intermediates between plastids and the cytosol has been demonstrated based on the application of MEP and MVA pathway-specific inhibitors and the incorporation of stable-isotope precursors in primary and specialized terpenoid metabolites (e.g., [100, 111–117]. There is frequent evidence for trafficking of isoprenoid intermediates from the plastid to the cytosol in photosynthetic tissues (e.g., [113]). However, the contribution of the MVA pathway to the biosynthesis of plastidial isoprenoids can be substantial in the absence of light as was demonstrated by Opitz et al. [118] in roots of cotton seedlings or in dark-grown *Arabidopsis* seedlings [92].

To date, no specific transporters of isoprenoid precursors have been identified in the plastid membrane. The export of IPP from plastids to the cytosol was suggested to proceed by a plastidial proton symport system [119]. Studies by Flügge and Gao [120] indicated that IPP is not transported by plastidic phosphate translocators but depends on phosphorylated counter-substrates. In addition to the transport of IPP, there is evidence that longer prenyl diphosphates such as GPP and FPP are moved from plastids to the cytosol in tomato [121], the grape berry exocarp [122], and glandular trichomes of *Stevia rebaudiana* [116]. Genomic and proteomic analyses of single cells such as trichomes could be a promising approach to identify the isoprenoid transporter machinery between both compartments.

Despite some degree of exchange of isoprenoid intermediates between the plastid and the cytosol, the spatial separation of terpenoid biosynthetic pathways has been of benefit for the engineering of terpenoid end products. Expression and targeting of an FPP synthase and sesquiterpene synthase to plastids in tobacco did prevent carbon flux competition with sterol biosynthesis in the cytosol and promoted sesquiterpenoid yields by a thousandfold [123]. The same approach was successfully applied to produce high levels of the triterpene squalene in plastids and in tobacco trichomes although the latter case came at the cost of severely reduced growth [124]. Efforts have also been made to insert the entire MVA pathway in the tobacco chloroplast genome resulting in increased levels of mevalonate and carotenoids, but also squalene and sterols [125].

As mentioned above, expression of the MEP and MVA pathway genes is coordinately regulated by external stimuli. Other interdependent mechanisms of regulation between the pathways have been detected at posttranslational levels. Recent studies in tobacco demonstrated that blocking MEP pathway-dependent protein geranylgeranylation by treatment with the monoterpene *S*-carvone suppresses signaling to induce the MVA pathway-dependent formation of the sesquiterpene phytoalexin capsidiol [126]. Other possible roles of multicompartment networks in regulating the MVA pathway have been addressed by Verbitskiy et al. [127]. Work by these authors on proteins involved in RNA editing suggests that retrograde signaling between mitochondria and the cytosol might modify MVA pathway activity and, according to Tang et al. [128], this interaction seems to involve the mitochondrial respiratory pathway. Most notably, the MEP pathway intermediate, MEcPP, was found to function as a retrograde signaling molecule between plastids and the nucleus. MEcPP elicits the expression of stress-responsive nuclear-encoded plastidial proteins which suggests that the MEP pathway functions in stress sensing and coordinating stress-induced nuclear genes [129].

3 Isomerization and Condensation of the C5 Building Blocks

The construction of terpenoids with more than five carbons requires a sufficient supply of IPP and its more reactive, electrophilic isomer DMAPP. Therefore, IPP derived from the MVA pathway needs to be converted to DMAPP by the activity of an IPP isomerase (IDI; Figs. 1 and 2). Type I IPP isomerase isoenzymes in plants have been localized to mitochondria and plastids and shorter isoforms have been predicted to remain in the cytosol [130]. In analogy to mammalian cells, an alternative localization of IPP isomerases in peroxisomes has been discussed [108] but additional evidence for the role of peroxisomes in plant isoprenoid metabolism is needed. Although the formation of DMAPP from IPP derived from the MVA pathway is essential for downstream reactions in the cytosol and mitochondria, IPP isomerization seems less important in plastids where both C5 building blocks are produced by the MEP pathway. However, plastidial IPP isomerase activity might be necessary to produce an optimal ratio of IPP and DMAPP for the downstream condensation reactions and to provide precursors for a possible transport to the cytosol.

In the second major stage of terpenoid biosynthesis, IPP and DMAPP units are fused by the catalytic activity of prenyltransferases (isoprenyl diphosphate synthases) to form prenyl diphosphates as the linear central precursors of all terpenoids (Fig. 2). The initial reaction catalyzed by a prenyltransferase is a head-to tail (1′–4) condensation of IPP with the allylic cosubstrate DMAPP based on an ionization–condensation–elimination mechanism to produce a C10-allylic diphosphate. Additional rounds of head-to-tail condensation of the allylic product with more IPP units lead to the formation of short-chain (C15–C25), medium-chain (C30–C35), and long-chain (C40–Cn) prenyl diphosphates. The *cis*- or *trans*-stereochemistry of the double bonds of the prenyl diphosphate product determines whether the enzyme operates as *cis*-prenyltransferase or *trans*-prenyltransferase, which belong to families of structurally unrelated enzymes [131]. Much knowledge has been gained on the biochemistry and evolution of short-chain *trans*-prenyltransferases, which synthesize C10-geranyl diphosphate (GPP), C15-*trans*,*trans*-farnesyl diphosphate ((*E*,*E*)-FPP), or C20-all-*trans*-geranylgeranyl diphosphate (all-*trans*-GGPP) as the main precursors in terpenoid metabolism, although more recent work has discovered similar roles of previously undetected short-chain *cis*-prenyltransferases (see below).

3.1 Geranyl Diphosphate Synthases

As a precursor in the biosynthesis of C10-monoterpenoids, GPP is synthesized from IPP and DMAPP by the activity of GPP synthase enzymes (GPSs), which are usually targeted to plastids (Fig. 2). Different classes of homodimeric and heterodi/tetrameric GPSs have been identified in plants [132–136] (Fig. 2). A heterotetrameric GPS from peppermint was the first GPS to be discovered in plants [137] and since then related heterodimeric proteins have been found in a variety of other species such as *Anthirrinum majus*, *Clarkia breweri*, and *Humulus lupulus* [137–139]. The enzymes consist of a large subunit (LSU), which has significant homology ($\sim$50 %) to GGPP synthases (GGPS, see below) and can exhibit GGPP synthase activity as a recombinant protein, and a small subunit (SSU I) that shares only $\sim$20 % sequence similarity with homomeric prenyltransferases and is functionally inactive. It is generally thought that binding of SSU I modifies the activity of the LSU to produce GPP. The importance of the physical interaction of both subunits to make GPP has been confirmed by structural analysis of the heterotetrameric GPS from peppermint [134]. In *Arabidopsis*, Wang and Dixon [139] identified a separate lineage of SSU (SSU II) genes encoding GGPS-related proteins (GGR). *Arabidopsis* GGR modifies the in vitro activity of GGPS 11 to produce GPP and contains two conserved CxxxC motifs that are essential for the interaction of both subunits [139]. In contrast to the role of SSU I-containing GPSs in monoterpene formation in peppermint or hops, the function of heterodimeric GPSs carrying SSU II subunits is less clear because of the absence of a tight correlation between protein expression and the biosynthesis of monoterpenes in different tissues [139].

Engineering of GPS activity has been achieved by the expression of GPS.SSU I from snapdragon in tobacco and tomato fruits. The expressed subunit recruits plastidial GGPS proteins to form functionally active heterodimeric GPS proteins [121, 140]. The study on tomato also revealed that GPP produced in plastids is exported to the cytosol, where it can be used for monoterpene biosynthesis [121]. However, the exchange of GPP between both compartments might be limited in the absence of engineered GPP pools as was shown for a bifunctional *Arabidopsis* monoterpene/sesquiterpene synthase (TPS02), which is located in the cytosol and produces sesquiterpenes but no monoterpenes in planta [42].

Homodimeric GPS enzymes have been described from angiosperms and gymnosperms [135, 136, 141, 142]. These proteins belong to different lineages and are evolutionarily related to GGPSs (see below). The existence of a homodimeric GPS in *Arabidopsis* has been discussed controversially. A single *GPS1* gene was originally identified to encode a functionally active GPS enzyme [143]; however, more recently the GPS1 protein has been characterized as a multiproduct medium-/long-chain prenyl diphosphate synthase. The latter activity was observed when IPP was supplied in excess to the allylic substrates DMAPP, GPP, and FPP and was supported by the structural analysis of an active-site cavity with sufficient size to accommodate the medium-/long-chain products [144]. The GPS1 protein (renamed by Hsieh et al. as polyprenyl di(pyro)phosphate synthase, PPS) is targeted to plastids [143] where IPP and DMAPP are produced at ratios of approximately 5:1 by the MEP pathway. Thus, it is possible that this enzyme exhibits a PPS activity in vivo.

3.2 Farnesyl Diphosphate Synthases

Trans-FPP synthases (FPSs) catalyze the formation of (*E,E*)-FPP as a central precursor in the biosynthesis of terpene primary metabolites (phytosterols, brassinosteroids, dolichols, ubiquinones), for protein prenylation, and in the production of specialized metabolites such as sesquiterpenoids and triterpenoids (Fig. 2). As type I (eukaryotic) FPSs, plant *trans*-FPSs build a superfamily of homodimeric enzymes that are often encoded by small species-specific gene families (e.g., [145–147]). FPS isozymes of different size that are produced as a result of differential gene transcription have been localized to the cytosol or the mitochondria where they produce FPP pools for the biosynthesis of cytosolic and mitochondrial downstream products [148] (Fig. 2). Targeting of FPSs to peroxisomes has been discussed based on YFP fusion experiments in *Catharantus roseus* cells [149]. However, no peroxisomal targeting has been demonstrated for fluorescent FPS fusion proteins in *Arabidopsis*, which is consistent with results from proteomic studies of the cytosol and purified peroxisomes [150, 151].

As with the isozymes of the MEP and MVA pathways, it has been a primary interest to elucidate the possible functional differences of prenyltransferase isoforms. In *Arabidopsis*, the two *FPS* paralogues, *FPS1* and *FPS2*, have overlapping expression patterns and can rescue each other's loss, whereas double mutants are impaired in male genetic transmission and arrested at early embryo development [152]. However, there is no complete functional redundancy between the two isozymes inasmuch as FPS2 is the predominantly expressed isozyme in mature seeds and early seedling development, and FPS1 appears to be only expressed in the maternal seed coat [153]. Consequently, seeds of *fps2* mutants have a reduced sterol content [152]. Keim et al. propose that the specific expression of FPS2 in mature seeds is related to its higher enzymatic activity and thermal stability. The authors further speculate that during early development of the embryo (in the absence of *FPS2* expression), FPP might be imported from the seed tissue where *FPS1* is expressed [153].

3.3 Geranylgeranyl Diphosphate Synthases

Similar to (*E,E*)-FPP, all-*trans*-GGPP synthesized by all-*trans*-GGPSs is a major branching point for several downstream terpenoid pathways in primary and specialized metabolism. These include the biosynthesis of carotenoids and their breakdown products (abscisic acid, strigolactones), chlorophylls, tocopherols, gibberellins, plastoquinones, and diterpenoids (all synthesized in plastids), geranylgeranylated proteins and poly-/oligoprenols (synthesized in the cytosol), and poly-/oligoprenols synthesized in the plastids and mitochondria (Fig. 2). Compared to FPSs, GPPS isozymes are represented by larger gene families. For example, the *Arabidopsis* genome contains 12 *GGPS* paralogues, of which 10 have been identified to encode functional GGPS proteins of most likely homodimeric architecture and with GGPP as the primary or sole product [154]. The different GGPS isozymes are located in the plastids, mitochondria, and the ER consistent with the subcellular compartmentalization of the diverse GGPP-dependent terpenoid pathways. With the exception of two of the *Arabidopsis* isozymes (GGPS1-mitochondrial, GGPS11-plastidial), which are expressed in the whole plant, the remaining family members exhibit distinct spatiotemporal expression patterns [154]. Seedling-lethal albino and embryo-lethal phenotypes are found in *ggps1* mutants, indicating that GGPS1 has essential functions in development and the chlorophyll biosynthetic pathway [155]. Although possible redundant or more specific functions of most of the GGPS isozymes are not well understood, it is apparent that the divergence in the *Arabidopsis GGPS* gene family is the result of functional specialization and fine-tuning of metabolic pathways in different cellular compartments and in tissues at different developmental stages or under different environmental conditions.

Both FPPS and GGPPS proteins have been expressed in modules with sesquiterpene synthases and diterpene synthases, respectively, to engineer the biosynthesis of sesquiterpenoids and diterpenoids in microbial systems and in planta [123, 156]. Specifically, the buildup of FPP pools in plastids improved the precursor supply and allowed for a substantial increase in yield of the desired sesquiterpene products [123]. Other strategies to improve pathway productivity include generating combinatorial mutations in prenyldiphosphate synthase and downstream terpene synthases. For example, prokaryotic expression of pathway variants of a GGPPS and a terpene synthase, which produces a levopimaradiene diterpene precursor in ginkgolide biosynthesis, led to a more than 2,000-fold increase in the levels of the levopimaradiene product thereby stressing the importance of protein engineering in these approaches [157].

3.4 Chain Length Regulation and Evolution of Prenyltransferases

Structural analysis combined with random or site-directed mutagenesis has provided substantial insight to the chain length regulation of short-chain prenyltransferase products [158]. Based on crystal structures of several homodimeric FPPs and GGPS from eukaryotes and prokaryotes [159–165], short-chain prenyltransferases share a common protein fold composed of 13 α-helices with 10 helices surrounding the active site cavity. IPP and the allylic substrate are bound by two highly conserved aspartate-rich regions, a first DDx2-4D motif (FARM) and a second DDxxD motif (SARM), which are positioned on opposite walls of the cavity. Product chain length is in part regulated by amino acid residues upstream of the FARM motif (position –4, –5), which change the size of the hydrophobic substrate binding or elongation pocket of the polyisoprenoid chain [166, 167]. According to this mechanism, type I FPSs such as *Arabidopsis* FPS1 and FPS2 have a smaller binding pocket because of the presence of "bulkier" aromatic amino acid residues. In type II GGPSs, which comprise eubacterial and plant GGPSs, these aromatic amino acids are replaced by smaller residues such as alanine, serine, and methionine allowing the formation of a longer C20 chain. Studies of yeast GGPS indicated that chain termination at C20 depends on residues located deeper in the catalytic cavity [162]. Poulter and colleagues recently employed a large-scale bioinformatics approach combined with experimental enzyme characterization, protein crystallization, and computational modeling to predict the chain length specificity of a large number of putative polyprenyl transferases [168]. The approach, which resulted in a high rate of correctly predicted functions, largely supported the notion that steric hindrance in the elongation cavity is the main criterion determining chain length specificity. It is important to note that the study also suggested a chain-length–determining effect of "second shell" residues that are positioned in the vicinity of the residues lining the elongation pocket. Depending on their size, these neighboring residues may or may not provide flexibility for bulkier aromatic residues that protrude into the cavity to be moved or displaced by the growing polyprenyl chain [168].

Phylogenetic analyses of prokaryotic and eukaryotic prenyltransferases place plant FPSs in a clade with other eukaryotic FPSs that is distinct from a cluster containing plant GPS and GGPS proteins [158]. A comprehensive phylogenetic study of GGPS and GPS homologues of land plants and green algae demonstrated a lineage and species-specific expansion of GGPS families indicating gene duplication events and functional divergence [169]. The phylogeny shows several evolutionary transitions from proteins with GGPS to GPS activity. For example, gymnosperm homodimeric GGPSs, which form a distinct clade among plant GGPSs, can produce shorter prenyldiphosphates or synthesize exclusively GPP [133, 142]. In comparison to GGPSs from green algae and mosses that possess the FARM and SARM motifs and a conserved CxxxC motif, the gymnosperm GGPSs have acquired a second CxxxS (bifunctional GGPS) or CxxxC (GPS) motif. The two CxxxC motifs are characteristic of most proteins that are associated with GPS activity. Thus, they are present in the SSU I and SSU II subunits of heteromeric GPS proteins and critical in binding the LSU. The binding of both subunits limits access to the elongation cavity and terminates chain elongation at the formation of a C10-product [134]. SSU I and SSU II proteins have lost both aspartate-rich motifs or carry a mutated SARM, respectively, which is associated with the loss of prenyl diphosphate activity [135, 169]. Interestingly, an earlier study reported a flower-specific GPS from orchids with similarity to SSU II [136]. This protein lacks the SARM but maintains GPS activity as a homodimeric enzyme (Fig. 2).

Several proteins with homology to *Arabidopsis* PPS (former GPS1) have been reported from other plants and designated as homodimeric GPSs (Fig. 2). These proteins do not carry the CxxxC motifs and it remains to be determined whether they function as true GPSs in vivo or may exhibit medium-chain or long-chain polyprenyl diphosphate activity as was shown for the *Arabidopsis* enzyme. For instance, GPS activity was demonstrated for a protein in tomato, but assays were performed at a low IPP/DMAPP ratio [141]. Furthermore, silencing or mutation of this enzyme and of PPS in *Arabidopsis* resulted in dwarfed or embryo lethal phenotypes, which could be related to promiscuous GGPS activity to produce GGPP for gibberellin biosynthesis or the synthesis of longer precursors in plastoquinone biosynthesis. The formation of longer chain products by *Arabidopsis* PPS is also supported by the absence of aromatic amino acids near the FARM. Computational predictions such as those presented by Wallrapp et al. [168] should facilitate determining the chain length specificity of PPS homologues. In summary, GPS activity appears to be the result of promiscuity and neofunctionalization of GGPS (or PPS?) proteins in conjunction with the evolutionary adaptation of individual plant lineages to produce monoterpenes as constituents of floral scent or for chemical defense.

3.5 Cis-Isoprenyl Diphosphate Synthases

One of the surprising findings in the field of terpene biosynthesis in the past five years was the identification of short-chain *cis*-prenyltransferases (CPTs) and the conversion of their *cis*-prenyl diphosphate products to terpenoids by the activity of terpene synthases (see below). Prior to this discovery, it was generally believed that *CPTs* synthesize prenyl diphosphate products with a chain length of more than 50 carbons by using all-*trans* short-chain prenyl diphosphates as allylic primer substrates [170]. Such prenyltransferases in plants include enzymes that produce C70–C120 dehydrodolichol diphosphates or natural rubber (>C10,000) from (*E*,*E*)-FPP by head-to-tail condensations in a *cis* orientation [170–172]. Functional genomics studies of terpene biosynthetic genes in glandular trichomes of wild tomato then revealed the presence of a short-chain (Z,Z)-FPP synthase that produces (Z,Z)-FPP [173] (Figs. 2 and 3). Characterization of a nine-member CPT family in cultivated tomato gave additional evidence for short-chain enzyme activity by the identification of three genes encoding a neryldiphosphate (NPP) synthase (NDPS1 or SlCPT1, expressed in trichomes), a (Z,Z)-FPP synthase (SlCPT6, expressed in root and fruit), and a nerylneryl diphosphate (NNPP) synthase (NNDPS or SlCPT2, expressed in the stem), respectively [174] (Figs. 2 and 3). All three proteins are targeted to plastids [174]. Notably, the Z,Z-FPP pool produced by (Z,Z)-FPP synthase in trichome-specific plastids in wild tomato is used by plastidic sesquiterpene terpene synthases (santalene/bergamotene sesquiterpene synthase [173] and 7-epizingiberene synthase [175, 176]), which are related to diterpene synthases. Engineering of (Z,Z)-FPP synthase and 7-epizingiberene synthase in trichomes of

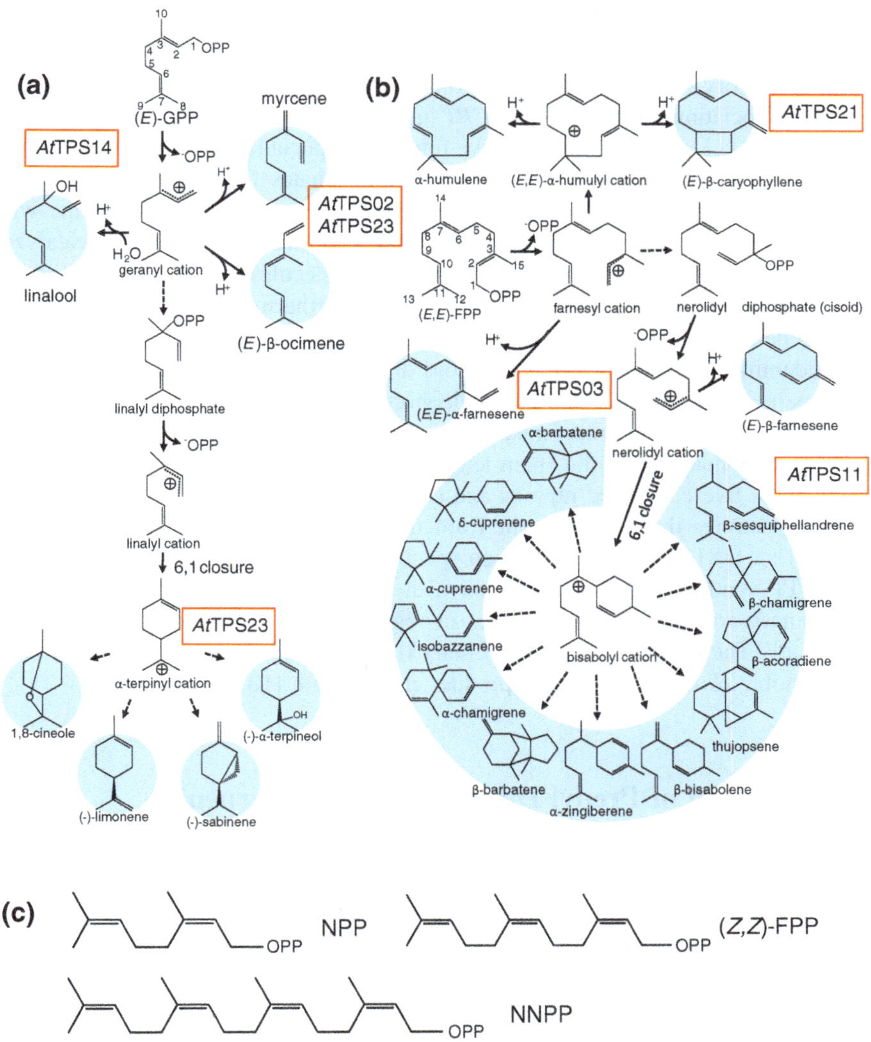

Fig. 3 Structural diversity of terpenoid specialized metabolites and their precursors. **a**, **b** Examples of monoterpenoids and sesquiterpenoids produced by different *Arabidopsis* terpene synthases (*At*TPS). **c** Structures of prenyl diphosphates produced by short-chain CPTs; NPP neryl diphosphate; NNPP, nerylneryl diphosphate (**a**, **b** from Tholl and Lee 2011) [217], thearabidopsisbook.org, Copyright American Society of Plant Biologists

cultivated tomato led to the production of 7-epizingiberene and increased resistance to herbivores [175]. NPP has been shown to be converted by a monoterpene synthase to β-phellandrene among other monoterpenes [177]. Consequently, coexpression of the NDPS1 enzyme with phellandrene synthase 1 was used successfully for metabolic engineering of monoterpene formation in tomato fruits [178]. Intriguingly, expression of NDPS1 alone led to the reduction of carotenoid levels in fruits because of feedback inhibition of GGPS by NPP. Based on these findings, it is plausible that NPP production is restricted primarily to trichomes to avoid inhibitory effects on carotenoid biosynthesis.

The association of the tomato *CPT* genes with terpenoid biosynthetic gene clusters [174] clearly indicates adaptive functional specialization in the tomato *CPT* gene family to provide short-chain prenyl diphosphates for different terpene biosynthetic pathways including trichome-specific terpene biosynthesis. In line with these findings, a *cis*-type prenyltransferase was identified in lavender that catalyzes the head-to-middle condensation of two DMAPP molecules to synthesize lavandulol diphosphate, the precursor of lavendulol [179]. Furthermore, in the nine-member *CPT* gene family of *Arabidopsis* a multiproduct prenyltransferase (AtCPT6) has been identified that makes polyisoprenoid diphosphates with six to eight isoprene units as precursors of polyisoprenoid alcohols in roots [180].

As with *trans*-prenyltransferases, efforts have been made to determine amino acid residues that control the chain length specificity of CPTs [170]. Sequences of CPTs share five conserved regions and employ residues for substrate binding and catalytic activity that are different from those of *trans*-prenyltransferases [170]. Kang et al. [181] exploited accession-specific sequence differences of NDPS and (*Z*,*Z*)-FPP synthase in tomato coupled with homology modeling and site-directed mutagenesis to identify four residues in region II that are important for product specificity. These residues are part of helix II, which, together with helix III, lines a hydrophobic cleft that influences product chain length [182, 183].

4 Conversion of Prenyl Diphosphates and Terpene Synthase Function and Regulation

Trans- and *cis*-prenyldiphosphates are the entry points to various downstream primary and specialized terpenoid biosynthetic routes in plastids, mitochondria, and the cytosol (summarized in Fig. 2). It is beyond the scope of this chapter to address all of these pathways and the reader is referred to other chapters in this series (e.g., carotenoid biosynthesis) or more specialized recent reviews in the field.

The tremendous diversity of terpenoids in specialized metabolism can to a large extent be attributed to the activity of terpene synthases (TPSs; Fig. 3a). TPS enzymes have, therefore, become a focus point of in planta and heterologous metabolic engineering of terpenoid end products with use as pharmaceuticals, flavors, biofuels, or plant chemical defenses [184] (see other chapters in this series). The TPS superfamily, which is divided into eight subfamilies (TPSa–h), comprises a large and still growing number of enzymes from almost all taxa in the plant kingdom [185]. TPSs convert acyclic C5 to C20 *cis*- or *trans*-prenyl diphosphate intermediates into C5-hemiterpenes such as isoprene, C10-monoterpenoids, C15-sesquiterpenoids, or C20-diterpenoids (Fig. 2). The primary enzymatic products are in most cases acyclic or cyclic hydrocarbons (Fig. 3a) that are frequently modified by secondary enzymatic reactions such as hydroxylation, peroxidation, methylation, acylation, glycosylation, or cleavage to produce biologically active end products of even larger structural diversity [186]. TPS enzymes facilitate adaptations of terpene metabolism to the changing environment because their promiscuous activity often results in the production of more than a single compound (e.g., [187]) and TPS proteins easily acquire new catalytic properties by minor structural changes [187–192].

Mechanistically, TPS proteins are divided into class I and class II enzymes. The enzymatic reaction catalyzed by class I TPSs starts with the ionization of the prenyl diphosphate substrate by a divalent cation-dependent subtraction of the diphosphate moiety. The produced carbocation intermediate then enters different reactions that can include cyclizations, hydride shifts, and rearrangements prior to a termination of the reaction by proton loss or the addition of a nucleophile such as water [193] (Fig. 3). By contrast, class II TPSs, which include oxidosqualene cyclases (see below) and diterpene synthases, catalyze the ionization of their substrate by adding a proton to an epoxide ring or via protonation at the 14,15-double bond of GGPP, respectively. Class II diterpene synthases that fall into this category are *ent*-copalyl diphosphate (CPP) synthases (CPSs), which are involved in gibberellin and phytoalexin biosynthesis [194] (Fig. 2). In the gibberellin biosynthetic pathway, CPSs catalyze a protonation-induced bicyclization of the substrate GGPP to form *ent*-CPP, which is further ionized and converted to ent-kaur-16-ene by a class I ent-kaurene synthase (KS) activity. Detailed genomic studies of land plants revealed that the gibberellin biosynthetic pathway gave rise to the biosynthesis of an array of specialized labdane-related diterpenoids largely by gene duplication and divergence of CPS and KS homologues [194]. The ability to produce kaurene arose early in land plant evolution as can be assumed from the identification of a bifunctional classII/I CPS/KS in the moss *Physcomitrella patens*, which catalyzes the formation of ent-kaurene (and 16-hydroxykaurene) via a CPP intermediate in the biosynthesis of kaurenoic acid [195, 196]. Similar class II/I diterpene synthases such as abietadiene synthase occur in gymnosperms and can be considered early diterpene synthases. These enzymes produce (+)-CPP from GGPP prior to an ionization-initiated cyclization of (+)-CPP to the diterpene product [197]. An interesting new view on the evolution of plant *TPS* genes comes from a genomic study of a large *TPS* gene family in the fern *Selaginella moellendorffii* [198]. Two distinct types of *TPS* genes were identified: a group of diterpene synthases that represent a new plant TPS-h subfamily, and, surprisingly, a group of monoterpene synthases and sesquiterpene synthases that are more closely related to microbial TPSs and may be the first indication for a horizontal gene transfer of *TPS* genes [198].

It should be noted here that, recently, a new mechanism for the enzymatic formation of cyclic terpenes was discovered in the iridoid monoterpene biosynthetic pathway [199]. Iridoids have pharmaceutical and antibacterial activities and are also produced by aphids as pheromones [200, 201]. The iridoid synthase from *Catharantus roseus* is a short-chain reductase that most likely generates a C5-iridoid ring in the linear monoterpene 10-oxogeranial substrate by coupling a reduction step with a cyclization step via a Diels–Alder cycloaddition or a Michael addition [199].

This exciting finding may open the way for future discovery of similar reductase-type terpene cyclases in plants and other organisms.

More insight to the evolution of "regular" TPS enzymes has been gained from the analysis of an increasing number of crystal structures including those from an isoprene synthase [202], monoterpene synthases [203–206], sesquiterpene synthases [207, 208], a class I diterpene synthase (taxadiene synthase [209]), a class II CPP synthase [210, 211], and a class II/I diterpene synthase (abietadiene synthase [212]). Comparisons of the assembly of a class I type α-domain and class II type β and γ domains led to the prediction of an evolutionary scenario according to which an ancestral bifunctional classII/classI diterpene synthase (consisting of all three domains with a functional α- and β-domain) similar to the CPS/KS enzyme of *P. patens* gave rise to class II type diterpene synthases (consisting of all three domains with a functionally active β-domain and an inactive α-domain) and class I type TPSs (consisting of a nonfunctional β-domain and a functionally active α-domain) [213, 214]. A functionally active class I α-domain carries the highly conserved aspartate-rich motif, DDxxD, and a less conserved NSE/DTE motif, which are located on opposite sides of the entrance of the catalytic side and help position the diphosphate substrate by binding of a trinuclear magnesium cluster [215]. By contrast, functional class II β-domains carry a conserved DxDD motif, which is required for protonation-initiated carbocation formation [213].

Although TPS enzymes may convert more than one prenyl diphosphate substrate in vitro, their function in vivo is largely determined by the substrate pool that is available in the respective cellular compartment. In this regard, TPS enzymes localized in plastids generally produce monoterpenoids or diterpenoids from plastidial GPP and all-*trans*-GGPP, respectively, whereas TPSs in the cytosol primarily convert (*E*,*E*)-FPP to sesquiterpenes (or squalene in the biosynthesis of C30 terpenes). However, this general rule has recently been challenged by the discovery of plastidial (*Z*,*Z*)-FPSs and sesquiterpene synthases in tomato, the latter of which are more closely related to kaurene synthases in the TPS-e subfamily [177].

The existence of medium-size to large TPS families in *Arabidopsis* and many other plant species strongly supports the notion that *TPS* genes evolve by gene duplication and neofunctionalization [185, 216, 217]. Such duplication events combined with relocation in the genome can include other genes that encode modifying enzymes such as cytochrome P450s, and thus lead to the assembly of gene clusters. From the first discovery of a thalianol triterpene biosynthetic gene cluster in *Arabidopsis* [218], several such clusters have been found in the arabidiol, marneral, and avenacin triterpene biosynthetic pathways in *Arabidopsis* and oat, respectively [219, 220] (Sohrabi et al. in preparation), and for the biosynthesis of labdane-related diterpenoids in rice [221] or monoterpenoids and sesquiterpenoids in tomato [222]. The triterpene biosynthetic clusters carry genes for oxidosqualene cyclases (OSCs), which catalyze the cyclization of oxidosqualene to one or more cyclic triterpene alcohols via formation of a carbocationic intermediate [223, 224]. Coexpression with other cluster genes (e.g., P450s, desaturase, acyltransferase) in an operon-like manner then allows a consecutive derivatization of the triterpene precursor [218–220]. The evolutionary forces driving this coordinated gene cluster assembly are believed to be twofold. Clustering of genes for pathway building facilitates the regulation of multiple genes at the level of chromatin and/or prevents the accumulation of possible cytotoxic products [219, 225, 226]. However, a strict coregulation of gene expression does not seem to be the case in all clusters as was shown for a diterpene biosynthetic cluster in rice containing P450s that are differentially regulated and function in two different pathways [194].

Clusters that exhibit a coordinated expression of their genes have allowed the identification of putative key regulators such as in the case of the basic leucine zipper transcription factor, OsTGAP1, which is involved in regulating a diterpenoid biosynthetic gene cluster in rice [227]. Another transcription factor that was identified previously to regulate terpene biosynthetic genes positively is a WRKY transcription factor in cotton, GaWRKY1, which regulates the transcription of a sesquiterpene synthase gene in the gossypol biosynthesis pathway [228]. More recent studies on *Artemisia annua* suggest that APETALA2/ethylene-response factors (AP2/ERF) are positive regulators of biosynthetic genes in the formation of the sesquiterpene artemisinin, an insect deterrent and antimalaria drug produced in leaf glandular trichomes [229]. However, these studies thus far do not place the identified transcription factors into regulatory networks related to development and cell specification.

A better understanding of the regulatory networks controlling terpene volatile formation has been gained in the process of flower maturation in *Arabidopsis*. Two R2R3 MYB transcription factors, MYB21 and MYB24, were identified that promote gynoecium growth and nectary development and positively affect expression of the major floral (*E*)-β-caryophyllene sesquiterpene synthase *TPS21* [230]. Both MYB TFs respond positively to jasmonic acid (JA), the levels of which are induced by the auxin response factor 6 (ARF6) and ARF8, both master regulators of flower maturation. TPS21 and the second floral sesquiterpene synthase, TPS11 [187], also respond more directly to JA by the direct binding of their promoters to the bHLH transcription factor MYC2 [231], which is a central regulator of the JA signaling pathway in developmental and stress responses [232, 233]. In addition, *TPS21* and *TPS11* gene expression is indirectly regulated by gibberellins through the binding of DELLA proteins (gibberellin signaling repressors) [231].

Similar to the tissue-specificity of terpene formation in flowers, terpene-specialized metabolism in roots appears to be a highly coordinated cell type-specific process. Genes of the thalianol and marneral triterpene biosynthetic gene clusters are coexpressed primarily in the root epidermis [218, 219]. Likewise, 14 genes of the *Arabidopsis TPS* family are expressed in different root tissues. For example, a recently identified rhizathalene diterpene synthase (TPS08; Fig. 4) was found to be primarily expressed in the root stele (see below) [234]. In addition, two 1,8-cineole monoterpene synthase genes are constitutively expressed in the stele of the root elongation zone and differentiation/maturation zone and in the epidermis and cortex of more mature roots; a similar expression pattern has been observed for two closely related (*Z*)-γ-bisabolene sesquiterpene synthases [188, 235]. However, no networks of temporal and spatial regulation have yet been defined for these root-specific genes.

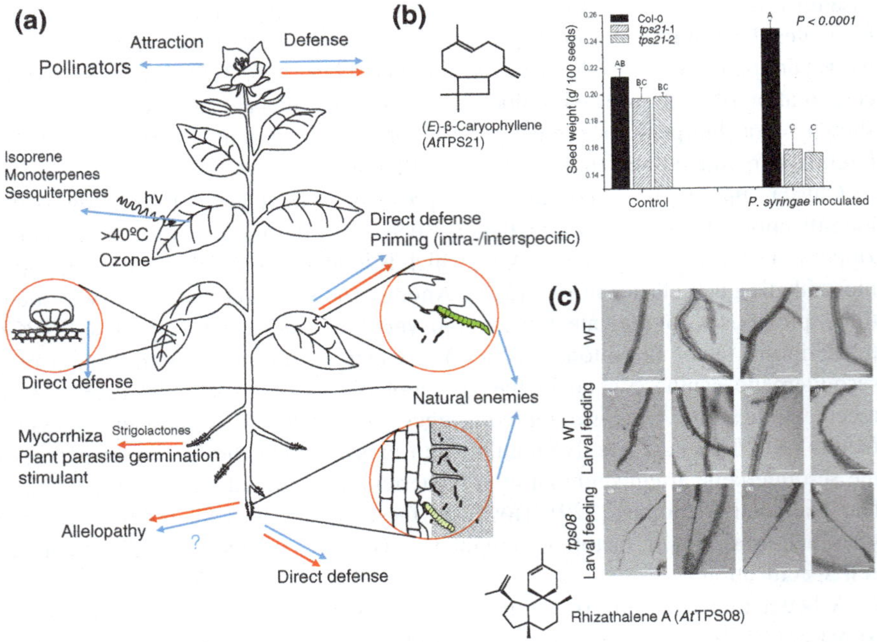

Fig. 4 Biological functions of plant terpenoids. **a** Functions of volatile terpenoids (*blue arrows*) and nonvolatile terpenoids (*red arrows*) in the interactions of plants with their environment. **b** Defensive activity of the volatile sesquiterpene, (*E*)-β-caryophyllene, against infection of *Arabidopsis* flowers by the microbial pathogen *Pseudomonas syringae*. Mutants deficient in (*E*)-β-caryophyllene biosynthesis in the floral stigmatic tissue (*tps21*) produce lighter seeds post inoculation of flowers with *P. syringae*. **c** Antifeedant activity of the semivolatile diterpenoid, rhizathalene, in *Arabidopsis* roots. Light microscopic pictures of roots of wild-type and rhizathalene biosynthetic mutants (*tps08*) with and without feeding by *Bradysia* (fungus gnat) larvae. Increased feeding damage is observed in the absence of the diterpenoid compound (**b**, **c** from Huang et al. 2012 [245] and Vaughan et al. 2013 [234], www.plantcell.org, Copyright American Society of Plant Biologists partially modified)

5 Multifunctionality of Plant Terpenoids

Although terpenoids serve important primary functions as photosynthetic pigments (carotenoids), electron carriers (side-chains of ubiquinone and plastoquinone), regulators of growth and development (gibberellins, abscisic acid, strigolactones, brassinosteroids, cytokinins), in protein glycosylation (dolichols), or as elements of membrane structure and function (phytosterols), specialized terpenoid metabolites (covered here), in particular, have been recognized for an array of biological roles. Volatile or semivolatile, low-molecular–weight terpenoids, which include isoprene, monoterpenoids, sesquiterpenoids, and diterpenoids, are implicated in the protection of plants against abiotic stress and in various biotic interactions above- and below-ground [236] (Fig. 4a). The substantial emissions of isoprene and monoterpenes from various vascular and nonvascular plants have been associated with the protection against thermal stress. This process is presumably based on an intercalation of the volatile compounds with the photosynthetic membranes and thereby enhances membrane functionality [237–239]. Moreover, transgenic approaches in tobacco and poplar support a role of isoprene in oxidative stress protection [240–243] and are addressed in a separate chapter by Vickers et al.

Volatile terpenoids as constituents of floral scent are implicated in mutualistic interactions with plant pollinators. For instance, choice tests with bumblebees have indicated a role of monoterpenoids emitted by monkeyflowers in pollinator attraction [244]. Nevertheless, distinct evidence for a specific role of terpenoids in pollinator attraction by the use of biosynthetic mutants is still missing, but it can be assumed that attractive effects depend on mixtures of volatiles rather than individual compounds. The notion that floral volatile terpenoids serve multiple functions has been supported by their role in the defense of floral tissues against microbial pathogens. This interaction was demonstrated in flowers of *Arabidopsis* mutants, which lack the emission of (*E*)-β-caryophyllene from their stigmatic tissue. The mutant flowers were more susceptible to infection by *P. syringae*, which resulted in lighter and often misshaped seeds suggesting reduced plant fitness [245] (Fig. 4b). Similar findings were made by Junker et al. [246] demonstrating that floral volatiles play roles in the structuring of bacterial communities that colonize flower petals by providing compound-dependent niches.

Volatile terpenoids also serve important functions as constitutive or pathogen- and herbivore-induced compounds in the defense of photosynthetic tissues. For example, repellent activities have been reported for monoterpene volatiles that are emitted by leaves of *Chrysanthemum morifolium* and, notably, herbivore-deterrent effects have been observed for isoprene [247, 248]. Furthermore, volatile terpenoids that accumulate in glandular trichomes function as insect repellents as was, for example, found for the activity of sesquiterpenes in trichomes of wild tomato against white flies [249] (Fig. 4a). In conifers, the production of terpenoid oleoresin and terpenoid volatile emissions constitute an important chemical defense system [250]. In a search for resistance factors, the monoterpene (+)-3-carene was found to be associated with resistance of Sitka spruce (*Picea sitchensis*) to white pine weevil (*Pissodes strobi*) [251]. Variation of the (+)-3-carene production in resistant and susceptible trees was demonstrated to depend on the copy number of a (+)-3-carene *TPS* gene, differences in gene transcript and protein levels, and variation in catalytic efficiencies. Similarly, in *Arabidopsis*, ecotype-specific variation of the herbivore-induced volatiles, (*E*)-beta-ocimene and (*E*,*E*)-alpha-farnesene, is controlled by allelic variation and differences in subcellular targeting of the two terpene synthases, TPS02 and TPS03 [42].

The role of herbivore-induced volatile blends in the attraction of natural enemies of herbivores (Fig. 4a) and at higher trophic levels has been investigated in numerous studies (reviewed by [252, 253]). Work with transgenic *Arabidopsis* provided strong evidence for the role of volatile terpenes in these interactions [254–256]. However, as indicated for floral scent, the effect of these compounds has to be considered in the context of the entire herbivore-induced volatile blend, and actual fitness benefits to the plant host under natural conditions are still debated [252, 257, 258]. Indirect defense responses mediated by volatile compounds also occur upon insect oviposition [259]. For example, egg deposition on the foliage of European field elm (*Ulmus minor*) by the elm leaf beetle (*Xanthogaleruca luteola*) leads to the emission of volatiles including the irregular homoterpene, (*E*)-4,8-dimethyl-1,3,7-nonatriene (DMNT), which play a role in the attraction of the specialist egg parasitoid, *Oomyzus gallerucae* [260].

In addition to their function in the interaction with herbivores and their enemies, constitutive and induced volatile mixtures (including volatile terpenes such as homoterpenes) can serve as interspecific, intraspecific, and intraplant "alarm" signals to prime or induce defense responses in neighboring plants or in unattacked tissues of the same plant [261–264]. In these interactions, volatiles may not necessarily need to enter the leaf tissue of the neighboring plant but remain on the leaf surface. This effect was observed for sesquiterpenoids that are emitted by rhododendron leaves and adsorbed on the leaves of birch trees, where they exhibit direct herbivore-repellent activities [265]. Moreover, terpenoids were suggested to be involved in parasitic plant interactions, specifically, in the attraction of the parasitic plant *Cuscuta pentagona* (dodder) to establish contact with tomato as its host [266]. The molecular mechanism of host plant detection in this response as in other volatile-mediated plant–plant interactions is still poorly understood.

The described functions of volatile terpenoids in aboveground plant defense are complemented by nonvolatile terpenoids. As an example, glycosides of geranyllinalool serve as potent antifeedants in the wild tobacco, *Nicotiana obtusifolia* [267], and recently detected *ent*-kaurane–related diterpenoids in maize named kauralexins as well as acidic sesquiterpenes called zealexins function as pathogen-inducible phytoalexins [268, 269]. Similarly, metabolomics studies of *Barbarea vulgaris* revealed that triterpene saponins contribute to resistance against flea beetle attack [270].

An increased interest in the role of specialized metabolites belowground has shown that terpenoids serve functions similar to those aboveground. Recent studies in *Arabidopsis* roots discovered semivolatile diterpene hydrocarbons with an unusual tricyclic spiro-hydrindane structure called rhizathalenes [234] (Fig. 4c). These compounds are produced in the root stele, from where they diffuse through the surrounding cell layers to function as local antifeedants by reducing root herbivore damage on these cell layers [234] (Fig. 4c). The role of volatile terpenes in belowground indirect defense has been well established based on studies in maize showing that the sesquiterpene, (*E*)-β-caryophyllene, which is emitted from roots upon attack by the Western corn root worm *Diabrotica virgifera*, attracts entomopathogenic nematodes [271, 272]. These findings prompted attempts to engineer (*E*)-β-caryophyllene production in nonemitting American maize cultivars, which resulted in an increased attraction of nematodes and higher resistance to corn root worm attack [273]. However, constitutive emissions of (*E*)-β-caryophyllene were found to have additional costs inasmuch as they compromise seed germination, plant growth, and yield [274]. Therefore, more fine-tuned engineering strategies considering herbivore-induced emissions may have to be developed to circumvent these cost effects.

Nonvolatile terpenoids can be exuded from roots into the rhizosphere and the surrounding soil environment where they are involved in different defense responses. Studies using rice mutants convincingly demonstrated that labdane-related diterpenoids named momilactones exhibit allelopathic effects on barnyard grass competitors [275]. Moreover, avenacins, which are triterpene saponins exuded by the roots of oat, are known for their potent activity as phytoalexins [276]. Excitingly, a recent study by Osbourn and colleagues revealed that common triterpene precursors have additional signaling functions in root development. Specifically, it was demonstrated that β-amyrin is involved with determining the patterns of epidermal root hair cells [277]. These findings indicate that the roles of specialized metabolites in biotic interactions and potential "primary" functions become increasingly blurred. Signaling functions have also been demonstrated for the abietane diterpenoid, dehydroabietinal, which is produced at picomolar concentrations in *Arabidopsis* leaf tissue and serves as a vascular signaling compound and potent activator of systemic acquired resistance [278]. This activity seems to depend on the association of dehydroabietinal with vascular sap proteins.

Finally, it should be noted that strigolactones have become an exciting model for the multifunctionality of small molecules. As carotenoid-derived compounds (reviewed by [279]), strigolactones have important roles as exogenous signals by recruiting arbuscular mycorrhizal fungi in the rhizosphere [280]. Parasitic plants such as *Striga lutea* (witchweed) eavesdrop on these compounds by using them as germination signals [281]. As internal signals, strigolactones function as growth and developmental hormones that suppress shoot branching [282, 283]. Other processes that involve strigolactone signaling functions include root growth and development, stem elongation, secondary growth, leaf expansion and senescence, and responses to drought and salinity [279, 284]. Rapid progress has been made in understanding the perception of strigolactones but many open questions remain about downstream targets and the role of strigolactone-related compounds [279].

6 Outlook

In the past years, research in terpenoid metabolism has received a boost from developments in synthetic biology to generate engineering platforms for the production of high-value terpenoid products. Production systems in microbes have been developed to result in substantial yields [285], however, engineering of terpenoids in plants still faces challenges because of the complexity of metabolic and regulatory networks. Nevertheless, strategies to avoid metabolic flux competition by targeting biosynthetic modules to different cellular compartments have proved to be promising. Likewise, establishing pathways in specialized cells such as trichomes helps avoid metabolic competition and phytotoxic effects that could negatively affect growth and yield. The discovery of trichome-specific CPTs and TPS enzymes with substrate specificity for *cis*-prenyl diphosphates most likely will facilitate the engineering efforts in these tissues and provide new gene tools for building synthetic modules. Despite the successful use of distinct organelles such as plastids as "mini" subcellular factories, more efforts need to be made to understand the compartmentalization of the core terpenoid pathway. Especially, additional work should be performed to clarify the putative localization of the MVA pathway and prenyltransferase enzymes in peroxisomes, which would add yet another dimension to the compartmental complexity of terpenoid metabolism in plants. Genomics-based efforts to better understand the regulation of the early terpenoid pathways and terpenoid biosynthetic gene clusters are on their way and will be essential to gain a better understanding of the regulatory networks and epigenetic factors coordinating terpenoid metabolic routes in space and time. Finally, our knowledge of the biological roles of terpenoids is still far from complete. The recent findings of overlapping activities of terpenoids such as strigolactones or triterpenoids in biotic interactions and as internal signals indicate a need to use advanced mutant-based approaches for elucidating the multifunctionality of plant terpenoid compounds.

Acknowledgments This work was supported by the National Science Foundation MCB grant 0950865.

References

1. Kubeczka KH (2010) History and sources of essential oil research. In: Baser KHC, Buchbauer G (eds) Handbook of essential oils: science, technology, and applications. CRC Press/Taylor & Francis, Boca Raton, pp 3–38
2. Zhang FZ, Rodriguez S, Keasling JD (2011) Metabolic engineering of microbial pathways for advanced biofuels production. Curr Opin Biotechnol 22:775–783
3. Lange BM, Ahkami A (2013) Metabolic engineering of plant monoterpenes, sesquiterpenes and diterpenes-current status and future opportunities. Plant Biotechnol J 11:169–196
4. Hemmerlin A, Harwood JL, Bach TJ (2012) A raison d'etre for two distinct pathways in the early steps of plant isoprenoid biosynthesis? Prog Lip Res 51:95–148
5. Hemmerlin A (2013) Post-translational events and modifications regulating plant enzymes involved in isoprenoid precursor biosynthesis. Plant Sci 203:41–54
6. Vranova E, Coman D, Gruissem W (2013) Network analysis of the MVA and MEP pathways for isoprenoid synthesis. In: Merchant SS (ed) Ann Rev Plant Biol, vol 64, pp 665–700
7. Vranova E, Coman D, Gruissem W (2012) Structure and dynamics of the isoprenoid pathway network. Mol Plant 5:318–333
8. Caelles C, Ferrer A, Balcells L, Hegardt FG, Boronat A (1989) Isolation and structural characterization of a cDNA-encoding *Arabidopsis thaliana* 3-hydroxy-3-methylglutaryl coenzyme A reductase. Plant Mol Biol 13:627–638
9. Enjuto M, Balcells L, Campos N, Caelles C, Arro M, Boronat A (1994) *Arabidopsis thaliana* contains 2 differentially expressed 3-hydroxy-3-methylglutaryl-CoA reductase genes, which encode microsomal forms of the enzyme. Proc Natl Acad Sci USA 91:927–931
10. Campos N, Boronat A (1995) Targeting and topology in the membrane of plant 3-hydroxy-3-methylglutaryl coenzyme a reductase. Plant Cell 7:2163–2174

11. Denbow CJ, Lang S, Cramer CL (1995) Targeting and membrane orientation of tomato 3-hydroxy-3-methylglutaryl coenzyme A reductases. Plant Physiol 108:144
12. Re EB, Brugger S, Learned M (1997) Genetic and biochemical analysis of the transmembrane domain of *Arabidopsis* 3-hydroxy-3-methylglutaryl coenzyme A reductase. J Cell Biochem 65:443–459
13. Vollack KU, Dittrich B, Ferrer A, Boronat A, Bach TJ (1994) Two radish genes for 3-hydroxy-3-methylglutaryl-CoA reductase isozymes complement mevalonate auxotrophy in a yeast mutant and yield membrane-bound active enzyme. J Plant Physiol 143:479–487
14. Re EB, Jones D, Learned RM (1995) Coexpression of native and introduced genes reveals cryptic regulation of HMG CoA reductase expression in *Arabidopsis*. Plant J 7:771–784
15. Holmberg N, Harker M, Wallace AD, Clayton JC, Gibbard CL, Safford R (2003) Co-expression of N-terminal truncated 3-hydroxy-3-methylglutaryl CoA reductase and C24-sterol methyltransferase type 1 in transgenic tobacco enhances carbon flux towards end-product sterols. Plant J 36:12–20
16. Ohyama K, Suzuki M, Masuda K, Yoshida S, Muranaka T (2007) Chemical phenotypes of the *hmg1* and *hmg2* mutants of *Arabidopsis* demonstrate the *in-planta* role of HMG-CoA reductase in triterpene biosynthesis. Chem Pharm Bull (Tokyo) 55:1518–1521
17. Suzuki M, Kamide Y, Nagata N, Seki H, Ohyama K, Kato H, Masuda K, Sato S, Kato T, Tabata S, Yoshida S, Muranaka T (2004) Loss of function of 3-hydroxy-3-methylglutaryl coenzyme A reductase 1 (HMG1) in *Arabidopsis* leads to dwarfing, early senescence and male sterility, and reduced sterol levels. Plant J 37:750–761
18. Manzano D, Fernandez-Busquets X, Schaller H, Gonzalez V, Boronat A, Arro M, Ferrer A (2004) The metabolic imbalance underlying lesion formation in *Arabidopsis thaliana* overexpressing farnesyl diphosphate synthase (isoform 1S) leads to oxidative stress and is triggered by the developmental decline of endogenous HMGR activity. Planta 219:982–992
19. Chappell J, Vonlanken C, Vogeli U (1991) Elicitor-inducible 3-hydroxy-3-methylglutaryl coenzyme A reductase activity is required for sesquiterpene accumulation in tobacco cell-suspension cultures. Plant Physiol 97:693–698
20. Chappell J, Wolf F, Proulx J, Cuellar R, Saunders C (1995) Is the reaction catalyzed by 3-hydroxy-3-methylglutaryl coenzyme A reductase a rate-limiting step for isoprenoid biosynthesis in plants? Plant Physiol 109:1337–1343
21. Choi D, Bostock RM, Avdiushko S, Hildebrand DF (1994) Lipid-derived signals that discriminate wound-responsive and pathogen-responsive isoprenoid pathways in plants—methyl jasmonate and the fungal elicitor arachidonic acid induce different 3-hydroxy-3-methylglutaryl coenzyme A reductase genes and antimicrobial isoprenoids in *Solanum tuberosum* L. Proc Natl Acad Sci USA 91:2329–2333
22. Lichtenthaler HK (1998) The plant 1-deoxy-D-xylulose-5-phosphate pathway for biosynthesis of isoprenoids. Fett-Lipid 100:128–138
23. Rohmer M (1999) The discovery of a mevalonate-independent pathway for isoprenoid biosynthesis in bacteria, algae and higher plants. Nat Prod Rep 16:565–574
24. Boucher Y, Doolittle WF (2000) The role of lateral gene transfer in the evolution of isoprenoid biosynthesis pathways. Mol Microbiol 37:703–716
25. Lange BM, Rujan T, Martin W, Croteau R (2000) Isoprenoid biosynthesis: the evolution of two ancient and distinct pathways across genomes. Proc Natl Acad Sci USA 97:13172–13177
26. Bouvier F, d'Harlingue A, Suire C, Backhaus RA, Camara B (1998) Dedicated roles of plastid transketolases during the early onset of isoprenoid biogenesis in pepper fruits. Plant Physiol 117:1423–1431
27. Lange BM, Wildung MR, McCaskill D, Croteau R (1998) A family of transketolases that directs isoprenoid biosynthesis via a mevalonate-independent pathway. Proc Natl Acad Sci USA 95:2100–2104
28. Chahed K, Oudin A, Guivarc'h N, Hamdi S, Chenieux JC, Rideau M, Clastre M (2000) L-Deoxy-D-xylulose 5-phosphate synthase from periwinkle: cDNA identification and induced gene expression in terpenoid indole alkaloid-producing cells. Plant Physiol Biochem 38: 559–566

29. Lois LM, Rodriguez-Concepcion M, Gallego F, Campos N, Boronat A (2000) Carotenoid biosynthesis during tomato fruit development: regulatory role of 1-deoxy-D-xylulose 5-phosphate synthase. Plant J 22:503–513
30. Walter MH, Fester T, Strack D (2000) Arbuscular mycorrhizal fungi induce the non-mevalonate methylerythritol phosphate pathway of isoprenoid biosynthesis correlated with accumulation of the 'yellow pigment' and other apocarotenoids. Plant J 21:571–578
31. Estevez JM, Cantero A, Reindl A, Reichler S, Leon P (2001) 1-Deoxy-D-xylulose-5-phosphate synthase, a limiting enzyme for plastidic isoprenoid biosynthesis in plants. J Biol Chem 276:22901–22909
32. Mandel MA, Feldmann KA, HerreraEstrella L, RochaSosa M, Leon P (1996) CLA1, a novel gene required for chloroplast development, is highly conserved in evolution. Plant J 9:649–658
33. Estevez JM, Cantero A, Romero C, Kawaide H, Jimenez LF, Kuzuyama T, Seto H, Kamiya Y, Leon P (2000) Analysis of the expression of *CLA1*, a gene that encodes the 1-deoxyxylulose 5-phosphate synthase of the 2-C-methyl-D-erythritol-4-phosphate pathway in *Arabidopsis*. Plant Physiol 124:95–103
34. Araki N, Kusumi K, Masamoto K, Niwa Y, Iba K (2000) Temperature-sensitive *Arabidopsis* mutant defective in 1-deoxy-D-xylulose 5-phosphate synthase within the plastid non-mevalonate pathway of isoprenoid biosynthesis. Physiol Plant 108:19–24
35. Schwender J, Muller C, Zeidler J, Lichlenthaler HK (1999) Cloning and heterologous expression of a cDNA encoding 1-deoxy-D-xylulose-5-phosphate reductoisomerase of *Arabidopsis thaliana*. FEBS Lett 455:140–144
36. Carretero-Paulet L, Ahumada I, Cunillera N, Rodriguez-Concepcion M, Ferrer A, Boronat A, Campos N (2002) Expression and molecular analysis of the *Arabidopsis DXR* gene encoding 1-deoxy-D-xylulose 5-phosphate reductoisomerase, the first committed enzyme of the 2-C-methyl-D-erythritol 4-phosphate pathway. Plant Physiol 129:1581–1591
37. Kuzuyama T, Shimizu T, Takahashi S, Seto H (1998) Fosmidomycin, a specific inhibitor of 1-deoxy-D-xylulose 5-phosphate reductoisomerase in the nonmevalonate pathway for terpenoid biosynthesis. Tetrahedron Lett 39:7913–7916
38. Jomaa H, Wiesner J, Sanderbrand S, Altincicek B, Weidemeyer C, Hintz M, Turbachova I, Eberl M, Zeidler J, Lichtenthaler HK, Soldati D, Beck E (1999) Inhibitors of the nonmevalonate pathway of isoprenoid biosynthesis as antimalarial drugs. Science 285:1573–1576
39. Steinbacher S, Kaiser J, Eisenreich W, Huber R, Bacher A, Rohdich F (2003) Structural basis of fosmidomycin action revealed by the complex with 2-C-methyl-D-erythritol 4-phosphate synthase (IspC)—implications for the catalytic mechanism and anti-malaria drug development. J Biol Chem 278:18401–18407
40. Zeidler J, Schwender J, Muller C, Wiesner J, Weidemeyer C, Beck E, Jomaa H, Lichtenthaler HK (1998) Inhibition of the non-mevalonate 1-deoxy-D-xylulose-5-phosphate pathway of plant isoprenoid biosynthesis by fosmidomycin. Z fur Naturforsch C- J Biosci 53:980–986
41. Rodriguez-Concepcion M, Ahumada I, Diez-Juez E, Sauret-Gueto S, Lois LM, Gallego F, Carretero-Paulet L, Campos N, Boronat A (2001) 1-Deoxy-D-xylulose 5-phosphate reductoisomerase and plastid isoprenoid biosynthesis during tomato fruit ripening. Plant J 27:213–222
42. Huang MS, Abel C, Sohrabi R, Petri J, Haupt I, Cosimano J, Gershenzon J, Tholl D (2010) Variation of herbivore-induced volatile terpenes among *Arabidopsis* ecotypes depends on allelic differences and subcellular targeting of two terpene synthases, TPS02 and TPS03. Plant Physiol 153:1293–1310
43. Xing SF, Miao J, Li SA, Qin GJ, Tang S, Li HN, Gu HY, Qu LJ (2010) Disruption of the 1-deoxy-D-xylulose-5-phosphate reductoisomerase (*DXR*) gene results in albino, dwarf and defects in trichome initiation and stomata closure in *Arabidopsis*. Cell Res 20:688–700
44. Rohdich F, Wungsintaweekul J, Fellermeier M, Sagner S, Herz S, Kis K, Eisenreich W, Bacher A, Zenk MH (1999) Cytidine 5'-triphosphate-dependent biosynthesis of isoprenoids: YgbP protein of *Escherichia coli* catalyzes the formation of 4-diphosphocytidyl-2-C-methylerythritol. Proc Natl Acad Sci USA 96:11758–11763

45. Rohdich F, Wungsintaweekul J, Eisenreich W, Richter G, Schuhr CA, Hecht S, Zenk MH, Bacher A (2000) Biosynthesis of terpenoids: 4-Diphosphocytidyl-2C-methyl-D-erythritol synthase of *Arabidopsis thaliana*. Proc Natl Acad Sci USA 97:6451–6456
46. Lange BM, Croteau R (1999) Isopentenyl diphosphate biosynthesis via a mevalonate-independent pathway: Isopentenyl monophosphate kinase catalyzes the terminal enzymatic step. Proc Natl Acad Sci USA 96:13714–13719
47. Luttgen H, Rohdich F, Herz S, Wungsintaweekul J, Hecht S, Schuhr CA, Fellermeier M, Sagner S, Zenk MH, Bacher A, Eisenreich W (2000) Biosynthesis of terpenoids: YchB protein of *Escherichia coli* phosphorylates the 2-hydroxy group of 4-diphosphocytidyl-2C-methyl-D-erythritol. Proc Natl Acad Sci USA 97:1062–1067
48. Rohdich F, Wungsintaweekul J, Luttgen H, Fischer M, Eisenreich W, Schuhr CA, Fellermeier M, Schramek N, Zenk MH, Bacher A (2000) Biosynthesis of terpenoids: 4-diphosphocytidyl-2-C-methyl-D-erythritol kinase from tomato. Proc Natl Acad Sci USA 97:8251–8256
49. Rohdich F, Hecht S, Gärtner K, Adam P, Krieger C, Amslinger S, Arigoni D, Bacher A, Eisenreich W (2002) Studies on the nonmevalonate terpene biosynthetic pathway: metabolic role of IspH (LytB) protein. Proc Natl Acad Sci USA 99:1158–1163
50. Rohdich F, Zepeck F, Adam P, Hecht S, Kaiser J, Laupitz R, Grawert T, Amslinger S, Eisenreich W, Bacher A, Arigoni D (2003) The deoxyxylulose phosphate pathway of isoprenoid biosynthesis: studies on the mechanisms of the reactions catalyzed by IspG and IspH protein. Proc Natl Acad Sci USA 100:1586–1591
51. Tritsch D, Hemmerlin A, Bach TJ, Rohmer M (2010) Plant isoprenoid biosynthesis via the MEP pathway: in vivo IPP/DMAPP ratio produced by (*E*)-4-hydroxy-3-methylbut-2-enyl diphosphate reductase in tobacco BY-2 cell cultures. FEBS Lett 584:129–134
52. Hsieh MH, Goodman HM (2006) Functional evidence for the involvement of *Arabidopsis* IspF homolog in the nonmevalonate pathway of plastid isoprenoid biosynthesis. Planta 223:779–784
53. Hsieh MH, Chang CY, Hsu SJ, Chen JJ (2008) Chloroplast localization of methylerythritol 4-phosphate pathway enzymes and regulation of mitochondrial genes in *ispD* and *ispE* albino mutants in *Arabidopsis*. Plant Mol Biol 66:663–673
54. Gutierrez-Nava MDL, Gillmor CS, Jimenez LF, Guevara-Garcia A, Leon P (2004) Chloroplast biogenesis genes act cell and noncell autonomously in early chloroplast development. Plant Physiol 135:471–482
55. Hsieh MH, Goodman HM (2005) The *Arabidopsis* IspH homolog is involved in the plastid nonmevalonate pathway of isoprenoid biosynthesis. Plant Physiol 138:641–653
56. Gil MJ, Coego A, Mauch-Mani B, Jorda L, Vera P (2005) The *Arabidopsis csb3* mutant reveals a regulatory link between salicylic acid-mediated disease resistance and the methyl-erythritol 4-phosphate pathway. Plant J 44:155–166
57. Wang H, Nagegowda DA, Rawat R, Bouvier-Nave P, Guo DJ, Bach TJ, Chye ML (2012) Overexpression of *Brassica juncea* wild-type and mutant HMG-CoA synthase 1 in *Arabidopsis* up-regulates genes in sterol biosynthesis and enhances sterol production and stress tolerance. Plant Biotechnol J 10:31–42
58. Alex D, Bach TJ, Chye ML (2000) Expression of *Brassica juncea* 3-hydroxy-3-methylglutaryl CoA synthase is developmentally regulated and stress-responsive. Plant J 22:415–426
59. Nagegowda DA, Ramalingam S, Hemmerlin A, Bach TJ, Chye ML (2005) *Brassica juncea* HMG-CoA synthase: localization of mRNA and protein. Planta 221:844–856
60. Ishiguro S, Nishimori Y, Yamada M, Saito H, Suzuki T, Nakagawa T, Miyake H, Okada K, Nakamura K (2010) The Arabidopsis *FLAKY POLLEN1* gene encodes a 3-hydroxy-3-methylglutaryl-coenzyme A synthase required for development of tapetum-specific organelles and fertility of pollen grains. Plant Cell Physiol 51:896–911
61. Choi D, Ward BL, Bostock RM (1992) Differential induction and suppression of potato 3-hydroxy-3-methylglutaryl coenezyme A reductase genes in response to *Phytophthora infestans* and to its elicitor arachidonic acid. Plant Cell 4:1333–1344

62. Rodriguez-Concepcion M, Gruissem W (1999) Arachidonic acid alters tomato HMG expression and fruit growth and induces 3-hydroxy-3-methylglutaryl coenzyme A reductase-independent lycopene accumulation. Plant Physiol 119:41–48
63. Suzuki H, Xia YJ, Cameron R, Shadle G, Blount J, Lamb C, Dixon RA (2004) Signals for local and systemic responses of plants to pathogen attack. J Exp Bot 55:169–179
64. Suzuki M, Nakagawa S, Kamide Y, Kobayashi K, Ohyama K, Hashinokuchi H, Kiuchi R, Saito K, Muranaka T, Nagata N (2009) Complete blockage of the mevalonate pathway results in male gametophyte lethality. J Exp Biol 60:2055–2064
65. Walter MH, Hans J, Strack D (2002) Two distantly related genes encoding 1-deoxy-D-xylulose 5-phosphate synthases: differential regulation in shoots and apocarotenoid-accumulating mycorrhizal roots. Plant J 31:243–254
66. Paetzold H, Garms S, Bartram S, Wieczorek J, Uros-Gracia EM, Rodriguez-Concepcion M, Boland W, Strack D, Hause B, Walter MH (2010) The isogene 1-deoxy-D-xylulose 5-phosphate synthase 2 controls isoprenoid profiles, precursor pathway allocation, and density of tomato trichomes. Mol Plant 3:904–916
67. Brooker JD, Russell DW (1975) Properties of microsomal 3-hydroxy-3-methylglutaryl coenzyme A reductase from *Pisum sativum* seedlings. Arch Biochem Biophys 167:723–729
68. Soto G, Stritzler M, Lisi C, Alleva K, Pagano ME, Ardila F, Mozzicafreddo M, Cuccioloni M, Angeletti M, Ayub ND (2011) Acetoacetyl-CoA thiolase regulates the mevalonate pathway during abiotic stress adaptation. J Exp Bot 62:5699–5711
69. Schulte AE, van der Heijden R, Verpoorte R (2000) Purification and characterization of mevalonate kinase from suspension-cultured cells of *Catharanthus roseus* (L.) G. Don. Arch Biochem Biophys 378:287–298
70. Banerjee A, Wu Y, Banerjee R, Li Y, Yan HG, Sharkey TD (2013) Feedback inhibition of deoxy-D-xylulose-5-phosphate synthase regulates the methylerythritol 4-phosphate pathway. J Biol Chem 288:16926–16936
71. Ghirardo A, Wright LP, Bi Z, Rosenkranz M, Pulido P, Rodriguez-Concepcion M, Niinemets U, Brueggemann N, Gershenzon J, Schnitzler J-P (2014) Metabolic flux analysis of plastidic isoprenoid biosynthesis in poplar leaves emitting and nonemitting isoprene. Plant Physiol 165:37–51
72. Liao P, Wang H, Wang M, Hsiao A-S, Bach TJ, Chye M-L (2014) Transgenic tobacco overexpressing *Brassica juncea* HMG-CoA Synthase 1 shows increased plant growth, pod size, and seed yield. Plos One 9:e98264
73. Jin HN, Song ZH, Nikolau BJ (2012) Reverse genetic characterization of two paralogous acetoacetyl CoA thiolase genes in *Arabidopsis* reveals their importance in plant growth and development. Plant J 70:1015–1032
74. Holmberg N, Harker M, Gibbard CL, Wallace AD, Clayton JC, Rawlins S, Hellyer A, Safford R (2002) Sterol C-24 methyltransferase type 1 controls the flux of carbon into sterol biosynthesis in tobacco seed. Plant Physiol 130:303–311
75. Wentzinger LF, Bach TJ, Hartmann MA (2002) Inhibition of squalene synthase and squalene epoxidase in tobacco cells triggers an up-regulation of 3-hydroxy-3-methylglutaryl coenzyme A reductase. Plant Physiol 130:334–346
76. Ahn CS, Pai HS (2008) Physiological function of *ispE*, a plastid MEP pathway gene for isoprenoid biosynthesis, in organelle biogenesis and cell morphogenesis in *Nicotiana benthamiana*. Plant Mol Biol 66:503–517
77. Jung KH, Lee J, Dardick C, Seo YS, Cao P, Canlas P, Phetsom J, Xu X, Ouyang S, An K, Cho YJ, Lee GC, Lee Y, An G, Ronald PC (2008) Identification and functional analysis of light-responsive unique genes and gene family members in rice. PLoS Genet 4
78. Wille A, Zimmermann P, Vranova E, Furholz A, Laule O, Bleuler S, Hennig L, Prelic A, von Rohr P, Thiele L, Zitzler E, Gruissem W, Buhlmann P (2004) Sparse graphical Gaussian modeling of the isoprenoid gene network in *Arabidopsis thaliana*. Genome Biol 5

79. Ghassemian M, Lutes J, Tepperman JM, Chang HS, Zhu T, Wang X, Quail PH, Lange BM (2006) Integrative analysis of transcript and metabolite profiling data sets to evaluate the regulation of biochemical pathways during photomorphogenesis. Arch Biochem Biophys 448:45–59
80. Meier S, Tzfadia O, Vallabhaneni R, Gehring C, Wurtzel ET (2011) A transcriptional analysis of carotenoid, chlorophyll and plastidial isoprenoid biosynthesis genes during development and osmotic stress responses in *Arabidopsis thaliana*. BMC Syst Biol 5
81. Flores-Perez U, Perez-Gila J, Closa M, Wright LP, Botella-Pavia P, Phillips MA, Ferrer A, Gershenzon J, Rodriguez-Concepcion M (2010) PLEIOTROPIC REGULATORY LOCUS 1 (PRL1) integrates the regulation of sugar responses with isoprenoid metabolism in *Arabidopsis*. Mol Plant 3:101–112
82. Dale S, Arro M, Becerra B, Morrice NG, Boronat A, Hardie DG, Ferrer A (1995) Bacterial expression of the catalytic domain of 3-hydroxy-3-methylglutaryl-CoA reductase (isoform HMGR1) from *Arabidopsis thaliana*, and its inactivation by phosphorylation at Ser577 by *Brassica oleracea* 3-hydroxy-3-methylglutaryl-CoA reductase kinase. Eur J Biochem 233:506–513
83. Leivar P, Antolin-Llovera M, Ferrero S, Closa M, Arro M, Ferrer A, Boronat A, Camposa N (2011) Multilevel control of *Arabidopsis* 3-hydroxy-3-methylglutaryl coenzyme A reductase by protein phosphatase 2A. Plant Cell 23:1494–1511
84. Yoshioka H, Miyabe M, Hayakawa Y, Doke N (1996) Expression of genes for phenylalanine ammonia-lyase and 3-hydroxy-3-methylglutaryl CoA reductase in aged potato tubers infected with *Phytophthora infestans*. Plant Cell Physiol 37:81–90
85. Nieto B, Fores O, Arro M, Ferrer A (2009) *Arabidopsis* 3-hydroxy-3-methylglutaryl-CoA reductase is regulated at the post-translational level in response to alterations of the sphingolipid and the sterol biosynthetic pathways. Phytochemistry 70:53–59
86. Kang JH, McRoberts J, Shi F, Moreno JE, Jones AD, Howe GA (2014) The flavonoid biosynthetic enzyme chalcone isomerase modulates terpenoid production in glandular trichomes of tomato. Plant Physiol 164:1161–1174
87. Pourcel L, Irani NG, Koo AJK, Bohorquez-Restrepo A, Howe GA, Grotewold E (2013) A chemical complementation approach reveals genes and interactions of flavonoids with other pathways. Plant J 74:383–397
88. Sahu NK, Balbhadra SS, Choudhary J, Kohli DV (2012) Exploring pharmacological significance of chalcone scaffold: a review. Curr Med Chem 19:209–225
89. Saslowsky DE, Warek U, Winkel BSJ (2005) Nuclear localization of flavonoid enzymes in *Arabidopsis*. J Biol Chem 280:23735–23740
90. Ben Zvi MM, Shklarman E, Masci T, Kalev H, Debener T, Shafir S, Ovadis M, Vainstein A (2012) PAP1 transcription factor enhances production of phenylpropanoid and terpenoid scent compounds in rose flowers. New Phytol 195:335–345
91. Cordoba E, Salmi M, Leon P (2009) Unravelling the regulatory mechanisms that modulate the MEP pathway in higher plants. J Exp Bot 60:2933–2943
92. Rodriguez-Concepcion M, Fores O, Martinez-Garcia JF, Gonzalez V, Phillips MA, Ferrer A, Boronat A (2004) Distinct light-mediated pathways regulate the biosynthesis and exchange of isoprenoid precursors during *Arabidopsis* seedling development. Plant Cell 16:144–156
93. Toledo-Ortiz G, Huq E, Rodriguez-Concepcion M (2010) Direct regulation of phytoene synthase gene expression and carotenoid biosynthesis by phytochrome-interacting factors. Proc Natl Acad Sci USA 107:11626–11631
94. Wiberley AE, Donohue AR, Westphal MM, Sharkey TD (2009) Regulation of isoprene emission from poplar leaves throughout a day. Plant Cell Environ 32:939–947
95. Mongelard G, Seemann M, Boisson AM, Rohmer M, Bligny R, Rivasseau C (2011) Measurement of carbon flux through the MEP pathway for isoprenoid synthesis by P-31-NMR spectroscopy after specific inhibition of 2-C-methyl-D-erythritol 2,4-cyclodiphosphate reductase. Effect of light and temperature. Plant Cell Environ 34:1241–1247

96. Kim YJ, Lee OR, Oh JY, Jang MG, Yang DC (2014) Functional analysis of 3-hydroxy-3-methylglutaryl coenzyme A reductase encoding genes in triterpene saponin-producing ginseng. Plant Physiol 165:373–387
97. Mannen K, Matsumoto T, Takahashi S, Yamaguchi Y, Tsukagoshi M, Sano R, Suzuki H, Sakurai N, Shibata D, Koyama T, Nakayama T (2014) Coordinated transcriptional regulation of isopentenyl diphosphate biosynthetic pathway enzymes in plastids by phytochrome-interacting factor 5. Biochem Biophys Res Commun 443:768–774
98. Flores-Perez U, Sauret-Gueto S, Gas E, Jarvis P, Rodriguez-Concepcion M (2008) A mutant impaired in the production of plastome-encoded proteins uncovers a mechanism for the homeostasis of isoprenoid biosynthetic enzymes in *Arabidopsis* plastids. Plant Cell 20:1303–1315
99. Fukushima A, Kusano M, Nakamichi N, Kobayashi M, Hayashi N, Sakakibara H, Mizuno T, Saito K (2009) Impact of clock-associated *Arabidopsis* pseudo-response regulators in metabolic coordination. Proc Natl Acad Sci USA 106:7251–7256
100. Dudareva N, Andersson S, Orlova I, Gatto N, Reichelt M, Rhodes D, Boland W, Gershenzon J (2005) The nonmevalonate pathway supports both monoterpene and sesquiterpene formation in snapdragon flowers. Proc Natl Acad Sci USA 102:933–938
101. Lemaire SD, Guillon B, Le Marechal P, Keryer E, Miginiac-Maslow M, Decottignies P (2004) New thioredoxin targets in the unicellular photosynthetic eukaryote *Chlamydomonas reinhardtii*. Proc Natl Acad Sci USA 101:7475–7480
102. Balmer Y, Koller A, del Val G, Manieri W, Schurmann P, Buchanan BB (2003) Proteomics gives insight into the regulatory function of chloroplast thioredoxins. Proc Natl Acad Sci USA 100:370–375
103. Seemann M, Bui BTS, Wolff M, Miginlac-Maslow M, Rohmer M (2006) Isoprenoid biosynthesis in plant chloroplasts via the MEP pathway: direct thylakoid/ferredoxin-dependent photoreduction of GcpE/IspG. FEBS Lett 580:1547–1552
104. Vogeli U, Chappell J (1988) Induction of sesquiterpene cyclase and suppression of squalene synthase activities in plant cell cultures treated with fungal elicitor. Plant Physiol 88:1291–1296
105. Kim CY, Zhang SQ (2004) Activation of a mitogen-activated protein kinase cascade induces WRKY family of transcription factors and defense genes in tobacco. Plant J 38:142–151
106. Jin HL, Liu YD, Yang KY, Kim CY, Baker B, Zhang SQ (2003) Function of a mitogen-activated protein kinase pathway in N gene-mediated resistance in tobacco. Plant J 33:719–731
107. Kevei Z, Lougnon G, Mergaert P, Horvath GV, Kereszt A, Jayaraman D, Zaman N, Marcel F, Regulski K, Kiss GB, Kondorosi A, Endre G, Kondorosi E, Ane JM (2007) 3-hydroxy-3-methylglutaryl coenzyme A reductase1 interacts with NORK and is crucial for nodulation in *Medicago truncatula*. Plant Cell 19:3974–3989
108. Sapir-Mir M, Mett A, Belausov E, Tal-Meshulam S, Frydman A, Gidoni D, Eyal Y (2008) Peroxisomal localization of *Arabidopsis* isopentenyl diphosphate isomerases suggests that part of the plant isoprenoid mevalonic acid pathway is compartmentalized to peroxisomes. Plant Physiol 148:1219–1228
109. Simkin AJ, Guirimand G, Papon N, Courdavault V, Thabet I, Ginis O, Bouzid S, Giglioli-Guivarc'h N, Clastre M (2011) Peroxisomal localisation of the final steps of the mevalonic acid pathway in planta. Planta 234:903–914
110. Guirimand G, Guihur A, Phillips MA, Oudin A, Glevarec G, Melin C, Papon N, Clastre M, St-Pierre B, Rodriguez-Concepcion M, Burlat V, Courdavault V (2012) A single gene encodes isopentenyl diphosphate isomerase isoforms targeted to plastids, mitochondria and peroxisomes in *Catharanthus roseus*. Plant Mol Biol 79:443–459
111. Lichtenthaler HK (1999) The 1-deoxy-D-xylulose-5-phosphate pathway of isoprenoid biosynthesis in plants. Annu Rev Plant Physiol Plant Mol Biol 50:47–65
112. Hemmerlin A, Hoeffler JF, Meyer O, Tritsch D, Kagan IA, Grosdemange-Billiard C, Rohmer M, Bach TJ (2003) Cross-talk between the cytosolic mevalonate and the plastidial methylerythritol phosphate pathways in tobacco bright yellow-2 cells. J Biol Chem 278:26666–26676

113. Laule O, Furholz A, Chang HS, Zhu T, Wang X, Heifetz PB, Gruissem W, Lange BM (2003) Crosstalk between cytosolic and plastidial pathways of isoprenoid biosynthesis in *Arabidopsis thaliana*. Proc Natl Acad Sci USA 100:6866–6871
114. Bartram S, Jux A, Gleixner G, Boland W (2006) Dynamic pathway allocation in early terpenoid biosynthesis of stress-induced lima bean leaves. Phytochemistry 67:1661–1672
115. Chaurasiya ND, Sangwan NS, Sabir F, Misra L, Sangwan RS (2012) Withanolide biosynthesis recruits both mevalonate and DOXP pathways of isoprenogenesis in Ashwagandha *Withania somnifera* L. (Dunal). Plant Cell Rep 31:1889–1897
116. Woelwer-Rieck U, May B, Lankes C, Wuest M (2014) Methylerythritol and mevalonate pathway contributions to biosynthesis of mono-, sesqui-, and diterpenes in glandular trichomes and leaves of *Stevia rebaudiana* Bertoni. J Agric Food Chem 62:2428–2435
117. Zhao S, Wang L, Liu L, Liang Y, Sun Y, Wu J (2014) Both the mevalonate and the non-mevalonate pathways are involved in ginsenoside biosynthesis. Plant Cell Rep 33:393–400
118. Opitz S, Nes WD, Gershenzon J (2014) Both methylerythritol phosphate and mevalonate pathways contribute to biosynthesis of each of the major isoprenoid classes in young cotton seedlings. Phytochemistry 98:110–119
119. Bick JA, Lange BM (2003) Metabolic cross talk between cytosolic and plastidial pathways of isoprenoid biosynthesis: unidirectional transport of intermediates across the chloroplast envelope membrane. Arch Biochem Biophys 415:146–154
120. Flügge UI, Gao W (2005) Transport of isoprenoid intermediates across chloroplast envelope membranes. Plant Biol 7:91–97
121. Gutensohn M, Orlova I, Nguyen TTH, Davidovich-Rikanati R, Ferruzzi MG, Sitrit Y, Lewinsohn E, Pichersky E, Dudareva N (2013) Cytosolic monoterpene biosynthesis is supported by plastid-generated geranyl diphosphate substrate in transgenic tomato fruits. Plant J 75:351–363
122. May B, Lange BM, Wuest M (2013) Biosynthesis of sesquiterpenes in grape berry exocarp of *Vitis vinifera* L.: evidence for a transport of farnesyl diphosphate precursors from plastids to the cytosol. Phytochemistry 95:135–144
123. Wu SQ, Schalk M, Clark A, Miles RB, Coates R, Chappell J (2006) Redirection of cytosolic or plastidic isoprenoid precursors elevates terpene production in plants. Nat Biotechnol 24:1441–1447
124. Wu S, Jiang Z, Kempinski C, Nybo SE, Husodo S, Williams R, Chappell J (2012) Engineering triterpene metabolism in tobacco. Planta 236:867–877
125. Kumar S, Hahn FM, Baidoo E, Kahlon TS, Wood DF, McMahan CM, Cornish K, Keasling JD, Daniell H, Whalen MC (2012) Remodeling the isoprenoid pathway in tobacco by expressing the cytoplasmic mevalonate pathway in chloroplasts. Metab Eng 14:19–28
126. Huchelmann A, Gastaldo C, Veinante M, Zeng Y, Heintz D, Tritsch D, Schaller H, Rohmer M, Bach TJ, Hemmerlin A (2014) *S*-Carvone suppresses cellulase-induced capsidiol production in *Nicotiana tabacum* by interfering with protein isoprenylation. Plant Physiol 164:935–950
127. Verbitskiy D, Zehrmann A, van der Merwe JA, Brennicke A, Takenaka M (2010) The PPR protein encoded by the Lovastatin Insensitive 1 gene is involved in RNA editing at three sites in mitochondria of *Arabidopsis thaliana*. Plant J 61:446–455
128. Tang JW, Kobayashi K, Suzuki M, Matsumoto S, Muranaka T (2010) The mitochondrial PPR protein LOVASTATIN INSENSITIVE 1 plays regulatory roles in cytosolic and plastidial isoprenoid biosynthesis through RNA editing. Plant J 61:456–466
129. Xiao YM, Savchenko T, Baidoo EEK, Chehab WE, Hayden DM, Tolstikov V, Corwin JA, Kliebenstein DJ, Keasling JD, Dehesh K (2012) Retrograde signaling by the plastidial metabolite MEcPP regulates expression of nuclear stress-response genes. Cell 149:1525–1535
130. Phillips MA, D'Auria JC, Gershenzon J, Pichersky E (2008) The *Arabidopsis thaliana* type I isopentenyl diphosphate isomerases are targeted to multiple subcellular compartments and have overlapping functions in isoprenoid biosynthesis. Plant Cell 20:677–696

131. Kharel Y, Koyama T (2003) Molecular analysis of cis-prenyl chain elongating enzymes. Nat Prod Rep 20:111–118
132. Burke C, Croteau R (2002) Interaction with the small subunit of geranyl diphosphate synthase modifies the chain length specificity of geranylgeranyl diphosphate synthase to produce geranyl diphosphate. J Biol Chem 277:3141–3149
133. Schmidt A, Gershenzon J (2008) Cloning and characterization of two different types of geranyl diphosphate synthases from Norway spruce (*Picea abies*). Phytochemistry 69:49–57
134. Chang TH, Hsieh FL, Ko TP, Teng KH, Liang PH, Wang AHJ (2010) Structure of a heterotetrameric geranyl pyrophosphate synthase from mint (*Mentha piperita*) reveals intersubunit regulation. Plant Cell 22:454–467
135. Rai A, Smita SS, Singh AK, Shanker K, Nagegowda DA (2013) Heteromeric and homomeric geranyl diphosphate synthases from *Catharanthus roseus* and their role in monoterpene indole alkaloid biosynthesis. Mol Plant 6:1531–1549
136. Hsiao YY, Jeng MF, Tsai WC, Chuang YC, Li CY, Wu TS, Kuoh CS, Chen WH, Chen HH (2008) A novel homodimeric geranyl diphosphate synthase from the orchid *Phalaenopsis bellina* lacking a DD(X)(2-4)D motif. Plant J 55:719–733
137. Burke CC, Wildung MR, Croteau R (1999) Geranyl diphosphate synthase: cloning, expression, and characterization of this prenyltransferase as a heterodimer. Proc Natl Acad Sci USA 96:13062–13067
138. Tholl D, Kish CM, Orlova I, Sherman D, Gershenzon J, Pichersky E, Dudareva N (2004) Formation of monoterpenes in *Antirrhinum majus* and *Clarkia breweri* flowers involves heterodimeric geranyl diphosphate synthases. Plant Cell 16:977–992
139. Wang GD, Dixon RA (2009) Heterodimeric geranyl(geranyl)diphosphate synthase from hop (*Humulus lupulus*) and the evolution of monoterpene biosynthesis. Proc Natl Acad Sci USA 106:9914–9919
140. Orlova I, Nagegowda DA, Kish CM, Gutensohn M, Maeda H, Varbanova M, Fridman E, Yamaguchi S, Hanada A, Kamiya Y, Krichevsky A, Citovsky V, Pichersky E, Dudareva N (2009) The small subunit of snapdragon geranyl diphosphate synthase modifies the chain length specificity of tobacco geranylgeranyl diphosphate synthase in planta. Plant Cell 21:4002–4017
141. van Schie CCN, Ament K, Schmidt A, Lange T, Haring MA, Schuurink RC (2007) Geranyl diphosphate synthase is required for biosynthesis of gibberellins. Plant J 52:752–762
142. Schmidt A, Wachtler B, Temp U, Krekling T, Seguin A, Gershenzon J (2010) A bifunctional geranyl and geranylgeranyl diphosphate synthase is involved in terpene oleoresin formation in *Picea abies*. Plant Physiol 152:639–655
143. Bouvier F, Suire C, d'Harlingue A, Backhaus RA, Camara B (2000) Molecular cloning of geranyl diphosphate synthase and compartmentation of monoterpene synthesis in plant cells. Plant J 24:241–252
144. Hsieh F-L, Chang T-H, Ko T-P, Wang AHJ (2011) Structure and mechanism of an *Arabidopsis* medium/long-chain-length prenyl pyrophosphate synthase. Plant Physiol 155:1079–1090
145. Cunillera N, Arro M, Delourme D, Karst F, Boronat A, Ferrer A (1996) *Arabidopsis thaliana* contains two differentially expressed farnesyl-diphosphate synthase genes. J Biol Chem 271:7774–7780
146. Gaffe J, Bru JP, Causse M, Vidal A, Stamitti-Bert L, Carde JP, Gallusci P (2000) LEFPS1, a tomato farnesyl pyrophosphate gene highly expressed during early fruit development. Plant Physiol 123:1351–1362
147. Hemmerlin A, Rivera SB, Erickson HK, Poulter CD (2003) Enzymes encoded by the farnesyl diphosphate synthase gene family in the big sagebrush *Artemisia tridentata* ssp *spiciformis*. J Biol Chem 278:32132–32140
148. Cunillera N, Boronat A, Ferrer A (1997) The *Arabidopsis thaliana FPS1* gene generates a novel mRNA that encodes a mitochondrial farnesyl-diphosphate synthase isoform. J Biol Chem 272:15381–15388

149. Thabet I, Guirimand G, Courdavault V, Papon N, Godet S, Dutilleul C, Bouzid S, Giglioli-Guivarc'h N, Clastre M, Simkin AJ (2011) The subcellular localization of periwinkle farnesyl diphosphate synthase provides insight into the role of peroxisome in isoprenoid biosynthesis. J Plant Physiol 168:2110–2116
150. Ito J, Batth TS, Petzold CJ, Redding-Johanson AM, Mukhopadhyay A, Verboom R, Meyer EH, Millar AH, Heazlewood JL (2011) Analysis of the *Arabidopsis* cytosolic proteome highlights subcellular partitioning of central plant metabolism. J Proteome Res 10:1571–1582
151. Reumann S, Quan S, Aung K, Yang P, Manandhar-Shrestha K, Holbrook D, Linka N, Switzenberg R, Wilkerson CG, Weber APM, Olsen LJ, Hu J (2009) In-depth proteome analysis of *Arabidopsis* leaf peroxisomes combined with in vivo subcellular targeting verification indicates novel metabolic and regulatory functions of peroxisomes. Plant Physiol 150:125–143
152. Closa M, Vranova E, Bortolotti C, Bigler L, Arro M, Ferrer A, Gruissem W (2010) The *Arabidopsis thaliana* FPP synthase isozymes have overlapping and specific functions in isoprenoid biosynthesis, and complete loss of FPP synthase activity causes early developmental arrest. Plant J 63:512–525
153. Keim V, Manzano D, Fernandez FJ, Closa M, Andrade P, Caudepon D, Bortolotti C, Vega MC, Arro M, Ferrer A (2012) Characterization of *Arabidopsis* FPS isozymes and *FPS* gene expression analysis provide insight into the biosynthesis of isoprenoid precursors in seeds. Plos One 7
154. Beck G, Coman D, Herren E, Ruiz-Sola M, Rodriguez-Concepcion M, Gruissem W, Vranova E (2013) Characterization of the GGPP synthase gene family in *Arabidopsis thaliana*. Plant Mol Biol 82:393–416
155. Ruppel NJ, Kropp KN, Davis PA, Martin AE, Luesse DR, Hangarter RP (2013) Mutations in geranylgeranyl diphosphate synthase 1 affect chloroplast development in *Arabidopsis thaliana* (*Brassicaceae*). Am J Bot 100:2074–2084
156. Dai ZB, Liu Y, Huang LQ, Zhang XL (2012) Production of miltiradiene by metabolically engineered *Saccharomyces cerevisiae*. Biotechnol Bioeng 109:2845–2853
157. Leonard E, Ajikumar PK, Thayer K, Xiao WH, Mo JD, Tidor B, Stephanopoulos G, Prather KLJ (2010) Combining metabolic and protein engineering of a terpenoid biosynthetic pathway for overproduction and selectivity control. Proc Natl Acad Sci USA 107:13654–13659
158. Vandermoten S, Haubruge E, Cusson M (2009) New insights into short-chain prenyltransferases: structural features, evolutionary history and potential for selective inhibition. Cell Mol Life Sci 66:3685–3695
159. Tarshis LC, Yan MJ, Poulter CD, Sacchettini JC (1994) Crystal structure of recombinant farnesyl diphosphate synthase at 2.6 Angstrom resolution. Biochemistry 33:10871–10877
160. Koyama T, Gotoh Y, Nishino T (2000) Intersubunit location of the active site of farnesyl diphosphate synthase: reconstruction of active enzymes by hybrid-type heteromeric dimers of site-directed mutants. Biochemistry 39:463–469
161. Hosfield DJ, Zhang YM, Dougan DR, Broun A, Tari LW, Swanson RV, Finn J (2004) Structural basis for bisphosphonate-mediated inhibition of isoprenoid biosynthesis. J Biol Chem 279:8526–8529
162. Chang TH, Guo RT, Ko TP, Wang AHJ, Liang PH (2006) Crystal structure of type-III geranylgeranyl pyrophosphate synthase from *Saccharomyces cerevisiae* and the mechanism of product chain length determination. J Biol Chem 281:14991–15000
163. Gabelli SB, McLellan JS, Montalvetti A, Oldfield E, Docampo R, Amzel LM (2006) Structure and mechanism of the farnesyl diphosphate synthase from *Trypanosoma cruza*: implications for drug design. Proteins Struct Funct Bioinform 62:80–88
164. Kavanagh KL, Dunford JE, Bunkoczi G, Russell RGG, Oppermann U (2006) The crystal structure of human geranylgeranyl pyrophosphate synthase reveals a novel hexameric arrangement and inhibitory product binding. J Biol Chem 281:22004–22012
165. Kloer DP, Welsch R, Beyer P, Schulz GE (2006) Structure and reaction geometry of geranylgeranyl diphosphate synthase from *Sinapis alba*. Biochemistry 45:15197–15204

166. Ohnuma S, Hirooka K, Hemmi H, Ishida C, Ohto C, Nishino T (1996) Conversion of product specificity of archaebacterial geranylgeranyl- diphosphate synthase. Identification of essential amino acid residues for chain length determination of prenyltransferase reaction. J Biol Chem 271:18831–18837
167. Ohnuma SI, Nakazawa T, Hemmi H, Hallberg AM, Koyama T, Ogura K, Nishino T (1996) Conversion from farnesyl diphosphate synthase to geranylgeranyl diphosphate synthase by random chemical mutagenesis. J Biol Chem 271:10087–10095
168. Wallrapp FH, Pan J-J, Ramamoorthy G, Almonacid DE, Hillerich BS, Seidel R, Patskovsky Y, Babbitt PC, Almo SC, Jacobson MP, Poulter CD (2013) Prediction of function for the polyprenyl transferase subgroup in the isoprenoid synthase superfamily. Proc Natl Acad Sci USA 110:E1196–E1202
169. Coman D, Altenhoff A, Zoller S, Gruissem W, Vranova E (2014) Distinct evolutionary strategies in the GGPPS family from plants. Front Plant Sci 5
170. Takahashi S, Koyama T (2006) Structure and function of *cis*-prenyl chain elongating enzymes. Chem Rec 6:194–205
171. Surmacz L, Swiezewska E (2011) Polyisoprenoids—secondary metabolites or physiologically important superlipids? Biochem Biophys Res Commun 407:627–632
172. Schmidt T, Lenders M, Hillebrand A, van Deenen N, Munt O, Reichelt R, Eisenreich W, Fischer R, Prufer D, Gronover CS (2010) Characterization of rubber particles and rubber chain elongation in *Taraxacum koksaghyz*. BMC Biochem 11
173. Sallaud C, Rontein D, Onillon S, Jabes F, Duffe P, Giacalone C, Thoraval S, Escoffier C, Herbette G, Leonhardt N, Causse M, Tissier A (2009) A novel pathway for sesquiterpene biosynthesis from *Z,Z*-farnesyl pyrophosphate in the wild tomato *Solanum habrochaites*. Plant Cell 21:301–317
174. Akhtar TA, Matsuba Y, Schauvinhold I, Yu G, Lees HA, Klein SE, Pichersky E (2013) The tomato *cis*-prenyltransferase gene family. Plant J 73:640–652
175. Bleeker PM, Mirabella R, Diergaarde PJ, VanDoorn A, Tissier A, Kant MR, Prins M, de Vos M, Haring MA, Schuurink RC (2012) Improved herbivore resistance in cultivated tomato with the sesquiterpene biosynthetic pathway from a wild relative. Proc Natl Acad Sci USA 109:20124–20129
176. Gonzales-Vigil E, Hufnagel DE, Kim J, Last RL, Barry CS (2012) Evolution of TPS20-related terpene synthases influences chemical diversity in the glandular trichomes of the wild tomato relative *Solanum habrochaites*. Plant J 71:921–935
177. Schilmiller AL, Schauvinhold I, Larson M, Xu R, Charbonneau AL, Schmidt A, Wilkerson C, Last RL, Pichersky E (2009) Monoterpenes in the glandular trichomes of tomato are synthesized from a neryl diphosphate precursor rather than geranyl diphosphate. Proc Natl Acad Sci USA 106:10865–10870
178. Gutensohn M, Nguyen TTH, McMahon RD III, Kaplan I, Pichersky E, Dudareva N (2014) Metabolic engineering of monoterpene biosynthesis in tomato fruits via introduction of the non-canonical substrate neryl diphosphate. Metab Eng 24:107–116
179. Demissie ZA, Erland LAE, Rheault MR, Mahmoud SS (2013) The biosynthetic origin of irregular monoterpenes in *Lavandula*: isolation and biochemical characterization of a novel *cis*-prenyl diphosphate synthase gene, lavendulyl diphosphate synthase. J Biol Chem 288:6333–6341
180. Surmacz L, Plochocka D, Kania M, Danikiewicz W, Swiezewska E (2014) cis-Prenyltransferase AtCPT6 produces a family of very short-chain polyisoprenoids in planta. Biochim Biophys Acta, Mol Cell Biol Lipids 1841:240–250
181. Kang J-H, Gonzales-Vigil E, Matsuba Y, Pichersky E, Barry CS (2014) Determination of residues responsible for substrate and product specificity of *Solanum habrochaites* short-chain *cis*-prenyltransferases. Plant Physiol 164:80–91
182. Kharel Y, Takahashi S, Yamashita S, Koyama T (2006) Manipulation of prenyl chain length determination mechanism of *cis*-prenyltransferases. FEBS J 273:647–657

183. Noike M, Katagiri T, Nakayama T, Koyama T, Nishino T, Hemmi H (2008) The product chain length determination mechanism of type II geranylgeranyl diphosphate synthase requires subunit interaction. FEBS J 275:3921–3933
184. Bohlmann J, Keeling CI (2008) Terpenoid biomaterials. Plant J 54:656–669
185. Chen F, Tholl D, Bohlmann J, Pichersky E (2011) The family of terpene synthases in plants: a mid-size family of genes for specialized metabolism that is highly diversified throughout the kingdom. Plant J 66:212–229
186. Degenhardt J, Kollner TG, Gershenzon J (2009) Monoterpene and sesquiterpene synthases and the origin of terpene skeletal diversity in plants. Phytochemistry 70:1621–1637
187. Tholl D, Chen F, Petri J, Gershenzon J, Pichersky E (2005) Two sesquiterpene synthases are responsible for the complex mixture of sesquiterpenes emitted from *Arabidopsis* flowers. Plant J 42:757–771
188. Chen F, Ro D-K, Petri J, Gershenzon J, Bohlmann J, Pichersky E, Tholl D (2004) Characterization of a root-specific *Arabidopsis* terpene synthase responsible for the formation of the volatile monoterpene 1,8-cineole. Plant Physiol 135:1956–1966
189. Kollner TG, Schnee C, Gershenzon J, Degenhardt J (2004) The sesquiterpene hydrocarbons of maize (*Zea mays*) form five groups with distinct developmental and organ-specific distribution. Phytochemistry 65:1895–1902
190. Xu MM, Wilderman PR, Peters RJ (2007) Following evolution's lead to a single residue switch for diterpene synthase product outcome. Proc Natl Acad Sci USA 104:7397–7401
191. Keeling CI, Weisshaar S, Lin RPC, Bohlmann J (2008) Functional plasticity of paralogous diterpene synthases involved in conifer defense. Proc Natl Acad Sci USA 105:1085–1090
192. Greenhagen BT, O'Maille PE, Noel JP, Chappell J (2006) Identifying and manipulating structural determinates linking catalytic specificities in terpene synthases. Proc Natl Acad Sci USA 103:9826–9831
193. Davis EM, Croteau R (2000) Cyclization enzymes in the biosynthesis of monoterpenes, sesquiterpenes, and diterpenes. In: Leeper FJ, Vederas JC (eds) Topics in current chemistry: biosynthesis-aromatic polyketides, isoprenoids, alkaloids. Springer, Heidelberg, pp 53–95
194. Zi J, Mafu S, Peters RJ (2014) To gibberellins and beyond! Surveying the evolution of (di) terpenoid metabolism. Annu Rev Plant Biol 65:259–286
195. Hayashi K, Kawaide H, Notomi M, Sakigi Y, Matsuo A, Nozaki H (2006) Identification and functional analysis of bifunctional ent-kaurene synthase from the moss *Physcomitrella patens*. FEBS Lett 580:6175–6181
196. Anterola A, Shanle E, Mansouri K, Schuette S, Renzaglia K (2009) Gibberellin precursor is involved in spore germination in the moss *Physcomitrella patens*. Planta 229:1003–1007
197. Peters RJ, Carter OA, Zhang Y, Matthews BW, Croteau RB (2003) Bifunctional abietadiene synthase: mutual structural dependence of the active sites for protonation-initiated and ionization-initiated cyclizations. Biochemistry 42:2700–2707
198. Li GL, Kollner TG, Yin YB, Jiang YF, Chen H, Xu Y, Gershenzon J, Pichersky E, Chen F (2012) Nonseed plant *Selaginella moellendorfii* has both seed plant and microbial types of terpene synthases. Proc Natl Acad Sci USA 109:14711–14715
199. Geu-Flores F, Sherden NH, Courdavault V, Burlat V, Glenn WS, Wu C, Nims E, Cui Y, O'Connor SE (2012) An alternative route to cyclic terpenes by reductive cyclization in iridoid biosynthesis. Nature 492:138–142
200. Tundis R, Loizzo MR, Menichini F, Statti GA, Menichini F (2008) Biological and pharmacological activities of iridoids: recent developments. Mini-Rev Med Chem 8:399–420
201. Dewhirst SY, Pickett JA, Hardie J (2010) Aphid pheromones. In: Litwack G (ed) Vitamins and hormones: pheromones, pp 551–574
202. Koksal M, Zimmer I, Schnitzler JP, Christianson DW (2010) Structure of isoprene synthase illuminates the chemical mechanism of teragram atmospheric carbon emission. J Mol Biol 402:363–373
203. Whittington DA, Wise ML, Croteau R, Christianson DW (2002) Insights into monoterpene cyclization reactions in biology: crystal structure of (+)-bornyl diphosphate synthase. Biochemistry 41:8973

204. Whittington DA, Wise ML, Urbansky M, Coates RM, Croteau RB, Christianson DW (2002) Bornyl diphosphate synthase: structure and strategy for carbocation manipulation by a terpenoid cyclase. Proc Natl Acad Sci USA 99:15375–15380
205. Hyatt DC, Youn BY, Zhao YX, Santhamma B, Coates RM, Croteau RB, Kang CH (2007) Structure of limonene synthase, a simple model for terpenoid cyclase catalysis. Proc Natl Acad Sci USA 104:5360–5365
206. Kampranis SC, Ioannidis D, Purvis A, Mahrez W, Ninga E, Katerelos NA, Anssour S, Dunwell JM, Degenhardt J, Makris AM, Goodenough PW, Johnson CB (2007) Rational conversion of substrate and product specificity in a *Salvia* monoterpene synthase: structural insights into the evolution of terpene synthase function. Plant Cell 19:1994–2005
207. Starks CM, Back KW, Chappell J, Noel JP (1997) Structural basis for cyclic terpene biosynthesis by tobacco 5-epi-aristolochene synthase. Science 277:1815–1820
208. Gennadios HA, Gonzalez V, Di Costanzo L, Li AA, Yu FL, Miller DJ, Allemann RK, Christianson DW (2009) Crystal structure of (+)-delta-cadinene synthase from *Gossypium arboreum* and evolutionary divergence of metal binding motifs for catalysis. Biochemistry 48:6175–6183
209. Koksal M, Jin YH, Coates RM, Croteau R, Christianson DW (2011) Taxadiene synthase structure and evolution of modular architecture in terpene biosynthesis. Nature 469:116–120
210. Koksal M, Potter K, Peters RJ, Christianson DW (2014) 1.55 angstrom-resolution structure of *ent*-copalyl diphosphate synthase and exploration of general acid function by site-directed mutagenesis. Biochim Biophys Acta, Gen Sub 1840:184–190
211. Koeksal M, Hu H, Coates RM, Peters RJ, Christianson DW (2011) Structure and mechanism of the diterpene cyclase *ent*-copalyl diphosphate synthase. Nat Chem Biol 7:431–433
212. Zhou K, Gao Y, Hoy JA, Mann FM, Honzatko RB, Peters RJ (2012) Insights into diterpene cyclization from structure of bifunctional abietadiene synthase from *Abies grandis*. J Biol Chem 287:6840–6850
213. Cao R, Zhang YH, Mann FM, Huang CC, Mukkamala D, Hudock MP, Mead ME, Prisic S, Wang K, Lin FY, Chang TK, Peters RJ, Odfield E (2010) Diterpene cyclases and the nature of the isoprene fold. Proteins: Struct Funct Bioinform 78:2417–2432
214. Gao Y, Honzatko RB, Peters RJ (2012) Terpenoid synthase structures: a so far incomplete view of complex catalysis. Nat Prod Rep 29:1153–1175
215. Christianson DW (2006) Structural biology and chemistry of the terpenoid cyclases. Chem Rev 106:3412–3442
216. Aubourg S, Lecharny A, Bohlmann J (2002) Genomic analysis of the terpenoid synthase (*AtTPS*) gene family of *Arabidopsis thaliana*. Mol Genet Genomics 267:730–745
217. Tholl D, Lee S (2011) Terpene specialized metabolism in *Arabidopsis thaliana*. The Arabidopsis Book 9:e0143
218. Field B, Osbourn AE (2008) Metabolic diversification-independent assembly of operon-like gene clusters in different plants. Science 320:543–547
219. Field B, Fiston-Lavier AS, Kemen A, Geisler K, Quesneville H, Osbourn AE (2011) Formation of plant metabolic gene clusters within dynamic chromosomal regions. Proc Natl Acad Sci USA 108:16116–16121
220. Mugford ST, Louveau T, Melton R, Qi XQ, Bakht S, Hill L, Tsurushima T, Honkanen S, Rosser SJ, Lomonossoff GP, Osbourn A (2013) Modularity of plant metabolic gene clusters: a trio of linked genes that are collectively required for acylation of triterpenes in oat. Plant Cell 25:1078–1092
221. Wilderman PR, Xu MM, Jin YH, Coates RM, Peters RJ (2004) Identification of *syn*-pimara-7,15-diene synthase reveals functional clustering of terpene synthases involved in rice phytoalexin/allelochemical biosynthesis. Plant Physiol 135:2098–2105
222. Falara V, Akhtar TA, Nguyen TTH, Spyropoulou EA, Bleeker PM, Schauvinhold I, Matsuba Y, Bonini ME, Schilmiller AL, Last RL, Schuurink RC, Pichersky E (2011) The tomato terpene synthase gene family. Plant Physiol 157:770–789
223. Segura MJR, Jackson BE, Matsuda SPT (2003) Mutagenesis approaches to deduce structure-function relationships in terpene synthases. Nat Prod Rep 20:304–317

224. Phillips DR, Rasbery JM, Bartel B, Matsuda SPT (2006) Biosynthetic diversity in plant triterpene cyclization. Curr Opin Plant Biol 9:305–314
225. Wegel E, Koumproglou R, Shaw P, Osbourn A (2009) Cell type-specific chromatin decondensation of a metabolic gene cluster in oats. Plant Cell 21:3926–3936
226. Mylona P, Owatworakit A, Papadopoulou K, Jenner H, Qin B, Findlay K, Hill L, Qi X, Bakht S, Melton R, Osbourn A (2008) Sad3 and Sad4 are required for saponin biosynthesis and root development in oat. Plant Cell 20:201–212
227. Yamane H (2013) Biosynthesis of phytoalexins and regulatory mechanisms of it in rice. Biosci Biotech Biochem 77:1141–1148
228. Xu YH, Wang JW, Wang S, Wang JY, Chen XY (2004) Characterization of GaWRKY1, a cotton transcription factor that regulates the sesquiterpene synthase gene (+)-delta-cadinene synthase-A. Plant Physiol 135:507–515
229. Lu X, Zhang L, Zhang FY, Jiang WM, Shen Q, Zhang LD, Lv ZY, Wang GF, Tang KX (2013) AaORA, a trichome-specific AP2/ERF transcription factor of *Artemisia annua*, is a positive regulator in the artemisinin biosynthetic pathway and in disease resistance to *Botrytis cinerea*. New Phytol 198:1191–1202
230. Reeves PH, Ellis CM, Ploense SE, Wu MF, Yadav V, Tholl D, Chetelat A, Haupt I, Kennerley BJ, Hodgens C, Farmer EE, Nagpal P, Reed JW (2012) A regulatory network for coordinated flower maturation. PLoS Genet 8
231. Hong GJ, Xue XY, Mao YB, Wang LJ, Chen XY (2012) *Arabidopsis* MYC2 interacts with DELLA proteins in regulating sesquiterpene synthase gene expression. Plant Cell 24:2635–2648
232. Dombrecht B, Xue GP, Sprague SJ, Kirkegaard JA, Ross JJ, Reid JB, Fitt GP, Sewelam N, Schenk PM, Manners JM, Kazan K (2007) MYC2 differentially modulates diverse jasmonate-dependent functions in *Arabidopsis*. Plant Cell 19:2225–2245
233. Yadav V, Mallappa C, Gangappa SN, Bhatia S, Chattopadhyay S (2005) A basic helix-loop-helix transcription factor in *Arabidopsis*, MYC2, acts as a repressor of blue light-mediated photomorphogenic growth. Plant Cell 17:1953–1966
234. Vaughan MM, Wang Q, Webster FX, Kiemle D, Hong YJ, Tantillo DJ, Coates RM, Wray AT, Askew W, O'Donnell C, Tokuhisa JG, Tholl D (2013) Formation of the unusual semivolatile diterpene rhizathalene by the *Arabidopsis* class I terpene synthase TPS08 in the root stele is involved in defense against belowground herbivory. Plant Cell 25:1108–1125
235. Ro DK, Ehlting J, Keeling CI, Lin R, Mattheus N, Bohlmann J (2006) Microarray expression profiling and functional characterization of *AtTPS* genes: duplicated *Arabidopsis thaliana* sesquiterpene synthase genes At4g13280 and At4g13300 encode root-specific and wound-inducible (*Z*)-*Y*-bisabolene synthases. Arch Biochem Biophys 448:104–116
236. Loreto F, Dicke M, Schnitzler JP, Turlings TCJ (2014) Plant volatiles and the environment. Plant Cell Environ 37:1905–1908
237. Behnke K, Ehlting B, Teuber M, Bauerfeind M, Louis S, Hasch R, Polle A, Bohlmann J, Schnitzler JP (2007) Transgenic, non-isoprene emitting poplars don't like it hot. Plant J 51:485–499
238. Sharkey TD, Yeh SS (2001) Isoprene emission from plants. Annu Rev Plant Physiol Plant Mol Biol 52:407–436
239. Velikova V, Ghirardo A, Vanzo E, Merl J, Hauck SM, Schnitzler J-P (2014) Genetic manipulation of isoprene emissions in poplar plants remodels the chloroplast proteome. J Proteome Res 13:2005–2018
240. Behnke K, Kleist E, Uerlings R, Wildt J, Rennenberg H, Schnitzler JP (2009) RNAi-mediated suppression of isoprene biosynthesis in hybrid poplar impacts ozone tolerance. Tree Physiol 29:725–736
241. Loreto F, Schnitzler JP (2010) Abiotic stresses and induced BVOCs. Trends Plant Sci 15:154–166
242. Vickers CE, Possell M, Laothawornkitkul J, Ryan AC, Hewitt CN, Mullineaux PM (2011) Isoprene synthesis in plants: lessons from a transgenic tobacco model. Plant Cell Environ 34:1043–1053

243. Schnitzler JP, Louis S, Behnke K, Loivamaki M (2010) Poplar volatiles—biosynthesis, regulation and (eco)physiology of isoprene and stress-induced isoprenoids. Plant Biol 12:302–316
244. Byers K, Bradshaw HD, Riffell JA (2014) Three floral volatiles contribute to differential pollinator attraction in monkeyflowers (*Mimulus*). J Exp Biol 217:614–623
245. Huang M, Sanchez-Moreiras AM, Abel C, Sohrabi R, Lee S, Gershenzon J, Tholl D (2012) The major volatile organic compound emitted from *Arabidopsis thaliana* flowers, the sesquiterpene (*E*)-β-caryophyllene, is a defense against a bacterial pathogen. New Phytol 193:997–1008
246. Junker RR, Loewel C, Gross R, Dötterl S, Keller A, Blüthgen N (2011) Composition of epiphytic bacterial communities differs on petals and leaves. Plant Biol 13:918–924
247. Wang H, Guo WF, Zhang PJ, Wu ZY, Liu SS (2008) Experience-induced habituation and preference towards non-host plant odors in ovipositing females of a moth. J Chem Ecol 34:330–338
248. Laothawornkitkul J, Paul ND, Vickers CE, Possell M, Taylor JE, Mullineaux PM, Hewitt CN (2008) Isoprene emissions influence herbivore feeding decisions. Plant Cell Environ 31:1410–1415
249. Bleeker PM, Diergaarde PJ, Ament K, Schutz S, Johne B, Dijkink J, Hiemstra H, de Gelder R, de Both MTJ, Sabelis MW, Haring MA, Schuurink RC (2011) Tomato-produced 7-epizingiberene and *R*-curcumene act as repellents to whiteflies. Phytochemistry 72:68–73
250. Zulak KG, Bohlmann J (2010) Terpenoid biosynthesis and specialized vascular cells of conifer defense. J Integr Plant Biol 52:86–97
251. Hall DE, Robert JA, Keeling CI, Domanski D, Quesada AL, Jancsik S, Kuzyk MA, Hamberger B, Borchers CH, Bohlmann J (2011) An integrated genomic, proteomic and biochemical analysis of (+)-3-carene biosynthesis in Sitka spruce (*Picea sitchensis*) genotypes that are resistant or susceptible to white pine weevil. Plant J 65:936–948
252. Gols R (2014) Direct and indirect chemical defences against insects in a multitrophic framework. Plant Cell Environ 37:1741–1752
253. Pierik R, Ballare CL, Dicke M (2014) Ecology of plant volatiles: taking a plant community perspective. Plant Cell Environ 37:1845–1853
254. Schnee C, Kollner TG, Held M, Turlings TCJ, Gershenzon J, Degenhardt J (2006) The products of a single maize sesquiterpene synthase form a volatile defense signal that attracts natural enemies of maize herbivores. Proc Natl Acad Sci USA 103:1129–1134
255. Fontana A, Held M, Fantaye CA, Turlings TC, Degenhardt J, Gershenzon J (2011) Attractiveness of constitutive and herbivore-induced sesquiterpene blends of maize to the parasitic wasp *Cotesia marginiventris* (Cresson). J Chem Ecol 37:582–591
256. Kappers IF, Aharoni A, van Herpen TWJM, Luckerhoff LLP, Dicke M, Bouwmeester HJ (2005) Genetic engineering of terpenoid metabolism attracts bodyguards to *Arabidopsis*. Science 309:2070–2072
257. McCormick AC, Unsicker SB, Gershenzon J (2012) The specificity of herbivore-induced plant volatiles in attracting herbivore enemies. Trends Plant Sci 17:303–310
258. Kessler A, Heil M (2011) The multiple faces of indirect defences and their agents of natural selection. Funct Ecol 25:348–357
259. Hilker M, Meiners T (2006) Early herbivore alert: insect eggs induce plant defense. J Chem Ecol 32:1379–1397
260. Buchel K, Malskies S, Mayer M, Fenning TM, Gershenzon J, Hilker M, Meiners T (2011) How plants give early herbivore alert: volatile terpenoids attract parasitoids to egg-infested elms. Basic Appl Ecol 12:403–412
261. Arimura G, Ozawa R, Shimoda T, Nishioka T, Boland W, Takabayashi J (2000) Herbivory-induced volatiles elicit defence genes in lima bean leaves. Nature 406:512–515
262. Frost CJ, Appel M, Carlson JE, De Moraes CM, Mescher MC, Schultz JC (2007) Within-plant signalling via volatiles overcomes vascular constraints on systemic signalling and primes responses against herbivores. Ecol Lett 10:490–498

263. Heil M, Karban R (2010) Explaining evolution of plant communication by airborne signals. Trends Ecol Evol 25:137–144
264. Heil M (2014) Herbivore-induced plant volatiles: targets, perception and unanswered questions. New Phytol. doi:10.1111/nph.12977
265. Himanen SJ, Blande JD, Klemola T, Pulkkinen J, Heijari J, Holopainen JK (2010) Birch (*Betula* spp.) leaves adsorb and re-release volatiles specific to neighbouring plants—a mechanism for associational herbivore resistance? New Phytol 186:722–732
266. Runyon JB, Mescher MC, De Moraes CM (2006) Volatile chemical cues guide host location and host selection by parasitic plants. Science 313:1964–1967
267. Jassbi AR, Zamanizadehnajari S, Baldwin IT (2010) 17-Hydroxygeranyllinalool glycosides are major resistance traits of *Nicotiana obtusifolia* against attack from tobacco hornworm larvae. Phytochemistry 71:1115–1121
268. Schmelz EA, Kaplan F, Huffaker A, Dafoe NJ, Vaughan MM, Ni XZ, Rocca JR, Alborn HT, Teal PE (2011) Identity, regulation, and activity of inducible diterpenoid phytoalexins in maize. Proc Natl Acad Sci USA 108:5455–5460
269. Huffaker A, Kaplan F, Vaughan MM, Dafoe NJ, Ni XZ, Rocca JR, Alborn HT, Teal PEA, Schmelz EA (2011) Novel acidic sesquiterpenoids constitute a dominant class of pathogen-induced phytoalexins in maize. Plant Physiol 156:2082–2097
270. Kuzina V, Ekstrom CT, Andersen SB, Nielsen JK, Olsen CE, Bak S (2009) Identification of defense compounds in *Barbarea vulgaris* against the herbivore *Phyllotreta nemorum* by an ecometabolomic approach. Plant Physiol 151:1977–1990
271. Rasmann S, Kollner TG, Degenhardt J, Hiltpold I, Toepfer S, Kuhlmann U, Gershenzon J, Turlings TCJ (2005) Recruitment of entomopathogenic nematodes by insect-damaged maize roots. Nature 434:732–737
272. Turlings TCJ, Hiltpold I, Rasmann S (2012) The importance of root-produced volatiles as foraging cues for entomopathogenic nematodes. Plant Soil 358:47–56
273. Degenhardt J, Hiltpold I, Kollner TG, Frey M, Gierl A, Gershenzon J, Hibbard BE, Ellersieck MR, Turlings TCJ (2009) Restoring a maize root signal that attracts insect-killing nematodes to control a major pest. Proc Natl Acad Sci USA 106:13213–13218
274. Robert CAM, Erb M, Hiltpold I, Hibbard BE, Gaillard MDP, Bilat J, Degenhardt J, Cambet-Petit-Jean X, Turlings TCJ, Zwahlen C (2013) Genetically engineered maize plants reveal distinct costs and benefits of constitutive volatile emissions in the field. Plant Biotechnol J 11:628–639
275. Xu MM, Galhano R, Wiemann P, Bueno E, Tiernan M, Wu W, Chung IM, Gershenzon J, Tudzynski B, Sesma A, Peters RJ (2012) Genetic evidence for natural product-mediated plant–plant allelopathy in rice (*Oryza sativa*). New Phytol 193:570–575
276. Thimmappa R, Geisler K, Louveau T, O'Maille P, Osbourn A (2014) Triterpene biosynthesis in plants. Annu Rev Plant Biol 65:225–257
277. Kemen AC, Honkanen S, Melton RE, Findlay KC, Mugford ST, Hayashi K, Haralampidis K, Rosser SJ, Osbourn A (2014) Investigation of triterpene synthesis and regulation in oats reveals a role for beta-amyrin in determining root epidermal cell patterning. Proc Natl Acad Sci USA 111:8679–8684
278. Chaturvedi R, Venables B, Petros RA, Nalam V, Li MY, Wang XM, Takemoto LJ, Shah J (2012) An abietane diterpenoid is a potent activator of systemic acquired resistance. Plant J 71:161–172
279. Waldie T, McCulloch H, Leyser O (2014) Strigolactones and the control of plant development: lessons from shoot branching. Plant J 79:607–622
280. Akiyama K, Matsuzaki K, Hayashi H (2005) Plant sesquiterpenes induce hyphal branching in arbuscular mycorrhizal fungi. Nature 435:824–827
281. Matusova R, Rani K, Verstappen FWA, Franssen MCR, Beale MH, Bouwmeester HJ (2005) The strigolactone germination stimulants of the plant-parasitic *Striga* and *Orobanche* spp. are derived from the carotenoid pathway. Plant Physiol 139:920–934

282. Gomez-Roldan V, Fermas S, Brewer PB, Puech-Pages V, Dun EA, Pillot JP, Letisse F, Matusova R, Danoun S, Portais JC, Bouwmeester H, Becard G, Beveridge CA, Rameau C, Rochange SF (2008) Strigolactone inhibition of shoot branching. Nature 455:189–U122
283. Domagalska MA, Leyser O (2011) Signal integration in the control of shoot branching. Nat Rev Mol Cell Biol 12:211–221
284. Brewer PB, Koltai H, Beveridge CA (2013) Diverse roles of strigolactones in plant development. Mol Plant 6:18–28
285. Rodriguez S, Kirby J, Denby CM, Keasling JD (2014) Production and quantification of sesquiterpenes in *Saccharomyces cerevisiae*, including extraction, detection and quantification of terpene products and key related metabolites. Nat Protoc 9:1980–1996

Adv Biochem Eng Biotechnol (2015) 148: 107–140
DOI 10.1007/10_2014_301

Published Online: 31 January 2015

Cytochromes P450 for Terpene Functionalisation and Metabolic Engineering

Irini Pateraki, Allison Maree Heskes and Björn Hamberger

Abstract Plants have evolved the capacity to produce a striking array of specialised metabolites. Terpenoids are the oldest and most diverse class of such compounds and have attracted interest for industrial and pharmaceutical applications. The development of biotechnological alternatives for their production is the focus of intense research. Photosynthetic systems provide new strategies for autotrophic metabolic engineering. Focusing on cytochromes P450, involved in the functionalisation of the core terpene molecules, this review highlights the latest approaches in this field and looks towards recent discoveries that have the potential to shape the future of terpenoid bioengineering.

Keywords Cytochromes P450 · Biotechnology · Terpenoids · Production hosts · Pathway engineering · Terpenoid biosynthesis

Abbreviations

P450	Cytochrome P450 dependent mono-oxygenase
POR	NADPH dependent cytochrome P450 oxidoreductase
CYP71A1 to CYP71 clan	Example of classification of P450s into subfamilies to clans

I. Pateraki · A.M. Heskes
Department of Plant and Environmental Sciences, Faculty of Science,
University of Copenhagen, Thorvaldsensvej 40, 1871 Copenhagen, Denmark

I. Pateraki · A.M. Heskes · B. Hamberger
Center for Synthetic Biology bioSYNergy, Copenhagen, Denmark

I. Pateraki · A.M. Heskes · B. Hamberger
Copenhagen Plant Sciences Centre, University of Copenhagen, Copenhagen, Denmark

B. Hamberger (✉)
Plant Biochemistry Laboratory, Department of Plant and Environmental Sciences,
University of Copenhagen, Thorvaldsensvej 40, Frederiksberg C, 1871 Copenhagen,
Denmark
e-mail: bjoernh@plen.ku.dk

EST	Expressed sequence tag
Fsp^3	Fraction of carbon atoms in compound with sp^3 hybridisation
ER	Endoplasmatic reticulum

Contents

1 Introduction

Plant-derived natural products have been utilised by humans to improve health and nutrition throughout history. Currently, two thirds of the global population still rely on traditional medicines consisting mostly of plant crude extracts. At the same time, the annual spending on pharmaceuticals by the established market economies (representing one fifth of the world's population) is projected to grow by US $130 billion over the next five years (e.g., [116]). Interestingly, it seems that these industrialised, synthetic pharmaceuticals still have their origins in the natural world. Of the over 900 small-molecule drugs approved between 1981 and 2010, approximately two thirds mimic or are analogues of natural compounds, carry a natural pharmacophore or are naturally inspired or derived [88].

Terpenoids represent prime examples of natural bioactive molecules. The earliest evidence of use and consumption of plant species rich in terpenoids dates back 50,000 years to the Palaeolithic era, with two independent lines of research indicating that Neanderthal hominids self-medicated with what are today recognised medicinal plants (i.e., yarrow, chamomile; [46, 67]). Terpenoids fulfil two key criteria that are associated with compounds that have good potential as pharmaceuticals [73]. First, the C_{15}, C_{20}, and C_{30} scaffolds of sesqui-, di-, and triterpenes carry an exceptionally high fraction of carbons in tetrahedral sp^3—versus planar sp^2—and linear sp-hybridisation (Fsp[3]). Second, they contain a high number of chiral centres (Fig. 1). In assays for inhibition of enzymes of the human liver, used as a proxy for toxicity of drugs due to off-target effects, the presence of these structural features is positively correlated with target specificity and low toxicity [72]. Yet, when compared with compounds of the alkaloid class, there is a discrepancy between the number of known plant terpenoids and those recognised in one of the 247 categories defined as medically active compounds in the Dictionary of Natural Products (Table 1; DNP23.1, accessed

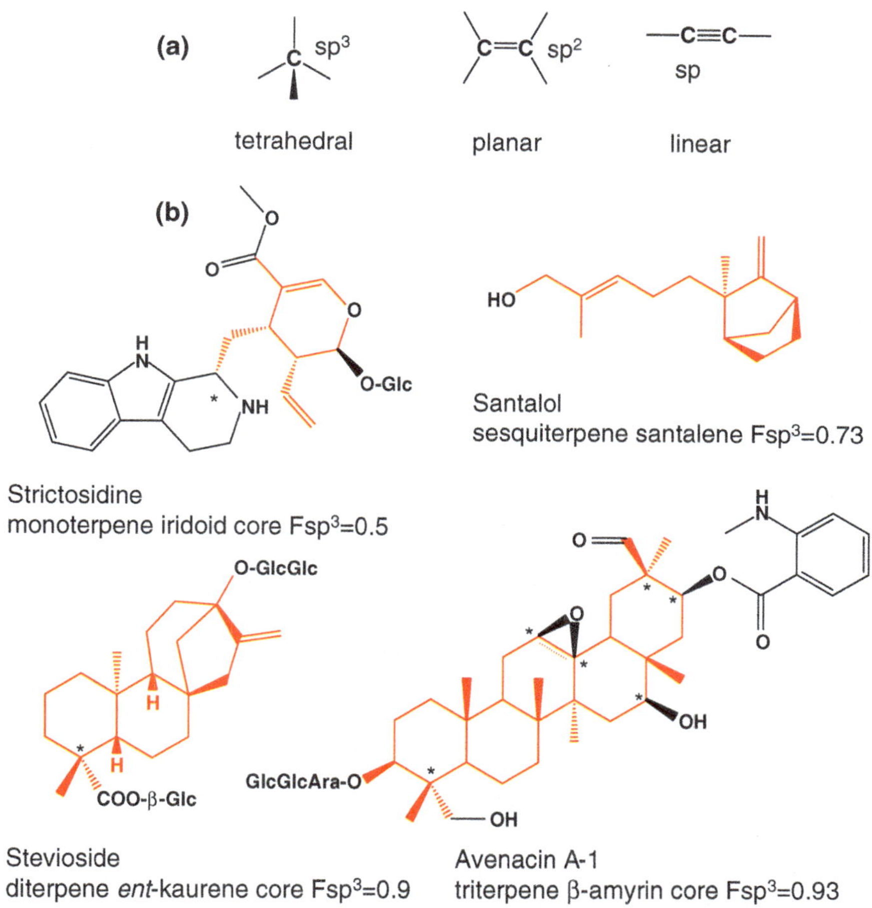

Fig. 1 The impact of carbon hybridisation on potential structural diversity. a Hybridisation of carbon orbitals and resulting geometry. *Note* Only sp[3] hybridised carbon atoms allow stereoisomers. **b** Selected examples of structurally complex terpenoids with increasing fraction of sp[3] hybridised carbon atoms (Fsp[3]). In *red* Iridoid core, two stereocenters; santalene sesquiterpene, three stereocenters; *ent*-kaurene diterpene, six stereocenters; beta-amyrin triterpene, eight stereocenters; an *asterisk* indicates stereocenters in the structures formed during functionalisation. *Dashed line* indicates a double bond in the beta-amyrin backbone, replaced by an epoxide in avenacin. Examples given are: secologanin/monoterpene iridoid core in the terpenoalkaloid seco-iridoid strictosidine [77]; (E)-epi-β-santalol [29]; Steviol in the terpene glycoside Stevioside [91]; 12,13β-epoxy-16β-hydroxy-β-amyrin in the compound antifungal Avenacin A-1 [38]

Table 1 Numbers and types of terpenoids currently deposited in the dictionary of natural products

Type of terpenoid[a]	Compounds (DNP 23.1)	Classified as drug	Fraction with 2+ oxygen[b] (%)
Monoterpenoid	4,129	28	92.2
Sesquiterpenoid	13,981	97	89.5
Diterpenoid	12,505	52	95.7
Triterpenoid	23,051	155	95.1
Tetraterpenoid	827	5	90.1
Total	54,493	337	93.5

[a] Including terpene-derived products within each class, such as apo- and nor-terpenoids

[b] Up to two oxygen atoms can be the result of water access and quenching of the carbocation during formation of the terpene backbone by terpene synthases

July 2014). Among the plant-based drugs approved, or with clinical trials launched in the period from 2000 to 2010, terpenoid-related compounds are dominated by only three scaffolds: Taxol (or paclitaxel), artemisinin, and vinblastine [14, 88, 103]. The key for the successful development of these scaffolds in clinical applications and establishment as treatment has been securing supply of the compounds through intense efforts in biotechnology or semi- or full chemical synthesis [23, 93, 128].

In addition to their potential as pharmaceuticals, bioactive terpenoids have attracted substantial interest as neutraceuticals, flavours and fragrances, and for their biopesticidal and antiherbivore activities (Fig. 1). This has led bioactive terpenoids to be considered as high-value natural products, setting this class of compounds apart from terpenoids that are used as fuels or feedstock for the chemical industry.

Regardless of the area of application, the availability of terpenoids for industrial purposes is, in general, hampered by laborious extraction and purification from their plant sources or limited economic competitiveness of chemical synthesis. Recently, biotechnological production has emerged as an alternative source of terpenoids for industrial applications. This technology involves the reconstruction of terpenoid biosynthetic pathways in host organisms that are suitable for large-scale cultivation of cell factories and holds the promise of being a biosustainable alternative to conventional production methods. The major requirement for successful engineering of the metabolic routes to complex target terpenoids in production hosts is knowledge of the enzymes involved in their biosynthesis.

Terpenoid biosynthesis involves two main classes of enzymes. The formation of the complex, often multicyclic scaffolds is catalysed by enzymes of the terpene synthase families [21]. The resulting scaffolds are then oxidatively functionalised through the action of cytochromes P450 (P450s). These enzymes carry out stereospecific hydroxylations, regularly contributing novel chiralities to the molecule, and which serve as molecular handles for further modifications, that is, linkage of sugar residues, alkylations, or esterifications. Further oxidation of hydroxyl groups to carbonyl and carboxyl groups is also a typical function of P450s. It should also be noted that introduction of up to two oxygen-containing functional groups into the terpenoid scaffold can be the result of water-quenching of the carbocation in the active site during formation by terpene synthases (e.g., [16, 131]).

This review highlights recent efforts and changes in the strategies for the discovery of plant enzymes with a focus on the extremely diverse family of P450s involved in oxidative functionalisation of high-value terpenoids. It also discusses advances in heterologous expression and metabolic engineering of selected pathways in the biotechnological hosts, *E. coli* and yeast, and also some key plant species. Finally, a brief future perspective is provided in which new discoveries which may shape the emerging field are discussed.

2 Genetic Diversity of Plant P450s as Driver of Terpenoid Diversity

Cytochromes P450 constitute the largest and functionally most versatile enzyme superfamily found in nature [48]. They are heme-containing enzymes, found in all kingdoms of life, noted for their extremely diverse amino acid sequences. They participate in many different biological processes, ranging from general biosynthetic functions (e.g., steroid biosynthesis) to specialised metabolism (plant defence molecules) and detoxification of xenobiotic chemicals such as drugs and pollutants [86]. The primary reaction they catalyse is the introduction of atmospheric oxygen into nonactivated carbon–hydrogen bonds [62]. For their catalytic activity, P450s require reducing equivalents, which are usually provided by NADPH or NADH, depending on the specific P450 enzyme. Inasmuch as P450s cannot receive electrons directly from these cofactors, different redox partners are engaged, such as membrane-bound FMN/FAD containing NADPH-dependent P450 oxidoreductases (POR) in the case of eukaryotic enzymes [62].

P450 enzymes are classified according to their sequence identity and not necessarily to their specific activity. Hence, the nomenclature reflects evolutionary relationships of the sequences [86]. Generally, P450s with a sequence identity higher than 40 % constitute members of one family, and sequence identity over 55 % defines members of a subfamily. In an effort to categorise the different P450 families in a higher order and make their nomenclature more comprehensive, the term ‘clan’ has been introduced [83, 84]. Clans are defined as groups of CYP families which segregate together on phylogenetic trees with members considered to have evolved from a common ancestor [86].

The diversity of P450 genes is vast, yet little is known about the evolutionary mechanisms driving the emergence of novel families. Due to their ubiquitous distribution, ancient P450s of terpenoid general metabolism are considered the genetic raw material for evolution. More than six decades after the first discovery of a P450, 127 families have been established in the plant kingdom, compared to 67 families in insects and 19 in vertebrates [85]. And within those families, over 100 subfamilies now carry functionally characterised P450s, of which 46 subfamilies, distributed over 8 of the 11 known plant clans, are involved in both general and specialised metabolism of terpenoids (see Appendix table for an update on details on plant species and references).

A characteristic of rapidly evolving families and subfamilies of P450s with roles in specialised metabolism is the occurrence of species- or genus-specific expansions called 'blooms' [33, 113]. This is particularly apparent in the CYP71 clan, where 20 out of 21 subfamilies are multimembered and have proliferated to such an extent that the previous definitions of family and subfamily have become problematic [85]. David Nelson coined the term 'tribe' for larger clades of subfamilies, in analogy to the rank of tribe between family and genus in taxonomy. This terminology had already been applied to the large group (i.e., tribe) of CYP71D found in the Euphorbiaceae, that consists of several phylogenetically distinct clusters of CYP71D and CYP726A subfamilies (Fig. 2; [132]).

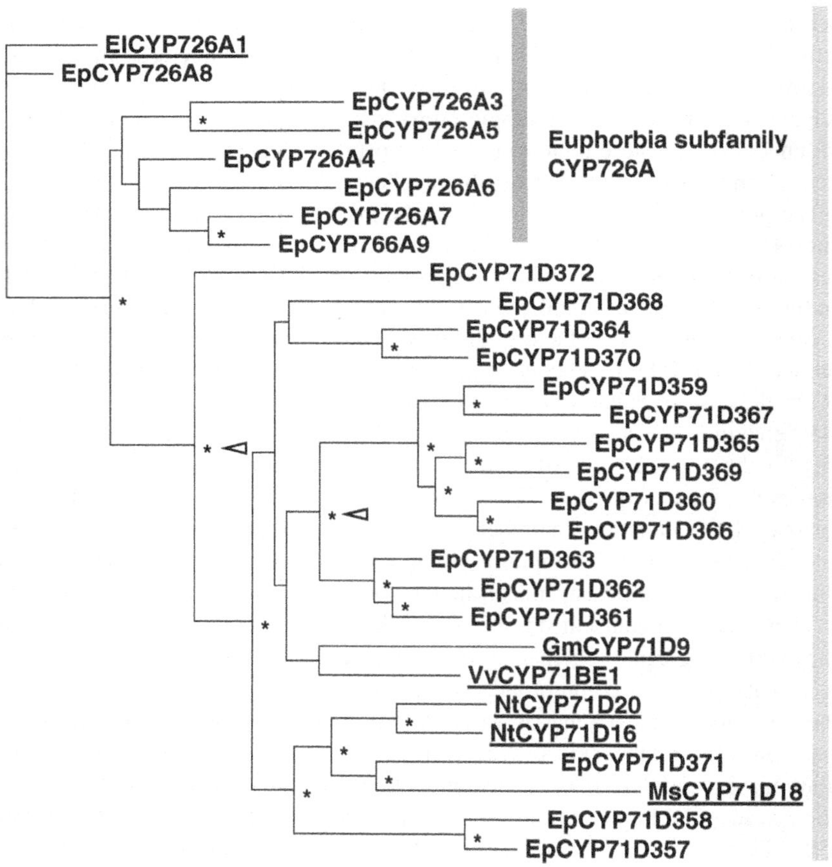

Fig. 2 Species and plant family specific blooms, resulting in P450 multigene families (modified after: [132]; copyright American Society of Plant Biologists, www.plantphysiol.org, permission obtained). El, *Euphorbia lagascae*; Ep, *E. peplus*; Gm, *Glycine max*; Vv, *Vitis vinifera*; Nt, *Nicotiana tabacum*; Ms, *Mentha spicata*. Asterisks indicate bootstrap values equal to, or above 80 %; triangles indicate phylogenetically distinct clades in *E. peplus*

The origin of such expansions of P450s is largely unknown, although it has been suggested that the fitness gain resulting from chemical diversification is one force driving the evolution of new P450s, with the original genes being recruited from the general metabolism [42]. Evidence of such a process can be observed in the rice genome, where five CYP701A homologues constitute a small multigene family, whereas only one, CYP701A6, is required for gibberellic acid biosynthesis. Its homologue CYP701A8, generated after duplication from general metabolism, has undergone neofunctionalisation and is now involved in the biosynthesis of rice antifungal phytoalexins [124]. A second example from rice is the grass-specific CYP51H subfamily, which contains nine members [87]. CYP51s are an ancestral P450 group typically found in eukaryotes in a single copy that catalyse a three-step demethylation of C_{30} sterol constituents of membranes. Little is known about the function of these additional CYP51 members in rice, but one of the members from the subfamily in oat has been functionally characterised as the first oxidative enzyme in the biosynthetic pathway of avenacin saponins, glycosylated triterpene-derived defence compounds [38].

3 P450s Involved in Plant Mono- to Triterpenoid Metabolism

The discovery of an enzymatic route to a given compound is critical if its biotechnological production is to be achieved. The development and maturation of next-generation sequencing techniques over the last decade, pioneered by 454 sequencing [76], has significantly increased our ability to perform this task. Several genomics and functional genomics projects focused on medicinal plants accumulating bioactive terpenoids have released vast nonmodel plant sequencing resources to the public, including the Medicinal Plant Genomics Resource (http://medicinalplantgenomics.msu.edu/), the Canadian PhytoMetaSyn Initiative (http://www.phytometasyn.com/), and the 1KP Plant project (http://www.onekp.com/project.html). In the wake of this radically increased availability of sequence data, creative and efficient approaches are emerging to identify the fraction of genes relevant for terpenoid biosynthesis, particularly in regard to the identification of P450s as these exist in high numbers in plant genomes. The following selected illustrative approaches have succeeded in the identification of specific P450s by integrating these sequencing data with proteomic and metabolomic data, performing deep comparative transcriptomics (e.g., coexpression analyses) or by exploiting the fact that genes in terpenoid biosynthetic routes may occur physically clustered at the genomic level. Considering such clustering recently prompted a genomewide scan for physically linked pairs of terpene synthases and P450s in 17 monocots and eudicots. In several plant species colocalisation of terpene synthases was found, with significant bias towards specific families of P450s, providing a strong incentive for functional investigation [12]. Generally, following identification, enzymes are functionally characterised in heterologous systems, such as the microbes

E. coli and yeast, but also with the classical higher eukaryotic ovarian-derived insect *Spodoptera frugiperda* 9 (*Sf*9)-cell line/Baculovirus system and increasingly through transient expression in the plant *Nicotiana benthamiana* using *Agrobacterium* infiltrations.

3.1 Monoterpenoids

Monoterpene indole alkaloids are a large group of plant-specialised metabolites with broad bioactivities ranging from anticancer and anti-inflammatory to antimicrobial properties. Biosynthesis of the central metabolite strictosidine involves oxidation of the initially volatile monoterpene geraniol to 8-hydroxygeraniol, cyclisation of the aldehyde 8-oxogeranial to an iridoid, and later formation of loganic acid, which is converted to loganin through an O-methyl transferase reaction. After oxidative ring-opening of loganin yielding secologanin, strictosidine is finally formed by coupling with tryptamine. The full biosynthetic pathway is suggested to include approximately 10 enzymes that catalyse the oxidation, reduction, glycosylation, and methylation reactions [77].

To identify missing steps in this complex biosynthetic pathway, deep transcriptome resources were developed for the key monoterpene-indole alkaloid-producing species, Madagascar periwinkle (*Catharanthus roseus*), the source of the anticancer monoterpene-indole alkaloid vinblastine (e.g., [39, 77]). Databases of annotated transcripts from various monoterpene-indole alkaloid-producing plant species were searched to identify candidate genes. Virus-induced gene silencing in plants of selected candidates and targeted metabolite analysis identified CYP72A224 by the accumulation of the presumed precursor of loganic acid, 7-deoxyloganic acid and a reduction of more complex monoterpene-indole alkaloids, indicating a function of this gene early in the pathway [104].

The last four unknown steps in the pathway to strictosidine were finally elucidated in 2014 using a gene coexpression analysis approach in which 23 transcriptomes were compared from different tissues, cell types, and treatments along with the proteomes of epidermal and mesophyll protoplasts isolated from *C. roseus* leaves [77]. By comparing different tissues, this approach was able to take into account cell-specific and spatially distant localisation of branches of the pathway. Biochemical characterisation of candidates resulted in identification of the enzymes catalysing formation of the iridoid-precursor, 8-oxogeranial, oxidation of the iridoid to 7-deoxyloganetic acid followed by glycosylation to loganic acid, and final oxidation to loganic acid (Fig. 3). Functional characterisation of the recombinant enzymes in *E. coli* established an 8-hydroxygeraniol dehydrogenase of the zinc-containing alcohol dehydrogenase family and a UDP-glucose-dependent glycosyltransferase involved in glycosylation of 7-deoxyloganetic acid to 7-deoxyloganic acid. Coexpression in yeast of CYP76A26 with the Arabidopsis POR1 established this CYP71 clan member as a multifunctional enzyme catalysing successive hydroxylation/dehydration steps en route to 7-deoxyloganetic acid. The mechanism

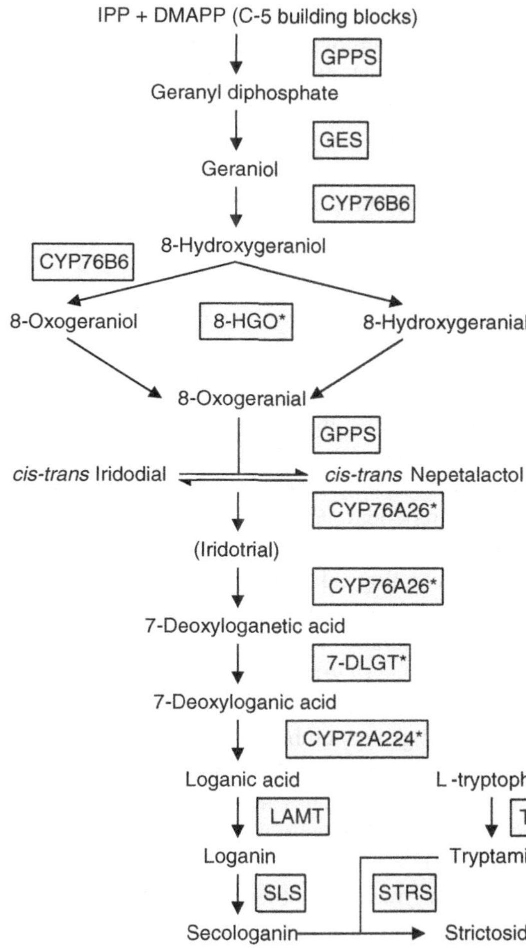

Fig. 3 Heterologous expression of the strictosidine biosynthetic pathway in *N. benthamiana* (modified after: [77]). Enzymes specifically discussed in the text are marked by an *asterisk*. IPP, isopentenyl diphosphate; DMAPP, dimethylallyl diphosphate; GPPS, geranyl diphosphate synthase; GES, geraniol synthase; CYP76B6 (*C. roseus*), geraniol 8-oxidase; 8-HGO, 8-hydroxygeraniol oxidoreductase; IS, iridoid synthase; CYP76A26 (*C. roseus*) iridoid oxidase; 7-DLGT, 7-deoxyloganetic acid glucosyl transferase; CYP72A224 (*C. roseus*), 7-deoxyloganic acid hydroxylase; LAMT, loganic acid O-methyltransferase; SLS, secologanin synthase; STRS, strictosidine synthase; TDC, tryptophan decarboxylase

resembles the oxidation of *ent*-kaurene to *ent*-kaurenoic acid in gibberellin biosynthesis by members of the subfamily CYP701A of general metabolism, typically rooting the CYP71 clan.

Functional characterisation of the previously identified CYP72A224 [104], by expression in yeast and transient expression in *N. benthamiana* following *Agrobacterium* infiltration showed that this enzyme converted 7-deoxyloganic acid, but not the aglycone derivative of 7-deoxyloganic acid (7-deoxyloganetic acid) into loganic acid, and indicated that the glycosylation precedes the 7-hydroxylation. From the perspective of pathway discovery, this example of recruitment of the indole alkaloid pathway genes from three highly divergent families, zinc-containing alcohol dehydrogenases and P450s of the divergent clans CYP71 and CYP72, shows that enzymatic steps of core-pathways are not necessarily encoded by related members of defined subfamilies. It remains to be tested whether other members of the relevant subfamilies can further increase chemical diversity of the individual steps. For functional characterisation of the enzymes involved, it also highlights the benefit of complementary host systems for heterologous expression. Finally, the availability of the enzymes catalysing all individual steps allowed the team to reconstitute the entire pathway to strictosidine transiently in the *N. benthamiana* system, demonstrating feasibility of biotechnological production [77].

3.2 Sesquiterpenoids

(+)-Nootkatone is a volatile sesquiterpenoid found in traces in grapefruit epicarp (flavedo) and as the main component of Nootka cypress heartwood essential oil (*Cupressus nootkatensis*). It is considered to be a high-value compound due to its use as an aroma compound and its potential as an environmentally safe biopesticide and insect repellent [95]. Hence, the biosynthesis of nootkatone has attracted substantial attention over the last years, leading to several stepwise improvements in its biotechnological production. Nootkatone carries three stereocenters introduced by its bicyclic hydrocarbon scaffold, valencene, and a keto-function, indicating a potentially simple biosynthetic route. As a terpene synthase from 'Valencia' orange (*Citrus sinensis* cv. Valencia) had been functionally characterised over a decade ago as an efficient converter of farnesyl diphosphate to valencene [114], several groups embarked on the quest to discover enzymes for the oxidative conversion to nootkatone. Chicory is well known to accumulate a diverse spectrum of oxidised sesquiterpenes in its root. The finding that microsomal enzymes from chicory root could metabolise valencene led to the discovery of the CYP71 clan member CYP76AV8 in a root transcriptome, based on homology to terpenoid-oxidising enzymes from other species. Biochemical characterisation of the enzyme showed high substrate promiscuity and conversion of valencene, resulting in predominant accumulation of the alcohol nootkatol, with traces of nootkatone [17]. During further testing of enzymes of the CYP71 clan from species known to accumulate oxidised sesquiterpenoids, a P450 catalysing efficient oxidation of valencene towards nootkatone was finally discovered in a targeted deep transcriptome derived from Nootka cypress wood tissue, consistent with detection of nootkatone synthesis in this tissue. The founding member of a new P450 subfamily, CYP706M1, was shown to catalyse the formation of nootkatone (see the section on yeast engineering for details). Interestingly, the degree of amino acid identity of CYP706M1 compared to the previously identified P450s with valencene oxidase activity did not exceed 30 % [18], indicating that the evolution of specific sesquiterpene oxidases may occur through independent events. Furthermore, the limited success in replacement of the required specific two-step oxidation to nootkatone with P450s nonnative to this pathway highlights the advantages of dedicated species- and tissue-specific transcriptome libraries for P450 discovery.

The recent identification of a group of P450s responsible for the oxidation of sesquiterpenes in sandalwood provides another example of the utility of tissue-specific transcriptome mining for P450 gene discovery. Sandalwood (*Santalum album*) is a hemiparasitic tree species of commercial interest for the fragrance industry due to the accumulation of highly prized sesquiterpene-based sandalwood oil in its xylem tissue. Mining of the transcriptome of the xylem of *S. album* led to identification of ten P450s constituting two distinct clades of the CYP76F subfamily. Consistent with a function in biosynthesis of sandalwood oil, the genes were found to be highly coexpressed in *S. album* xylem tissue [29]. Functional characterisation in yeast of nine of the ten recombinant P450s showed a distinct but overlapping product profile for each clade, with some oxidised sesquiterpenes alcohols detected in both. The produced bergamotol and santalol sesquiterpenes were all monohydroxylated at the terminal carbon of the isoprenyl side-chain, and the overall profile was shown to match the composition of the sandalwood oil, yet with different proportions of the individual compounds [29].

3.3 Diterpenoids

In general, the increasing molecular weight of terpenoids towards the C_{20}-based diterpenoids and the resulting loss of volatility of the core skeleton and changes in physicochemical properties leads to molecules with altered pharmaceutical, neutraceutical, and other industrially relevant properties [11]. One example of a group of high-value diterpenoids is the conifer diterpene resin acids, which are key components of high-end inks and tackifiers, used in formulating adhesives. To elucidate conifer diterpene resin acid biosynthesis, six divergent P450s from loblolly pine were selected based on remote similarity to families involved in triterpenoid phytohormones brassinosteroid metabolism (CYP90), or metabolism of Taxol (CYP725) of the CYP85 clan. Functional testing in yeast of all candidates allowed partial characterisation of CYP720B1 as a multisubstrate enzyme catalysing two consecutive oxidation steps leading from diterpene alcohols via the corresponding aldehyde to four structurally different resin acids [101]. Subsequent development and mining of large-scale transcriptome resources of nearly 1,000,000 pine and spruce ESTs revealed an extraordinary degree of multiplicity in the CYP720B subfamily specific for conifers. Recombinant Sitka spruce CYP720B4 was functionally characterised through complementary approaches in vitro, in vivo, and through RNA interference in planta and the enzyme was found to catalyse up to three oxidative steps of a panel of 24 diterpene olefins, alcohols, and aldehydes to varying degree [43].

To address the general lack of a standardised reference system for discovery and rapid functional annotation of genes involved in diterpenoid specialised metabolism in nonmodel plant systems, Zerbe and coworkers selected specific organs, tissues, or cell types for deep transcriptome sequencing, guided by targeted metabolomics, [132]. Feasibility of the generalised strategy was demonstrated by mining transcriptomes of 10 plant species accumulating high-value diterpenes with custom-built libraries for diterpene synthases and P450s. Several hundred sequences for terpenoid-related P450s were initially discovered in close to 1,000,000 predicted unique transcript reads. Phylogenetic analyses of these genes for blooms resulting specifically in novel P450 multigene subfamilies restricted to target plant orders, families, or species, effectively reduced the number of candidate genes. A particularly striking example of such a bloom is the Euphorbiaceae specific expansion in the CYP71 family. Based on the expression of the 22 members of this group they were implicated in the biosynthesis of the complex macrocyclic diterpenes of the lathyrane- and ingenane-type characteristic of the *Euphorbia* species [132]. This hypothesis received independent support through analysis of genomic clustering of sequence homologues in the genome of another member of the spurge family, castor bean (*Ricinus communis*), and indications that the P450s are involved in oxidative decoration of simple macrocyclic diterpenoids in that species [58].

Within model plant species, the metabolism of complex diterpenoids is perhaps best understood in rice (*Oryza sativa*), where functional roles of structurally divergent families of diterpenoids as phytoalexins, phytoanticipants, and allelochemicals are well established in plant–microbe and plant–plant interactions [56, 97]. In contrast to nonmodel species where genomic resources are lacking, the keys to the rapid and comprehensive elucidation of diterpenoid pathways in rice have been the physical organisation of groups of biosynthetic genes in clusters and their coordinated expression in response to environmental stimuli or elicitation. For example, four expanded subfamilies of the CYP71 clan are found in clusters (chromosome 2, CYP76M, CYP71Z; chromosome 4 CYP99A, and chromosome 6, CYP701A) with two clusters also carrying diterpene synthases (reviewed in [107]). This distribution of the functional enzymes of rice diterpene metabolism over several subfamilies indicates independent evolutionary events leading to their emergence. It was suggested that the observed pattern arose early in the grass family of plants (Poaceae; [107]), consistent also with clustering of the founding member *S. bicolor* CYP99A1 [7] on chromosome 6, and a CYP99A-paralogue on chromosome 1 with sequences potentially encoding terpene synthases (http://phytozome.jgi.doe.gov, Phytozome10, Sorghum bicolor v2.1 genome).

3.4 Triterpenoids

The oxidative functionalisation of triterpenoids involves evolutionary distant P450 enzymes found across four of the known eleven plant P450 clans: CYP51, CYP71, CYP72, and CYP85. In both oat (*Avena strigosa*) and liquorice (*Glycyrrhiza* spp.), discovery of duplications and divergence of genes originating in the metabolism of triterpene sterols (CYP51G) and diterpene gibberellins (CYP88A) indicated potential neofunctionalisation of newly emerged subfamilies. Indeed, liquorice CYP88D6 of the CYP85 clan was found to catalyse C-11 oxidation of β-amyrin, representing the first committed step in the biosynthesis of the natural sweetener and pharmacologically active glycyrrhizin. It was initially discovered by narrowing down unique ESTs encoding candidate P450s by transcript profiling in the glycyrrhizin-accumulating underground organs (stolons and roots) and the aboveground glycyrrhizin-free organs (leaves and stems). The less obvious and evolutionary distant member of the CYP72 clan, CYP72A154, performing the subsequent oxidation of C-30, was later identified by expression profiling to find genes whose expression matched the organ-specific accumulation of glycyrrhizin and testing of identified candidates for enzymatic activity against 11-oxo-β-amyrin. Both CYP88D6 and CYP72A154 were characterised in vitro using microsomes from recombinant *Sf9* insect cells, complemented by in vivo coexpression of the β-amyrin synthase, cytochrome P450 reductase (POR) with the P450s in yeast [111, 112].

Oat accumulates avenacins, glycosylated β-amyrin–derived triterpenoids conferring resistance to microbes. Even with limited biochemical knowledge, or evidence of the types of enzymes involved, linkage of biosynthetic genes had already been cautiously suggested 15 years ago based on genetic evidence of mutant lines, making this biosynthetic cluster an archetype for terpenoid biosynthesis [96]. The first gene involved in the oxidative decoration of avenacins, and encoding a P450, was indeed identified in a cluster with the triterpene β-amyrin synthase [98]. Both genes were shown to be coexpressed specifically in the epidermal cells of oat root tips, consistent with the localised accumulation of avenacin. Heterologous expression in yeast of CYP51H10 demonstrated catalytic activity converting β-amyrin to an oxidised triterpene. Yet, the conclusive structural elucidation of 12,13β-epoxy-16β-hydroxy-β-amyrin as intermediate of avenacin was only achieved after transient coexpression of the β-amyrin synthase with CYP51H10 in *N. benthamiana* following *Agrobacterium* infiltration and purification of the product for structure determination by NMR [38, 61].

Triterpene saponins from plants such as ginseng and liquorice have attracted interest due to their pharmacological properties. Substantial understanding of the biosynthetic routes of these compounds was gained over recent years from the model legume *Medicago truncatula*, a rich source of diverse saponins. A coexpression analysis based on methyl-jasmonate elicitation of triterpenoid biosynthesis, and correlation with expression of β-amyrin synthase was used to identify candidate P450s, which were reduced to only a few candidates by homology to P450s involved in terpenoid metabolism [82]. Indeed, *M. truncatula* CYP716A12, identified through homology with the conifer CYP720B, became the first functionally characterised member of this subfamily [35]. After initial demonstration of a functional enzyme in the microsomal fraction when expressed in the *Sf9* system, heterologous expression of CYP716A12 in yeast together with a POR and three distinct triterpene synthases yielded oleanolic acid, ursolic acid, and betulinic acid from β-amyrin, α-amyrin, and lupeol, respectively. Similar activity of CYP716A homologues from grape towards triterpenes was also demonstrated, indicating a possible conserved functionality within the CYP716A subfamily. This groundbreaking work spurred a burst of discoveries of novel triterpene-oxidising members in the CYP716 family, such as in ginseng (*Panax* spp.), the African shrub *Maesa lanceolata*, or the medicinal plant *Bupleurum falcatum* from the parsley family (Apiaceae; [44, 79, 80]) and also opened the door to combinatorial biosynthesis of triterpenes (see below in the section on perspectives).

4 Expression Systems: From Functional Testing of Recombinant P450s to Production Platforms

The biosynthesis of plant terpenoids by microbes for commercial production features several advantages over extraction from natural sources. These include: simple compound purification due to a limited chemical background, well-established genetic tools for metabolic engineering, and rapid generation of stable lines for scalable fermentation. However, the heterologous expression of functional terpene synthases and P450s, balancing of metabolic pathways, compartmentalisation of biochemical pathways, and metabolic channelling, are factors that currently need consideration and optimisation for every novel pathway and target compound where robust production lines and cost-effective biosynthesis are required. For example, plant mono- and diterpene synthases are plastidial enzymes in plants. Expression in *E. coli* or yeast hosts, which do not carry plastids, generally requires removal of the N-terminal plastid targeting peptide sequence of the native protein to yield a pseudomature form, as it could possibly interfere with proper protein folding and optimal expression [25, 106].

The expression of plant P450s in microbial hosts presents a specific set of challenges. Plant P450s are membrane-anchored enzymes localised at the endoplasmatic reticulum (ER) of the cell, making functional expression in prokaryotes problematic as these hosts do not contain this cellular compartment. Typical approaches to this problem involve removal or modifications of the N-terminal membrane anchor [8, 59, 66, 109]. Additionally, efficient electron transfer to recombinant P450s is required for function and this is normally achieved with the partner POR enzyme, colocalised and anchored at the ER [52]. A range of strategies for optimisations has been considered (summarised in [100]).

The following section provides examples of the successful expression of plant P450s in microbial hosts and a discussion of how arising problems have been overcome.

4.1 Bacterial Hosts

An example illustrating some of the difficulties encountered during expression of P450s in bacterial hosts is the semisynthetic artemisinin project [93]. Artemisinin is a plant natural product of the sesquiterpene lactones class, produced in the trichomes of *Artemisia annua* [9]. Despite its designation as the most potent antimalarial drug and its effective use in antimalarial combination therapies (ACTs), the supply was unstable due to the unreliable growth of *A. annua* crops. Thus, efforts have been undertaken over the last 10 years to develop an alternative biotechnologically produced source of artemisinin. The first microbial choice was *E. coli* where high titres (up to 27.4 g L^{-1}) of amorpha-4,11-diene (the sesquiterpene olefin precursor of artemisinin) were achieved by expression of the *A. annua* amorphadiene synthase (ADS) and a heterologous mevalonate pathway [121]. However, the biosynthetic pathway of artemisinin contains a single P450, CYP71AV1, which converts amorpha-4,11-diene to artemisinic acid or dihydroartemisinic acid through three consecutive oxidations [102, 118, 120]. Expression of the *A. annua* CYP71AV1 native enzyme in *E. coli* resulted in no detectable activity (in vivo or in vitro; [20]), creating a major obstacle for the successful use of *E. coli* as a production host.

To achieve activity of CYP71AV1 in *E. coli* multiple modifications and adjustments were required including: codon optimisation, manipulation of the N-terminal transmembrane domain, use of the native POR enzyme from *A. annua*, adjustment of the expression vectors, and also optimisation of the *E. coli* strain and culture conditions. Further efforts were undertaken to improve the yield, such as replacement of the anchor of the plant P450 with that of a heterologous CYP from an organism phylogenetically closer to bacteria (e.g., Candida tropicalis; [20]), or modifications of the anchor mimicking that of a well-expressing P450 [8]. Finally, with these changes, a significant amount of oxidation products could be detected (553 mg L^{-1}); however, only a fraction of these was artemisinic acid.

A different approach for the synthesis of artemisinic acid was followed by Dietrich et al. [30] with the hope of overcoming the low yields and the formation of by-products observed by Chang and coworkers [20]. Instead of the *A. annua* enzyme, a soluble substrate-promiscuous $P450_{BM3}$ from *Bacillus megaterium* (CYP102A1) was used to generate a new route towards artemisinic acid synthesis [30]. The wild-type $P450_{BM3}$ enzyme catalyses the hydroxylation of long-chain saturated fatty acids and has a very high turnover rate compared to other known P450s [90]. The wild-type enzyme was not active against amorphadiene and required several modifications for the enzyme to accept amorphadiene as a substrate. With the help of computer modelling, specific mutations were designed to increase the size of the active site binding pocket, to improve the enzyme specificity and for increasing the product titres (saturation mutagenesis). This semibiosynthetic strategy resulted in 250 mg L^{-1} of artemisinic epoxide which can be converted through dihydroartemisinic acid to artemisinin [30]. The titres achieved by this method were higher than the previous ones using the native enzyme (CYP71AV1) expressed in the same host. This was the first time that the properties of a $P450_{BM3}$ were modified by directed mutagenesis and it seems that this strategy holds promise for different substrates and systems.

An important advantage of $P450_{BM3}$ is that it is composed of a fusion between a FAD–FMN-containing NADPH oxidoreductase and a functional P450, forming an entire monooxygenase system in one soluble 119-kDa polypeptide [81]. Indeed, the enzyme exhibited the highest catalytic activity ever detected in a P450, likely due to more efficient electron transfer [130]. Additionally, $P450_{BM3}$ has proven amenable to targeted mutagenesis that can result in conformational changes and substrate specificity enabling novel P450 reactions for a diverse range of pathways and compounds [15, 22]. This case highlights the possibility of using existing enzymes for the catalysis of reactions leading to commercially interesting products whose pathways are not yet discovered or for the synthesis of new-to-nature products.

A further example where a bacterial enzyme has been used efficiently for the hydroxylation of a plant natural compound is in the pathway towards perillyl alcohol, an anticancer agent generated by the hydroxylation of limonene, a monoterpenoid found in several plants. Alonso-Gutierrez and coworkers [2] produced approximately 100 mg L^{-1} perillyl alcohol from an *E. coli* strain engineered to produce high levels of the precursor limonene via a heterologous mevalonate pathway and the corresponding terpene synthase [2]. The hydroxylation of limonene was achieved by a well-characterised P450 isolated from Mycobacterium HXN-1500, coupled with a ferredoxin, and a ferredoxin reductase [122]. The specific P450 was identified after the screening of 1,800 bacterial strains for their ability to hydroxylate L-limonene at the 7 position [122].

Inspired by progress with the $P450_{BM3}$ enzyme, several artificial POR–P450 fusions are known today. One of these examples regards one of the most noteworthy success stories of the terpenoid world, Taxol. Taxol, a potent anticancer agent [34], is naturally found in the cork of the Pacific yew tree in very small amounts (more than 500,000 mature Pacific yew trees were felled for extraction of 130 kg of Taxol, required for preclinical studies; [69]), and today it is produced from plant cell cultures or by chemically converting deacetylbaccatin III (isolated from *Taxus baccata* needles) into Taxol. Although production today is more stable and efficient, the price of the drug still remains high [69]. Thus, several efforts have been implemented to approach biotechnological solutions for the synthesis of this drug. It is predicted that Taxol biosynthesis requires 19 distinct enzymatic steps from GGPP [24], with approximately half of these reactions being catalysed by P450s [55]. A significant number of the enzymes participating in the pathway have been characterised, however, there still remain several enzymatic steps missing including some potentially catalysed by P450s. Although the majority of known P450s involved in Taxol biosynthesis have been expressed in yeast or insect cells for functional characterisation [49–51, 108], there are relatively few reports regarding expression of Taxol-related P450s in microorganisms for production purposes [53, 68].

The work of Ajikumar and coworkers [1] describes the optimisation of the MEP pathway in *E. coli* for the synthesis of high titres of taxadiene (1 g L^{-1} was achieved in fed-batch bioreactor fermentation) and the expression of the first P450 of the pathway, taxadien-5a-hydroxylase, which oxidises taxadiene to taxadien-5a-ol [51]. A codon-optimised taxadien-5a-hydroxylase was expressed, after N-terminal transmembrane domain engineering, as a fusion with the *Taxus* POR enzyme [1]. A number of fused enzymes (taxadien-5a-hydroxylase with *Taxus* POR) had been generated using different versions of the P450 transmembrane domain. One of the generated chimeric enzymes was able to convert 98 % of taxadiene to taxadien-5a-ol at a yield of 58 mg L^{-1} and equal amounts of the by-product 5(12)-Oxa-3(11)-cyclotaxane. Although taxadien-5a-hydroxylase was suggested to be a bottleneck of the pathway

[28], the applied strategy made it possible to express the recombinant enzyme in *E. coli* functionally and to produce relatively high titres of its product. Despite the complications encountered regarding P450 expression, the observed titres were approximately 2,400-fold higher than the ones observed previously from yeast [28]; for further details refer to the relevant chapter in this volume). As all Taxol-related P450s have high sequence similarity, a refined approach inspired by this progress could potentially lead to functional expression of the next biosynthetic steps in *E. coli* [55].

CYP97 is a family of plant P450s that has been successfully expressed in microbial hosts. Enzymes from this family are involved in the hydroxylation of carotenoids for the synthesis of lutein (from α-carotene) or zeaxanthin (from β-carotene; [99, 119]), and are some of the few P450s natively localised to the plastids. In Arabidopsis, eleven P450s have been predicted/identified as plastidial targeted including CYP97A3 (carotene β-ring hydroxylase), CYP97B3 (carotene β-ring hydroxylase), and CYP97C1 (carotene ε-ring hydroxylase; [57, 99, 110]). Homologous enzymes have also been found in rice, soy bean, and pea [119]. As lutein is an important ingredient (as antioxidant and natural colorant) in a number of pharmaceuticals, food supplements, or cosmetics, efforts have been made to produce it biotechnologically.

The group of Eleanore Wurtzel (City University of New York, USA) succeeded in producing 500 μg of lutein per gram of pelleted cells by overexpression of the CYP97A4 and CYP97C2 enzymes from rice in *E. coli* cells harbouring the pathways for α- and β-carotene [127]. In plant plastids, the nonheme, diiron enzyme, HYD4, performs exactly the same carotene β-ring hydroxylation reaction as the CYP97A. To test the redundancy of these two enzymes, attempted replacement of CYP97A with the HYD4 in planta, but also in the bacterial cells, failed in lutein production, which was only possible by the coexpression of CYP97A and CYP97C. Thus, it was assumed that for the efficient flux of the pathway and the synthesis of lutein, the observed protein–protein interaction between the CYP97 enzymes may be critical, whereas HYD4 did not seem to interact with the CYP97s [99, 127]. It is worth noting that for the plastidial CYP97 subfamily modifications were not necessary to facilitate expression in *E. coli*. This can probably be explained by the similarity of the plant cell plastids with the prokaryotic progenitors. CYP97s seem to be weakly associated with the envelope membrane of the plastids and they are not predicted to have any transmembrane domain [99].

4.2 Yeast Systems

Yeast has been used successfully for the expression of complex plant pathways including the P450s, in part due to the genomic tools available for this organism. Yeast has provided an alternative host for the functional characterisation of plant P450s recalcitrant to expression in bacterial systems [36, 129].

One of the most pronounced achievements of synthetic biology in high-value terpenoids in yeast is the synthesis of high titres of artemisinic acid with Paddon and coworkers achieving titres of 25 g L^{-1} [94]. To make this possible, multiple optimisation steps were required for efficient activity of the CYP71AV1 including the balancing of expression levels of CYP71AV1 in relation to the *A. annua* POR, and coexpression of an additional reductase, *A. annua* cytochrome *b*5. These changes increased the viability of the yeast cells, but the important developments that led to the high titres mentioned above were the discovery of the entire biosynthetic pathway of artemisinin including an *A. annua* alcohol dehydrogenase and artemisinic aldehyde dehydrogenase. The expression of these enzymes resulted in increased production of artemisinic acid from 2.4 to 8.1 g L^{-1}. Further improvements in fermentation tools and procedures then led to the final high titres [94]. These results show not only the remarkable potentials of synthetic biology and metabolic engineering for the synthesis of valuable compounds from microorganisms, but also provide a future strategy that can be applied to the production of other valuable terpenoids. A detailed description of this work can be found in the relevant chapter of this volume.

An interesting example demonstrating the potentials and also pitfalls of yeast bioengineering is the work regarding (+)-nootkatone synthesis. The development of several production platforms has been attempted using a range of hosts from bacteria to plants, yet to date the achieved titres of nootkatone are not sufficient for commercial applications [125]. Nootkatone can be produced from (+)-valencene, abundant in nature, through an enzymatic reaction assigned to a P450 enzyme. Due to the lack of a native plant enzyme at that time, several bacterial P450s were tested instead without success [37]. CYP71AV8 from chicory as well as the related tobacco CYP71D51v2 were expressed in yeast together with the Arabidopsis POR1 [17, 37]. In vitro results showed that both enzymes were able to catalyse the conversion of valencene to nootkatol efficiently but only to a minor degree to nootkatone. Although very small amounts of nootkatone were produced from yeast microsomes expressing CYP71D51v2, in vivo bioconversion (using the same yeast strain) of valencene and β-nootkatol to nootkatone proved to be much more efficient, yielding up to 3 mg L^{-1} of nootkatone. However, this production was found subsequently to be independent of CYP71D51v2, and likely catalysed by a yeast endogenous enzyme. Although increasing the substrate concentration led to the decrease of the final product due to toxicity issues, 6 mg L^{-1} of β-nootkatol and nootkatone could finally be produced from 200 mg L^{-1} of valencene [37].

A P450 from the Solanaceae family resulted in more promising results. Henbane (*Hyoscyamus muticus*) CYP71D55 is an enzyme previously shown to oxidise various decalin-ring sesquiterpenoids, including valencene [117]. The methylotrophic *Pichia pastoris* was used as host and has been shown to be tolerant for heterologous expression of specific P450s [125]. A variant of CYP71D55, mutated in two amino acids, V482I and A484I, improved the catalytic efficiency of the enzyme for nootkatol biosynthesis up to fivefold, but nevertheless failed to convert it to nootkatone [117]. Despite this observed lack of function of CYP71D55, the *P. pastoris* strain expressing the valencene synthase and CYP71D55/POR accumulated low amounts of nootkatone. It was shown that an endogenous *P. pastoris* alcohol dehydrogenase was responsible for this conversion. Overexpression of this enzyme, together with the remaining enzymes of the pathway (valencene synthase, CYP71D55, POR) and a truncated version of yeast HMG1 resulted in the production of 208 mg L^{-1} of nootkatone in fed-batch cultures. This production titre reaches the range of commercially sustainable production in grams per litre, and awaits further strain improvements [125].

With identification of CYP706M1, the native P450 from cypress catalysing formation of nootkatone from valencene, a turning point was reached [18]. In contrast to previous assays, expression of CYP706M1 in yeast cells resulted in nearly quantitative oxidation of valencene to nootkatone. Microsomal assays of yeast cells expressing CYP706M1 resulted mainly in the synthesis of nootkatone (82 % of total products) whereas only minor amounts of *trans*-nootkatol were observed. This indicates that CYP706M1 efficiently performs a two-step oxidation. Despite the high specificity of CYP706M1, engineering of a yeast strain coexpressing the Arabidopsis POR1 and the cypress valencene synthase resulted in 144 μg L^{-1} of nootkatone, whereas the control yeast strain lacking the P450 accumulated almost tenfold higher levels of valencene. This indicates that a bottleneck of the pathway is likely at the level of CYP706M1, making it a prime target for optimisation by enzyme engineering [18].

Triterpenoid saponins are a group of specialised metabolites found in many plant species and they are known for their diverse biological activities [4]. Because of their potential use as pharmaceutical compounds, large-scale biotechnological synthesis of these metabolites is desirable. Fukushima and coworkers showed that CYP716A12 and CYP93E2 are enzymes involved in the first step of the pathway to bioactive saponins, functionalising β-amyrin towards hemolytic and nonhemolytic sapogenins in the model legume plant *M. truncatula* [35]. Subsequently, the next step of oxidations catalysed by CYP72A68v2 and CYP72A61v2 was defined. In yeast strains coexpressing the *M. truncatula* P450s pairs CYP716A12/CYP72A68v2 and CYP93E2/CYP72A61v2 with the *Lotus japonicus* β-amyrin synthase and POR, gypsogenic acid and soyasapogenol B, respectively, were detected. Further, coexpression of novel combinations of P450s resulted in an array of novel terpenoids not yet detected in *M. truncatula*, including 4-epi-hederagenin, queretaroic acid, oleanolic acid, and β-amyrin derivatives oxidised at different carbon molecules [36]. Thus, it seems possible to combine P450 enzymes in yeast that do not co-occur in their native host for the synthesis of novel compounds with possible original biological functions.

One of the few examples of coexpression of multiple P450s in yeast aims at the synthesis of ginsenoside aglycons by Dai and coworkers [26]. Ginsenosides are the main bioactive phytochemicals of ginseng belonging to the triterpenoids class. The three basic aglycons of ginsenosides are protopanaxadiol, protopanaxatriol, and oleanolic acid. They are synthesised through cyclisation of 2,3-oxidosqualene by β-amyrin and dammarenediol-II synthase. The conversion of β-amyrin to oleanolic acid and of dammarenediol-II to protopanaxadiol and consequently to protopanaxatriol is catalysed by a number of previously identified P450 enzymes [35, 45]. The enzymes responsible for the synthesis of oleanolic acid, protopanaxadiol and protopanaxatriol were coexpressed in yeast by genomic integration together with enzymes from early steps of the pathway (a truncated HMG1, squalene synthase, and squalene epoxidase from yeast). The origin of the oleanolic acid synthase (CYP716A12) was *M. truncatula* and the protopanaxadiol (CYP716A47) and protopanaxatriol (CYP716A53v2) synthases were derived from *P. ginseng*. After coupling with *Arabidopsis* POR1 the P450s were functional in yeast without further modification yielding 17.2 mg L^{-1} protopanaxadiol, 15.9 mg L^{-1} protopanaxatriol, and 21.4 mg L^{-1} oleanolic acid in nonoptimised flask cultures [26]. These results may encourage further attempts at coexpression and assembly of pathways utilising P450s from multiple species.

4.3 Plant Systems: Nicotiana Benthamiana/Agrobacterium

The use of plant hosts for the biotechnological production of terpenoids appears to be an obvious approach because the isoprene precursors, cofactors, and critical enzymes, such as P450-compatible PORs required for terpenoid biosynthesis, are endogenously present. Furthermore, the protein translation machinery, posttranslational modifications, cellular compartments, and the use of transit peptides for protein targeting, are conserved among plants, eliminating some of the problems encountered when attempting to express plant genes in microbial hosts. Yet, plant systems remain largely underexplored. Possible reasons for this are that plants present some specific complications as hosts, notably the long turnover times for generation of transgenic lines expressing new proteins, the risk for transgene-silencing, highly differentiated cell types, and endogenous terpenoid biochemistry. *N. benthamiana* is one of the few examples of a plant expression system successfully used at an industrial scale (although its major use is still the production of pharmaceutically active peptides). Prominent examples of molecules made using the transient *N. benthamiana/Agrobacterium* expression system are vaccines for the treatment of malaria, influenza, and Ebola [32, 63, 92]. An advantage of this system is the short time between infiltration and harvest of typically four to six days, as well as its industrial scalability [41], and distinct and controllable glycosylation machinery [19, 75]. Concerning terpenoid production, *N. benthamiana* is well established and the robustness of the system has facilitated both high-throughput testing of enzymes and accumulation of substantial amounts of terpenes [5, 13].

The sesquiterpene lactone costunolide is of industrial interest due to inherent biological activities and as starting material for a broad range of high-value compounds such as the pharmaceutical parthenolide found in the medicinal plant feverfew (Tanacetum parthenium; [71]). The pathway to the precursor molecule, costunolide, is known and involves the formation of germacrene A by germacrene A synthase, oxidation at C-12 to germacrene A acid by CYP71AV (germacrene A oxidase; [70]), and then hydroxylation at C-6 by CYP71BL subfamily members. The product of this final reaction autocatalytically rearranges to costunolide [27, 47, 70]. With this pathway elucidated, transient expression in *N. benthamiana* as a potential production host was approached [70]. To optimise production of the first intermediate, several germacrene A synthases were evaluated and the most active enzyme from feverfew was subsequently targeted to the mitochondria using the targeting sequence of cytochrome oxidase subunit IV (CoxIV) leading to substantially improved production of germacrene A. Coexpression of both P450s, and the mitochondrially targeted germacrene A synthase, all driven by the strong constitutive Rubisco-promoter, resulted in production of up to 60 ng g^{-1} FW of costunolide. Nontargeted metabolomics showed the accumulation of considerable amounts of glutathione and cysteine conjugates of costunolide indicating that potential pathways for detoxification are active in planta. Although conjugation could present a potential problem in synthetic biology, it was on the other hand suggested as an opportunity to improve production and storage of terpenoids due to increased water solubility [70]. In plants, conjugation has the dual effect of masking reactive functional groups and increasing solubility, possibly targeting the molecules for further metabolisation and/or sequestration in the vacuole.

With formation of costunolide established, Liu and coworkers approached the final step, oxidation of costunolide to parthenolide. Coexpression of the established pathway together with feverfew CYP71CA1 and a soluble *Arabidopsis* HMG-CoA reductase, to increase substrate availability, yielded 2.05 ng g^{-1} FW of free parthenolide. Again, both parthenolide–cysteine and parthenolide–glutathione were found to constitute more than 90 % of the total amount of sesquiterpenoid produced [71]. A similar observation of effective conjugation of produced terpenoids was made when the model pathway to artemisinic acid was introduced into *N. benthamiana*. Here, predominant accumulation of approximately 40 ng g^{-1} FW artemisinic acid-12-β-diglycoside accompanied by absence of the nonconjugated terpenoid indicated efficient conversion by the endogenous conjugation machinery [123]. When reconstructing parts of the artemisinin pathway, LC-MS analysis of extracts revealed the presence of at least 19 conjugated forms of these functionalised sesquiterpenoids. Compounds displayed varying levels of glycosylation as well as conjugation with malonate and glutathione [120]. For construction of multistep pathways, it was suggested that the problem of conjugation could be alleviated through efficient and balanced expression of pathway enzymes to facilitate fast channelling of activated metabolic intermediates to downstream products [100].

Clearly, the conjugation of mono- and sesquiterpenoids oxidatively functionalised with hydroxyl or carboxyl groups by plant hosts needs to be considered in biotechnological production. However, in the few recent examples of *N. benthamiana*-based functional testing of P450 containing pathways to di- and triterpenoids, such as functional expression of the *Lotus japonicus* CYP71D353 which catalyses the formation of 20-hydroxybetulinic acid in a sequential three-step oxidation of 20-hydroxylupeol [60], no evidence was found indicating that this phenomenon may not play a significant role for larger and nonvolatile terpenoids.

Even though *N. benthamiana* is now routinely used for functional analysis of terpenoid biosynthetic pathways, including P450s, these few examples above indicate a promising future potential for targeted production of terpenoids. One limitation, though, which will need to be addressed in the future, is the production capacity of plants, with solutions possibly in storage of produced terpenoids in specialised structures such as glandular trichomes and regulation of the pathways supplying the universal C5 building blocks.

4.4 Plant Systems: Physcomitrella patens

The moss *Physcomitrella patens* represents a plant system uniquely suited to high-value terpenoid production due to a natural tolerance to accumulation of terpenoids, a simple endogenous terpenoid profile, and established genetic tools for genome editing by homologous recombination [6, 133]. In contrast to higher plant species, *P. patens* lacks gibberellin diterpene phytohormones but accumulates *ent*-kaurene intermediates of the gibberellin pathway. To date, there are at least two examples reported of the successful biosynthesis of nonnative terpenes by this moss. Constitutive expression of the taxadiene diterpene synthase from Pacific yew (*Taxus brevifolia*) resulted in the accumulation of up to 0.05 % tissue fresh weight of taxa-4(5),11(12)-diene, without significantly changing the accumulation or pattern of endogenous diterpenoids [3]. The *P. patens* genome encodes only a single diterpene synthase (copalyl-diphosphate/*ent*-kaurene synthase), which can be disrupted without phenotypic consequences in the relevant life stages. This was exploited for stable introduction of the single-step terpenoid biosynthetic pathways to the sesquiterpenoids patchoulol and santalene [133]. Although there are currently no published examples of the heterologous expression of functionally characterised P450s in *P. patens*, it is expected that this plant host will provide the same benefits as *N. benthamiana* such as the presence of the required subcellular compartments, partner enzymes, and cofactors, but with the added benefits of homologous recombination, short regeneration times, and scalable production in bioreactors.

5 Outlook and Perspectives

Significant technological advances have increased our abilities to discover, characterise, and manipulate P450s and this has led to breakthroughs in the biotechnological production of complex terpenoids (e.g., the production of artemisinic acid in yeast [94]). It is clear that major challenges remain.

The ability to combine enzymes from different terpenoid pathways and, to a degree, design new P450s [22] opens up the possibility of generating new to nature compounds with novel or specific bioactivities. Several groups have already demonstrated that this approach is feasible with novel triterpenes being generated from nonnatural combinations of plant-derived triterpene synthases and P450s expressed in yeast [26, 36, 78].

As photosynthesis in the chloroplast provides energy in the form of ATP, reducing equivalents, and a carbon source for isoprenoid C_5 building blocks, photosynthetic production systems may offer unique opportunities for development of new efficient P450-based pathways. Recent work has shown that it is possible to relocate P450s to the thylakoid membrane of the chloroplast and divert electrons directly from active photosynthesis to drive P450-mediated reactions, without the requirement of POR enzymes [64, 89]. Light-driven activity of P450s may therefore relieve current limitations of biotechnological applications, in particular the requirement of the membrane-anchored POR and stoichiometric amounts of NADPH. Consequently, this technology has a high potential for engineering of pathways already native to the plastids (i.e., mono- and diterpenoids), or those engineered for plastid localisation (i.e., sesqui- and triterpenoids; e.g., [105, 126]). Furthermore, recent progress in the development of tools for plastid genome transformation enables high-level expression of heterologous enzymes in chloroplasts due to the lack of transgene silencing and epigenetic mechanisms [10]. However, despite the encouraging successes and the inherent potential, there are currently no reports of engineering plastids or their autonomous equivalents, cyanobacteria such as *Synechocystis* or *Synechococcus* which harbour the pathways for C_5 isoprene building blocks [31], with P450s involved in terpenoid biosynthesis.

As it has become increasingly feasible to identify pathways of specialised metabolism, research focus is shifting towards understanding how the enzymes of these pathways interact and how these interactions affect pathway activity. In the case of terpenoid metabolism, this question relates to how the various involved P450s, PORs, and other ER or cytosolic enzymes, such as glycosyl-transferases and acyl-transferases, interact. There is growing evidence that the enzymes involved in a given pathway form dynamic multienzyme complexes (termed 'metabolons'; [54]). Metabolon formation may enable channelling of substrates, thought to result in more efficient and directed pathways, as well as provide an important means to regulate pathway activity [65]. In the case of plant-specialised metabolism, metabolon formation could be an important mechanism to reduce exposure of potentially toxic or labile intermediates to the cellular environment [65]. Currently our understanding of the factors enabling and controlling the formation and dissociation of metabolons is quite limited, but knowledge in this field is expected to expand with available new technologies. If metabolon formation does play an important role in pathway channelling, understanding how to control this phenomenon could provide significant efficiency gains for biotechnological production of terpenoids.

Acknowledgments Financial support was provided by the UNIK [Universitetsforskningens Investeringskapital (Investment Capital for University Research)] Center for Synthetic Biology "bioSYNergy" at the University of Copenhagen, funded by the Research Initiative of the Danish Ministry of Science, Technology and Innovation (BH), and the Novo Nordisk Foundation (BH). Additional support was provided by a personal postdoctoral stipend awarded by the FP7 PEOPLE MARIE CURIE ACTIONS Intra-European Fellowships (IP).

A.1 6 Appendix

See Table 2.

Table 2 P450s involved in terpenoid biosynthesis identified during 2013–2014

Clan	Subfamily	Metabolism	Class of metabolites	Function	Plant species	References
CYP51	CYP51H	Specialised	Triterpenoid	Triterpene avenacin biosynthesis (epoxydase); As CYP51H10 is a multifunctional P450 capable of modifying both the C and D rings of the pentacyclic triterpene scaffold to give 12,13β-epoxy-3β,16β-dihydroxy-oleanane (12,13β-epoxy-16β-hydroxy-β-amyrin)	Monocotelydons; oat (*Avena* spp.)	Geisler et al. [38]
CYP71	CYP71CA1	Specialised	Sesquiterpenoid	Epoxidation of costunolide to parthenolide	Feverfew (*Tanacetum parthenium*)	Liu et al. [71]
	CYP71D	Specialised	Mixed, including: sesquiterpenoid, diterpenoid, indole alkaloid, flavonoid	5-epi-aristolochene and 1-deoxycapsidiol to capsidiol in vitro, cembratrienol hydroxylase, limonene 3- and 6-hydroxylase, menthol hydroxylase, thymol and carvacrol, two-step production of (+)-trans-piperitenol, successive reactions at carbon 2 (C-2) of premnaspirodiene primarily to nootkatol in vitro, cembratriene-ol do CBT-diol, flavonoid hydroxylase, tabersonine hydroxylase	Tobacco (*Nicotiana tabacum*), mint, *Catharantus roseus*, *Hyoscyamus muticus*, *Ocimum* spp.	Gavira et al. [37]
	CYP76AH	Specialised	Diterpenoid	CYP76AH hydroxylate miltiradiene/abietatriene	Lamiaceae	Guo et al. [40], Zi et al. [134]
	CYP76F	Specialised	Sesquiterpenoid	A series of nine P450 was functionally characterized using in vitro and yeast in vivo assays and shown to encode santalene/bergamotene oxidases and bergamotene oxidases	Sandalwoold	Diaz-Chavez et al. [29]
	CYP93E	Specialised	Triterpenoid	C-24 hydroxylation of β-amyrin: hydroxylates triterpenes (β-amyrin or sophoradiol) to olean-12-ene-3β,24-diol and soyasapogenol B in both soybean and licorice (G. uralensis);	soybean (*Glycine max* L.); licorice (Glycyrrhiza); Medicago truncatula;	Fukushima et al. [36]
	CYP706M	Specialised	Sesquiterpenoid	Valencene oxidase producing trans-nootkatol and (+)-nootkatone in yeast	Alaska cedar (*Callitropsis nootkatensis*)	Cankar et al. [18]
	CYP726A	Specialised	Diterpenoid	Oxidation of macrocyclic diterpenes casbene, cembrene	Euphorbiaceae	King et al. [58]

(continued)

Table 2 (continued)

Clan	Subfamily	Metabolism	Class of metabolites	Function	Plant species	References
CYP72	CYP714A	General	Diterpenoid phytohormone GA	Inactivation of early GA intermediates; CYP714A1 catalyzes conversion of GA12 to 16-carboxylated GA12 (16-carboxy-16β,17-dihydro GA12), with GA12 12α-hydroxy GA12 (GA111) was produced; CYP714A2 catalyzed conversion of *ent*-kaurenoic acid into steviol (*ent*-13-hydroxy kaurenoic acid)	Arabidopsis	Nomura et al. [91]
	CYP714B	General	Diterpenoid phytohormone GA	Gibberellin 13-oxidases that reduce gibberellin activity; Transgenic Arabidopsis plants that overexpress CYP714B1 or CYP714B2 show semidwarfism	Rice (*Oryza sativa*)	Magome et al. [74]
	CYP72A	Specialised	Monoterpene indole alkaloid, triterpene saponins	Secologanin synthesis, catalyzing the conversion of loganin to secologanin and from 7-dexyloganic acid to loganic acid; ring-opening of cyclopentane ring through C–C cleavage; CYP72A154 catalyzed three sequential oxidation steps at C-30 of 11-oxo-β-amyrin to glycyrrhetinic acid; CYP72A63 catalyzes C-30 oxidation of β-amyrin	*Catharanthus roseus*, licorice (*Glycyrrhiza*), *Medicago truncatula*	Fukushima et al. [36], Salim et al. [104]
CYP85	CYP716A	Specialised	Triterpenoid	Multifunctional with β-amyrin 28-oxidase, α-amyrin 28-oxidase, and lupeol 28-oxidase activities; oxidation of β-amyrin and erythrodiol at C-28 position, yielding oleanolic acid. CYP716A47 and CYP716A53v2 catalyze the formation of dammarene-type triterpenesd in ginseng (CYP716A47 catalyzes formation of protopanaxadiol from dammarenediol-II and CYP716A53v2 catalyzes formation of protopanaxatriol from protopanaxadiol)	*Medicago truncatula*, grape, ginseng (Panax ginseng), genus Bupleurum	Fukushima et al. [36], Moses et al. [78]
CYP97	CYP97C	Specialised	Acyclic diterpenoid	Biosynthesis of plaunotol (18-OH geranylgeraniol)	*Croton stellatopilosus*	Sintupachee et al. [115]

See Hamberger and Bak [42] for a comprehensive list

References

1. Ajikumar PK, Xiao W-H, Tyo KEJ, Wang Y, Simeon F, Leonard E, Mucha O, Phon TH, Pfeifer B, Stephanopoulos G (2010) Isoprenoid Pathway Optimization for Taxol Precursor Overproduction in *Escherichia coli*. Science 330(6000):70–74
2. Alonso-Gutierrez J, Chan R, Batth TS, Adams PD, Keasling JD, Petzold CJ, Lee TS (2013) Metabolic engineering of *Escherichia coli* for limonene and perillyl alcohol production. Metab Eng 19:33–41
3. Anterola A, Shanle E, Perroud PF, Quatrano R (2009) Production of taxa-4(5),11(12)-diene by transgenic *Physcomitrella patens*. Transgenic Res 18(4):655–660
4. Augustin JM, Kuzina V, Andersen SB, Bak S (2011) Molecular activities, biosynthesis and evolution of triterpenoid saponins. Phytochemistry 72(6):435–457
5. Bach SS, Bassard JE, Andersen-Ranberg J, Moldrup ME, Simonsen HT, Hamberger B (2014) High-throughput testing of terpenoid biosynthesis candidate genes using transient expression in *Nicotiana benthamiana*. Methods Mol Biol 1153:245–255
6. Bach SS, King BC, Zhan X, Simonsen HT, Hamberger B (2014) Heterologous stable expression of terpenoid biosynthetic genes using the moss *Physcomitrella patens*. Methods Mol Biol 1153:257–271
7. Bak S, Kahn RA, Nielsen HL, Moller BL, Halkier BA (1998) Cloning of three A-type cytochromes P450, CYP71E1, CYP98, and CYP99 from *Sorghum bicolor* (L.) Moench by a PCR approach and identification by expression in *Escherichia coli* of CYP71E1 as a multifunctional cytochrome P450 in the biosynthesis of the cyanogenic glucoside dhurrin. Plant Mol Biol 36(3):393–405
8. Barnes HJ, Arlotto MP, Waterman MR (1991) Expression and enzymatic activity of recombinant cytochrome P450 17 alpha-hydroxylase in *Escherichia coli*. Proc Natl Acad Sci 88(13):5597–5601
9. Bertea CM, Freije JR, van der Woude H, Verstappen FW, Perk L, Marquez V, De Kraker JW, Posthumus MA, Jansen BJ, de Groot A, Franssen MC, Bouwmeester HJ (2005) Identification of intermediates and enzymes involved in the early steps of artemisinin biosynthesis in *Artemisia annua*. Planta Med 71(1):40–47
10. Bock R (2014) Genetic engineering of the chloroplast: novel tools and new applications. Curr Opin Biotechnol 26:7–13
11. Bohlmann J, Keeling CI (2008) Terpenoid biomaterials. Plant J Cell Mol Biol 54(4):656–669
12. Boutanaev AM, Moses T, Zi J, Nelson DR, Mugford ST, Peters RJ, Osbourn A (2014) Investigation of terpene diversification across multiple sequenced plant genomes. Proc Natl Acad Sci
13. Bruckner K, Tissier A (2013) High-level diterpene production by transient expression in *Nicotiana benthamiana*. Plant Methods 9(1):46
14. Butler MS (2008) Natural products to drugs: natural product-derived compounds in clinical trials. Nat Prod Rep 25(3):475–516
15. Butler CF, Peet C, McLean KJ, Baynham MT, Blankley RT, Fisher K, Rigby SE, Leys D, Voice MW, Munro AW (2014) Human P450-like oxidation of diverse proton pump inhibitor drugs by 'gatekeeper' mutants of flavocytochrome P450 BM3. Biochem J 460(2):247–259. LID—210.1042/BJ20140030 [doi]
16. Caniard A, Zerbe P, Legrand S, Cohade A, Valot N, Magnard JL, Bohlmann J, Legendre L (2012) Discovery and functional characterization of two diterpene synthases for sclareol biosynthesis in *Salvia sclarea* (L.) and their relevance for perfume manufacture. BMC Plant Biol 12:119
17. Cankar K, van Houwelingen A, Bosch D, Sonke T, Bouwmeester H, Beekwilder J (2011) A chicory cytochrome P450 mono-oxygenase CYP71AV8 for the oxidation of (+)-valencene. FEBS Lett 585(1):178–182

18. Cankar K, van Houwelingen A, Goedbloed M, Renirie R, de Jong RM, Bouwmeester H, Bosch D, Sonke T, Beekwilder J (2014) Valencene oxidase CYP706M1 from Alaska cedar (*Callitropsis nootkatensis*). FEBS Lett 588(6):1001–1007
19. Castilho A, Bohorova N, Grass J, Bohorov O, Zeitlin L, Whaley K, Altmann F, Steinkellner H (2011) Rapid high yield production of different glycoforms of Ebola virus monoclonal antibody. PLoS ONE 6(10):e26040
20. Chang MCY, Eachus RA, Trieu W, Ro D-K, Keasling JD (2007) Engineering *Escherichia coli* for production of functionalized terpenoids using plant P450s. Nat Chem Biol 3(5):274–277
21. Chen F, Tholl D, Bohlmann J, Pichersky E (2011) The family of terpene synthases in plants: a mid-size family of genes for specialized metabolism that is highly diversified throughout the kingdom. Plant J 66(1):212–229
22. Coelho PS, Brustad EM, Kannan A, Arnold FH (2013) Olefin cyclopropanation via carbene transfer catalyzed by engineered cytochrome P450 enzymes. Science 339(6117):307–310
23. Cragg GM (1998) Paclitaxel (Taxol): a success story with valuable lessons for natural product drug discovery and development. Med Res Rev 18(5):315–331
24. Croteau R, Ketchum RB, Long R, Kaspera R, Wildung M (2006) Taxol biosynthesis and molecular genetics. Phytochem Rev 5(1):75–97
25. Cyr A, Wilderman PR, Determan M, Peters RJ (2007) A modular approach for facile biosynthesis of labdane-related diterpenes. J Am Chem Soc 129(21):6684–6685
26. Dai Z, Wang B, Liu Y, Shi M, Wang D, Zhang X, Liu T, Huang L, Zhang X (2014) Producing aglycons of ginsenosides in bakers' yeast. Sci Rep 4:3698. doi:10.1038/srep03698
27. de Kraker JW, Franssen MC, Joerink M, de Groot A, Bouwmeester HJ (2002) Biosynthesis of costunolide, dihydrocostunolide, and leucodin. Demonstration of cytochrome p450-catalyzed formation of the lactone ring present in sesquiterpene lactones of chicory. Plant Physiol 129(1):257–268
28. DeJong JM, Liu Y, Bollon AP, Long RM, Jennewein S, Williams D, Croteau RB (2006) Genetic engineering of taxol biosynthetic genes in *Saccharomyces cerevisiae*. Biotechnol Bioeng 93(2):212–224
29. Diaz-Chavez ML, Moniodis J, Madilao LL, Jancsik S, Keeling CI, Barbour EL, Ghisalberti EL, Plummer JA, Jones CG, Bohlmann J (2013) Biosynthesis of sandalwood oil: *Santalum album* CYP76F cytochromes P450 produce santalols and bergamotol. PLoS ONE 8(9): e75053
30. Dietrich JA, Yoshikuni Y, Fisher KJ, Woolard FX, Ockey D, McPhee DJ, Renninger NS, Chang MCY, Baker D, Keasling JD (2009) A novel semi-biosynthetic route for artemisinin production using engineered substrate-promiscuous P450BM3. ACS Chem Biol 4(4):261–267
31. Englund E, Pattanaik B, Ubhayasekera SJ, Stensjo K, Bergquist J, Lindberg P (2014) Production of squalene in *Synechocystis* sp. PCC 6803. PLoS ONE 9(3):e90270
32. Feller T, Thom P, Koch N, Spiegel H, Addai-Mensah O, Fischer R, Reimann A, Pradel G, Fendel R, Schillberg S, Scheuermayer M, Schinkel H (2013) Plant-based production of recombinant *Plasmodium* surface protein pf38 and evaluation of its potential as a vaccine candidate. PLoS ONE 8(11):e79920
33. Feyereisen R (2011) Arthropod CYPomes illustrate the tempo and mode in P450 evolution. Biochim Biophys Acta 1814(1):19–28
34. Frense D (2007) Taxanes: perspectives for biotechnological production. Appl Microbiol Biotechnol 73(6):1233–1240
35. Fukushima EO, Seki H, Ohyama K, Ono E, Umemoto N, Mizutani M, Saito K, Muranaka T (2011) CYP716A subfamily members are multifunctional oxidases in triterpenoid biosynthesis. Plant Cell Physiol 52(12):2050–2061
36. Fukushima EO, Seki H, Sawai S, Suzuki M, Ohyama K, Saito K, Muranaka T (2013) Combinatorial biosynthesis of legume natural and rare triterpenoids in engineered yeast. Plant Cell Physiol 54(5):740–749

37. Gavira C, Höfer R, Lesot A, Lambert F, Zucca J, Werck-Reichhart D (2013) Challenges and pitfalls of P450-dependent (+)-valencene bioconversion by *Saccharomyces cerevisiae*. Metab Eng 18:25–35
38. Geisler K, Hughes RK, Sainsbury F, Lomonossoff GP, Rejzek M, Fairhurst S, Olsen C-E, Motawia MS, Melton RE, Hemmings AM, Bak S, Osbourn A (2013) Biochemical analysis of a multifunctional cytochrome P450 (CYP51) enzyme required for synthesis of antimicrobial triterpenes in plants. Proc Natl Acad Sci 110(35):E3360–E3367
39. Gongora-Castillo E, Childs KL, Fedewa G, Hamilton JP, Liscombe DK, Magallanes-Lundback M, Mandadi KK, Nims E, Runguphan W, Vaillancourt B, Varbanova-Herde M, Dellapenna D, McKnight TD, O'Connor S, Buell CR (2012) Development of transcriptomic resources for interrogating the biosynthesis of monoterpene indole alkaloids in medicinal plant species. PLoS ONE 7(12):e52506
40. Guo J et al (2013) PNAS 110(29):12108–12113
41. Hahn S, Giritch A, Bartels D, Bortesi L, Gleba Y (2014) A novel and fully scalable *Agrobacterium* spray-based process for manufacturing cellulases and other cost-sensitive proteins in plants. Plant Biotechnol J. doi:10.1111/pbi.12299 Early online version [Epub ahead of print]
42. Hamberger B, Bak S (2013) Plant P450s as versatile drivers for evolution of species-specific chemical diversity. Philos Trans R Soc B: Biol Sci 368(1612):20120426. doi:10.1098/rstb.2012.0426
43. Hamberger B, Ohnishi T, Hamberger B, Séguin A, Bohlmann J (2011) Evolution of diterpene metabolism: sitka spruce CYP720B4 catalyzes multiple oxidations in resin acid biosynthesis of conifer defense against insects. Plant Physiol 157(4):1677–1695
44. Han JY, Kim HJ, Kwon YS, Choi YE (2011) The Cyt P450 enzyme CYP716A47 catalyzes the formation of protopanaxadiol from dammarenediol-II during ginsenoside biosynthesis in *Panax ginseng*. Plant Cell Physiol 52(12):2062–2073
45. Han J-Y, Hwang H-S, Choi S-W, Kim H-J, Choi Y-E (2012) Cytochrome P450 CYP716A53v2 catalyzes the formation of protopanaxatriol from protopanaxadiol during ginsenoside biosynthesis in *Panax Ginseng*. Plant Cell Physiol 53(9):1535–1545
46. Hardy K, Buckley S, Collins MJ, Estalrrich A, Brothwell D, Copeland L, Garcia-Tabernero A, Garcia-Vargas S, de la Rasilla M, Lalueza-Fox C, Huguet R, Bastir M, Santamaria D, Madella M, Wilson J, Cortes AF, Rosas A (2012) Neanderthal medics? Evidence for food, cooking, and medicinal plants entrapped in dental calculus. Naturwissenschaften 99(8):617–626
47. Ikezawa N, Göpfert JC, Nguyen DT, Kim S-U, O'Maille PE, Spring O, Ro D-K (2011) Lettuce costunolide synthase (CYP71BL2) and its homolog (CYP71BL1) from sunflower catalyze distinct regio- and stereoselective hydroxylations in sesquiterpene lactone metabolism. J Biol Chem 286(24):21601–21611
48. Jackson CJ, Lamb DC, Marczylo TH, Warrilow AGS, Manning NJ, Lowe DJ, Kelly DE, Kelly SL (2002) A novel sterol 14α-demethylase/ferredoxin fusion protein (MCCYP51FX) from *Methylococcus capsulatus* represents a new class of the cytochrome P450 superfamily. J Biol Chem 277(49):46959–46965
49. Jennewein S, Rithner CD, Williams RM, Croteau RB (2001) Taxol biosynthesis: taxane 13α-hydroxylase is a cytochrome P450-dependent monooxygenase. Proc Natl Acad Sci 98(24):13595–13600
50. Jennewein S, Rithner CD, Williams RM, Croteau R (2003) Taxoid metabolism: taxoid 14β-hydroxylase is a cytochrome P450-dependent monooxygenase. Arch Biochem Biophys 413(2):262–270
51. Jennewein S, Long RM, Williams RM, Croteau R (2004) Cytochrome P450 taxadiene 5α-hydroxylase, a mechanistically unusual monooxygenase catalyzing the first oxygenation step of taxol biosynthesis. Chem Biol 11(3):379–387
52. Jensen K, Møller BL (2010) Plant NADPH-cytochrome P450 oxidoreductases. Phytochemistry 71(2–3):132–141
53. Jiang M, Stephanopoulos G, Pfeifer BA (2012) Toward biosynthetic design and implementation of *Escherichia coli*-derived paclitaxel and other heterologous polyisoprene compounds. Appl Environ Microbiol 78(8):2497–2504

54. Jorgensen K, Rasmussen AV, Morant M, Nielsen AH, Bjarnholt N, Zagrobelny M, Bak S, Moller BL (2005) Metabolon formation and metabolic channeling in the biosynthesis of plant natural products. Curr Opin Plant Biol 8(3):280–291
55. Kaspera R, Croteau R (2006) Cytochrome P450 oxygenases of taxol biosynthesis. Phytochem Rev 5(2–3):433–444
56. Kato-Noguchi H, Peters RJ (2013) The role of momilactones in rice allelopathy. J Chem Ecol 39(2):175–185
57. Kim J-E, Cheng KM, Craft NE, Hamberger B, Douglas CJ (2010) Over-expression of *Arabidopsis thaliana* carotenoid hydroxylases individually and in combination with a β-carotene ketolase provides insight into in vivo functions. Phytochemistry 71(2–3):168–178
58. King AJ, Brown GD, Gilday AD, Larson TR, Graham IA (2014) Production of bioactive diterpenoids in the euphorbiaceae depends on evolutionarily conserved gene clusters. Plant Cell 26(8):3286–3298
59. Kirby J, Keasling JD (2009) Biosynthesis of plant isoprenoids: perspectives for microbial engineering. Annu Rev Plant Biol 60(1):335–355
60. Krokida A, Delis C, Geisler K, Garagounis C, Tsikou D, Pena-Rodriguez LM, Katsarou D, Field B, Osbourn AE, Papadopoulou KK (2013) A metabolic gene cluster in *Lotus japonicus* discloses novel enzyme functions and products in triterpene biosynthesis. New Phytol 200 (3):675–690
61. Kunii M, Kitahama Y, Fukushima EO, Seki H, Muranaka T, Yoshida Y, Aoyama Y (2012) b-Amyrin oxidation by oat CYP51H10 expressed heterologously in yeast cells: the first example of CYP51-dependent metabolism other than the 14-demethylation of sterol precursors. Biol Pharm Bull 35(5):801–804
62. Lamb DC, Waterman MR (2013) Unusual properties of the cytochrome P450 superfamily. Philos Trans R Soc B: Biol Sci 368(1612):20120434. doi:10.1098/rstb.2012.0434
63. Landry N, Pillet S, Favre D, Poulin JF, Trepanier S, Yassine-Diab B, Ward BJ (2014) Influenza virus-like particle vaccines made in *Nicotiana benthamiana* elicit durable, poly-functional and cross-reactive T cell responses to influenza HA antigens. Clin Immunol 154 (2):164–177
64. Lassen LM, Nielsen AZ, Olsen CE, Bialek W, Jensen K, Moller BL, Jensen PE (2014) Anchoring a plant cytochrome P450 via PsaM to the thylakoids in *Synechococcus* sp. PCC 7002: evidence for light-driven biosynthesis. PLoS ONE 9(7):e102184
65. Laursen T, Møller BL, Bassard J-E (2014) Plasticity of specialized metabolism as mediated by dynamic metabolons. Trends Plant Sci. doi:10.1016/j.tplants.2014.11.002 [Epub ahead of print]
66. Leonard E, Koffas MAG (2007) Engineering of artificial plant cytochrome P450 enzymes for synthesis of isoflavones by *Escherichia coli*. Appl Environ Microbiol 73(22):7246–7251
67. Leroi-Gourhan A (1975) The flowers found with Shanidar IV, a neanderthal burial in Iraq. Science 190(4214):562–564
68. Li Y, Pfeifer BA (2014) Heterologous production of plant-derived isoprenoid products in microbes and the application of metabolic engineering and synthetic biology. Curr Opin Plant Biol 19:8–13
69. Liu T, Khosla C (2010) A balancing act for taxol precursor pathways in *E. coli*. Science 330 (6000):44–45
70. Liu Q, Majdi M, Cankar K, Goedbloed M, Charnikhova T, Verstappen FW, de Vos RC, Beekwilder J, van der Krol S, Bouwmeester HJ (2011) Reconstitution of the costunolide biosynthetic pathway in yeast and *Nicotiana benthamiana*. PLoS ONE 6(8):e23255
71. Liu Q, Manzano D, Tanic N, Pesic M, Bankovic J, Pateraki I, Ricard L, Ferrer A, de Vos R, van de Krol S, Bouwmeester H (2014) Elucidation and in planta reconstitution of the parthenolide biosynthetic pathway. Metab Eng 23:145–153
72. Lovering F (2013) Escape from Flatland 2: complexity and promiscuity. Med Chem Comm 4 (3):515–519
73. Lovering F, Bikker J, Humblet C (2009) Escape from Flatland: increasing saturation as an approach to improving clinical success. J Med Chem 52(21):6752–6756
74. Magome H et al (2013) PNAS 110(5):1947–1952

75. Mamedov T, Ghosh A, Jones RM, Mett V, Farrance CE, Musiychuk K, Horsey A, Yusibov V (2012) Production of non-glycosylated recombinant proteins in *Nicotiana benthamiana* plants by co-expressing bacterial PNGase F. Plant Biotechnol J 10(7):773–782
76. Margulies M, Egholm M, Altman WE, Attiya S, Bader JS, Bemben LA, Berka J, Braverman MS, Chen YJ, Chen Z, Dewell SB, Du L, Fierro JM, Gomes XV, Godwin BC, He W, Helgesen S, Ho CH, Irzyk GP, Jando SC, Alenquer ML, Jarvie TP, Jirage KB, Kim JB, Knight JR, Lanza JR, Leamon JH, Lefkowitz SM, Lei M, Li J, Lohman KL, Lu H, Makhijani VB, McDade KE, McKenna MP, Myers EW, Nickerson E, Nobile JR, Plant R, Puc BP, Ronan MT, Roth GT, Sarkis GJ, Simons JF, Simpson JW, Srinivasan M, Tartaro KR, Tomasz A, Vogt KA, Volkmer GA, Wang SH, Wang Y, Weiner MP, Yu P, Begley RF, Rothberg JM (2005) Genome sequencing in microfabricated high-density picolitre reactors. Nature 437(7057):376–380
77. Miettinen K, Dong L, Navrot N, Schneider T, Burlat V, Pollier J, Woittiez L, van der Krol S, Lugan R, Ilc T, Verpoorte R, Oksman-Caldentey KM, Martinoia E, Bouwmeester H, Goossens A, Memelink J, Werck-Reichhart D (2014) The seco-iridoid pathway from *Catharanthus roseus*. Nat Commun 5:3606
78. Moses T, Pollier J, Almagro L, Buyst D, Van Montagu M, Pedreno MA, Martins JC, Thevelein JM, Goossens A (2014) Combinatorial biosynthesis of sapogenins and saponins in Saccharomyces cerevisiae using a C-16alpha hydroxylase from Bupleurum falcatum. Proc Natl Acad Sci U S A 111(4):1634–1639
79. Moses T, Pollier J, Almagro L, Buyst D, Van Montagu M, Pedreño MA, Martins JC, Thevelein JM, Goossens A (2014) Combinatorial biosynthesis of sapogenins and saponins in *Saccharomyces cerevisiae* using a C-16α hydroxylase from *Bupleurum falcatum*. Proc Natl Acad Sci 111(4):1634–1639
80. Moses T, Pollier J, Faizal A, Apers S, Pieters L, Thevelein JM, Geelen D, Goossens A (2014) Unravelling the triterpenoid saponin biosynthesis of the african shrub *Maesa lanceolata*. Mol Plant 8(1):122−135. doi:10.1016/j.molp.2014.11.004
81. Munro AW, Leys DG, McLean KJ, Marshall KR, Ost TWB, Daff S, Miles CS, Chapman SK, Lysek DA, Moser CC, Page CC, Dutton PL (2002) P450 BM3: the very model of a modern flavocytochrome. Trends Biochem Sci 27(5):250–257
82. Naoumkina MA, Modolo LV, Huhman DV, Urbanczyk-Wochniak E, Tang Y, Sumner LW, Dixon RA (2010) Genomic and coexpression analyses predict multiple genes involved in triterpene saponin biosynthesis in *Medicago truncatula*. Plant Cell 22(3):850–866
83. Nelson DR (1998) Metazoan cytochrome P450 evolution. Comp Biochem Physiol C Pharmacol Toxicol Endocrinol 121(1–3):15–22
84. Nelson DR (1999) Cytochrome P450 and the individuality of species. Arch Biochem Biophys 369(1):1–10
85. Nelson D, Werck-Reichhart D (2011) A P450-centric view of plant evolution. Plant J 66 (1):194–211
86. Nelson DR, Koymans L, Kamataki T, Stegeman J, Feyereisen R, Waxman D, Waterman M, Gotoh O, Coon M, Estabrook R, Gunsalus I, Nebert D (1996) P450 superfamily: update on new sequences, gene mapping, accession numbers and nomenclature. Pharmacogenetics 6 (1):1–42
87. Nelson DR, Schuler MA, Paquette SM, Werck-Reichhart D, Bak S (2004) Comparative genomics of rice and arabidopsis. analysis of 727 cytochrome P450 genes and pseudogenes from a monocot and a dicot. Plant Physiol 135(2):756–772
88. Newman DJ, Cragg GM (2012) Natural products as sources of new drugs over the 30 years from 1981 to 2010. J Nat Prod 75(3):311–335
89. Nielsen AZ, Ziersen B, Jensen K, Lassen LM, Olsen CE, Møller BL, Jensen PE (2013) Redirecting photosynthetic reducing power toward bioactive natural product synthesis. ACS Synth Biol 2(6):308–315
90. Noble MA, Miles CS, Chapman SK, Lysek DA, MacKay AC, Reid GA, Hanzlik RP, Munro AW (1999) Roles of key active-site residues in flavocytochrome P450 BM3. Biochem J 339:371–379

91. Nomura T, Magome H, Hanada A, Takeda-Kamiya N, Mander LN, Kamiya Y, Yamaguchi S (2013) Functional analysis of arabidopsis CYP714A1 and CYP714A2 reveals that they are distinct gibberellin modification enzymes. Plant Cell Physiol 54(11):1837–1851
92. Olinger GG Jr, Pettitt J, Kim D, Working C, Bohorov O, Bratcher B, Hiatt E, Hume SD, Johnson AK, Morton J, Pauly M, Whaley KJ, Lear CM, Biggins JE, Scully C, Hensley L, Zeitlin L (2012) Delayed treatment of Ebola virus infection with plant-derived monoclonal antibodies provides protection in rhesus macaques. Proc Natl Acad Sci U S A 109 (44):18030–18035
93. Paddon CJ, Keasling JD (2014) Semi-synthetic artemisinin: a model for the use of synthetic biology in pharmaceutical development. Nat Rev Micro 12(5):355–367
94. Paddon CJ, Westfall PJ, Pitera DJ, Benjamin K, Fisher K, McPhee D, Leavell MD, Tai A, Main A, Eng D, Polichuk DR, Teoh KH, Reed DW, Treynor T, Lenihan J, Jiang H, Fleck M, Bajad S, Dang G, Dengrove D, Diola D, Dorin G, Ellens KW, Fickes S, Galazzo J, Gaucher SP, Geistlinger T, Henry R, Hepp M, Horning T, Iqbal T, Kizer L, Lieu B, Melis D, Moss N, Regentin R, Secrest S, Tsuruta H, Vazquez R, Westblade LF, Xu L, Yu M, Zhang Y, Zhao L, Lievense J, Covello PS, Keasling JD, Reiling KK, Renninger NS, Newman JD (2013) High-level semi-synthetic production of the potent antimalarial artemisinin. Nature 496 (7446):528–532
95. Panella NA, Dolan MC, Karchesy JJ, Xiong Y, Peralta-Cruz J, Khasawneh M, Montenieri JA, Maupin GO (2005) Use of novel compounds for pest control: insecticidal and acaricidal activity of essential oil components from heartwood of Alaska yellow cedar. J Med Entomol 42(3):352–358
96. Papadopoulou K, Melton RE, Leggett M, Daniels MJ, Osbourn AE (1999) Compromised disease resistance in saponin-deficient plants. Proc Natl Acad Sci U S A 96(22): 12923–12928
97. Peters RJ (2006) Uncovering the complex metabolic network underlying diterpenoid phytoalexin biosynthesis in rice and other cereal crop plants. Phytochemistry 67(21): 2307–2317
98. Qi X, Bakht S, Leggett M, Maxwell C, Melton R, Osbourn A (2004) A gene cluster for secondary metabolism in oat: implications for the evolution of metabolic diversity in plants. Proc Natl Acad Sci U S A 101(21):8233–8238
99. Quinlan RF, Shumskaya M, Bradbury LMT, Beltrán J, Ma C, Kennelly EJ, Wurtzel ET (2012) Synergistic interactions between carotene ring hydroxylases drive lutein formation in plant carotenoid biosynthesis. Plant Physiol 160(1):204–214
100. Renault H, Bassard JE, Hamberger B, Werck-Reichhart D (2014) Cytochrome P450-mediated metabolic engineering: current progress and future challenges. Curr Opin Plant Biol 19:27–34
101. Ro D-K, Arimura G-I, Lau SYW, Piers E, Bohlmann J (2005) Loblolly pine abietadienol/abietadienal oxidase PtAO (CYP720B1) is a multifunctional, multisubstrate cytochrome P450 monooxygenase. Proc Natl Acad Sci 102(22):8060–8065
102. Ro D-K, Paradise EM, Ouellet M, Fisher KJ, Newman KL, Ndungu JM, Ho KA, Eachus RA, Ham TS, Kirby J, Chang MCY, Withers ST, Shiba Y, Sarpong R, Keasling JD (2006) Production of the antimalarial drug precursor artemisinic acid in engineered yeast. Nature 440(7086):940–943
103. Saklani A, Kutty SK (2008) Plant-derived compounds in clinical trials. Drug Discov Today 13(3–4):161–171
104. Salim V, Yu F, Altarejos J, De Luca V (2013) Virus-induced gene silencing identifies *Catharanthus roseus* 7-deoxyloganic acid-7-hydroxylase, a step in iridoid and monoterpene indole alkaloid biosynthesis. Plant J 76(5):754–765
105. Saxena B, Subramaniyan M, Malhotra K, Bhavesh NS, Potlakayala SD, Kumar S (2014) Metabolic engineering of chloroplasts for artemisinic acid biosynthesis and impact on plant growth. J Biosci 39(1):33–41

106. Schalk M, Pastore L, Mirata MA, Khim S, Schouwey M, Deguerry F, Pineda V, Rocci L, Daviet L (2012) Toward a biosynthetic route to sclareol and amber odorants. J Am Chem Soc 134(46):18900–18903
107. Schmelz EA, Huffaker A, Sims JW, Christensen SA, Lu X, Okada K, Peters RJ (2014) Biosynthesis, elicitation and roles of monocot terpenoid phytoalexins. Plant J 79(4):659–678
108. Schoendorf A, Rithner CD, Williams RM, Croteau RB (2001) Molecular cloning of a cytochrome P450 taxane 10β-hydroxylase cDNA from taxus and functional expression in yeast. Proc Natl Acad Sci 98(4):1501–1506
109. Schuler MA, Werck-Reichhart D (2003) Functional genomics of P450s. Annu Rev Plant Biol 54(1):629–667
110. Schuler M, Duan H, Bilgin M, Ali S (2006) Arabidopsis cytochrome P450s through the looking glass: a window on plant biochemistry. Phytochem Rev 5(2–3):205–237
111. Seki H, Ohyama K, Sawai S, Mizutani M, Ohnishi T, Sudo H, Akashi T, Aoki T, Saito K, Muranaka T (2008) Licorice beta-amyrin 11-oxidase, a cytochrome P450 with a key role in the biosynthesis of the triterpene sweetener glycyrrhizin. Proc Natl Acad Sci U S A 105 (37):14204–14209
112. Seki H, Sawai S, Ohyama K, Mizutani M, Ohnishi T, Sudo H, Fukushima EO, Akashi T, Aoki T, Saito K, Muranaka T (2011) Triterpene functional genomics in licorice for identification of CYP72A154 involved in the biosynthesis of glycyrrhizin. Plant Cell 23 (11):4112–4123
113. Sezutsu H, Le Goff G, Feyereisen R (2013) Origins of P450 diversity. Philos Trans R Soc B: Biol Sci 368(1612):20120428. doi:10.1098/rstb.2012.0428
114. Sharon-Asa L, Shalit M, Frydman A, Bar E, Holland D, Or E, Lavi U, Lewinsohn E, Eyal Y (2003) Citrus fruit flavor and aroma biosynthesis: isolation, functional characterization, and developmental regulation of Cstps1, a key gene in the production of the sesquiterpene aroma compound valencene. Plant J 36(5):664–674
115. Sintupachee S et al (2014) Plant Sci 229:131–141
116. Sullivan R, Behncke I, Purushotham A (2010) Why do we love medicines so much? An evolutionary perspective on the human love of pills, potions and placebo. EMBO Rep 11 (8):572–578
117. Takahashi S, Yeo Y-S, Zhao Y, O'Maille PE, Greenhagen BT, Noel JP, Coates RM, Chappell J (2007) Functional characterization of premnaspirodiene oxygenase, a cytochrome P450 catalyzing regio- and stereo-specific hydroxylations of diverse sesquiterpene substrates. J Biol Chem 282(43):31744–31754
118. Teoh KH, Polichuk DR, Reed DW, Nowak G, Covello PS (2006) *Artemisia annua* L. (Asteraceae) trichome-specific cDNAs reveal CYP71AV1, a cytochrome P450 with a key role in the biosynthesis of the antimalarial sesquiterpene lactone artemisinin. FEBS Lett 580 (5):1411–1416
119. Tian L, Musetti V, Kim J, Magallanes-Lundback M, DellaPenna D (2004) The arabidopsis LUT1 locus encodes a member of the cytochrome P450 family that is required for carotenoid ε-ring hydroxylation activity. Proc Natl Acad Sci 101(1):402–407
120. Ting HM, Wang B, Ryden AM, Woittiez L, van Herpen T, Verstappen FW, Ruyter-Spira C, Beekwilder J, Bouwmeester HJ, van der Krol A (2013) The metabolite chemotype of *Nicotiana benthamiana* transiently expressing artemisinin biosynthetic pathway genes is a function of CYP71AV1 type and relative gene dosage. New Phytol 199(2):352–366
121. Tsuruta H, Paddon CJ, Eng D, Lenihan JR, Horning T, Anthony LC, Regentin R, Keasling JD, Renninger NS, Newman JD (2009) High-level production of amorpha-4,11-diene, a precursor of the antimalarial agent artemisinin, in *Escherichia coli*. PLoS ONE 4(2):e4489
122. van Beilen JB, Holtackers R, Lüscher D, Bauer U, Witholt B, Duetz WA (2005) Biocatalytic production of perillyl alcohol from limonene by using a novel *Mycobacterium* sp. Cytochrome P450 alkane hydroxylase expressed in *Pseudomonas putida*. Appl Environ Microbiol 71 (4):1737–1744

123. van Herpen TW, Cankar K, Nogueira M, Bosch D, Bouwmeester HJ, Beekwilder J (2010) *Nicotiana benthamiana* as a production platform for artemisinin precursors. PLoS ONE 5 (12):e14222
124. Wang Q, Hillwig ML, Wu Y, Peters RJ (2012) CYP701A8: a rice ent-kaurene oxidase paralog diverted to more specialized diterpenoid metabolism. Plant Physiol 158(3): 1418–1425
125. Wriessnegger T, Augustin P, Engleder M, Leitner E, Müller M, Kaluzna I, Schürmann M, Mink D, Zellnig G, Schwab H, Pichler H (2014) Production of the sesquiterpenoid (+)-nootkatone by metabolic engineering of *Pichia pastoris*. Metab Eng 24:18–29
126. Wu S, Jiang Z, Kempinski C, Eric Nybo S, Husodo S, Williams R, Chappell J (2012) Engineering triterpene metabolism in tobacco. Planta 236(3):867–877
127. Wurtzel ET, Quinlan R (2013) Cells and methods for producing lutein. WO2013119552 A2, PCT/US2013/024746. Aug 15, 2013
128. Yokoshima S, Tokuyama H, Fukuyama T (2010) Total synthesis of (+)-vinblastine: control of the stereochemistry at C18′. Chem Rec 10(2):101–118
129. Yu F, Okamoto S, Harada H, Yamasaki K, Misawa N, Utsumi R (2011) *Zingiber zerumbet* CYP71BA1 catalyzes the conversion of α-humulene to 8-hydroxy-α-humulene in zerumbone biosynthesis. Cell Mol Life Sci 68(6):1033–1040
130. Yun C-H, Kim K-H, Kim D-H, Jung H-C, Pan J-G (2007) The bacterial P450 BM3: a prototype for a biocatalyst with human P450 activities. Trends Biotechnol 25(7):289–298
131. Zerbe P, Chiang A, Yuen M, Hamberger B, Hamberger B, Draper JA, Britton R, Bohlmann J (2012) Bifunctional cis-abienol synthase from Abies balsamea discovered by transcriptome sequencing and its implications for diterpenoid fragrance production. J Biol Chem 287 (15):12121–12131
132. Zerbe P, Hamberger B, Yuen MMS, Chiang A, Sandhu HK, Madilao LL, Nguyen A, Hamberger B, Bach SS, Bohlmann J (2013) Gene discovery of modular diterpene metabolism in nonmodel systems. Plant Physiol 162(2):1073–1091
133. Zhan X, Han LA, Zhang Y-H, Chen D-F, Simonsen HT (2014) Metabolic engineering of the moss *Physcomitrella patens* to produce the sesquiterpenoids patchoulol and α/β-santalene. Front Plant Sci 5:636. doi: 10.3389/fpls.2014.00636 eCollection 2014
134. Zi J et al (2013) Org Biomol Chem 11:7650–7652

Part II
Fundamental Technologies for Metabolic, Enzyme and Process Engineering of Isoprenoids

Adv Biochem Eng Biotechnol (2015) 148: 143–160
DOI 10.1007/10_2014_284

Published Online: 31 December 2014

Enabling Technologies to Advance Microbial Isoprenoid Production

Yun Chen, Yongjin J. Zhou, Verena Siewers and Jens Nielsen

Abstract Microbial production of isoprenoids provides an attractive alternative to biomass extraction and chemical synthesis. Although widespread research aims for isoprenoid biosynthesis, it is still in its infancy in terms of delivering commercial products. Large barriers remain in realizing a cost-competitive process, for example, developing an optimal microbial cell factory. Here, we summarize the many tools and methods that have been developed in the metabolic engineering of isoprenoid production, with the advent of systems biology and synthetic biology, and discuss how these technologies advance to accelerate the design–build–test engineering cycle to obtain optimum microbial systems. It is anticipated that innovative combinations of new and existing technologies will continue to emerge, which will enable further development of microbial cell factories for commercial isoprenoid production.

Keywords Terpenoids · Sesquiterpenes · Metabolic engineering · Enabling technologies · *E. coli* · *S. cerevisiae*

Contents

Y. Chen · Y.J. Zhou · V. Siewers · J. Nielsen (✉)
Systems and Synthetic Biology, Department of Chemical and Biological Engineering, Chalmers University of Technology, Kemivägen 10, SE-412 96 Gothenburg, Sweden
e-mail: nielsenj@chalmers.se

1 Introduction

Isoprenoids constitute a large and diverse group of natural products, encompassing more than 40,000 structures known thus far [1]. This extremely diverse array of chemical structures is probably a reflection of their many varied biological functions, which have great attraction for traditional and modern human exploitation, for instance, as pharmaceuticals, flavors, fragrances, dietary supplements, food ingredients, biomaterials, and biofuels [2, 3]. From environmental and economic concerns, microbial production of isoprenoids has gained more and more attention as most isoprenoids are present only in low quantities in their natural plant resources and are structurally complex, making it difficult to produce them by chemical synthesis.

All isoprenoids are derived from the universal C_5 monomers isopentenyl diphosphate (IPP) and dimethylallyl diphosphate (DMAPP; Fig. 1). Two distinct biosynthetic pathways, the mevalonate (MVA) pathway (Fig. 1a) and the methylerythritol phosphate (MEP) pathway (Fig. 1b) have evolved for the biosynthesis of these two vital building blocks [2, 4]. Condensations of DMAPP with one or more molecules of IPP result in the linear prenyl diphosphate precursors geranyl diphosphate (GPP), farnesyl diphosphate (FPP), and geranylgeranyl diphosphate (GGPP), as well as larger units. These skeletons normally pass through a series of reactions, such as cyclization, methylation, acetylation, phenylation, and rearrangements to form a diversity of isoprenoid families.

Owing to the excellent progress in metabolic engineering with the advent of synthetic biology and systems biology, much success has been seen in the field of microbial production of isoprenoids. The production of artemisinic acid, a chemical precursor for the production of the important antimalarial drug artemisinin, represents a milestone achievement in this area [5]. However, microbial production of most other isoprenoids is still far behind and will require further development before they reach the commercial stage. One of the key factors is that construction and improvements of microbial cell factories are time consuming and need massive endeavors. Herein, we attempt to cover the recent developments in metabolic engineering of bacterial and yeast microbial platform strains for isoprenoid production with a focus on applied technologies.

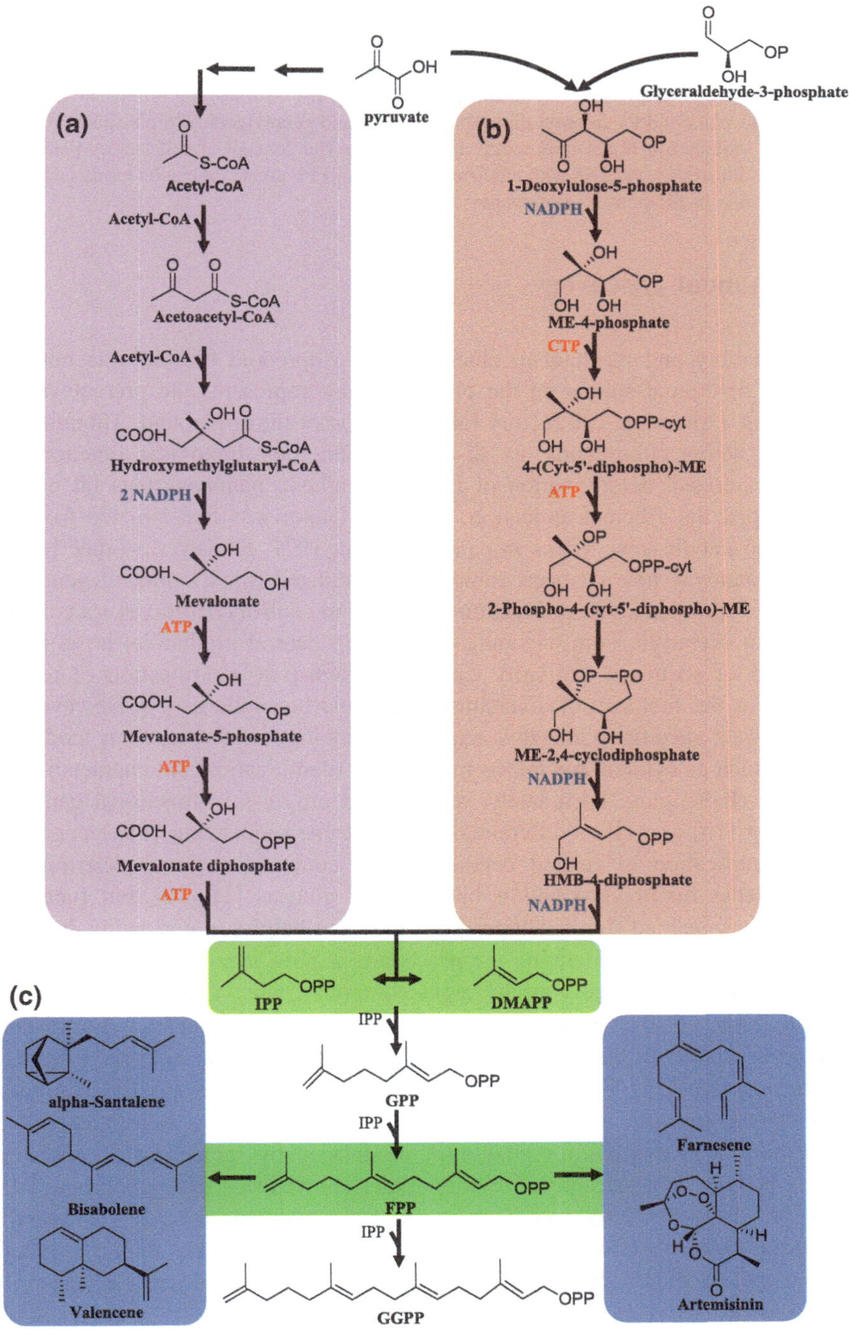
pyruvate
Glyceraldehyde-3-phosphate
(a)
Acetyl-CoA
Acetyl-CoA
Acetoacetyl-CoA
Acetyl-CoA
Hydroxymethylglutaryl-CoA
2 NADPH
Mevalonate
ATP
Mevalonate-5-phosphate
ATP
Mevalonate diphosphate
ATP
(b)
1-Deoxylulose-5-phosphate
NADPH
ME-4-phosphate
CTP
4-(Cyt-5'-diphospho)-ME
ATP
2-Phospho-4-(cyt-5'-diphospho)-ME
ME-2,4-cyclodiphosphate
NADPH
HMB-4-diphosphate
NADPH
IPP
DMAPP
(c)
IPP
GPP
IPP
FPP
IPP
GGPP
alpha-Santalene
Bisabolene
Valencene
Farnesene
Artemisinin

◂**Fig. 1** Biosynthesis of isoprenoids. Complementary routes: **a** the mevalonate (MVA) pathway and **b** the methylerythritol phosphate (MEP) pathway are responsible for the biosynthesis of two vital precursors, isopentenyl diphosphate (IPP) and dimethylallyl diphosphate (DMAPP). Condensations of DMAPP with one or more molecules of IPP result in the immediate precursor geranyl diphosphate (GPP), farnesyl diphosphate (FPP), and geranylgeranyl diphosphate (GGPP), from which various functional isoprenoids are formed. **c** The concept of cell factory platform is simplified as illustrated. A strain with efficient provision of FPP precursor can be easily adapted to produce a range of different sesquiterpenes

2 Component Discovery

Gene discovery and mechanism elucidation in isoprenoid biosynthesis not only facilitates functional studies in the plant, but also represents the prerequisite for engineering a microbial cell factory for production of the compound. Though more and more active compounds are discovered and their biological functions are well-characterized, determination of their biosynthetic pathways lags far behind. For example, the efficient anticancer terpenoid Taxol was first isolated from the bark of the Pacific yew, *Taxus brevifolia* Nutt in 1971. However, to date its biosynthetic pathway has not been completely elucidated [6]. Cloning of genes differentially expressed under production compared to control conditions is a common strategy for identifying the biosynthesis genes in natural production hosts [7, 8]. Evaluation of sequence similarity can further assist in identification of terpene synthases as the first step of diversified isoprenoid biosynthesis [9]. However, the high substrate specificity and low sequence homology of downstream modifying enzymes such as cytochrome P450s makes their identification and characterization extremely challenging. Fortunately, recent rapid progress in functional genomics sequencing [10] as well as differential transcriptome sequencing [11] has enabled faster identification of related candidates. Of course, functional determination strongly relies on in vitro and/or biochemical analysis [11, 12], but functional genomics has enabled significant progress and been helpful in narrowing down the candidate genes and accelerating the procedure of gene identification.

Sometimes, knowledge gaps in target compound biosynthesis will hinder the reconstruction of the entire heterologous biosynthetic pathway. Fortunately, directed evolution provides a feasible strategy to modify an enzyme towards a specific function [13]. For instance, evolution of a methyl ketone-preferring aminotransferase towards acceptance of prositagliptin ketone as substrate enabled establishing a bioprocess for sitagliptin production by increasing the substrate binding pocket [14]. Recently, exciting studies showed that the cytochrome P450 P450BM3 from *Bacillus megaterium* could be evolved towards catalyzing a carbene transfer reaction for olefin cyclopropanation, rather than its natural monooxygenation [15, 16]. Directed evolution not only produces enzymes with novel specific activities, but may also increase catalytic efficiency. For example, directed evolution of a sesquiterpene cadinene synthase led to increased terpene–

synthase–cyclization activity [17]. Here, an efficient high-throughput screening strategy is essential for rapid enzyme selection during the evolution procedure.

Exploiting enzyme catalytic promiscuity, such as condition promiscuity, substrate promiscuity, and catalytic promiscuity, provides another possibility of gap repairing for constructing artificial synthesis pathways that are currently not available [18]. For example, a substrate-promiscuous P450BM3 from *B. megaterium* was engineered and incorporated into the semibiosynthetic route for artemisinin production in *Escherichia coli*, which enabled efficient conversion of its precursor amorphadiene to artemisinic-11S,12-epoxide, at titers of 250 mg/L [19]. Similarly, a D-lactate dehydrogenase (encoded by *d-ldh* from *Lactobacillus pentosus*) and a hydroxylase complex (encoded by *hpaBC* from *E. coli*) were shown to have enough promiscuity to catalyze the consecutive reactions for salvianic acid A production from 4-hydroxyphenylpyruvate in *E. coli* [20].

3 Cell Factory Construction

3.1 Host Selection

Saccharomyces cerevisiae and *E. coli* are commonly used hosts for isoprenoid production due to the well-characterized genetic background and excellent accessibility of molecular techniques. *E. coli* showed great performance in terpene production due to its fast growth [21, 22]. However, the biosynthesis of many functional terpenoids, especially phytoterpenoids, involves several modification steps catalyzed by complex enzymes including membrane-bound cytochrome P450 enzymes, which can be difficult to establish in bacteria including *E. coli* due to differences in translation and membrane structure. Though engineering of these enzymes can be helpful for functional expression in *E. coli*, their catalytic efficiency usually remains low. Actually, CYP71AV1 expressing *E. coli* just produced artemisinic acid at a titer of 105 mg/L even though CYP71AV1 expression was systematically optimized including codon-optimization, N-terminal modification, and reductase balancing [23], whereas the corresponding engineered *S. cerevisiae* produced artemisinic acid at up to 25 g/L [5]. Alternative prokaryotic hosts such as *Corynebacterium glutamicum* [24] and *Bacillus subtilis* [25] have also been engineered for isoprenoid production, but with relatively low efficiency. To date, the *S. cerevisiae* has been showing good performance in producing a number of functional isoprenoids including artemisinic acid [5], tanshinone precursor [11] and ginsenosides [26]. Nevertheless, its endogenous MVA pathway awaits further improvement for high-level production. The red yeast *Rhodosporidium toruloides*, which has efficient lipid biosynthesis, is well suited for isoprenoid biosynthesis, and its genome, transcriptome, and proteome have been systematically analyzed [27], which makes it a potential efficient host for functional isoprenoid production after establishing a genetic manipulation platform.

3.2 Genetic Implementation

Rapid and efficient DNA assembly and genome engineering are the prerequisites for fast pathway construction and optimization. Yeast homologous recombination-based DNA assembler [28] and scarless DNA assembly via ligase cycling reaction (LCR) [29] both enabled assembly of up to 12 DNA parts with 60–100 % of individual clones being correct, and provided fast and reliable workflow for assembly of DNA constructs from PCR amplification or DNA synthesis. In yeast efficient seamless gene disruption [30] enables rapid blocking of by-product formation and one-step multigene genomic integration [31] is helpful for constructing genetic stable strains. Similarly, in *E. coli* ePathBrick enables rapid assembly of up to 7 genes (around 9 kB in total) [32] and chromosomal gene disruption using PCR primers as the homologue to the targeted gene is well established [33]. However, genetic engineering of unconventional microorganisms such as *R. toruloides* is still difficult. Any endeavor, on establishing a genetic engineering platform for this, is helpful for constructing alternative microbial isoprenoid cell factories [34].

4 Rational Design and Optimization

4.1 Precursor Supply

High-level production of isoprenoids using either *E. coli* or *S. cerevisiae* requires engineering a strain for efficient provision of the universal C_5 monomer IPP or its isomer DMAPP, as all isoprenoids are assembled from these monomers. Two pathways exist for the biosynthesis of these precursors (Fig. 1). One approach to increase the precursor flux is to employ a heterologous pathway to supplement the native pathway, which may bypass endogenous regulation and alternative dissipative pathways. Importing the MVA pathway from *S. cerevisiae* into *E. coli* substantially increased the availability of isoprenoid precursors [35]. However, it should be noted that this procedure may also cause undesired effects such as the toxicity of pathway intermediates to the host. In this case, the accumulation of hydroxymethylglutaryl-CoA (HMG-CoA) imposed through heterologous expression of the MVA pathway was shown to be toxic for *E. coli* [36]. In addition, the host may lack the appropriate machinery to functionalize the heterologous enzyme. In a study, it was found that the difficulty functionally to express iron sulfur cluster proteins in yeast, which are essential for the last two reactions in the MEP pathway, and this led to failure in establishing a functional MEP pathway in *S. cerevisiae* [37, 38].

Another approach to increase the precursor pool is to manipulate the metabolic flux and regulation of the native pathway. Extensive endeavors have focused on metabolic engineering of the MVA pathway for enhanced supply of IPP/DMAPP, and further FPP, which is the universal precursor for sesquiterpenes. Except for the well-elucidated targets for engineering the MVA pathway in yeast such as overexpression of a truncated Hmg1, transcriptional factor mutant Upc2-1, and FPP synthase Erg20 combined with down-regulation of squalene synthase gene *ERG9* [5, 12, 39–41] additional strategies have been developed to increase the flux to FPP. These include, for example, overexpression of all genes in the MVA pathway either by integration of an additional copy [42] or replacing the promoters of all relevant MVA pathway genes [43], or integration of one more copy of *tHMG*1 with strong promoter (P_{TDH3}) into the chromosome [42]. Owing to all efforts made, a thoroughly optimized strain was developed and combined with the discovery and implementation of plant dehydrogenase and cytochromes; this enabled an efficient biosynthetic route in *S. cerevisiae* to artemisinic acid with fermentation titers of 25 g/L [5].

Taking advantage of a developed platform strain that has a high flux towards FPP, a commercial process was developed for the production of farnesene [44], another sesquiterpene derived from FPP. In addition, the flexibility of the FPP overproducing platform allowed a rapid switch from the overproduction of amorphadiene to that of bisabolene, the immediate precursor to bisabolane, which is a novel biosynthetic alternative to D2 diesel fuel [45]. These examples well elucidate the concept of a cell factory platform (Fig. 1c), where the cellular metabolism has been optimized towards formation of universal precursors, which can then be used by "plug-and-play" for the production of different types of related products.

Similar to this concept, a supply of precursor acetyl-CoA is essential for high-level production of isoprenoids via the MVA pathway and many other commercially interesting compounds as well [46]. However, acetyl-CoA metabolism in yeast is quite complex and this metabolite is present in three different compartments and is not directly interchangeable among these compartments. Therefore, the metabolism of acetyl-CoA in *S. cerevisiae* has been thoroughly investigated [47]. Based on this, a combined pull–push–block strategy has been recently developed to increase the availability of acetyl-CoA in the cytoplasm [46]. The strategy involved overexpression of endogenous alcohol dehydrogenase Adh2, aldehyde dehydrogenase Ald6, and expression of acetyl-CoA synthetase variant from *Salmonella enterica*. The use of this variant had been previously demonstrated to redirect flux from acetaldehyde to acetyl-CoA successfully and to lead to improved production of the sesquiterpene amorphadiene [48]. Furthermore, through combining these with deleting acetyl-CoA consuming reactions, such as peroxisomal citrate synthase Cit2 or cytosolic malate synthase Mls1, the developed acetyl-CoA platform improved the production of the sesquiterpene santalene fourfold [46]. The platform host was also shown to have a substantially improved production of 1-butanol by eightfold [49], and polyhydroxybutyrate by 18-fold [50]. However, these strategies exclusively depend on the native pyruvate dehydrogenase (PDH) bypass pathway, which is highly ATP-consuming for provision of acetyl-CoA making this route unfavorable. Alternative pathways for cytosolic acetyl-CoA synthesis such as replacement of native acetyl-CoA synthetases (ACSs) by acetylating acetaldehyde dehydrogenase or pyruvate formate lyase [51], may overcome the constraints on the maximum yields and increase the production of acetyl-CoA–derived isoprenoids and other compounds.

4.2 Pathway Balance

It is important to increase the flux to important precursors and the final products, however, it is also essential that the relative levels of the enzymes are coordinated in such a way that no intermediates are accumulated, as these metabolites may be toxic to the host and/or affect further product formation. In order to balance the multigene pathways, an excellent example by using an approach named "multivariate modular pathway engineering" was demonstrated to improve production of the diterpene taxadiene successfully [52]. This technique redefines the metabolic network as a collection of distinct modules. Four key enzymes (encoded by *dxs*, *ispD*, *ispF*, and *idi*) in the MEP pathway were grouped into one module, and two enzymes catalyzing the synthesis of GGPP and taxadiene were grouped into another module. The expression of the different modules in *E. coli* was sequentially and systematically tuned by, for example, varying plasmid copy numbers and altering promoter strengths. Consequently, the highest producer among these strains reached production levels of 1 g/L taxadiene, a nearly 15,000-fold improvement compared to the control strain [52]. This framework allows for the assessment and elimination of regulatory and pathway bottlenecks, offers the opportunity to evaluate a complicated system in a more tractable way, and is particularly useful for experimentation lacking high-throughput screens. The module-balancing strategy has also been proven to be useful for the production of different types of products, such as the diterpene miltiradiene [53], fatty acids [54], and the flavonoid pinocembrin [55].

4.3 Spatial Optimization of Metabolism

To obtain the desired cell factories, it is crucial to coordinate the performances of many different enzymes and multiple cellular pathways. In addition to modular optimization, spatial coupling of complex metabolic reactions at multiple scales has become an attractive strategy to improve pathway function. These optimization approaches function at many scales and are responsive to external signals, ensuring that the cell does not waste metabolic resources by producing unnecessary enzymes and channeling flux through appropriate metabolic routes.

One strategy is organizing pathway enzymes as macromolecular complexes by generation of unnatural fusion proteins. Direct fusion of FPP synthase with a heterologous sesquiterpene synthase, patchoulol synthase in *S. cerevisiae* was found successful in diverting the flux towards formation of the desired product and resulted in increased patchoulol production by about twofold [56]. Physical fusion of diterpene synthases SmCPS (labdadienyl/copalyl diphosphate synthase, from *Salvia miltiorrhiza*) and SmKSL (kaurene synthase-like, from *S. miltiorrhiza*) as well as the fusion of GGPP and FPP synthase led to significantly improved diterpene miltiradiene production and reduced by-product formation in *S. cerevisiae* [53]. Although a fusion of two normally monomeric enzymes is straightforward, fusion of two normally multimeric or more than two enzymes may be problematic due to misfolding or proteolysis of the fusion proteins.

As an alternative to artificial protein fusions, synthetic scaffolds with the assistance of proteins, DNA, or RNA have been recently developed to colocalize multiple enzymes in a designable manner. Dueber et al. [57] expressed scaffolds from the interaction domains of metazoan signaling proteins to assemble three mevalonate biosynthetic enzymes in optimal stoichiometry, and achieved a 77-fold improvement in mevalonate production with lower enzyme expression levels and reduced metabolic load in *E. coli*. Similar to protein scaffolds, RNA aptamer-based scaffolds are recruited to control the spatial organization of two metabolic enzymes involved in biological hydrogen production expressed in *E. coli*, resulting in about 50-fold enhancements in hydrogen production [58]. Unlike protein-based approaches, the use of RNA scaffolds permits the assembly of complex multidimensional architecture with nanometer precision. However, these techniques have not been exploited in application of isoprenoid production. Along similar lines, DNA-based scaffolds were used in combination with chimeras between target biosynthetic enzymes and zinc fingers (ZFs) to increase mevalonate production up to threefold with optimal scaffold designs in *E. coli* [59].

In addition to the linear design and channeling of metabolic pathways, intracellular spatial organization of cellular functions to membrane-surrounded organelles offers great benefits such as concentrating reactions and pathways that need special conditions (e.g., pH) or may be toxic to other organelles. Mitochondrial targeting of a heterologous FPP synthase and a sesquiterpene synthase improved the production of sesquitrepene valencene and amorphadiene by eight- and 20-fold, respectively, in yeast coengineered with a truncated *HMG1* [60]. Targeting the Ehrlich pathway into yeast mitochondria increased isobutanol production more than twofold, whereas overexpression of the same pathway in the cytoplasm only improved yields by 10 % [61]. These successes are not only due to higher concentrations of enzymes, substrates, and cofactors, but might also result from removing the need to transport intermediates between organelles and from reducing the loss of intermediates to competing pathways.

4.4 Cofactor Engineering

Cofactors such as NAD(P)H, ATP, and CoA are among the most highly connected metabolites and influence a broad range of cellular functions. Manipulation of cofactor availability not only allows overcoming bottlenecks in cofactor-dependent biosynthesis pathways, but might also have a widespread impact on cellular metabolism.

Pathways for synthesis of isoprenoids often involve oxidation-reduction reactions catalyzed by enzymes using cofactors NADH and NADPH that are as critical for catalysis as the enzymes themselves. In native cell metabolism NADH is the major redox product of catabolism, and NADPH is predominantly required in anabolism where it serves as a reducing agent for amino acid, lipid, and nucleotide biosynthesis. Thus, pathway enzymes such as, for instance, HMG-CoA reductase, which is a key player in the MVA pathway and requires NADPH as a reducing agent, will increase the demand of NADPH aimed for isoprenoid production and often have to compete with the requirements of other anabolic pathways for this essential cofactor.

One approach is altering cofactor requirements by enzyme substitution. Ma et al. [62] characterized various HMG-CoA reductases to substitute the corresponding component in the yeast MVA pathway that was heterologously expressed in *E. coli*. A NADH-dependent HMG-CoA reductase from *Delftia acidovorans* improved production of the sesquiterpene amorphadiene by 50 % over the yeast enzyme. A further enhanced performance was observed by increasing intracellular NADH availability using a NAD^+-dependent formate dehydrogenase from *Candida boidinii* along with formate supplementation.

Another approach is targeting native host pathways that can bring about cofactor recycling. In order to restore the NADH/NADPH balance impaired by installing a sesquiterpene santalene biosynthesis pathway in *S. cerevisiae*, the ammonium assimilation pathway was modified to increase the availability of the reductive cofactor NADPH. This was achieved by deleting the NADPH-consuming reaction of glutamate dehydrogenase encoded by *GDH*1, and simultaneously overexpressing the NAD^+-dependent glutamate dehydrogenase encoded by *GDH*2. This strategy was found to improve the production of the sesquiterpene cubebol by 85 % [63], and together with other strategies the production of the sesquiterpene santalene could be increased fourfold [64].

Altering cofactor specificity via protein engineering is another approach as demonstrated in the enhanced synthesis of butanediol [65], in which the coenzyme specificity of butanediol dehydrogenase was altered from NAD(H) to NADP(H).

5 Computational and Omics-Aided Engineering and Optimization

5.1 Computational Tool-Driven Optimization

The identification of gene targets for strain improvement outside the actual biosynthetic pathway is a difficult task as it needs comprehensive knowledge about all pathways and genes functioning in the cell and their interactions with each other. Increasing availability of stoichiometric models for a range of organisms with developed metabolic optimization algorithms enables researchers to identify such gene targets in silico, and may thereby accelerate the process of metabolic engineering for strain improvement towards the enhanced production of desired products.

A gene deletion strategy using minimization of metabolic adjustments (MOMA) allowed the successful identification of genes to be deleted among a large number of gene candidates, and improved the biosynthesis of lycopene in the engineered *E. coli* strain by nearly 40 % [66]. This was further proved to be useful by using OptGene [67] as a modeling framework and MOMA as an objective function to identify gene knockouts in *S. cerevisiae* and led to enhanced production of the sesquiterpene cubebol by 85 % [63].

On the other hand, it has been difficult to identify gene overexpression targets using similar techniques. One strategy called flux scanning based on enforced objective flux (FSEOF) has been successfully used to select genomewide gene amplification targets in *E. coli*, and improved lycopene production [68]. These successes point out examples where genome-scale metabolic models can be used effectively. However, there are areas where models that incorporate regulatory constraints and signaling should be used in combination with additional tools, for example, constraint-based reconstruction and analysis (COBRA) [69], which would allow for more precise target identification.

5.2 *Omics-Based Technique Aided Optimization*

The advancements of omics techniques have enabled a rapid acceleration in the ability to obtain system-level information for the cell, capturing cell physiology in a holistic manner from several different levels. These high-throughput omics data, integrated with computational approaches now serve biomolecular network analysis and model-based prediction in the field of systems biology [70, 71].

Transcription analysis, as the most widely implemented omics technology in metabolic engineering, has enabled quantitative measurements of the dynamic expression of mRNAs between different states at the genome scale. This allows integration of genome-scale metabolic models using different approaches to, for instance, map the transcriptional and metabolic regulation in metabolic networks [72, 73]. This valuable information enabled uncovering global regulatory mechanisms and complex metabolic networks. By exploiting whole-genome transcription analysis, a dynamic stress–response promoter strategy was developed to control accumulation of toxic intermediates in *E. coli* and improve the final titers of the sesquiterpene amorphadiene twofold over those from inducible or constitutive promoters [74].

Because mRNA levels do not necessarily correlate with protein levels, there is also much interest in the use of proteomics in metabolic engineering, to quantify protein levels or characterize states of proteins. A targeted proteomics approach via selected-reaction monitoring (SRM) mass spectrometry was employed to measure protein levels of the MVA pathway heterologously expressed in *E. coli* to produce the sesquiterpene amorphadiene [75]. The analysis revealed two proteins, mevalonate kinase (MK) and phosphomevalonate kinase (PMK), as potential bottlenecks. Manipulating the expression of these proteins led to a threefold improved final titer of amorphadiene. Nowadays, advancements in mass spectrometry combined with liquid chromatography have expanded the number of proteins that can be quantified in a cell lysate by several orders of magnitude. Also, other methods, such as stable isotope labeling-based isotope coded affinity tags (ICAT), isobaric tags for relative and absolute quantification (iTRAQ), or label-free comparative quantitative proteomics have enabled more accurate proteome measurements. However, these methods need to be employed to monitor abundance of proteins in metabolic engineering.

Similar to proteomics, the advent of mass spectrometry also permits measurements of intra-and extracellular metabolites for high-throughput metabolome studies [76]. However, some challenges still lie in sample preparation, as it is difficult to obtain quantitative information at the metabolite levels whereas relative levels can be measured [77]. Instead of looking at all metabolites simultaneously, target profiling compares the spectrum of interest to a library of reference spectra of pure compounds. In one study, targeted metabolite profiling coupled with transcriptome analysis was employed to identify the cytotoxic effects of accumulated HMG-CoA in an engineered *E. coli* strain expressing the MVA pathway [78]. It was revealed that HMG-CoA accumulation inhibits the biosynthesis of fatty acids. Based on these results a new fermentation strategy was suggested including supplementation of palmitic acid or oleic acid, which markedly improved the growth of the host strain.

As high-throughput omics analyses contribute to a better understanding of the biological system, this kind of analysis is expected to be more frequently employed for isoprenoid production to identify metabolic engineering targets.

6 Novel Fermentation Process Design and Optimization

A fundamental challenge to the microbial production of isoprenoids is product (and/or intermediate) tolerance; for instance, monoterpenes are generally highly toxic to microorganisms [79]. In addition to cellular engineering approaches such as transporter engineering [80] and modifying the components of the global transcription machinery [81] that can be used to overcome toxicity limitations, an alternative approach is novel fermentation process design. By adding an immiscible organic layer onto the cultivation medium, the product can be removed in situ, where the product is toxic or inhibits further product formation. A biphasic extractive system was demonstrated to reduce significantly toxicity of the monoterpene limonene to the host *S. cerevisiae*, with growth rates far beyond the ones being reached in the solvent-free system or after attempted adaptation in chemostat cultivation [79]. For example, using dibutyl phthalate as solvent, the minimum inhibitory concentration to limonene improved by about 700-fold compared to a solvent-free system.

The biphasic approach is also a suitable tool for recovery of the product. In an attempt to eliminate product loss through air-stripping, a two-phase partitioning fermentation process with a dodecane organic phase was used to trap the volatile sesquiterpene amorphadiene [82]. In another study, a gas-phase bioprocess was employed for isoprene production by using an extensively metabolically engineered *E. coli* strain reaching a high titer of over 60 g/L, a volumetric productivity of 2 g/L, and a yield of 11 % isoprene from glucose [22]. Numerous in situ product recovery techniques such as pervaporation, perstraction, and gas striping have been reported for the recovery of different kinds of compounds [83]. It can be expected that fermentation process design will gain more attention for microbial isoprenoid production in the future.

Optimization of the fermentation process is also a key factor for the viability of the overall bioprocess. Inasmuch as metabolic engineering of microorganisms is typically implemented in the process of using test tube or shake flask cultivation, a superior strain developed in this process does not necessarily perform well in a large-scale fermentation used in industrial production. Furthermore, an optimized fermentation process will further increase the performance of the metabolically engineered strains. Optimization of nitrogen delivery in the fermentation process using an engineered *E. coli* strain further enhanced the titers of the sesquiterpene amorphadiene, and up to 27 g/L amorphadiene were reached by using a dual restriction of carbon and nitrogen supply in the fed-batch process [21]. In another study, development of a fermentation process with ethanol feeding for the re-engineered yeast CEN.PK2 strain further enhanced amorphadiene production and led to the production of more than 40 g/L [43]. Moreover, during the fermentation optimization or process scale-up it might be necessary to use integrative analysis based on omics data to further identify metabolic engineering targets.

7 Outlook

Encouraged by recent advances, the production of isoprenoids by bacterial or yeast host platforms holds great promise. There are also an increasing number of research efforts on the molecular biochemistry and genetics of isoprenoid biosynthesis, and, to some extent, on their biological functions. This not only expands the resources for further exploitation, but also provides increasing opportunities for microbial production of isoprenoids. However, most of the catalytic diversity of plant enzymes is unexplored, because previous efforts have focused on a relatively small number of species. One limiting factor lies therefore in the elucidation of the biosynthetic pathways and identifying the appropriate genes. Of special note, terpene-modifying enzymes typically belonging to the P450s family pose great challenges in terms of both gene discovery and microbial expression. Advances and decreasing costs in sequencing technology will accelerate the rate of exploring new resources. Creating new enzymes by powerful protein engineering and directed evolution is another direction, as reported for an engineered cytochrome P450 that catalyzes a reaction that is fundamentally different from the one catalyzed by the natural enzyme [15]. This is expected to expand the range of isoprenoid products that can be synthesized by microbial cell factories.

The past few years have witnessed several successful examples of industrial production of isoprenoids in large-scale fermentation processes, such as isoprene, valencene, amorphadiene, and farnesene, by either *E. coli* or *S. cerevisiae* host platforms. However, expansion to many other isoprenoid products to become industrially relevant will depend on the continued development of technologies. Continuous efforts in systems biology to elucidate the complex regulatory and metabolic networks with novel algorithms emerging will advance the predictive potential of mathematical models, and thereby enhance the process for optimal microbial factory development. Moreover, the progress in design and construction of biological components in the field of synthetic biology will facilitate faster construction and reliable and desirable control of metabolic pathways. It can be predicted that new technologies and innovative combinations of new and existing technologies will continue emerging, which will enable further development of microbial factories for commercial isoprenoid production.

References

1. Bohlmann J, Keeling CI (2008) Terpenoid biomaterials. Plant J 54(4):656–669
2. Chang MC, Keasling JD (2006) Production of isoprenoid pharmaceuticals by engineered microbes. Nat Chem Biol 2(12):674–681
3. Tippmann S, Chen Y, Siewers V, Nielsen J (2013) From flavors and pharmaceuticals to advanced biofuels: production of isoprenoids in *Saccharomyces cerevisiae*. Biotechnol J 8 (12):1435–1444
4. Maury J, Asadollahi MA, Møller K, Clark A, Nielsen J (2005) Microbial isoprenoid production: an example of green chemistry through metabolic engineering. Adv Biochem Eng Biotechnol 100:19–51
5. Paddon CJ, Westfall PJ, Pitera DJ, Benjamin K, Fisher K, McPhee D, Leavell MD, Tai A, Main A, Eng D, Polichuk DR, Teoh KH, Reed DW, Treynor T, Lenihan J, Fleck M, Bajad S, Dang G, Dengrove D, Diola D, Dorin G, Ellens KW, Fickes S, Galazzo J, Gaucher SP, Geistlinger T, Henry R, Hepp M, Horning T, Iqbal T, Jiang H, Kizer L, Lieu B, Melis D, Moss N, Regentin R, Secrest S, Tsuruta H, Vazquez R, Westblade LF, Xu L, Yu M, Zhang Y, Zhao L, Lievense J, Covello PS, Keasling JD, Reiling KK, Renninger NS, Newman JD (2013) High-level semi-synthetic production of the potent antimalarial artemisinin. Nature 496 (7446):528–532
6. Guerra-Bubb J, Croteau R, Williams RM (2012) The early stages of taxol biosynthesis: an interim report on the synthesis and identification of early pathway metabolites. Nat Prod Rep 29(6):683–696
7. Gao W, Hillwig ML, Huang L, Cui G, Wang X, Kong J, Yang B, Peters RJ (2009) A functional genomics approach to tanshinone biosynthesis provides stereochemical insights. Org Lett 11(22):5170–5173
8. Jennewein S, Rithner CD, Williams RM, Croteau RB (2001) Taxol biosynthesis: Taxane 13 alpha-hydroxylase is a cytochrome P450-dependent monooxygenase. Proc Natl Acad Sci USA 98(24):13595–13600

9. Zerbe P, Hamberger B, Yuen MM, Chiang A, Sandhu HK, Madilao LL, Nguyen A, Hamberger B, Bach SS, Bohlmann J (2013) Gene discovery of modular diterpene metabolism in nonmodel systems. Plant Physiol 162(2):1073–1091
10. Chen S, Xu J, Liu C, Zhu Y, Nelson DR, Zhou S, Li C, Wang L, Guo X, Sun Y, Luo H, Li Y, Song J, Henrissat B, Levasseur A, Qian J, Li J, Luo X, Shi L, He L, Xiang L, Xu X, Niu Y, Li Q, Han MV, Yan H, Zhang J, Chen H, Lv A, Wang Z, Liu M, Schwartz DC, Sun C (2012) Genome sequence of the model medicinal mushroom Ganoderma lucidum. Nat Commun 3:913
11. Guo J, Zhou YJ, Hillwig ML, Shen Y, Yang L, Wang Y, Zhang X, Liu W, Peters RJ, Chen X, Zhao ZK, Huang L (2013) CYP76AH1 catalyzes turnover of miltiradiene in tanshinones biosynthesis and enables heterologous production of ferruginol in yeasts. Proc Natl Acad Sci USA 110(29):12108–12113
12. Ro DK, Paradise EM, Ouellet M, Fisher KJ, Newman KL, Ndungu JM, Ho KA, Eachus RA, Ham TS, Kirby J, Chang MC, Withers ST, Shiba Y, Sarpong R, Keasling JD (2006) Production of the antimalarial drug precursor artemisinic acid in engineered yeast. Nature 440 (7086):940–943
13. Brustad EM, Arnold FH (2011) Optimizing non-natural protein function with directed evolution. Curr Opin Chem Biol 15(2):201–210
14. Savile CK, Janey JM, Mundorff EC, Moore JC, Tam S, Jarvis WR, Colbeck JC, Krebber A, Fleitz FJ, Brands J, Devine PN, Huisman GW, Hughes GJ (2010) Biocatalytic asymmetric synthesis of chiral amines from ketones applied to sitagliptin manufacture. Science 329 (5989):305–309
15. Coelho PS, Brustad EM, Kannan A, Arnold FH (2013) Olefin cyclopropanation via carbene transfer catalyzed by engineered cytochrome P450 enzymes. Science 339(6117):307–310
16. Coelho PS, Wang ZJ, Ener ME, Baril SA, Kannan A, Arnold FH, Brustad EM (2013) A serine-substituted P450 catalyzes highly efficient carbene transfer to olefins in vivo. Nat Chem Biol 9(8):485–487
17. Lauchli R, Rabe KS, Kalbarczyk KZ, Tata A, Heel T, Kitto RZ, Arnold FH (2013) High-throughput screening for terpene-synthase-cyclization activity and directed evolution of a terpene synthase. Angew Chem Int Ed 52(21):5571–5574
18. Hult K, Berglund P (2007) Enzyme promiscuity: mechanism and applications. Trends Biotechnol 25(5):231–238
19. Dietrich JA, Yoshikuni Y, Fisher KJ, Woolard FX, Ockey D, McPhee DJ, Renninger NS, Chang MCY, Baker D, Keasling JD (2009) A novel semi-biosynthetic route for artemisinin production using engineered substrate-promiscuous P450(BM3). ACS Chem Biol 4(4):261–267
20. Yao YF, Wang CS, Qiao J, Zhao GR (2013) Metabolic engineering of *Escherichia coli* for production of salvianic acid A via an artificial biosynthetic pathway. Metab Eng 19:79–87
21. Tsuruta H, Paddon CJ, Eng D, Lenihan JR, Horning T, Anthony LC, Regentin R, Keasling JD, Renninger NS, Newman JD (2009) High-level production of amorpha-4,11-diene, a precursor of the antimalarial agent artemisinin, in *Escherichia coli*. PLoS ONE 4(2):e4489
22. Whited GM, Feher FJ, Benko DA, Cervin MA, Chotani GK, McAuliffe JC, LaDuca RJ, Ben-Shoshan EA, Sanford KJ (2010) Technology update: development of a gas-phase bioprocess for isoprene-monomer production using metabolic pathway engineering. Ind Biotechnol 6 (3):152–163
23. Chang MC, Eachus RA, Trieu W, Ro DK, Keasling JD (2007) Engineering *Escherichia coli* for production of functionalized terpenoids using plant P450s. Nat Chem Biol 3(5):274–277
24. Heider SA, Peters-Wendisch P, Wendisch VF (2012) Carotenoid biosynthesis and overproduction in *Corynebacterium glutamicum*. BMC Microbiol 12:198
25. Zhou K, Zou R, Zhang C, Stephanopoulos G, Too HP (2013) Optimization of amorphadiene synthesis in bacillus subtilis via transcriptional, translational, and media modulation. Biotechnol Bioeng 110(9):2556–2561
26. Dai Z, Liu Y, Zhang X, Shi M, Wang B, Wang D, Huang L, Zhang X (2013) Metabolic engineering of *Saccharomyces cerevisiae* for production of ginsenosides. Metab Eng 20:146–156

27. Zhu Z, Zhang S, Liu H, Shen H, Lin X, Yang F, Zhou YJ, Jin G, Ye M, Zou H, Zhao ZK (2012) A multi-omic map of the lipid-producing yeast *Rhodosporidium toruloides*. Nat Commun 3:1112
28. Shao Z, Zhao H, Zhao H (2009) DNA assembler, an in vivo genetic method for rapid construction of biochemical pathways. Nucleic Acids Res 37(2):e16
29. Sd Kok, Stanton LH, Slaby T, Durot M, Holmes VF, Patel KG, Platt D, Shapland EB, Serber Z, Dean J, Newman JD, Chandran SS (2014) Rapid and reliable DNA assembly via ligase cycling reaction. ACS Synthetic Biology 3(2):97–106
30. Akada R, Kitagawa T, Kaneko S, Toyonaga D, Ito S, Kakihara Y, Hoshida H, Morimura S, Kondo A, Kida K (2006) PCR-mediated seamless gene deletion and marker recycling in *Saccharomyces cerevisiae*. Yeast 23(5):399–405
31. Kuijpers NG, Chroumpi S, Vos T, Solis-Escalante D, Bosman L, Pronk JT, Daran JM, Daran-Lapujade P (2013) One-step assembly and targeted integration of multigene constructs assisted by the I-SceI meganuclease in *Saccharomyces cerevisiae*. FEMS Yeast Res 13(8):769–781
32. Xu P, Vansiri A, Bhan N, Koffas MAG (2012) EPathBrick: a synthetic biology platform for engineering metabolic pathways in *E. coli*. ACS Synth Biol 1(7):256–266
33. Datsenko KA, Wanner BL (2000) One-step inactivation of chromosomal genes in *Escherichia coli* K-12 using PCR products. Proc Natl Acad Sci USA 97(12):6640–6645
34. Lin X, Wang Y, Zhang S, Zhu Z, Zhou YJ, Yang F, Sun W, Wang X, Zhao ZK (2014) Functional integration of multiple genes into the genome of the oleaginous yeast *Rhodosporidium toruloides*. FEMS Yeast Res doi:10.1111/1567-1364.12140
35. Martin VJ, Pitera DJ, Withers ST, Newman JD, Keasling JD (2003) Engineering a mevalonate pathway in *Escherichia coli* for production of terpenoids. Nat Biotechnol 21(7):796–802
36. Pitera DJ, Paddon CJ, Newman JD, Keasling JD (2007) Balancing a heterologous mevalonate pathway for improved isoprenoid production in *Escherichia coli*. Metab Eng 9(2):193–207
37. Maury J, Asadollahi MA, Møller K, Schalk M, Clark A, Formenti LR, Nielsen J (2008) Reconstruction of a bacterial isoprenoid biosynthetic pathway in *Saccharomyces cerevisiae*. FEBS Lett 582(29):4032–4038
38. Partow S, Siewers V, Daviet L, Schalk M, Nielsen J (2012) Reconstruction and evaluation of the synthetic bacterial MEP pathway in *Saccharomyces cerevisiae*. PLoS ONE 7(12):e52498
39. Asadollahi MA, Maury J, Moller K, Nielsen KF, Schalk M, Clark A, Nielsen J (2008) Production of plant sesquiterpenes in Saccharomyces cerevisiae: effect of ERG9 repression on sesquiterpene biosynthesis. Biotechnol Bioeng 99(3):666–677
40. Asadollahi MA, Maury J, Schalk M, Clark A, Nielsen J (2010) Enhancement of farnesyl diphosphate pool as direct precursor of sesquiterpenes through metabolic engineering of the mevalonate pathway in *Saccharomyces cerevisiae*. Biotechnol Bioeng 106(1):86–96
41. Paradise EM, Kirby J, Chan R, Keasling JD (2008) Redirection of flux through the FPP branch-point in *Saccharomyces cerevisiae* by down-regulating squalene synthase. Biotechnol Bioeng 100(2):371–378
42. Keasling J, Newman J, Pitera D (2010) Method for enhancing production of isoprenoid compounds. US 7670825 B2
43. Westfall PJ, Pitera DJ, Lenihan JR, Eng D, Woolard FX, Regentin R, Horning T, Tsuruta H, Melis DJ, Owens A, Fickes S, Diola D, Benjamin KR, Keasling JD, Leavell MD, McPhee DJ, Renninger NS, Newman JD, Paddon CJ (2012) Production of amorphadiene in yeast, and its conversion to dihydroartemisinic acid, precursor to the antimalarial agent artemisinin. Proc Natl Acad Sci USA 109(3):E111–E118
44. Renninger NS, McPhee DJ (2008) Fuel compositions comprising farnesane and farnesane derivatives and method of making and using same. United States Patent
45. Peralta-Yahya PP, Ouellet M, Chan R, Mukhopadhyay A, Keasling JD, Lee TS (2011) Identification and microbial production of a terpene-based advanced biofuel. Nat Commun 2:483
46. Chen Y, Daviet L, Schalk M, Siewers V, Nielsen J (2013) Establishing a platform cell factory through engineering of yeast acetyl-CoA metabolism. Metab Eng 15:48–54

47. Chen Y, Siewers V, Nielsen J (2012) Profiling of cytosolic and peroxisomal acetyl-CoA metabolism in *Saccharomyces cerevisiae*. PLoS ONE 7(8):e42475
48. Shiba Y, Paradise EM, Kirby J, Ro DK, Keasing JD (2007) Engineering of the pyruvate dehydrogenase bypass in *Saccharomyces cerevisiae* for high-level production of isoprenoids. Metab Eng 9(2):160–168
49. Krivoruchko A, Serrano-Amatriain C, Chen Y, Siewers V, Nielsen J (2013) Improving biobutanol production in engineered *Saccharomyces cerevisiae* by manipulation of acetyl-CoA metabolism. J Ind Microbiol Biotechnol 40(9):1051–1056
50. Kocharin K, Chen Y, Siewers V, Nielsen J (2012) Engineering of acetyl-CoA metabolism for the improved production of polyhydroxybutyrate in *Saccharomyces cerevisiae*. AMB Express 2(1):52
51. Kozak BU, van Rossum HM, Benjamin KR, Wu L, Daran J-MG, Pronk JT, van Maris AJA (2014) Replacement of the *Saccharomyces cerevisiae* acetyl-CoA synthetases by alternative pathways for cytosolic acetyl-CoA synthesis. Metab Eng 21:46–59
52. Ajikumar PK, Xiao WH, Tyo KE, Wang Y, Simeon F, Leonard E, Mucha O, Phon TH, Pfeifer B, Stephanopoulos G (2010) Isoprenoid pathway optimization for Taxol precursor overproduction in *Escherichia coli*. Science 330(6000):70–74
53. Zhou YJ, Gao W, Rong Q, Jin G, Chu H, Liu W, Yang W, Zhu Z, Li G, Zhu G, Huang L, Zhao ZK (2012) Modular pathway engineering of diterpenoid synthases and the mevalonic acid pathway for miltiradiene production. J Am Chem Soc 134(6):3234–3241
54. Xu P, Gu Q, Wang W, Wong L, Bower AG, Collins CH, Koffas MA (2013) Modular optimization of multi-gene pathways for fatty acids production in *E. coli*. Nat Commun 4:1409
55. Wu J, Du G, Zhou J, Chen J (2013) Metabolic engineering of *Escherichia coli* for (2S)-pinocembrin production from glucose by a modular metabolic strategy. Metab Eng 16(1):48–55
56. Albertsen L, Chen Y, Bach LS, Rattleff S, Maury J, Brix S, Nielsen J, Mortensen UH (2011) Diversion of flux toward sesquiterpene production in *Saccharomyces cerevisiae* by fusion of host and heterologous enzymes. Appl Environ Microbiol 77(3):1033–1040
57. Dueber JE, Wu GC, Malmirchegini GR, Moon TS, Petzold CJ, Ullal AV, Prather KL, Keasling JD (2009) Synthetic protein scaffolds provide modular control over metabolic flux. Nat Biotechnol 27(8):753–759
58. Delebecque CJ, Lindner AB, Silver PA, Aldaye FA (2011) Organization of intracellular reactions with rationally designed RNA assemblies. Science 333(6041):470–474
59. Conrado RJ, Wu GC, Boock JT, Xu H, Chen SY, Lebar T, Turnek J, Tomšič N, Avbelj M, Gaber R, Koprivnjak T, Mori J, Glavnik V, Vovk I, Beninča M, Hodnik V, Anderluh G, Dueber JE, Jerala R, Delisa MP (2012) DNA-guided assembly of biosynthetic pathways promotes improved catalytic efficiency. Nucleic Acids Res 40(4):1879–1889
60. Farhi M, Marhevka E, Masci T, Marcos E, Eyal Y, Ovadis M, Abeliovich H, Vainstein A (2011) Harnessing yeast subcellular compartments for the production of plant terpenoids. Metab Eng 13(5):474–481
61. Avalos JL, Fink GR, Stephanopoulos G (2013) Compartmentalization of metabolic pathways in yeast mitochondria improves the production of branched-chain alcohols. Nat Biotech 31 (4):335–341
62. Ma SM, Garcia DE, Redding-Johanson AM, Friedland GD, Chan R, Batth TS, Haliburton JR, Chivian D, Keasling JD, Petzold CJ, Lee TS, Chhabra SR (2011) Optimization of a heterologous mevalonate pathway through the use of variant HMG-CoA reductases. Metab Eng 13(5):588–597
63. Asadollahi MA, Maury J, Patil KR, Schalk M, Clark A, Nielsen J (2009) Enhancing sesquiterpene production in *Saccharomyces cerevisiae* through in silico driven metabolic engineering. Metab Eng 11(6):328–334
64. Scalcinati G, Partow S, Siewers V, Schalk M, Daviet L, Nielsen J (2012) Combined metabolic engineering of precursor and co-factor supply to increase alpha-santalene production by *Saccharomyces cerevisiae*. Microb Cell Fact 11:117

65. Ehsani M, Fernández MR, Biosca JA, Dequin S (2009) Reversal of coenzyme specificity of 2,3-butanediol dehydrogenase from *Saccharomyces cerevisae* and in vivo functional analysis. Biotechnol Bioeng 104(2):381–389
66. Alper H, Jin Y-S, Moxley JF, Stephanopoulos G (2005) Identifying gene targets for the metabolic engineering of lycopene biosynthesis in *Escherichia coli*. Metab Eng 7(3):155–164
67. Patil KR, Rocha I, Forster J, Nielsen J (2005) Evolutionary programming as a platform for in silico metabolic engineering. BMC Bioinform 6:308
68. Choi HS, Lee SY, Kim TY, Woo HM (2010) In silico identification of gene amplification targets for improvement of lycopene production. Appl Environ Microbiol 76(10):3097–3105
69. Lewis NE, Nagarajan H, Palsson BO (2012) Constraining the metabolic genotype-phenotype relationship using a phylogeny of in silico methods. Nat Rev Microbiol 10(4):291–305
70. Nielsen J, Jewett MC (2008) Impact of systems biology on metabolic engineering of *Saccharomyces cerevisiae*. FEMS Yeast Res 8(1):122–131
71. Tyo KE, Kocharin K, Nielsen J (2010) Toward design-based engineering of industrial microbes. Curr Opin Microbiol 13(3):255–262
72. Bordel S, Agren R, Nielsen J (2010) Sampling the solution space in genome-scale metabolic networks reveals transcriptional regulation in key enzymes. PLoS Comput Biol 6(7):e1000859
73. Patil KR, Nielsen J (2005) Uncovering transcriptional regulation of metabolism by using metabolic network topology. Proc Natl Acad Sci USA 102(8):2685–2689
74. Dahl RH, Zhang F, Alonso-Gutierrez J, Baidoo E, Batth TS, Redding-Johanson AM, Petzold CJ, Mukhopadhyay A, Lee TS, Adams PD, Keasling JD (2013) Engineering dynamic pathway regulation using stress-response promoters. Nat Biotechnol 31(11):1039–1046
75. Redding-Johanson AM, Batth TS, Chan R, Krupa R, Szmidt HL, Adams PD, Keasling JD, Soon Lee T, Mukhopadhyay A, Petzold CJ (2011) Targeted proteomics for metabolic pathway optimization: application to terpene production. Metab Eng 13(2):194–203
76. Fuhrer T, Heer D, Begemann B, Zamboni N (2011) High-throughput, accurate mass metabolome profiling of cellular extracts by flow injection-time-of-flight mass spectrometry. Anal Chem 83(18):7074–7080
77. Canelas AB, Harrison N, Fazio A, Zhang J, Pitkanen JP, van den Brink J, Bakker BM, Bogner L, Bouwman J, Castrillo JI, Cankorur A, Chumnanpuen P, Daran-Lapujade P, Dikicioglu D, van Eunen K, Ewald JC, Heijnen JJ, Kirdar B, Mattila I, Mensonides FI, Niebel A, Penttila M, Pronk JT, Reuss M, Salusjarvi L, Sauer U, Sherman D, Siemann-Herzberg M, Westerhoff H, de Winde J, Petranovic D, Oliver SG, Workman CT, Zamboni N, Nielsen J (2010) Integrated multilaboratory systems biology reveals differences in protein metabolism between two reference yeast strains. Nat Commun 1(9):145
78. Kizer L, Pitera DJ, Pfleger BF, Keasling JD (2008) Application of functional genomics to pathway optimization for increased isoprenoid production. Appl Environ Microbiol 74 (10):3229–3241
79. Brennan TCR, Turner CD, Kromer JO, Nielsen LK (2012) Alleviating monoterpene toxicity using a two-phase extractive fermentation for the bioproduction of jet fuel mixtures in *Saccharomyces cerevisiae*. Biotechnol Bioeng 109(10):2513–2522
80. Dunlop MJ, Dossani ZY, Szmidt HL, Chu HC, Lee TS, Keasling JD, Hadi MZ, Mukhopadhyay A (2011) Engineering microbial biofuel tolerance and export using efflux pumps. Mol Syst Biol 7:487
81. Alper H, Moxley J, Nevoigt E, Fink GR, Stephanopoulos G (2006) Engineering yeast transcription machinery for improved ethanol tolerance and production. Science 314 (5805):1565–1568
82. Newman JD, Marshall J, Chang M, Nowroozi F, Paradise E, Pitera D, Newman KL, Keasling JD (2006) High-level production of amorpha-4,11-diene in a two-phase partitioning bioreactor of metabolically engineered *Escherichia coli*. Biotechnol Bioeng 95(4):684–691
83. Stark D, von Stockar U (2003) In situ product removal (ISPR) in whole cell biotechnology during the last twenty years. Adv Biochem Eng Biotechnol 80:149–175

Adv Biochem Eng Biotechnol (2015) 148: 161–200
DOI 10.1007/10_2014_290

Published Online: 31 January 2015

Metabolic Engineering of Higher Plants and Algae for Isoprenoid Production

Chase Kempinski, Zuodong Jiang, Stephen Bell and Joe Chappell

Abstract Isoprenoids are a class of compounds derived from the five carbon precursors, dimethylallyl diphosphate, and isopentenyl diphosphate. These molecules present incredible natural chemical diversity, which can be valuable for humans in many aspects such as cosmetics, agriculture, and medicine. However, many terpenoids are only produced in small quantities by their natural hosts and can be difficult to generate synthetically. Therefore, much interest and effort has been directed toward capturing the genetic blueprint for their biochemistry and engineering it into alternative hosts such as plants and algae. These autotrophic organisms are attractive when compared to traditional microbial platforms because of their ability to utilize atmospheric CO2 as a carbon substrate instead of supplied carbon sources like glucose. This chapter will summarize important techniques and strategies for engineering the accumulation of isoprenoid metabolites into higher plants and algae by choosing the correct host, avoiding endogenous regulatory mechanisms, and optimizing potential flux into the target compound. Future endeavors will build on these efforts by fine-tuning product accumulation levels via the vast amount of available "-omic" data and devising metabolic engineering schemes that integrate this into a whole-organism approach. With the development of high-throughput transformation protocols and synthetic biology molecular tools, we have only begun to harness the power and utility of plant and algae metabolic engineering.

Keywords Isoprenoids · Terpenoids · Metabolism · Genetic engineering · Metabolic engineering · Plants · Algae · Sub-cellular · Tissue-specific · Mevalonic acid pathway · Methylerythritol phosphate pathway · Plastid · Mitochondria · Terpene synthase · Prenyl transferase · Synthetic biology · Genetically modified organisms · Natural products · Carbon neutral · Biotechnology

---

C. Kempinski · Z. Jiang · S. Bell · J. Chappell (✉)
Department of Pharmaceutical Sciences, University of Kentucky,
Lexington, KY 40536, USA
e-mail: chappell@uky.edu

Contents

1 Introduction

Isoprenoids constitute one of the most structurally diverse classes of compounds found in nature and are synthesized from two C_5 isomeric precursors: isopentenyl diphosphate (IPP) and dimethylallyl diphosphate (DMAPP). These compounds have many roles in eukaryotic biology with special importance in plants. These roles are both critical (i.e., hormones, aliphatic membrane anchors, maintaining membrane structure) and unique (i.e., defense compounds, insect/animal attractants). There are tens of thousands of specialized metabolites (metabolites that are nonessential for normal cell homeostasis) and a significant proportion of these are isoprenoids, with well over 55,000 examples [178, 189]. Many excellent reviews have summarized the biosynthesis and roles of these compounds; the general model for isoprenoid biosynthesis is formation of the above-mentioned C_5 precursors by either the mevalonate (MVA) pathway which is present in the cytoplasm/endoplasmic reticulum (ER) of plants (and is the route of isoprenoid biosynthesis ubiquitous in all eukaryotes) or the methylerythritol phosphate (MEP) pathway that is present in prokaryotes and the plastid compartment of plants (Fig. 1) [5, 12, 79, 161, 171, 189]. Briefly, these C_5 precursors are then conjugated together in an additive fashion to generate prenyl diphosphates of varying chain lengths (in increments of five carbons) which are then cyclized and/or modified by terpene synthases, and can then be further decorated by downstream enzymes such as cytochrome P450s [80]. In plants, the MVA pathway is the primary pathway for the biosynthesis of sesqui- (C_{15}) and triterpenes (C_{30}), both generated from the C_{15} intermediate farnesyl diphosphate (FPP). The MEP pathway is the primary pathway for the biosynthesis of hemi- (C_5),

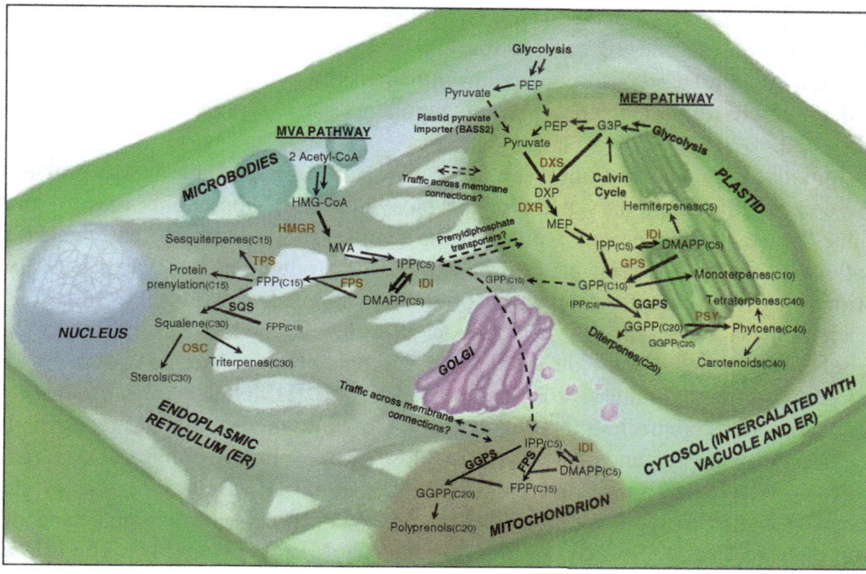

Fig. 1 Summary of the two isoprenoid biosynthetic routes in plant cells. The mevalonic acid (MVA) pathway is shown starting with two acetyl-coenzyme A (CoA) molecules and proceeds through two steps (not shown) in the cytosol at microbodies to produce 3-hydroxy-3-methylglutaryl-CoA (HMG-CoA) which is reduced by HMG-CoA reductase (*HMGR*) at the ER interface to produce MVA. Three more steps (not shown) produce IPP and it is interconverted to its isomer DMAPP by isopentenyl diphosphate isomerase (*IDI*). Two molecules of IPP and one molecule of DMAPP are condensed to produce farnesyl pyrophosphate (*FPP*) by farnesyl pyrophosphate synthase (*FPS*). FPP serves as a precursor for sesquiterpenes, which are formed by terpene synthases and can be decorated by other various enzymes. Two FPP molecules are condensed by squalene synthase (*SQS*) at the ER face to produce squalene the precursor for triterpenes and sterols, which are generated by oxidosqualene cyclases (*OSC*) and are modified by various downstream enzymes. The methylerythritol phosphate (MEP) pathway present in the plastid compartment initiates its isoprenoid biosynthesis with pyruvate and D-glyceraldehyde 3-phosphate (*G3P*) condensed into 1-deoxy-D-xylulose 5-phosphate (*DXP*) by DXP synthase (*DXS*). Glycolysis and the Calvin cycle provide the precursors for the MEP pathway and recent studies have elucidated plastid importers of pyruvate (*BASS2*) and phosphoenolpyruvate (*PEP*) [47, 52]. DXP is then converted to MEP by DXP reductoisomerase (*DXR*). Five steps (not shown) convert MEP into IPP, which is also interconverted to DMAPP by the plastid IDI enzyme. DMAPP/IPP are also the precursor for hemiterpenes. One molecule of IPP and one molecule of DMAPP are condensed to form geranyl pyrophosphate (*GPP*) by GPP synthase (*GPS*). GPP is the immediate precursor for monoterpenes. Further condensations of two molecules of IPP to GPP forms geranylgeranyl pyrophosphate (*GGPP*) by GGPP synthase (*GGPS*). GGPP is the precursor for diterpenes and 2 GGPPs can be condensed by phytoene synthase (*PSY*) to form phytoene, the immediate precursor for carotenoids and tetraterpenes. Mitochondria also contain nuclear encoded, catalytically active isoforms of IDI, GGPS, and FPS. Mitochondrial FPP is critical for ubiquinone. Transport of precursors and prenyl diphosphates between membranes and sites of biosynthesis are represented by dashed arrows, while direct catalytic reactions are depicted by bold arrows. Enzymes that catalyze important regulatory steps or are important engineering targets are indicated in bold dark red. Figure adapted from information in [20, 52, 57, 123]

mono- (C_{10}), di- (C_{20}), tetra- (C_{40}), and polyterpenes (>C_{50}) which are generated from additions to the 10 carbon starting molecule geranyl diphosphate (GPP). Figure 1 summarizes the isoprenoid biosynthetic pathways operating in higher plants and highlights important enzymes and regulatory steps within the typical higher plant cell.

Initially, the identification of the MVA pathway suggested that it was probably the conserved evolutionary path for the biosynthesis of isoprenoids in all organisms. However, pioneering work starting in the 1980s showed that the MEP pathway was the bacterial method of isoprenoid production and that it was also primarily responsible for isoprenoid production in plant plastids [94, 150, 151]. Each pathway's catalytic enzymes have been well described and several important regulatory elements are known [171]. Canonically, in the MVA pathway, 3-hydroxy-3-methylglutaryl coenzyme A reductase (HMGR) is one of the key regulatory steps controlling the flux of acetyl-CoA into the isoprenoid precursors in the MVA pathway, and 1-deoxy-D-xylulose 5-phosphate synthase (DXS) and 1-deoxy-D-xylulose 5-phosphate reductoisomerase (DXR) are key regulatory elements in the MEP pathway (Fig. 1) [62]. For some time evidence suggested the important role of DXS in controlling flux into the MEP pathway [42]. The role of DXS has been further underlined by several studies looking at MEP pathway carbon flux using ^{13}C labeling experiments [54, 179]. These studies corroborate the findings by Estévez et al. [42], showing that both increased and decreased DXS activity directly appears to influence MEP pathway flux. However, overexpression of *DXS* alone did not seem to increase downstream isoprenoid accumulation highly, indicating other critical regulatory steps at downstream enzymes. It is important to note that DXS activity does not directly correlate with increased *DXS* transcription and posttranscriptional regulation of DXS activity has also been described [134]. The regulatory system of DXS appears typical for both isoprenoid pathways: control of flux is complex, occurring at multiple points in each pathway, in response to many different types of stimuli, and through varying methods across all levels of regulation (i.e., transcriptionally, posttranscriptionally, and posttranslationally). Untangling these regulatory networks is further complicated by the presence of gene families for each coding enzyme of the plant isoprenoid pathways. Furthermore, these gene families have varying compositions from species to species with specialized functions (e.g., active for specific developmental regimes) for the various isozyme members in each [62].

This subcellular division of labor in higher plants is an important biochemical facet that can be capitalized upon using metabolic engineering that will become apparent. The division effectively partitions major biochemical control over two areas within the cell, each of which may be exploited to avoid the endogenous regulation that controls production in either the cytoplasm or plastid. During the elucidation of the plastid MEP pathway, several clues hinted at the independence of plastid isoprenoids from the cytosolic MVA pathway. Labeling studies using MVA precursors indicated they were not effectively incorporated into carotenoids, diterpenes, or monoterpenes. Furthermore, chlorophylls and carotenoids were not affected in plants treated with the MVA pathway inhibitor, mevinolin, and initially it was speculated that the inhibitor was not able to permeate through the plastid envelope. Interestingly though, chloroplasts were able to incorporate exogenous IPP [150]. Mitochondria are also able to import IPP; however, mitochondria are void of an endogenous pathway that synthesizes the basic five-carbon prenyl unit, and must obtain IPP from the cytosolic MVA pathway [35]. This is critical for the production of polyprenols such as those present in ubiquinone [62]. It is appreciated that flux through the MEP pathway is more robust than flux through the MVA pathway. This is due to the plastid synthesizing large amounts of carotenoids and chlorophyll polyprenyl chains primarily from photosynthetically obtained CO_2. The MVA pathway precursor, acetyl-CoA, is a common intermediate in several metabolic processes within the cell and is derived from the breakdown of carbohydrate (glycolysis), fatty acids (ß-oxidation), and amino acid recycling [62].

Each pathway appears to control its flux independently, however, they do have some basal level of intermediate sharing and this may be more prevalent during specific times of growth [63]. Continued study of various isoprenoid compounds indicates that the separation of these two pathways and the isolation of their respective prenyldiphosphate precursors do not follow any generalizable rules, and that noncanonical sharing of substrates across pathway boundaries does appear to occur. Whether these exchanges are facilitated by transporters or through another biophysical mechanism (such as intermediate sharing of tocopherol precursors as described by Mehrshahi et al. [111]) is not known, and may be a combination of methods [62].

Many of the isoprenoids produced by plants are of high value to humans. They have uses in medicine (e.g., the anticancer diterpene Taxol, the antimalarial sesquiterpene artemisinin), cosmetics (e.g., the triterpene squalene, the many fragrant monoterpenes), industry (e.g., the monoterpenes and diterpene resin acids for chemical feedstock), and agriculture (e.g., sesquiterpenes involved in complex ecological trophic interactions and isoprenoid-derived plant hormones such as the gibberellins [10]). However, many of these compounds are chemically complex with organic synthesis sometimes being prohibitively costly and complicated for large-scale production. Thus, much effort has been focused on producing these compounds in biological hosts using metabolic/genetic engineering technologies that have become possible within the past few decades. Although biological production has mostly focused on production using microbial fermentation systems, these require their own expensive infrastructure, feedstock (usually sugars), and purification methods. Compared to microbial systems, plants are especially appealing because they require a simple infrastructure, including water and fertilizer, are easily amenable to variable production scales, and are environmentally friendly by fixing atmospheric carbon into desired compounds.

However, plants are complex multicellular organisms, which have unique evolutionary adaptations that make them more difficult to engineer genetically when compared to microbes. Although efforts have indicated that metabolically engineering plants is feasible, further work will continue to refine and identify the most successful strategies in developing plant production platforms. The purpose of this chapter is to summarize important considerations, technologies, and strategies that

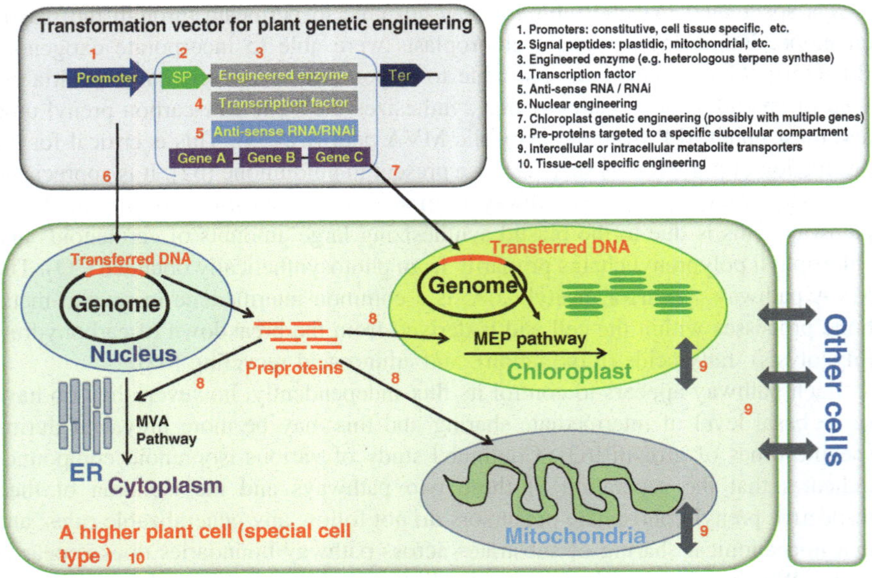

Fig. 2 A schematic blueprint illustrating important variables and approaches to metabolic engineering in plants. This figure shows a typical transgene design that can capitalize upon various permutations to yield the desired flux of carbon to the biosynthesis of unique end-products. The process(es) controlled by each arrow are defined within the inset figure legend

should be considered when designing a metabolic engineering effort in higher plants, and also touches on work and the state of metabolic engineering in green algae. These important considerations begin at the macroscale in deciding upon an appropriate plant host in which the desired engineering might prove most effective, and continue to discussions of how tissue-specific (and cell-specific) targeting can be used in particular circumstances. Illustrative case examples where these engineering approaches have been successful are highlighted. Microscale and molecular approaches are also considered, with specific examples presented when applicable. Specific molecular techniques and engineering approaches that have been successful, or may be successful in the future, are also discussed.

Overall, we stress three essential components that underwrite any engineering strategy: (1) manipulating gene(s) expression (e.g., using heterologous genes from a species unrelated to the host under a promoter that drives the desired temporal and/or spatial expression patterns, or altering endogenous gene expression to change flux towards the desired pathway); (2) avoiding endogenous regulation (e.g., using specific hosts and targeting the introduced pathway to specific tissues and subcellular compartments); and (3) combining the first two considerations in unique and novel ways with the available technologies (e.g., using metabolic models, synthetic biology to ensure enzyme cooperation, using transporters to sequester accumulated compounds). In addition to the choice of plant host, these are all variables that can be controlled by construct design and are best summarized in Fig. 2.

Before summarizing the technologies and efforts to engineer terpene metabolism in plants, an important distinction that arises from the literature needs to be mentioned: there is a difference between efforts to characterize genes, proteins, and biochemical pathways using transgenic plants as a tool for basic research versus efforts to introduce transgenes aimed towards the accumulation of large amounts of desired compound(s). As expected, many of the initial efforts to alter terpene metabolism in plants were proof-of-concept experiments [64, 90, 122, 163, 172, 190] and the first dedicated efforts to engineer particular isoprenoid compounds into plants soon followed [9, 12, 39, 73, 90, 100–102, 181]. Several recent review articles summarize many of these efforts in the field quite well [4, 87, 117, 160] and it is not the aim of this chapter to review comprehensively all the efforts to engineer isoprenoid compound accumulation in plants. Instead, we focus on the strategies important to obtain high levels of desired isoprenoid compounds.

2 Host Choice

The choice of host plant for metabolic engineering should be based upon relevant scientific criteria such as whether the endogenous flux of carbon through a desired pathway is high enough and flexible enough to accommodate the funneling of precursor metabolites towards a desired product. Additional criteria include how amenable the plant host is to genetic transformation, will the host produce enough biomass to obtain the desired amount of compound(s), and is the host devoid (or need to be made devoid) of endogenous enzymes that may modify or metabolize the desired end-product compound(s). Most of the current work has been done in model species that are amenable to transformation, have short life-spans (allowing for efficient genetic experiments), and for whom we generally have a more thorough understanding of their biochemistry (e.g., *Arabidopsis* and *Nicotiana*). Host choice should also consider the endogenous biochemical milieu of the target, especially if there exists the potential to modify the engineered compound chemically or sequester it via an undesired biochemical transformation. One way to avoid these latter concerns has been to choose a host for which the engineered metabolism is completely foreign (i.e., there is no endogenous amount of that compound produced naturally). However, this may not be possible either due to the compound's ubiquity (e.g., squalene [180]) or the desire to alter the balance of compounds natively produced in a particular species (e.g., tocopherols). However, even introduction of foreign metabolism may not be enough to prevent further compound derivatization, as seen in work by Lücker et al. [100], where they saw modification of engineered *S*-linalool to *S*-linalyl-β-D-glucopyranoside in engineered *Petunia*. In that study, control plants did not natively produce *S*-linalool, but transgenic plants did emit *S*-linalool into the headspace from leaves, yet did not have any detectable *S*-linalool in extracted leaf tissue. Upon treating transgenic tissue with hydrolyzing enzymes, free linalool was detected. Using HPLC-MS/MS and multidimensional gas chromatography–mass spectrometry with comparison to standards, they were able to elucidate the conjugated linalool as *S*-linalyl-β-D-glucopyranoside. However, the main problem with model species is that they generally do not produce much biomass (e.g., *Arabidopsis thaliana* or *Brachypodium distachyon*), and therefore are of little applied use, besides proving that the particular engineering can be accomplished at a genetic and biochemical level.

When thinking of substantial biomass we generally think of crop plants (e.g., maize, rice, soybean, sugarcane, etc.), many of which have proven to be more recalcitrant to genetic transformation and thus have fewer successful reports of engineering efforts. However, with recent developments in increasing the efficiency of transformation for previously low-throughput species (as in the work done by Sahoo et al. [156] and Zhang et al. [186]), it is reasonable to speculate that the introduction of transgenes will not be the limiting factor in conducting metabolic engineering experiments in plants in the years to come. Therefore, host selection should revolve around the desired engineering outcome and how a particular plant can match those needs (i.e., whether the plant displays substantial tissue biomass for tissue-specific accumulation, if desired). Examining host choice from a more applied and agronomic sense, it may prove worthwhile to consider the choice of a particular line in light of where the plant would actually be grown. Selecting a line having already been bred to display resistance to pathogens and/or particular climate nuances would undoubtedly produce higher product accumulation. Perhaps most importantly, the host to be chosen should be a high-performing line, meaning it should display robust growth characteristics in a field setting with an emphasis on yield of desired engineered tissues.

3 Organ- and Tissue-Specific Engineering

It's been well known that certain plant cells and tissues have unique chemical profiles, especially for the case of specialized metabolites (e.g., avenacins found in oat roots [106], alkaloids in the latex of poppy [43], and sesquiterpenes in glands of cotton Elzen et al. [40]) for which recent technologies and methods in metabolomics and transcriptomic studies have provided more understanding of the genes and enzymes involved in the production of unique metabolites [118, 119, 149, 161]. Often these unique metabolites are of special value and many of them are terpenoids (or terpenoid-derivatives). Therefore, it could be beneficial to hijack the biochemical flux through these specific cellular pathways without the potential disruption of plant homeostasis that often occurs when constitutive promoters are used. In an engineering strategy designed for large amounts of product accumulation, cell-/tissue-/organ-specific targeting would probably have the best results in one of four main cell/tissue/organ types: seeds, fruits, trichomes, and leaves. There are other cell/tissue types that are specific to certain plant species, such as resin glands in gymnosperms or laticifers in Hevea species and plants in the Asteraceae and Euphorbiaceae, which could also serve as excellent platforms in which to divert endogenous carbon flux [75]. However, engineering efforts into these cell types are limited by our incomplete understanding of the overall biology of these specialized cell types. The ability to target single cell types or even specific tissues is really limited by the number of promoters fully characterized for a specific pattern of gene expression/cell specificity. Future production platforms might also benefit from the development of synthetic promoters that could allow for more tailoring, tunable, optimal gene expression [71].

3.1 Leaves

With regard to organ specificity, leaf-specific targeting does not seem to be a goal towards which much effort has been expended. This is partly due to the historical use of constitutive promoters that drive ubiquitous expression and thus, leaf expression. Because leaves usually are the most accessible and easy to work with of plant tissues, they generally serve as the evaluation platform for testing new constructs or strategies using constitutive promoters to drive construct expression. Leaves are generally the default tissue engineered product accumulation is assayed within due to their biomass that they serve as the basic model in which plant biochemistry and metabolism are considered. This is intuitive because the leaf serves as the main site of photosynthesis and contains all of the primary metabolites derived from photosynthate that could be diverted into heterologous compounds. Presented here are a few cases showing how the separation of leaf-specific and constitutive expression most likely depends on the desired product outcome, and that it is difficult to anticipate the phenotypic effect that a particular compound may elicit *in planta*.

Plant isoprenoid production as a result of pathogen attack has been well documented [77, 128, 169, 170]. These elicited compounds can function to prevent further pathogen attack and recruit predators of the attacking pathogen. Recent studies have continued to describe the functions of isoprenoids in these roles [69]. Thus, engineering accumulation (or reduction) of these compounds to protect plants against pathogens is an intriguing proposition. Rodríguez et al. [147] engineered an antisense limonene synthase into oranges (under the *CaMV35S* promoter) and found reduced limonene (and other monoterpenes, sesquiterpenes, and monoterpene aldehydes), increased monoterpene alcohols, and most interestingly a marked resistance to infection by *Penicillium digitatum*, *Xanthomonas citri* subsp. *Citri*, and less attraction to the fruits by the insect *Ceratitis capitata*. Although introduction of the antisense limonene synthase changed the terpenoid profile of the orange fruits, the transgenic plants were phenotypically identical to wild-type (WT) plants. Engineering of maize plants with an oregano (*E*)-ß-caryophyllene (EbC) synthase resulted in an increase in aboveground herbivory during field growth, and olfactory assays showed a preference by *Spodoptera frugiperda* towards the transgenic lines [145]. These EbC-expressing lines did not repress *S. frugiperda* growth any more than WT lines. Thus, engineering the accumulation (or reduction) of the terpene constituents can affect the resistance characteristics of plants to pathogens in unexpected ways. Although, as previously reported, constitutive *EbC* expression in maize did decrease root damage by the economically relevant western corn rootworm, *Diabrotica virgifera virgifera* LeConte, probably through attraction of entomopathogenic nematodes [32, 145].

These *EbC*-expressing maize lines did show reduced root herbivory, however, the agronomic performance of the plants in a field setting was compromised. Although plant height and water content in the *EbC*-expressing lines were similar to WT, the leaf biomass was only 25 % of WT and yield was reduced compared to WT lines [145]. On average, the *EbC*-expressing maize plants emitted β-caryophyllene and α-humulene between 542–2,024 pg/h* plant and these values did not significantly change after attack by the *S. frugiperda* larvae [145]. These results suggest two possibilities for the reduced plant biomass: carbon flux into the engineered compounds is so robust that it is stealing isoprenoid precursors away from downstream products that are essential for plant growth (e.g., chlorophyll), or there is a certain amount of accumulating monoterpene that is directly harmful to plant cell growth. Inasmuch as tissue accumulation of β-caryophyllene or α-humulene was not presented in these studies [32, 145] the first possibility cannot be definitively ruled out. Other work has supported the latter possibility.

Work by Wu et al. [181] showed that plants engineered to produce patchoulol within the plastid accumulated high levels in leaf tissue and also emitted the sesquiterpene into the headspace (emission was at a higher level than the monoterpenes in maize presented above, 50–100 ng/h*g fresh weight, [FW]), and that some of these transgenic lines showed vein clearing and chlorosis of the leaves, indicating precursor funneling from plastid-produced pigments to the engineered terpene. Wu et al. [180] also showed that transgenic tobacco lines producing high amounts of triterpenes in the plastid (using a multigene approach) also had reduced biomass (Fig. 3). In addition, Aharoni et al. [2, 3] saw stunted growth with Arabidopsis engineered with a *S*-linalool/[3*S*]-*E*-nerolidol synthase (*FaNES1*) targeted to the plastid. Similarly, Aharoni et al. [4] saw stunted growth and leaf necrosis in potato plants engineered with *FaNES1* under control of the strong chrysantheum Rubisco small subunit promoter [127].

In contrast to these results, Cahoon et al. [12] engineered tocotrienol accumulation into *Arabidopsis* leaves using barley homogentisate geranylgeranyl transferase (HGGT) under the *CaMV35S* promoter, and reported elevated tocopherols and tocotrienols increased to 700–900 μg/g dry weight (DW) from <100 μg/g DW with tocotrienols accounting for ~85 % of the total content. These transgenic plants had tocopherol levels approximately equal to the WT plants, but no changes in the levels of chlorophyll or carotenoids. Sitbon and Jonsson [164] engineered *Nicotiana tabacum* with sterol methyltransferase genes one and two (*SMT1* and *SMT2*) under a *TET1/35S* promoter and saw reduced biomass in only *SMT2* engineered plants (Table 1). Interestingly, they did not observe changes in total sterol content in the transgenic lines versus WT, but did observe changes in the sterol profile of the transgenic lines versus the WT. These results suggest that unknown or unappreciated levels of metabolite control can act as compensatory mechanisms to maintain homeostasis and simultaneously allow high accumulation

Fig. 3 Phenotypic effects of high accumulation of plastid-targeted triterpene production in tobacco. Plants engineered to express plastid-targeted FPS and SQS under putative trichome specific promoters (*far left*) and constitutive promoters (*middle*) have varying degrees of compromised growth compared to WT (*far right*). Transgenic plants presented are homozygous (T2 generation). Taken from Wu et al. [180]

Table 1 Mean heights (±SD) of tobacco plants engineered with sterol methyltransferases

Genotype	Height (cm)	Height relative to control
Control	115 ± 4 (16)	
GmSMT1		
Line 6	114 ± 4 (16)	−1 %
Line 51	109 ± 4 (10)	−6 %
AtSMT		
Line 6	94 ± 4 (10)	−18 % (*)
Line 8	95 ± 3 (10)	−17 % (*)

Two lines from each transformation event are presented: *Glycine max* Sterol Methyltransferase 1 (GmSMT1) and *Arabidopsis thaliana* Sterol Methyltransferase 2 (AtSMT2). Heights of AtSMT2 plants (but not GmSMT1) plants had significantly reduced height (Student's *t*-test, $P < 0.001$). Taken from Sitbon and Jonsson [164] of the desired product. This also reiterates the point that the choice of the enzyme(s) in a pathway where attempts to push or pull carbon downstream towards the desired endpoint is of crucial importance.

3.2 Seeds

Although leaf biomass can be prolific in some plants (e.g., tobacco), seeds are probably the oldest target of selection/engineering, usually to increase biomass (yield) for animal/human nutritional value, and in oilseeds, to increase oil content for both industrial and nutritional value. There have been successful efforts in diverting carbon flux into terpenoid compounds in seeds, perhaps the best known being the introduction of β-carotene biosynthesis into rice endosperm, producing the titular "Golden Rice" of the engineered plant [183]. Ye et al. [183] did this by engineering in heterologous versions of PSY, phytoene desaturase, and lycopene β-cyclase. These enzymes were targeted to the plastid (the site of carotenoid biosynthesis) and segregating lines showed β-carotene accumulation of varying levels with the highest presented line producing 1.6 μg/g endosperm. However, expressing only the endogenous PSY under the *Brassica napus NAPIN1* (*BcNA1*) seed-specific promoter in *Arabidopsis* saw an increase in carotenoids and chlorophyll, with the largest change being a 43-fold increase in β-carotene (260 μg/g FW) compared to WT (6 μg/g FW). Fujisawa et al. [51] introduced seven genes into the seeds of *B. napus* for the production of ketocarotenoids, including a synthetic IDI to increase the availability of upstream IPP. Engineered *B. napus* seeds produced 19–30 times higher carotenoids (412–657 μg/g FW) compared to the WT, with 60–190 μg/g FW being ketocarotenoids. Importantly, they also noticed that introduction of heterologous genes altered the expression of endogenous carotenoid biosynthesis genes. Similar changes to endogenous carotenoid gene expression were also seen in Gutensohn et al. [57] when tomato fruits were engineered with a GPS small-subunit (GPS-SSU) from snapdragon.

Holmberg et al. [58, 66] showed elevated sterols when seeds were targeted to express heterologous sterol methyltransferases, one of the rate-limiting steps into sterol biosynthesis. Savidge et al. [159] showed elevated levels of total seed tocopherols, up to 60 % increases, when engineered with *Arabidopsis* homogentisate phytyl transferase (*HPT1*) under the expression of the *BcNA1* seed-specific promoter. Cahoon et al. [12] engineered tocotrienol accumulation into *Zea mays* embryos using barley HGGT under the embryo-specific *Zea mays* 16 kDa oleosin gene promoter and found increased whole seed tocotrienol production at least 20-fold compared to WT whole seeds (Fig. 4). These efforts indicate that altering terpenoid flux in developing seeds is possible and even more important, that they are generated in a natural storage repository that can be harvested/extracted at a later time. The fact that seeds can serve as storage for produced compounds with little biochemical alteration once the seed is dormant, is of special interest, especially if the site of production and extraction are physically separated or long-term storage is desired.

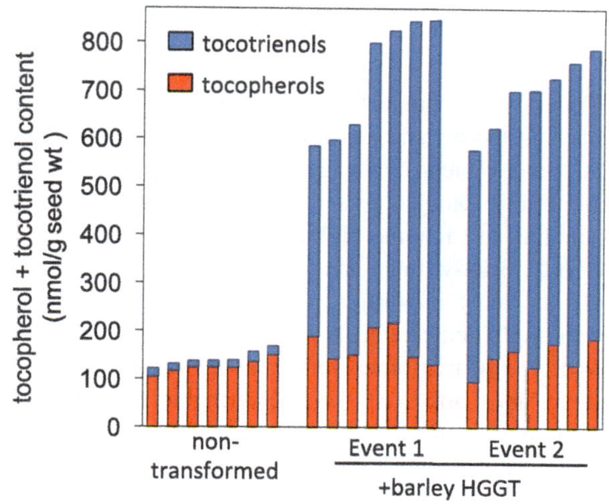

Fig. 4 *Zea mays* seeds engineered with barley HGGT under a seed-specific promoter. Tocotrienol production was increased at least 20-fold compared to WT. Modified from Cahoon et al. [12]

3.3 Fruits

Like seeds, fruits also have the potential to be natural storage tissues and targeting to fruits is desirable for compounds that have beneficial impacts on human nutrition or could deter pathogens (avoiding the application of pesticides that could be detrimental to the environment and human/animal health). Work has been done in both of these arenas to varying degrees of success. Like the efforts to increase the nutritional content of seeds by engineering in carotenoid accumulation, similar, early work was done in tomato fruits, which are naturally rich in carotenoids [48, 154]. Introduction of a *PSY* gene increased total tomato fruit carotenoid accumulation approximately 2-fold (over 2 mg/g DW in transgenic fruits vs. 1.3 mg/g DW for WT plants).

In contrast, when a GPS-SSU gene was introduced, it caused a 5-fold reduction in overall carotenoid accumulation in tomato fruits (~215 μg/g FW in WT vs. an average of ~41 μg/g FW in the transgenic lines). Heterologous expression of the snapdragon GPS-SSU was previously observed to control prenyl-chain length formation of endogenous GGPS enzymes in transgenic tobacco, resulting in more GPP and less GGPP accumulation [126]. Hence, the decrease in carotenoid formation is most likely due to the heterologous expressed GPS-SSU saturating the endogenous GGPS enzymes, causing a reduction in the biosynthesis of the carotenoid precursor GGPP. Interestingly, the concentrations of some of the lycopene-derived carotenoids (β-carotene, zeaxanthin, and lutein) were not significantly decreased in the GPS-SSU transgenic fruits [57]. This indicates a much more complex regulatory network controlling the flux of GGPP into carotenoid compounds, and suggests the existence of multiple branch pathways to specific carotenoid end-products.

3.4 Trichomes

Trichomes encompass a group of specialized cells that originate from the epidermis of plant tissues and are differentiated on the basis of their biochemical capabilities. A recent review by Lange and Turner [89] summarizes the current knowledge of isoprenoid biosynthesis in trichomes and touches upon how our current understanding in trichome biology might be harnessed for use in metabolic engineering efforts. Briefly, there are several types of trichomes but glandular trichomes (GTs) are the most important in terms of specialized metabolism capacity due to their seemingly dedicated role as metabolite production factories. The presence, number, and type of trichomes vary between plant species. Thus, the capacity to engineer them does not exist for every engineering project (although, this could be an important factor in choosing a particular production host). Equally important, glandular trichomes can secrete their products onto the leaf surface, facilitating collection, or the glandular head synthesized compounds may accumulate in cavities/sacs associated with the metabolically active cells. Trichomes secreting hydrophobic compounds including isoprenoids generally have interesting intracellular features such as an extensive smooth ER network that maintains contact with nonpigmented leucoplasts. The leucoplasts appear to have a nonuniform shape that could be implicated in increasing the contact surface area with the smooth ER [89]. These extensive connections between intracellular membrane networks may be critical for transport of large amounts of hydrophobic compounds. A recent study has also implicated a possible role of a lipid transfer protein in the export of isoprenoids from *Nicotiana* tall GTs [17].

The species in which isoprenoid biology of the trichomes has been best studied is the pathway leading to the production of (−)-menthol in *Mentha x piperita* (peppermint), which occurs exclusively in this plant's trichomes. The biosynthesis and subcellular organization of (−)-menthol production has been well described [24]. The production of this monoterpene through the MEP pathway (Fig. 1) illustrates the robust carbon flux through trichome plastids. Further support for a high carbon flux through the MEP pathway in trichomes is the high level of Δ^9-tetrahydrocannabinolic acid accumulation in *Cannabis sativa*. Tetrahydrocannabinolic acid contains an integrated GPP molecule attached to a phenolic precursor that allows for further carbon- and lactone-cyclization yielding the final product [89]. Trichome-specific promoters have also been identified that enable trichome-specific transgene targeting and novel biotechnology applications (summarized in [168]). The natural capacity for trichomes to function as specialized biochemical factories, in fact, makes them intriguing targets for the redirection of flux into heterologous, high-valued compounds [180].

3.5 Roots

Targeting compound accumulation in roots may pose special technical challenges as well as physical limitations on harvesting. However, compound accumulation in roots can be beneficial for defense compounds that prevent pathogen attack in the soil and for nutritional enhancement in tubers or other root vegetables. As noted above, Degenhardt et al. [32] and Robert et al. [145] showed that root emissions of β-caryophyllene and α-humulene attracted nematodes that helped prevent root herbivory by western corn rootworms.

Work by Diretto et al. [34] also demonstrated how the chemical profile of roots was amenable to manipulation by genetic engineering. Silencing the lycopene epsilon cyclase (*LCY-e*) increased β-carotene levels >13-fold and total carotenoids were increased >2-fold in transgenic tubers relative to WT. The highest accumulating tubers showed a mild yellow phenotype. Follow-up work by Diretto et al. [33] reported engineering PSY, phytoene desaturase/carotene isomerase, and lycopene β-cyclase, all expressed under tuber-specific promoters, caused a greater than 3600-fold increase in β-carotene (47.4 μg/g DW) compared to WT. Total tuber carotenoid content was also increased >19-fold (up to ~114 μg/g DW) in these lines. These high carotenoid producing transgenic lines showed a distinct yellow color (compare wild-type to engineered lines P-YBI-17 and P-YBI-30; Fig. 5). These same investigators also transformed potato plants with the same three bacterial genes, but only placed the *PSY* gene under a tuber-specific promoter, whereas the remaining two genes were driven by the constitutive *35S* promoter (K-YBI-41). Transgenic lines with expression of the *PSY* gene under the tuber-specific promoter definitely prevented severe chlorosis/growth retardation as previously reported by Fray et al. [50].

The transgenic lines constitutively expressing bacterial phytoene desturase/carotene isomerase and lycopene β-cyclase (but tuber-specific *PSY*) showed hampered regeneration of transgenic plantlets. The authors attributed this to a reduced ability to produce pigments necessary for photosynthesis, which was correlated with a chlorotic phenotype (compare middle leaf, to the WT and the tuber-specific expression line, P-YBI-17 in Fig. 5). These negative phenotypes were not observed when all three genes were placed under tuber-specific promoters. Indeed, these pK-YBI-41 plants showed approximately half the leaf chlorophyll accumulation of WT plants [33]. When these constructs were transformed into *Solanum tuberosum* L. cv. Désirée, which has an endogenous tuber carotenoid content of approximately 5.6 μg/g DW, the transgenic lines accumulated 6-fold more carotenoid, up to 35 μg/g DW. However, when the same bacterial *PSY* gene was transformed into the *S. tuberosum* cv. Mayan Gold (which has an endogenous tuber carotenoid content of approximately 20 μg/g DW), the highest carotenoid accumulating lines had nearly four times the amount of total carotenoid (78 μg/g DW) as found in the nontransgenic parental line [37]. This illustrates the importance of choosing a plant cultivar that naturally exhibits a robust metabolic flow through the desired pathway in order to generate the highest accumulating lines.

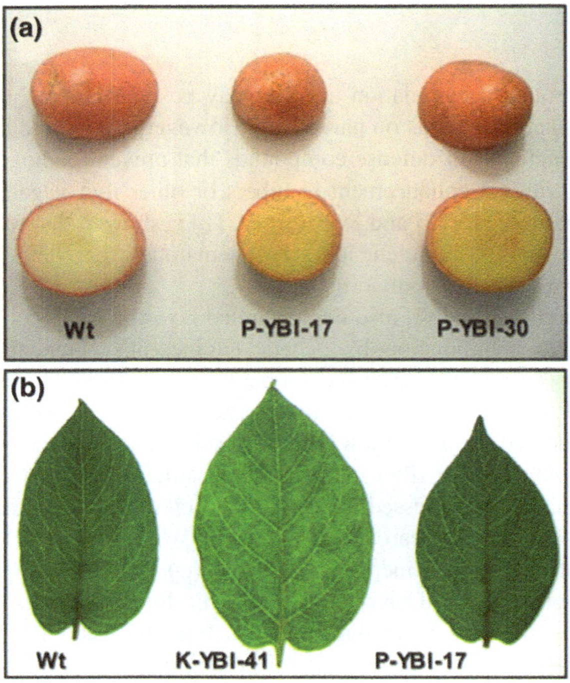

Fig. 5 Transgenic *Solanum tuberosum* engineered with multiple carotenoid biosynthetic genes. Plants were engineered with phytoene synthase, phytoene desaturase/carotene isomerase, and lycopene β-cyclase. P-YBI-17 and P-YBI-30 transgenic lines had all genes expressed under tuber-specific promoters. K-YBI-41 plants had only phytoene synthase under a tuber-specific promoter, while the other two genes were expressed constitutively under the direction of the 35S promoter. **a** Transgenic tubers show a yellow pigmentation. **b** Note the apparent decreased leaf pigmentation of line K-YBI-41 compared to the WT and the tuber-specific P-YBI-17 line. Taken from Diretto et al. [33]

Other work utilized the *Or* transcription factor, originally identified as the gene controlling orange pigmentation in cauliflower mutants by causing the formation of chromoplast structures and the increased accumulation of carotenoid content [98]. Transformation of the *Or* gene into *S. tuberosum* cv. Désirée under the control of the primarily tuber-specific starch synthase promoter resulted in orange pigmented potato tubers containing >5-fold higher carotenoid content than the nontransgenic parental line ($\sim$31 μg/g DW) [92, 97].

Altogether, these observations have demonstrated that carotenoid content can be manipulated by the overexpression of either genes encoding for biosynthetic enzymes or relevant trans-acting regulatory factors. An interesting question remaining unanswered is if overexpression of the *Or* gene in combination with the bacterial carotenoid biosynthetic genes could enhance carotenoid accumulation even more in potato than either of the separate engineering strategies?

3.6 Other Specialized Cell Types

Many plants contain specialized cells that serve as endogenous synthesis and storage compartments for specialized metabolites. Often these cells are utilized to sequester defense compounds that would otherwise impose their cytotoxic activities onto the host plant. An example of this cell type is the resin ducts formed in conifer species upon herbivore or fungal attack [76]. These cells produce prodigious amounts of monoterpenes and diterpene resin acids along with smaller amounts of sesquiterpenes [75]. The formation and biochemistry of these terpenoid factory cells is still a matter of investigation. In one study, for instance, introducing a *Pinus abies* bifunctional isoprenyl diphosphate synthase (IDS; IDS catalyzes the biosynthesis of GPP and GGPP) into *Pinus glauca* did not reveal an increase in any of the monoterpenes, sesquiterpenes, diterpene resin acids, sterols, or carotenoids in any tissue. The prenyl diphosphate intermediates, GPP and GGPP, were increased in needles but not bark, and higher levels of esters of geranylgeraniol with fatty acids were noted [120]. This was unexpected because IDS expression was driven by the constitutive *Zea mays* ubiquitin 1 promoter, and although transcript accumulation was much lower in the bark than in needles, no real metabolic changes were observed in the bark tissue. This is indicative of unique control mechanisms that might need further elucidation before transgenic conifers could be used to increase production of the valuable oleoresins.

In contrast to the induced formation of resin ducts in conifers, lactifers are reticulated, interconnected cells present in particular plant species and often in particular tissues and that generate and accumulate latex. Latex is a combination of many specialized compounds, including phenolics, proteins, alkaloids, terpenes, and the main constituent being polyisoprenoids, that is, natural rubber which can contain polyisoprene chains with >18,000 isoprenyl subunits [1, 8]. Human utilization of latex has been important for thousands of years, with an expanding role more recently for its use in industrial production of diverse rubber products. Interestingly, not all the steps in rubber biosynthesis have been elucidated [18], which has definitely limited the prospects for engineering greater rubber production capacities [8].

Specialized plant cell cultures might also serve as production platforms for important terpenoid compounds [11, 146]. These systems have been especially important for high-value pharmaceuticals that tend to accumulate to low levels *in planta*, such as providing intermediates from *Taxus baccata* for the semisynthetic production of Taxol, portending similar opportunities for terpene indole alkaloids (TIAs) from *Catharanthus roseus* and artemisinin from *Artemisia annua* cultures.

4 Engineering Isoprenoid Metabolism in Plant Organelles

As noted earlier, metabolic flux in isoprenoid biosynthesis revolves around a complex network that involves multiple subcellular compartments within the plant cell: the cytosolic MVA pathway, the plastid-localized MEP pathway, localization of many sequential isoprenoid biosynthetic enzymes arrayed on the ER membrane or targeted to the mitochondria and microbodies, and the possible participation of the vacuole as a storage/sequestration organelle (Fig. 1). Hence, targeting of enzymes and biosynthetic capacity to an appropriate subcellular location is a prerequisite for successful metabolic engineering in plants, which must overcome several layers of complicated regulation [61, 87]. Early metabolic engineering attempts achieved less success in elevating terpene production by introducing enzymes into the cytoplasm without considering the regulatory complexity of the MVA pathway [181]. Now, engineering terpene metabolism into organelles to overcome these innate regulatory constraints has become an important strategy, as illustrated by the following two examples.

It is well established that the biosynthesis of distinct classes of isoprenoids take place in distinct organelles (Fig. 1). For example, the major steps of monoterpene, diterpene, and carotenoid biosynthesis are known to operate in higher plant chloroplasts, so most of the engineering efforts to enhance biosynthesis of these compounds have used heterologous enzymes fused with a plastid-targeting signal peptide to direct them to their endogenous site of biosynthesis (or by introducing the transgene into the plastid genome directly). This type of "straightforward" genetic engineering has been conducted in various species of higher plants, extensively reviewed (6, 45, 49, 88, 7, 114, 116, 182), and is not reiterated in this section other than to note that this type of engineering utilizes pre-existing substrate biosynthesis, which is captured and diverted to produce compounds that may or may not be normally present within a particular organelle. Hence, this type of strategy does not completely avoid endogenous regulatory elements that may respond directly to the synthesized compound or indirectly through changes in intermediates and large amounts of target compound accumulation. It is also necessary to be aware of any inherent regulatory systems (e.g., allosteric sites) of the introduced enzymes that may inhibit their activity.

In order to avoid this, targeting introduced enzymes to particular organelles to divert available substrate(s) for the formation of a novel pathway that is foreign to the organelle is becoming a commonplace strategy. In an early study, a strawberry linalool/nerolidol synthase *FaNES1*, was targeted to mitochondria by fusing a mitochondrial targeting signal sequence to the amino terminus of the FaNES1 protein, leading to generation of two novel sesquiterpenes in transgenic *Arabidopsis*, (*3S*)-(*E*)-nerolidol and (*E*)-DMNT, neither of which is found in WT lines [73]. In another recent study [44], amorphadiene-4,11-diene synthase (ADS) targeted to the mitochondria was coexpressed with a cytochrome P450 (CYP71AV1), a cytochrome P450 reductase (CPR), an artemisinic aldehyde reductase (DBR2) from *A. annua*, and a truncated and deregulated HMGR from yeast in *N. tabacum*. This resulted in artemisinin accumulation of 5–7 μg/g DW, which is 8-fold more than when ADS was targeted to the cytoplasm [44]. These results provide strong evidence that mitochondrial targeted FaNES1 and ADS have access to FPP in the mitochondria and convert this intermediate into (3*S*)-(*E*)-nerolidol and amorpha-4,11-diene, respectively. These reaction intermediates were then were further converted by native (and unknown enzymes in *Arabidopsis*) or engineered enzymes (CYP71AV1, CPR, DBR2) present in cytosol to the final products (*E*)-DMNT and artemisinin, respectively. Although it is not clear how the intermediate product (3*S*)-(*E*)-nerolidol and amorpha-4,11-diene are shuttled between the mitochondria and cytosol, the results show that the heterologous terpene synthases can efficiently use the FPP pool in the mitochondria for novel sesquiterpene production. In contrast, free pools of FPP in chloroplasts and the cytoplasm must be less available in these species because simply introducing a FPP-dependent synthase does not yield appreciable novel terpene accumulation [3, 44, 73, 181].

Wu et al. [181] furthered this approach by applying it to chloroplasts. Accumulation greater than ~25 μg/g FW of the nonnative sesquiterpenes patchoulol and amorpha-4,11-diene, was achieved in transgenic *N. tabacum* when an avian FPS gene and nonnative sesquiterpene synthase, ADS or patchoulol synthase (PTS), genes were transformed into the nuclear genome with plastid targeting signal sequences appended to the amino-terminus of encoded proteins. Using the same strategy, up to ~500 μg/g FW of the triterpenes squalene and botryococcene were achieved when an avian FPS gene in combination with either a yeast squalene synthase (SQS) or an engineered algal botryococcene synthase (BS) were targeted to the chloroplast compartment [180] and Jiang et al., before unpublished). Conceptually, the plastid-targeted FPS diverts the IPP and DMAPP intermediates from the MEP pathway towards the accumulation of free FPP, the substrate for sesquiterpene and triterpene production. The biosynthesis of FPP, sesquiterpenes, and triterpenes are foreign to the chloroplast (Fig. 1), and introduction of these two-step biochemical pathways allows for elevated nonnative terpene production in the plastid compartment, sheltering these nonnative biosynthetic pathways from any native regulatory mechanisms and hence allowing for an unlimited flux of carbon to a desired terpene. Kumar et al. [84] corroborated this notion in an independent study where the entire yeast MVA pathway (a total of six enzymes) was introduced into the chloroplast genome of tobacco. The resulting homoplasmic transgenic lines accumulated multiple isoprenoid products, including mevalonate, carotenoids, sterols, squalene, and interestingly, triacylglycerides.

When engineering novel biosynthetic capacity into the chloroplast, there is an important choice to be made between plastidic transformation versus nuclear transformation. Plastidic transformation offers several advantages over nuclear transformation: homologous recombination methodology exists, expression of transgene operons could improve coordinated gene expression, transgene inheritance should only pass from maternal tissue, nuclear epigenetic effects should not present any difficulties, and the translation of expressed transgenes into protein is likely to be higher compared to nuclear transgene expression [27, 29, 84]. However, there are no direct comparisons of isoprenoid production by plastid genome transformations versus nuclear genome transformations expressing the same enzymes within or targeted to the chloroplast. Moreover, the higher level of protein expression does not always positively correlate with a higher level of terpene production and reflects the need to consider many factors, such as: protein (enzyme) activity, substrate availability, flux control within the pathway, and other regulatory complexities that may exist in the plastid. Thus, the best transformation strategy, nuclear versus plastidic, will probably vary on a case-by-case basis.

Metabolic engineering of higher plant chloroplasts offers additional potential for high yields of isoprenoid production: (1) there are many plastids within each higher plant cell (up to 50) that have polyploidy genomes (up to 1,000 copies per plastid); (2) because they are the site of photosynthesis, carbon flux in chloroplasts is robust and theoretically could provide for an unlimited supply of precursors; (3) chloroplasts may offer a good environment for exogenous protein folding, expression, and activity; (4) chloroplasts appear especially suitable for heterologous isoprenoid production due to lax endogenous regulation of the MEP pathway as compared to the MVA pathway operating in the cytosol; and (5) there is now the possibility to improve photosynthetic efficiency, which could increase production of engineered compounds [74].

A recent review by Heinig et al. [61] discussed the current challenges in conducting subcellular targeting in plant metabolic engineering. They suggested important considerations needed to be given to ensure substrate availability, whether this be overcome by co-introducing transporters, upstream catalytic enzymes to increase pathway flux, or suppressing endogenous pathways that bleed away precursors or cofactors. Although the most successful subcellular targeting efforts so far have used nuclear-encoded propeptides, which include transit peptides that direct the final protein to its intracellular destination, further development and optimization of plastid-encoded enzymes could allow for even higher titers of products. Engineering genetic constructs into the chloroplast has been possible since the late twentieth century [28, 82, 110], and high protein titers have been obtained, however, engineering chloroplast-encoded catalytically active enzymes has been less successful. Hence, there appears to be an inherent limitation in the chloroplast for the biosynthesis of high levels of catalytically competent enzymes, a problem that currently does not seem to be receiving considerable attention.

5 Altering Expression Patterns of Endogenous Genes

5.1 Downregulation

Downregulation or knock-out of endogenous biosynthetic genes is another important strategy commonly used to regulate/redirect metabolic pathway flux. These efforts attempt to suppress or abolish gene expression of a possible competing enzyme, thus flux can be redirected into the desired enzyme/pathway. However, this technique has not been used substantially and effectively in plant metabolic engineering compared with efforts in microbial systems. This is, in part, because of the difficulties in obtaining specific plant mutants and the low efficiency in obtaining appropriate amounts of downregulation of gene expression in plants (due mainly to large unknown genomes and/or redundant genes). Nevertheless, a number of studies have reported that antisense RNA and RNAi (RNA interference) techniques have been successfully used in manipulating plant terpene metabolism to increase terpenoid production.

Monoterpene essential oil production was elevated (61 % yield increase over WT plants) in transgenic peppermint expressing peppermint antisense (+)-menthofuran synthase (MFS) with simultaneous overexpression of DXR [88]. Downregulation of MFS alone was shown to decrease the level of side-product (+)-menthofuran, and redirected carbon flux to desirable monoterpene oil production, leading to an increased oil yield of roughly 35 % [104].

Tuber-specific expression of antisense fragments for either *LCY-e* or *β-carotene hydroxylase*, the genes encoding the enzymes that compete for lycopene and further metabolism of β-carotene, respectively, led to significantly increased levels of β-carotene (up to 14-fold and 38-fold, respectively) and total carotenoids (up to 2.5-fold and 4.5-fold, respectively) in potato tuber [33, 34]. When *LCY-e* was suppressed by an RNAi approach, increased carotenoid content in *B. napus* seeds was reported [185]. RNAi was also used in several studies to increase the content of artemisinin in *A. annua* by downregulating SQS and β-caryophyllene synthase, both enzymes that compete for FPP, which is a key intermediate in the artemisinin biosynthetic pathway [15, 46, 187]. When the *C. roseus* gene encoding 7-deoxyloganic acid 7-hydroxylase (*CrDL7H*), which is involved in secologanin biosynthesis, was suppressed by virus-induced gene silencing the accumulation of secologanin was reduced by at least 70 %. Critically, the accumulation of the intermediate, 7-deoxyloganic acid (the substrate for CrDL7H), was 4 mg/g FW in silenced plants whereas this compound is normally undetectable in WT plants [157].

Transgenic oranges with reduced levels of limonene caused by an antisense downregulation of the (+)-limonene synthase gene, were shown to be resistant to economically important pathogens [147]. Exactly how the reduced level of limonene in fruits activates a defense response has yet to be determined, but reduced limonene accumulation does correlate with increased levels of *GGPS* which could provide substrate for the formation of diterpene antimicrobial compounds that inhibit pathogen infection. The authors of this study suggested that the link between limonene accumulation and pathogen attack could play an ecological role in facilitating seed dispersal by allowing frugivores easier access to the fruit pulp [148].

5.2 Transcription Factors

Most specialized metabolites in plants accumulate when plants respond to acute developmental or environmental signals [14]. Therefore, it seems reasonable that their biosynthetic genes would be regulated in a coordinated manner by transcriptional factors. Although transcriptional regulation of the isoprenoid biosynthetic pathways is not well characterized, metabolic engineering of certain regulatory genes (the transcriptional factors) provides a novel approach to enhance terpene production in plants [129].

Recent studies have reported transcription factors in *A. annua* that appear to regulate artemisinin biosynthesis. AaWRKY1 was characterized as a transcription factor that regulates the native *ADS* gene in *A. annua*. Transient expression of

AaWRKY1 also led to increased transcript accumulation of a majority of artemisinin biosynthetic genes [103]. Two jasmonate-responsive (*AaERF1* and *AaERF2*) and a trichome-specific (*AaORA*) AP2/ERF transcription factors were also characterized as positive regulators for artemisinin biosynthesis in *A. annua*. Overexpressing either transcription factor resulted in increased accumulation of artemisinin and artemisinic acid [99, 184]. In contrast and contrary to expectations, constitutive expression of an *Arabidopsis* blue light receptor, *CRY1*, gene in *A. annua* increased the abundance of the *FPS*, *ADS*, and *CYP71AV1* transcripts, three important enzymes in artemisinin biosynthesis, and led to increased accumulation of artemisinin and anthocyanins [67].

Catharanthine accumulation was improved up to 6.5-fold higher than WT in *C. roseus* hairy roots by coexpression of the *ORCA3* transcription factor and the gene encoding for geraniol 10-hydroxylase, an enzyme involved in the TIA biosynthetic pathway [173]. Likewise, overexpression of the *Arabidopsis* transcription factor *Agamous-like 12* in *C. roseus* suspension cells promoted enhanced accumulation of ajmalicine, a TIA with antihypertensive properties [115]. *Arabidopsis* transcription factors MYC2 and MYB21 have also been reported to regulate the expression of sesquiterpene synthase genes *TPS11* and *TPS21*. When mutated, the *Arabidopsis myc2* and *myb21* mutants emit less sesquiterpene volatiles from their flowers than the WT plants, which was correlated with reduced levels of the *TPS11* and *TPS21* mRNAs [68, 142].

6 Understanding and Utilizing Compartmentalization

Understanding the flux of metabolites throughout the plant cell and attempts to use *in silico* algorithms to predict how specific changes in one metabolite may affect overall cell homeostasis has been an active area of interest. Efforts in creating integrative maps for comprehending the systemic biology of an organism is a rather new field that has become more accessible with the proliferation of available bioinformatics data and computing power. Full-scale system simulations are not available yet, partially due to our lack of understanding of full reaction dynamics and kinetics. Current genome-scale models have four main constituents: a list of enzymes and their reactions, the types and constraints on those reactions (e.g., reversibility, capacity), gene–protein–reaction linkages, and functions that dictate cellular activity, that is, the stipulations by which the listed reactions will be constrained within the cellular model [162]. Databases defining a biological system in terms of its biochemical, genetic, and genomic data have been generated for several types of species, including plants. Plants that have had genome-scale models generated for them include: *Arabidopsis* [16, 113, 125, 132, 137, 177], *Zea mays* [26, 155], *Sorghum bicolor* [26], *Saccharum officinarum* [26], *Hordeum vulgare* [56, 112, 152], *Oryza sativa* [85], and *B. napus* [59, 60, 131].

In the case of *Arabidopsis*, one of the metabolic networks incorporated 5,253 gene-reaction inputs with a total of 1,567 unique reactions and 1,748 metabolites, and it also considered the subcellular divisions of the typical plant cell (e.g., plastids, mitochondria, peroxisomes) but was limited in the consideration of cellular metabolic pathways that it models [26]. These models are still being refined and so far only consider primary metabolism, thus they have limited utility in understanding flux through the isoprenoid pathways especially when the system is altered by heterologous enzymes, although they can give insight into disconnects between available precursors and experimentally observed flux. However, recent mathematical models for understanding important regulatory components of essential oil production in peppermint have received experimental support [143, 144]. The reader is directed to two recent reviews discussing metabolic modeling for further details [124, 162].

One of the annotations that is important in generating genome-scale models across the subcellular locations of a plant cell, is the role of transporters in allowing accessibility of specific metabolites to the various enzymes of specialized cellular compartments. Metabolite transporters for terpenoid compounds probably exist to facilitate interorganelle sharing of intermediates (Fig. 1). However, identification of these transporters has been lacking. Doshi et al. [36] conducted analyses on bacterial homologues (including some mutated versions) of the ATP-binding cassette (ABC) exporter, MsbA, which is responsible for exporting the lipid A-moiety of lipopolysaccharide from the inner to outer leaflet of the *E. coli* inner membrane. They found that heterologous expression of some of these MsbA homologues allowed an increase in carotenoid excretion from *E. coli* cells also engineered for novel carotenoid biosynthesis. Although these transporters could export the hydrophobic carotenoid compounds, natural transporters for all types of isoprenoids have yet to be elucidated. There is ongoing research into this area and recently a plasma membrane ABC transporter from *N. tabacum*, NtPDR1, was described that is involved in diterpene export across the plasma membrane. Expression of *NtPDR1* was observed across several tissue types and induced in response to several stimuli [25]. ABC transporters for the secretion of hydrophobic cuticular waxes in epidermal cells have been identified [158] as well as a plastid-localized ABC transporter important for importing ER-synthesized lipids [153]. In light of this, it seems reasonable to speculate that specialized transporters for intra- and extracellular movement of isoprenoid compounds may exist. Alteration of isoprenoid exchange within the cell and deposition to the exterior of the cell could also be altered using engineered lipid transfer proteins (LTPs).

As stated previously, a recently identified tobacco trichome LTP showed increased trichome exudate of alkanes and diterpenes when overexpressed [17]. Other recent studies have identified additional putative lipid transfer proteins from *B. napus* and *Zea mays* which remain to be biochemically validated [93, 191]. However, even if these LTPs do not normally transport isoprenoid substrates, it may be possible to engineer them to recognize certain isoprenoid compounds. Using transporters within a metabolic engineering strategy could allow for sequestration of produced molecules within a specialized structure away from the site of synthesis, allowing for higher titers by removing the product from the site of synthesis, decreasing possible negative feedback on the biosynthetic enzyme(s), and promoting continued pathway flux via simple mass-action principles.

The notion of using synthetic biology to increase flux through a pathway has been a topic of considerable discussion. Synthetic protein scaffolds, where successive acting catalytic enzymes are tethered via specific epitopes to a protein scaffold, in theory allow for the creation of a synthetic metabolon in vivo. These synthetic metabolons can be designed so that the stoichiometry of the pathway components is optimized. This is exemplified by Dueber et al. [38] where a protein scaffold was used to tether three MVA pathway enzymes in a stoichiometrically optimized ratio, thus allowing for a 77-fold increase in MVA accumulation in scaffold-containing *E. coli* cells versus cells without the scaffold.

7 Algae Engineering

Algae are a polyphyletic group of water-dwelling, eukaryotic single and multicellular organisms with diverse morphological, physiological, and biochemical features. As do higher plants, algae fix CO_2 through photosynthesis and can ultimately sequester this carbon in numerous downstream products such as terpenoids. Mass algae cultures can produce large amounts of terpenoids that are used in nutraceutical [188], pharmaceutical [174], and renewable fuel [130] applications without competing for arable farmland used in food production. Moreover, algae can achieve higher cellular densities than plants, enabling them to produce more compound per unit of land area [53]. Several species are already cultivated on a large scale for production of carotenoids with useful antioxidant and pigmentation properties [58]. Higher titers of these and other terpenoids are always desirable and metabolic engineering offers one route to accomplish this using genetic engineering strategies [53, 55]. Currently, no one phylum can be labeled superior for terpenoid metabolic engineering; too little is known about the incredible diversity that exists. From a biotechnological standpoint members of chlorophyta, particularly *Chlamydomonas reinhardtii*, have received the greatest attention because they are the best studied "omically," physiologically, and biochemically, and because most terpenoid metabolic engineering efforts have been attempted in green algae.

Fundamental concepts for metabolic engineering of terpenoids in algae are largely built around strategies developed for higher plants; these include but are not limited to increasing substrate levels/access, deregulation of rate-limiting steps, reduction/elimination of competing pathways, and so on (see [31] for review). Accomplishing such feats first requires consideration of the pathway(s) that supply terpenoid precursors in the algae species of interest. Depending on phylogeny, the MVA and MEP biosynthetic pathways responsible for supplying IPP are differentially dispersed. Primary endosymbiotic algae such as glaucophyta and rhodophyta (red algae) typically possess both the MVA and MEP pathways whereas chlorophyta (green algae) only harbor the plastidic MEP pathway; secondary endosymbiotic algae phyla euglenophyta, chlorarachniophyta, and heterokontophyta generally utilize both pathways [96].

As noted for higher plants, metabolic engineering of terpenes in chlorophyta can be approached differentially by expressing transgenes from either the nuclear or chloroplast genome. Methodology exists for both types of genetic manipulations in *Chlamydomonas reinhardtii*, *Haematococcus pluvialis*, *Chlorella ellipsoidea*, *Chlorella sorokiniana*, *Chlorella kessleri*, *Volvox carteri*, and *Gonium pectorale* [133, 175]. Recombinant DNA is integrated randomly in the nuclear genome presumably through double-stranded chromosome breakage followed by nonhomologous end joining; homologous recombination has only been described in the nuclear genome of one nonchlorophyte algae, *Nannochloropsis* sp. [79]. As in higher plants, algal chloroplast transformations proceed through homologous recombination lending better genetic precision and yielding more reliable titers of protein, however, no significant manipulations of chloroplast terpenoid metabolism have been reported, possibly because of protein misfolding or the lack of posttranslational modifications that are needed for obtaining catalytically active enzyme [135]. Nuclear engineering offers alternative means for protein folding and posttranslation modifications. However, this approach has been plagued by low-level expression of transgenes [83, 139] due to transgene silencing [166] and positional effects [121]; hence, many independent transgenic lines per gene construct must be assessed before high producers are identified. Other important aspects of algae terpenoid metabolic engineering include promoter/terminator choice, intron inclusion, codon optimization, exogenous versus endogenous isoforms of catalytic enzymes, transformation method, and metabolite assessment (colorimetric, spectroscopic, chromatographic, cytometric) [53, 55, 70, 175]. However, despite all the advances in algal biotechnology [140], no precedent-setting examples have been published for algae metabolically engineered to produce high levels of terpenoids or any other specialized metabolites from nuclear encoded transgenes [21–23, 91, 95, 165].

Lackluster results for algal terpenoid engineering insinuate underlying biological issues not yet fully appreciated. One area that has been investigated extensively is construct design for nuclear transformation. Transgene promoter analyses suggest that nuclear gene silencing occurs via a chromatin-based, epigenetic mechanism (histone methylation/acylation) that can be bypassed if proper *cis*-elements are placed upstream of a strong constitutive promoter [166]. One study demonstrated successful nuclear transgene expression by using an endogenous nuclear promoter from the highly expressed, intronless *psaD* gene, but also noted the importance of a transcriptional terminator element as well [83]. These investigators also determined that transformation with linear DNA improved transformant recovery and transgene expression [83].

An alternative approach to promoter/terminator choice has been constructing inframe chimeras with transgenes fused to an antibiotic resistance enzyme via a 2A linker. When this chimera is expressed, two separate polypeptides are produced due to ribosomal skipping of a peptide bond in the 2A linker sequence. In effect, the cell is forced to express both proteins for survival because of the antibiotic present [139–141]. This has been utilized successfully to express active enzymes and could be used to target enzymes to various subcellular locales such the ER or chloroplasts [139–141].

In addition to construct design and transformation methodology, *C. reinhardtii* strains (UVM4 and UVM11) have been developed for higher efficiency expression of nonselectable transgenes [121]. The open-reading frame for *C. reinhardtii* SQS has been integrated into the nuclear genome of the UVM strains and although appreciable levels of the corresponding protein were detected, enzymatic activity was not assessed [81]. Thus, using the *C. reinhardtii* UVM strains to express nuclear transgenes inframe with the 2A linker and antibiotic resistance gene under control of the highly expressed, intronless *psaD* promoter and terminator may be beneficial in future attempts to engineer terpene metabolism in algae. Overcoming the nuclear transgene expression problem in *C. reinhardtii*, and presumably other green algae, is one hurdle that needs to be addressed before complex genetic engineering efforts such as expressing the entire MVA pathway in the cytosol of a green alga or adding extra copies of the MEP pathway enzymes can be contemplated.

Another key area for future algal metabolic engineering efforts will be to understand algae-specific oleaginous traits. New "omics" studies have suggested there are different genes and proteins contributing to the ability of a species to accumulate their respective lipids. Specifically, in the heterokont *Nannochloropsis* sp., gene dosage (13 copies of DGAT vs. 4 in *C. reinhardtii*) seems to be extremely important for the accumulation of high levels of TAGs [81]. Moreover, *B. braunii* race B constitutively expresses three distinct genes for DXS, presumably to promote high carbon flux through the MEP pathway and into the abundant triterpene molecules that it accumulates [107]. *In silica* predictions such as these are needed to guide the design of synthetic mechanisms and provide the testable hypotheses that will be necessary to uncover the fundamental principles controlling terpene metabolism in algae and higher plants [109].

High titers of naturally occurring algal terpenoids are possible as observed for *Dunaliella salina* [138] and *Botryococcus braunii* race B [41]. However, an overarching question is how do these algae accomplish this and what are the prospects for capturing these mechanisms and deploying them in other algae or plant species? *B. braunii* race B accumulates 30 % of its mass in triterpene oils, albeit at the cost of growth rate [41]. Increased growth rates can be achieved under various culture conditions while maintaining high levels of hydrocarbon production [78], but genetically engineered strains of this colony-forming algae have not been reported. Moreover, *B. braunii* hydrocarbon production and secretion into the extracellular matrix seem to be intimately linked with the cell cycle [167, 176], making prospects for the deployment of this mechanism into other algae more daunting. Thus, using *B. braunii* as a design template for terpenoid metabolic engineering approaches may not be as fruitful as other algae such as *Dunaliella salina*. This halophilic, marine green alga can accumulate 10–15 % of its mass in β-carotene stored in plastid globules; it can also form cytoplasmic globules of TAGs produced during stressful conditions [19, 30, 86, 136, 138]. Supposing that the carotenogenic enzymes could be knocked down/out in this alga, the available IPP/DMAPP in the chloroplast might be increased.; this could be bolstered by additional nuclear copies of DXS targeted to the plastid. Nuclear copies of isopentenyl diphosphate isomerase, prenyl phosphate synthase (such as FPS), and triterpene synthases including BS could then be targeted to the cytoplasm. Hypothetically, this could create an environment where the high levels of substrates produced in the chloroplast are drawn into the cytoplasm by mass-action effects and channeled into botryococcene (foreign molecule, no feedback regulation) that could be stored in pre-existing TAG cytoplasmic globules.

Currently, green algae are blank slates for testing new strategies to engineer higher terpenoid accumulations metabolically. Emerging information about specialized green algae metabolism will buttress systems/synthetic biology approaches where carbon fluxes are optimized among cellular pathways in order to achieve maximum growth and hydrocarbon production rates. Coupled with improved genetic engineering methodology and thorough bioinformatics, machine learning, and network flux analyses, ideal strategies will be revealed for manipulating terpenoid metabolism in green algae.

8 Conclusions and Future Directions

Several important considerations regarding metabolic engineering efforts in plants have been covered here, as well as a look at concerns that should be addressed before advanced higher plant and algae engineering can be successfully accomplished. Once a goal has been identified, one must approach the engineering variables from a macro- to microscale. However, the main choices are which host to engineer and how the genetic construct design will be designed. Construct design is where one can fine tune expression to specific cell/tissue/organ types or developmental stages using specific promoters, as well as direct the introduced protein(s) to discrete subcellular locales. A useful analogy for viewing internal metabolite flux is that of a plumbing system (Fig. 6). It is important to remember that the metabolic pathways within the cell are interconnected and that changing one parameter (such as swapping a connecting pipe within Fig. 6) will also alter flow through the various other pipes and pathways within the scheme. However, as represented by the open pipes within Fig. 6, we do not always know what and where certain metabolic pathways meet with each other nor how they physically interact. Certainly, a limitation in our knowledge base is that we can currently only obtain such information through empirical experimentation.

The future of metabolic engineering will need to embrace a holistic use of all available technologies presented. Early plant engineering efforts focused on proof-of-concept experiments using basic tools to ensure transgene expression and show that heterologous proteins can be expressed and be catalytically active (such as [64]). Now our primary challenges seem more dependent upon our ability to integrate vast amounts of genetic, proteomic, and metabolomic information to construct complex recombinant expression cassettes, and test these engineering designs in economically valuable hosts [72, 105, 108, 156, 186]. For instance, refinement of metabolic flux models will help to identify endogenous rate-limiting steps and control mechanisms that can be altered to increase desired product accumulation.

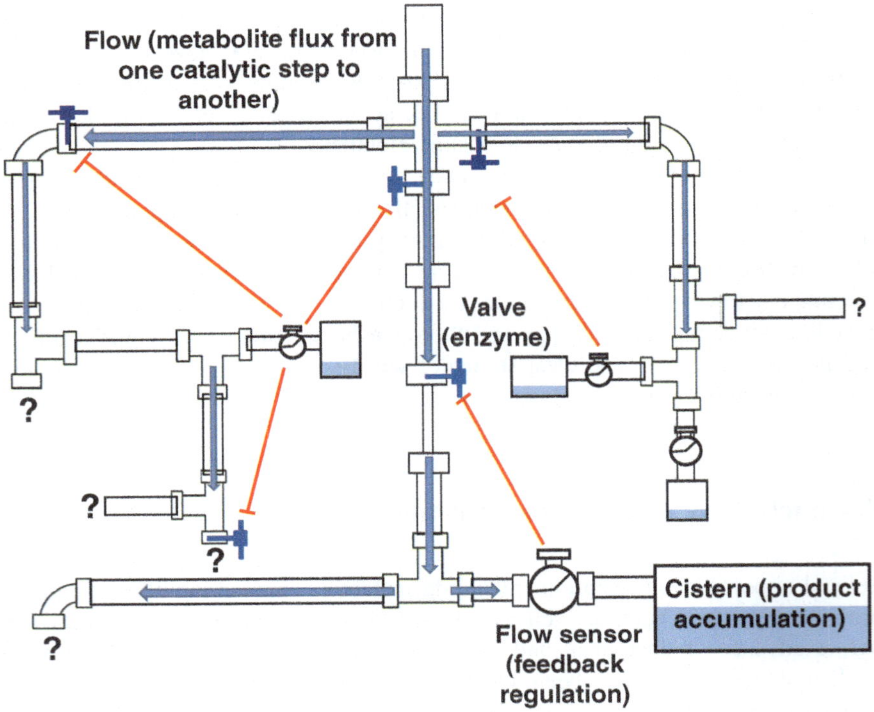

Fig. 6 Plumbing model and analogy for metabolic flux. One can imagine that carbon flux throughout the cell can be modeled as a system of connected pipes. The pipe junctions represent enzymes and the connecting pipes represent the reactions catalyzed by those enzymes—with thinner pipes representing rate-limiting steps. In this model the water collects in various cisterns which represent the final metabolic product. These cisterns have a certain maximum capacity (maximum amount of end product which can accumulate) and this is monitored by various sensors (feedback mechanisms), which can alter valves (blue tee-shaped objects) that represent regulatory proteins (*e.g.* transcription factors) controlling flux through the various connecting pipes or reactions. However, there are many steps designated by question marks for which we do not know how they might interact with our metabolic network

The continued improvement and success of future experiments will also rely on precise transgene tuning using synthetic biology components to introduce and integrate metabolic pathway(s) into seamless functional units and avoid unwanted feedback from homeostatic mechanisms. This may involve using synthetic promoters that allow transgene expression during an appropriate time in development and in coordination with suppression of endogenous pathway genes, using RNAi or repressive transcription factors, which would otherwise rob substrate from the introduced pathway. Engineering expression cassettes into specific genomic locales may also remove the necessity to screen large numbers of transgenic lines that have had transgenic DNA randomly inserted, and allow for a more rapid and direct comparison between engineering efforts in various cultivars of the same species.

Creating synthetic regulatory mechanisms that effectively partition precursor availability and provide robust flux to the desired product without allowing buildup of intermediates may be another way to utilize endogenous enzymes to boost overall productivity. Efforts that hope to generate plants applicable to actual commercial targets should also focus on obtaining the highest amount of product accumulation in high-biomass/yielding lines with low input requirements (e.g., fertilizer). Building better plant production platforms now only seems limited by our ability to assimilate all of the available biological and technological information into higher levels or more complex architectures taking into account the nuances of network interactions and connectivity.

Acknowledgments Work from the Chappell laboratory mentioned in this chapter has been supported by the USDA, DOE, NSF, and by Sapphire Energy, Inc. The authors are also grateful for the many insightful discussions with past and current members of the Chappell laboratory.

References

1. Agrawal A, Konno K (2009) Latex: a model for understanding mechanisms, ecology, and evolution of plant defense against herbivory. Annu Rev Ecol Evol Syst 40:311–331
2. Aharoni A, Giri A, Deuerlein S (2003) Terpenoid metabolism in wild-type and transgenic Arabidopsis plants. Plant Cell 15:2866–2884
3. Aharoni A, Giri AP, Deuerlein S, Griepink F, de Kogel W-J, Verstappen FWA, Verhoeven HA, Jongsma MA, Schwab W, Bouwmeester HJ (2003) Terpenoid metabolism in wild-type and transgenic Arabidopsis plants. Plant Cell 15:2866–2884
4. Aharoni A, Jongsma MA, Kim T-Y, Ri M-B, Giri AP, Verstappen FWA, Schwab W, Bouwmeester HJ (2006) Metabolic engineering of terpenoid biosynthesis in plants. Phytochem Rev 5:49–58
5. Bach TJ (1995) Some new aspects of isoprenoid biosynthesis in plants–a review. Lipids 30:191–202
6. Bai C, Twyman RM, Farre G, Sanahuja G, Christou P, Capell T, Zhu C (2011) A golden era–pro-vitamin A enhancement in diverse crops. In Vitro Cell Dev Biol 47:205–221
7. Beyer P (2010) Golden rice and 'golden' crops for human nutrition. New Biotechnol 27:478–481
8. Van Beilen JB, Poirier Y (2008) Production of renewable polymers from crop plants. Plant J 54:684–701
9. Besumbes O, Sauret-Güeto S, Phillips MA, Imperial S, Rodríguez-Concepción M, Boronat A (2004) Metabolic engineering of isoprenoid biosynthesis in Arabidopsis for the production of taxadiene, the first committed precursor of Taxol. Biotechnol Bioeng 88:168–175
10. Bohlmann J, Keeling CI (2008) Terpenoid biomaterials. Plant J 54:656–669
11. Bonfill M, Malik S, Mirjalili M, Goleniowski M, Cusido R, Palazón J (2013) Production and genetic engineering of terpenoids production in plant cell and organ cultures. In: Ramawat KG, Mérillon J-M (eds) Natural products. Springer, Berlin, pp 2761–2796
12. Cahoon EB, Hall SE, Ripp KG, Ganzke TS, Hitz WD, Coughlan SJ (2003) Metabolic redesign of vitamin E biosynthesis in plants for tocotrienol production and increased antioxidant content. Nat Biotechnol 21:1082–1087
13. Chappell J (1995) Biochemistry and molecular biology of the isoprenoid biosynthetic pathway in plants. Annu Rev Plant Physiol Plant Mol Biol 46:521–547

14. Chappell J, Nable R (1987) Induction of sesquiterpenoid biosynthesis in tobacco cell suspension cultures by fungal elicitor. Plant Physiol 85:469–473
15. Chen J-L, Fang H-M, Ji Y-P, Pu G-B, Guo Y-W, Huang L-L, Du Z-G, Liu B-Y, Ye H-C, Li G-F et al (2011) Artemisinin biosynthesis enhancement in transgenic *Artemisia annua* plants by downregulation of the β-caryophyllene synthase gene. Planta Med 77:1759–1765
16. Cheung CYM, Williams TCR, Poolman MG, Fell DA, Ratcliffe RG, Sweetlove LJ (2013) A method for accounting for maintenance costs in flux balance analysis improves the prediction of plant cell metabolic phenotypes under stress conditions. Plant J 75:1050–1061
17. Choi YE, Lim S, Kim H-J, Han JY, Lee M-H, Yang Y, Kim J-A, Kim Y-S (2012) Tobacco NtLTP1, a glandular-specific lipid transfer protein, is required for lipid secretion from glandular trichomes. Plant J 70:480–491
18. Chow K-S, Mat-Isa M-N, Bahari A, Ghazali A-K, Alias H, Mohd-Zainuddin Z, Hoh C-C, Wan K-L (2012) Metabolic routes affecting rubber biosynthesis in Hevea brasiliensis latex. J Exp Bot 63:1863–1871
19. Cifuentes AS, Gonzalez M, Conejeros M, Dellarossa V, Parra O (1992) Growth and carotenogenesis in eight strains of *Dunaliella saline* Teodoresco from Chile. J Appl Phycol 4:111–118
20. Closa M, Vranova E, Bortolotti C, Bigler L, Arro M, Ferrer A, Gruissem W (2010) The Arabidopsis thaliana FPP synthase isozymes have overlapping and specific functions in isoprenoid biosynthesis, and complete loss of FPP synthase activity causes early developmental arrest. Plant J 63:512–525
21. Cordero BF, Couso I, León R, Rodríguez H, Vargas MA (2011) Enhancement of carotenoids biosynthesis in *Chlamydomonas reinhardtii* by nuclear transformation using a phytoene synthase gene isolated from *Chlorella zofingiensis*. Appl Microbiol Biotechnol 91:341–351
22. Couso I, Cordero BF, Vargas MÁ, Rodríguez H (2012) Efficient heterologous transformation of *Chlamydomonas reinhardtii* npq2 mutant with the zeaxanthin epoxidase gene isolated and characterized from *Chlorella zofingiensis*. Mar Drugs 10:1955–1976
23. Couso I, Vila M, Rodriguez H, Vargas MA, León R (2011) Overexpression of an exogenous phytoene synthase gene in the unicellular alga *Chlamydomonas reinhardtii* leads to an increase in the content of carotenoids. Biotechnol Prog 27:54–60
24. Croteau RB, Davis EM, Ringer KL, Wildung MR (2005) (−)-Menthol biosynthesis and molecular genetics. Naturwissenschaften 92:562–577
25. Crouzet J, Roland J, Peeters E, Trombik T, Ducos E, Nader J, Boutry M (2013) NtPDR1, a plasma membrane ABC transporter from *Nicotiana tabacum*, is involved in diterpene transport. Plant Mol Biol 82:181–192
26. Dal'Molin CGDO, Quek L-E, Palfreyman RW, Brumbley SM, Nielsen LK (2010) C4GEM, a genome-scale metabolic model to study C4 plant metabolism. Plant Physiol 154:1871–1885
27. Daniell H (2006) Production of biopharmaceuticals and vaccines in plants via the chloroplast genome. Biotechnol J 1:1071–1079
28. Daniell H, Datta R, Varma S, Gray S, Lee SB (1998) Containment of herbicide resistance through genetic engineering of the chloroplast genome. Nat Biotechnol 16:345–348
29. Daniell H, Kumar S, Dufourmantel N (2005) Breakthrough in chloroplast genetic engineering of agronomically important crops. Trends Biotechnol 23:238–245
30. Davidi L, Katz A, Pick U (2012) Characterization of major lipid droplet proteins from Dunaliella. Planta 236:19–33
31. Davies FK, Jinkerson RE, Posewitz MC (2014) Toward a photosynthetic microbial platform for terpenoid engineering. Photosynth Res. doi:10.1007/s11120-014-9979-6
32. Degenhardt J, Hiltpold I, Köllner TG, Frey M, Gierl A, Gershenzon J, Hibbard BE, Ellersieck MR, Turlings TCJ (2009) Restoring a maize root signal that attracts insect-killing nematodes to control a major pest. Proc Natl Acad Sci USA 106:13213–13218
33. Diretto G, Al-Babili S, Tavazza R, Papacchioli V, Beyer P, Giuliano G (2007) Metabolic engineering of potato carotenoid content through tuber-specific overexpression of a bacterial mini-pathway. PLoS One 2:e350

34. Diretto G, Tavazza R, Welsch R, Pizzichini D, Mourgues F, Papacchioli V, Beyer P, Giuliano G (2006) Metabolic engineering of potato tuber carotenoids through tuber-specific silencing of lycopene epsilon cyclase. BMC Plant Biol 6:13
35. Disch A, Hemmerlin A, Bach TJ, Rohmer M (1998) Mevalonate-derived isopentenyl diphosphate is the biosynthetic precursor of ubiquinone prenyl side chain in tobacco BY-2 cells. Biochem J 331:615–621
36. Doshi R, Nguyen T, Chang G (2013) Transporter-mediated biofuel secretion. Proc Natl Acad Sci USA 110:7642–7647
37. Ducreux LJM, Morris WL, Hedley PE, Shepherd T, Davies HV, Millam S, Taylor MA (2005) Metabolic engineering of high carotenoid potato tubers containing enhanced levels of beta-carotene and lutein. J Exp Bot 56:81–89
38. Dueber JE, Wu GC, Malmirchegini GR, Moon TS, Petzold CJ, Ullal AV, Prather KLJ, Keasling JD (2009) Synthetic protein scaffolds provide modular control over metabolic flux. Nat Biotechnol 27:753–759
39. Van Eenennaam AL, Lincoln K, Durrett TP, Valentin HE, Shewmaker CK, Thorne GM, Jiang J, Baszis SR, Levering CK, Aasen ED et al (2003) Engineering vitamin E content: from Arabidopsis mutant to soy oil. Plant Cell 15:3007–3019
40. Elzen GW, Williams HJ, Alois AB, Stipanovic RD, Vinson SB (1985) Quantification of volatile terpenes of glanded and glandless Gossypium hirsutum L. cultivars and lines by gas chromatography. J Agric Food Chem 33:1079–1082
41. Eroglu E, Okada S, Melis A (2011) Hydrocarbon productivities in different Botryococcus strains: comparative methods in product quantification. J Appl Phycol 23:763–775
42. Estévez JM, Cantero A, Reindl A, Reichler S, León P (2001) 1-Deoxy-D-xylulose-5-phosphate synthase, a limiting enzyme for plastidic isoprenoid biosynthesis in plants. J Biol Chem 276:22901–22909
43. Facchini PJ, De Luca V (2008) Opium poppy and Madagascar periwinkle: model non-model systems to investigate alkaloid biosynthesis in plants. Plant J 54:763–784
44. Farhi M, Marhevka E, Ben-Ari J, Algamas-Dimantov A, Liang Z, Zeevi V, Edelbaum O, Spitzer-Rimon B, Abeliovich H, Schwartz B et al (2011) Generation of the potent anti-malarial drug artemisinin in tobacco. Nat Biotechnol 29:1072–1074
45. Farre G, Bai C, Twyman RM, Capell T, Christou P, Zhu C (2011) Nutritious crops producing multiple carotenoids–a metabolic balancing act. Trends Plant Sci 16:532–54
46. Feng L-L, Yang R-Y, Yang X-Q, Zeng X-M, Lu W-J, Zeng Q-P (2009) Synergistic re-channeling of mevalonate pathway for enhanced artemisinin production in transgenic *Artemisia annua*. Plant Sci 177:57–67
47. Fischer K, Kammerer B, Gutensohn M, Arbinger B, Weber A, Hausler R, Flugge U-I (1997) A new class of plastidic phosphate translocators: a putative link between primary and secondary metabolism by the phosphoenolpyruvate/phosphate antiporter. The Plant Cell 9:453–462
48. Fraser PD, Romer S, Shipton CA, Mills PB, Kiano JW, Misawa N, Drake RG, Schuch W, Bramley PM (2002) Evaluation of transgenic tomato plants expressing an additional phytoene synthase in a fruit-specific manner. Proc Natl Acad Sci USA 99:1092–1097
49. Fraser PD, Enfissi EMA, Bramley PM (2009) Genetic engineering of carotenoid formation in tomato fruit and the potential application of systems and synthetic biology approaches. Arch Biochem Biophys 483:196–204
50. Fray RG, Wallace A, Fraser PD, Valero D, Hedden P, Bramley PM, Grierson D (1995) Constitutive expression of a fruit phytoene synthase gene in transgenic tomatoes causes dwarfism by redirecting metabolites from the gibberellin pathway. Plant J 8:693–701
51. Fujisawa M, Takita E, Harada H, Sakurai N, Suzuki H, Ohyama K, Shibata D, Misawa N (2009) Pathway engineering of *Brassica napus* seeds using multiple key enzyme genes involved in ketocarotenoid formation. J Exp Bot 60:1319–1332
52. Furumoto T, Yamaguchi T, Ohshima-Ichie Y, Nakamura M, Tsuchida-Iwata Y, Shimamura M, Ohnishi J, Hata S, Gowik U, Westhoff P, Brautigam A, Weber APM, Izui K (2011) A plastidial sodium-dependent pyruvate transporter. Nature 476:472–475

53. Georgianna DR, Mayfield SP (2012) Exploiting diversity and synthetic biology for the production of algal biofuels. Nature 488:329–335
54. Ghirardo A, Wright LP, Bi Z, Rosenkranz M, Pulido P, Rodríguez-Concepción M, Niinemets Ü, Brüggemann N, Gershenzon J, Schnitzler J-P (2014) Metabolic flux analysis of plastidic isoprenoid biosynthesis in poplar leaves emitting and nonemitting isoprene. Plant Physiol 165:37–51
55. Gimpel JA, Specht EA, Georgianna DR, Mayfield SP (2013) Advances in microalgae engineering and synthetic biology applications for biofuel production. Curr Opin Chem Biol 17:489–495
56. Grafahrend-Belau E, Junker A, Eschenröder A, Müller J, Schreiber F, Junker BH (2013) Multiscale metabolic modeling: dynamic flux balance analysis on a whole-plant scale. Plant Physiol 163:637–647
57. Gutensohn M, Orlova I, Nguyen TTH, Davidovich-Rikanati R, Ferruzzi MG, Sitrit Y, Lewinsohn E, Pichersky E, Dudareva N (2013) Cytosolic monoterpene biosynthesis is supported by plastid-generated geranyl diphosphate substrate in transgenic tomato fruits. Plant J 75:351–363
58. Gómez PI, Inostroza I, Pizarro M, Pérez J (2013) From genetic improvement to commercial-scale mass culture of a Chilean strain of the green microalga *Haematococcus pluvialis* with enhanced productivity of the red ketocarotenoid astaxanthin. AoB Plants 5:plt026
59. Hay J, Schwender J (2011) Computational analysis of storage synthesis in developing *Brassica napus* L. (oilseed rape) embryos: flux variability analysis in relation to ^{13}C metabolic flux analysis. Plant J 67:513–525
60. Hay J, Schwender J (2011) Metabolic network reconstruction and flux variability analysis of storage synthesis in developing oilseed rape (*Brassica napus* L.) embryos. Plant J 67:526–541
61. Heinig U, Gutensohn M, Dudareva N, Aharoni A (2013) The challenges of cellular compartmentalization in plant metabolic engineering. Curr Opin Biotechnol 24:239–246
62. Hemmerlin A, Harwood JL, Bach TJ (2012) A raison d'être for two distinct pathways in the early steps of plant isoprenoid biosynthesis? Prog Lipid Res 51:95–148
63. Hemmerlin A, Hoeffler J-F, Meyer O, Tritsch D, Kagan IA, Grosdemange-Billiard C, Rohmer M, Bach TJ (2003) Cross-talk between the cytosolic mevalonate and the plastidial methylerythritol phosphate pathways in tobacco bright yellow-2 cells. J Biol Chem 278:26666–26676
64. Hohn TM, Ohlrogge JB (1991) Expression of a fungal sesquiterpene cyclase gene in transgenic tobacco. Plant Physiol 97:460–462
65. Holmberg N, Harker M, Wallace AD, Clayton JC, Gibbard CL, Safford R (2003) Co-expression of N-terminal truncated 3-hydroxy-3-methylglutaryl CoA reductase and C24-sterol methyltransferase type 1 in transgenic tobacco enhances carbon flux towards end-product sterols. Plant J 36:12–20
66. Holmberg N, Harker M, Gibbard CL, Wallace AD, Clayton JC, Rawlins S, Hellyer A, Safford R (2002) Sterol C-24 methyltransferase type 1 controls the flux of carbon into sterol biosynthesis in tobacco seed. Plant Physiol 130:303–311
67. Hong G-J, Hu W-L, Li J-X, Chen X-Y, Wang L-J (2009) Increased accumulation of Artemisinin and Anthocyanins in *Artemisia annua* expressing the arabidopsis blue light receptor CRY1. Plant Mol Biol Report 27:334–341
68. Hong G-J, Xue X-Y, Mao Y-B, Wang L-J, Chen X-Y (2012) Arabidopsis MYC2 interacts with DELLA proteins in regulating sesquiterpene synthase gene expression. Plant Cell 24:2635–2648
69. Huang M, Sanchez-Moreiras AM, Abel C, Sohrabi R, Lee S, Gershenzon J, Tholl D (2012) The major volatile organic compound emitted from *Arabidopsis thaliana* flowers, the sesquiterpene (E)-β-caryophyllene, is a defense against a bacterial pathogen. New Phytol 193:997–1008
70. Hyka P, Lickova S, Přibyl P, Melzoch K, Kovar K (2012) Flow cytometry for the development of biotechnological processes with microalgae. Biotechnol Adv 31:2–16

71. Jackson MA, Sternes PR, Mudge SR, Graham MW, Birch RG (2014) Design rules for efficient transgene expression in plants. Plant Biotechnol J 12:925–933
72. Jaganath B, Subramanyam K, Mayavan S, Karthik S, Elayaraja D, Udayakumar R, Manickavasagam M, Ganapathi A (2014) An efficient in planta transformation of *Jatropha curcas* (L.) and multiplication of transformed plants through in vivo grafting. Protoplasma 251:591–601
73. Kappers IF, Aharoni A, van Herpen TWJM, Luckerhoff LLP, Dicke M, Bouwmeester HJ (2005) Genetic engineering of terpenoid metabolism attracts bodyguards to Arabidopsis. Science 309(80):2070–2072
74. Kebeish R, Niessen M, Thiruveedhi K, Bari R, Hirsch H-J, Rosenkranz R, Stäbler N, Schönfeld B, Kreuzaler F, Peterhänsel C (2007) Chloroplastic photorespiratory bypass increases photosynthesis and biomass production in *Arabidopsis thaliana*. Nat Biotechnol 25:593–599
75. Keeling CI, Bohlmann J (2006) Diterpene resin acids in conifers. Phytochemistry 67:2415–2423
76. Keeling CI, Bohlmann J (2006) Genes, enzymes and chemicals of terpenoid diversity in the constitutive and induced defence of conifers against insects and pathogens. New Phytol 170:657–675
77. Kessler A, Baldwin IT (2001) Defensive function of herbivore-induced plant volatile emissions in nature. Science 291:2141–2144
78. Khatri W, Hendrix R, Niehaus T, Chappell J, Curtis WR (2014) Hydrocarbon production in high density *Botryococcus braunii* race B continuous culture. Biotechnol Bioeng 111:493–503
79. Kilian O, Benemann CSE, Niyogi KK, Vick B (2011) High-efficiency homologous recombination in the oil-producing alga Nannochloropsis sp. Proc Natl Acad Sci 108:21265–21269
80. Kirby J, Keasling JD (2009) Biosynthesis of plant isoprenoids: perspectives for microbial engineering. Annu Rev Plant Biol 60:335–355
81. Kong F, Yamasaki T, Ohama T (2014) Expression levels of domestic cDNA cassettes integrated in the nuclear genomes of various *Chlamydomonas reinhardtii* strains. J Biosci Bioeng 117:613–616
82. Kota M, Daniell H, Varma S, Garczynski SF, Gould F, Moar WJ (1999) Overexpression of the *Bacillus thuringiensis* (Bt) Cry2Aa2 protein in chloroplasts confers resistance to plants against susceptible and Bt-resistant insects. Proc Natl Acad Sci USA 96:1840–1845
83. Kumar A, Falcao VR, Sayre RT (2013) Evaluating nuclear transgene expression systems in *Chlamydomonas reinhardtii*. Algal Res 2:321–332
84. Kumar S, Hahn FM, Baidoo E, Kahlon TS, Wood DF, McMahan CM, Cornish K, Keasling JD, Daniell H, Whalen MC (2012) Remodeling the isoprenoid pathway in tobacco by expressing the cytoplasmic mevalonate pathway in chloroplasts. Metab Eng 14:19–28
85. Lakshmanan M, Zhang Z, Mohanty B, Kwon J-Y, Choi H-Y, Nam H-J, Kim D-I, Lee D-Y (2013) Elucidating rice cell metabolism under flooding and drought stresses using flux-based modeling and analysis. Plant Physiol 162:2140–2150
86. Lamers PP, van de Laak CCW, Kaasenbrood PS, Lorier J, Janssen M, De Vos RCH, Bino RJ, Wijffels RH (2010) Carotenoid and fatty acid metabolism in light-stressed *Dunaliella salina*. Biotechnol Bioeng 106:638–648
87. Lange BM, Ahkami A (2013) Metabolic engineering of plant monoterpenes, sesquiterpenes and diterpenes—current status and future opportunities. Plant Biotechnol J 11:169–196
88. Lange BM, Mahmoud SS, Wildung MR, Turner GW, Davis EM, Lange I, Baker RC, Boydston RA, Croteau RB (2011) Improving peppermint essential oil yield and composition by metabolic engineering. Proc Natl Acad Sci USA 108:16944–16949
89. Lange BM, Turner GW (2013) Terpenoid biosynthesis in trichomes–current status and future opportunities. Plant Biotechnol J 11:2–22
90. Lavy M, Zuker A, Lewinsohn E, Larkov O, Ravid U, Weiss D (2002) Linalool and linalool oxide production in transgenic carnation flowers expressing the *Clarkia breweri* linalool synthase gene. Mol Breed 9:103–112

91. León R, Couso I, Fernández E (2007) Metabolic engineering of ketocarotenoids biosynthesis in the unicelullar microalga *Chlamydomonas reinhardtii*. J Biotechnol 130:143–152
92. Li L, Van Eck J (2007) Metabolic engineering of carotenoid accumulation by creating a metabolic sink. Transgenic Res 16:581–585
93. Li J, Gao G, Xu K, Chen B, Yan G, Li F, Qiao J, Zhang T, Wu X (2014) Genome-wide survey and expression analysis of the putative non-specific lipid transfer proteins in *Brassica rapa* L. PLoS One 9:e84556
94. Lichtenthaler HK, Schwender J, Disch A, Rohmer M (1997) Biosynthesis of isoprenoids in higher plant chloroplasts proceeds via a mevalonate-independent pathway. FEBS Lett 400:271–274
95. Liu J, Gerken H, Huang J, Chen F (2013) Engineering of an endogenous phytoene desaturase gene as a dominant selectable marker for *Chlamydomonas reinhardtii* transformation and enhanced biosynthesis of carotenoids. Process Biochem 48:788–795
96. Lohr M, Schwender J, Polle JEW (2012) Isoprenoid biosynthesis in eukaryotic phototrophs: a spotlight on algae. Plant Sci 185–186:9–22
97. Lopez AB, Van Eck J, Conlin BJ, Paolillo DJ, O'Neill J, Li L (2008) Effect of the cauliflower or transgene on carotenoid accumulation and chromoplast formation in transgenic potato tubers. J Exp Bot 59:213–223
98. Lu S, Van Eck J, Zhou X, Lopez AB, O'Halloran DM, Cosman KM, Conlin BJ, Paolillo DJ, Garvin DF, Vrebalov J et al (2006) The cauliflower Or gene encodes a DnaJ cysteine-rich domain-containing protein that mediates high levels of beta-carotene accumulation. Plant Cell 18:3594–3605
99. Lu X, Zhang L, Zhang F, Jiang W, Shen Q, Zhang L, Lv Z, Wang G, Tang K (2013) AaORA, a trichome-specific AP2/ERF transcription factor of *Artemisia annua*, is a positive regulator in the artemisinin biosynthetic pathway and in disease resistance to *Botrytis cinerea*. New Phytol 198:1191–1202
100. Lücker J, Bouwmeester HJ, Schwab W, Blaas J, van der Plas LH, Verhoeven HA (2001) Expression of Clarkia S-linalool synthase in transgenic petunia plants results in the accumulation of S-linalyl-beta-D-glucopyranoside. Plant J 27:315–324
101. Lücker J, Schwab W, Franssen MCR, Van Der Plas LHW, Bouwmeester HJ, Verhoeven HA (2004) Metabolic engineering of monoterpene biosynthesis: two-step production of (+)-trans-isopiperitenol by tobacco. Plant J 39:135–145
102. Lücker J, Schwab W, Van Hautum B, Blaas J, Van Der Plas LHW, Bouwmeester HJ, Verhoeven HA (2004) Increased and altered fragrance of tobacco plants after metabolic engineering using three monoterpene synthases from lemon. Plant Physiol 134:510–519
103. Ma D, Pu G, Lei C, Ma L, Wang H, Guo Y, Chen J, Du Z, Wang H, Li G et al (2009) Isolation and characterization of AaWRKY1, an *Artemisia annua* transcription factor that regulates the amorpha-4,11-diene synthase gene, a key gene of artemisinin biosynthesis. Plant Cell Physiol 50:2146–2161
104. Mahmoud SS, Croteau RB (2001) Metabolic engineering of essential oil yield and composition in mint by altering expression of deoxyxylulose phosphate reductoisomerase and menthofuran synthase. Proc Natl Acad Sci USA 98:8915–8920
105. Mallón R, Vieitez AM, Vidal N (2013) High-efficiency Agrobacterium-mediated transformation in *Quercus robur*: selection by use of a temporary immersion system and assessment by quantitative PCR. Plant Cell Tissue Organ Cult 114:171–185
106. Mary W, Crombie L, Crombie L (1986) Distribution of avenacins A-1, A-2, B-1 and B-2 in oat roots: their fungicidal activity towards "take-all" fungus. Phytochemistry 25:2069–2073
107. Matsushima D, Jenke-Kodama H, Sato Y, Fukunaga Y, Sumimoto K, Kuzuyama T, Matsunaga S, Okada S (2012) The single cellular green microalga *Botryococcus braunii*, race B possesses three distinct 1-deoxy-D-xylulose 5-phosphate synthases. Plant Sci 185–186:309–320
108. Mayavan S, Subramanyam K, Arun M, Rajesh M, Kapil Dev G, Sivanandhan G, Jaganath B, Manickavasagam M, Selvaraj N, Ganapathi A (2013) *Agrobacterium tumefaciens*-mediated in planta seed transformation strategy in sugarcane. Plant Cell Rep 32:1557–1574

109. Mayfield S (2013) The green revolution 2.0: the potential of algae for the production of biofuels and bioproducts. Genome 56:551–555
110. McBride KE, Svab Z, Schaaf DJ, Hogan PS, Stalker DM, Maliga P (1995) Amplification of a chimeric Bacillus gene in chloroplasts leads to an extraordinary level of an insecticidal protein in tobacco. Biotechnology (NY) 13:362–365
111. Mehrshahi P, Stefano G, Andaloro JM, Brandizzi F, Froehlich JE, DellaPenna D (2013) Transorganellar complementation redefines the biochemical continuity of endoplasmic reticulum and chloroplasts. Proc Natl Acad Sci USA 110:12126–12131
112. Melkus G, Rolletschek H, Fuchs J, Radchuk V, Grafahrend-Belau E, Sreenivasulu N, Rutten T, Weier D, Heinzel N, Schreiber F et al (2011) Dynamic $^{13}C/^{1}H$ NMR imaging uncovers sugar allocation in the living seed. Plant Biotechnol J 9:1022–1037
113. Mintz-Oron S, Meir S, Malitsky S, Ruppin E, Aharoni A, Shlomi T (2012) Reconstruction of Arabidopsis metabolic network models accounting for subcellular compartmentalization and tissue-specificity. Proc Natl Acad Sci USA 109:339–344
114. Misawa N (2009) Pathway engineering of plants toward astaxanthin production. Plant Biotechnol 26:93–99
115. Montiel G, Breton C, Thiersault M, Burlat V, Jay-Allemand C, Gantet P (2007) Transcription factor Agamous-like 12 from Arabidopsis promotes tissue-like organization and alkaloid biosynthesis in *Catharanthus roseus* suspension cells. Metab Eng 9:125–132
116. Morandini P (2013) Control limits for accumulation of plant metabolites: brute force is no substitute for understanding. Plant Biotechnol J 11:253–267
117. Moses T, Pollier J, Thevelein JM, Goossens A (2013) Bioengineering of plant (tri) terpenoids: from metabolic engineering of plants to synthetic biology in vivo and in vitro. New Phytol 200:27–43
118. Murata J, De Luca V (2005) Localization of tabersonine 16-hydroxylase and 16-OH tabersonine-16-O-methyltransferase to leaf epidermal cells defines them as a major site of precursor biosynthesis in the vindoline pathway in *Catharanthus roseus*. Plant J 44:581–594
119. Murata J, Roepke J, Gordon H, De Luca V (2008) The leaf epidermome of *Catharanthus roseus* reveals its biochemical specialization. Plant Cell 20:524–542
120. Nagel R, Berasategui A, Paetz C, Gershenzon J, Schmidt A (2013) Overexpression of an isoprenyl diphosphate synthase in spruce leads to unexpected terpene diversion products that function in plant defense. Plant Physiol 164:555–569
121. Neupert J, Karcher D, Bock R (2009) Generation of Chlamydomonas strains that efficiently express nuclear transgenes. Plant J 57:1140–1150
122. Ohara K, Ujihara T, Endo T, Sato F, Yazaki K (2003) Limonene production in tobacco with Perilla limonene synthase cDNA. J Exp Bot 54:2635–2642
123. Okada K, Saito T, Nakagawa T, Kawamukai M, Kamiya Y (2000) Five geranylgeranyl diphosphate synthases expressed in different organs are localized into three subcellular compartments in Arabidopsis. Plant Physiol 122:1045–1056
124. De Oliveira Dal'Molin CG, Nielsen LK (2013) Plant genome-scale metabolic reconstruction and modelling. Curr Opin Biotechnol 24:271–277
125. De Oliveira Dal'Molin CG, Quek L-E, Palfreyman RW, Brumbley SM, Nielsen LK (2010) AraGEM, a genome-scale reconstruction of the primary metabolic network in Arabidopsis. Plant Physiol 152:579–589
126. Orlova I, Nagegowda DA, Kish CM, Gutensohn M, Maeda H, Varbanova M, Fridman E, Yamaguchi S, Hanada A, Kamiya Y et al (2009) The small subunit of snapdragon geranyl diphosphate synthase modifies the chain length specificity of tobacco geranylgeranyl diphosphate synthase in planta. Plant Cell 21:4002–4017
127. Outchkourov NS, Peters J, de Jong J, Rademakers W, Jongsma MA (2003) The promoter-terminator of chrysanthemum rbcS1 directs very high expression levels in plants. Planta 216:1003–1012
128. Pare PW, Tumlinson JH (1999) Plant volatiles as a defense against insect herbivores. Plant Physiol 121:325–331

129. Patra B, Schluttenhofer C, Wu Y, Pattanaik S, Yuan L (2013) Transcriptional regulation of secondary metabolite biosynthesis in plants. Biochim Biophys Acta 1829:1236–1247
130. Peralta-Yahya PP, Ouellet M, Chan R, Mukhopadhyay A, Keasling JD, Lee TS (2011) Identification and microbial production of a terpene-based advanced biofuel. Nat Commun 2:483
131. Pilalis E, Chatziioannou A, Thomasset B, Kolisis F (2011) An in silico compartmentalized metabolic model of *Brassica napus* enables the systemic study of regulatory aspects of plant central metabolism. Biotechnol Bioeng 108:1673–1682
132. Poolman MG, Miguet L, Sweetlove LJ, Fell DA (2009) A genome-scale metabolic model of Arabidopsis and some of its properties. Plant Physiol 151:1570–1581
133. Potvin G, Zhang Z (2010) Strategies for high-level recombinant protein expression in transgenic microalgae: a review. Biotechnol Adv 28:910–918
134. Pulido P, Toledo-Ortiz G, Phillips MA, Wright LP, Rodríguez-Concepción M (2013) Arabidopsis J-protein J20 delivers the first enzyme of the plastidial isoprenoid pathway to protein quality control. Plant Cell 25:4183–4194
135. Purton S, Szaub JB, Wannathong T, Young R, Economou CK (2013) Genetic engineering of algal chloroplasts: progress and prospects. Russ J Plant Physiol 60:491–499
136. Rabbani S, Beyer P, Lintig J, Hugueney P, Kleinig H (1998) Induced beta-carotene synthesis driven by triacylglycerol deposition in the unicellular alga *Dunaliella bardawil*. Plant Physiol 116:1239–1248
137. Radrich K, Tsuruoka Y, Dobson P, Gevorgyan A, Swainston N, Baart G, Schwartz J-M (2010) Integration of metabolic databases for the reconstruction of genome-scale metabolic networks. BMC Syst Biol 4:114
138. Ramos AA, Polle J, Tran D, Cushman JC, Jin E-S, Varela JC (2011) The unicellular green alga Dunaliella salina Teod. as a model for abiotic stress tolerance: genetic advances and future perspectives. Algae 26:3–20
139. Rasala BA, Barrera DJ, Ng J, Plucinak TM, Rosenberg JN, Weeks DP, Oyler GA, Peterson TC, Haerizadeh F, Mayfield SP (2013) Expanding the spectral palette of fluorescent proteins for the green microalga *Chlamydomonas reinhardtii*. Plant J 74:545–556
140. Rasala BA, Chao S-S, Pier M, Barrera DJ, Mayfield SP (2014) Enhanced genetic tools for engineering multigene traits into green algae. PLoS One 9:e94028
141. Rasala BA, Lee PA, Shen Z, Briggs SP, Mendez M, Mayfield SP (2012) Robust expression and secretion of Xylanase1 in *Chlamydomonas reinhardtii* by fusion to a selection gene and processing with the FMDV 2A peptide. PLoS One 7:e43349
142. Reeves PH, Ellis CM, Ploense SE, Wu M-F, Yadav V, Tholl D, Chételat A, Haupt I, Kennerley BJ, Hodgens C et al (2012) A regulatory network for coordinated flower maturation. PLoS Genet 8:e1002506
143. Rios-Estepa R, Lange I, Lee JM, Lange BM (2010) Mathematical modeling-guided evaluation of biochemical, developmental, environmental, and genotypic determinants of essential oil composition and yield in peppermint leaves. Plant Physiol 152:2105–2119
144. Rios-Estepa R, Turner GW, Lee JM, Croteau RB, Lange BM (2008) A systems biology approach identifies the biochemical mechanisms regulating monoterpenoid essential oil composition in peppermint. Proc Natl Acad Sci USA 105:2818–2823
145. Robert CAM, Erb M, Hiltpold I, Hibbard BE, Gaillard MDP, Bilat J, Degenhardt J, Cambet-Petit-Jean X, Turlings TCJ, Zwahlen C (2013) Genetically engineered maize plants reveal distinct costs and benefits of constitutive volatile emissions in the field. Plant Biotechnol J 11:628–639
146. Roberts SC (2007) Production and engineering of terpenoids in plant cell culture. Nat Chem Biol 3:387–395
147. Rodríguez A, San Andrés V, Cervera M, Redondo A, Alquézar B, Shimada T, Gadea J, Rodrigo MJ, Zacarías L, Palou L et al (2011) Terpene down-regulation in orange reveals the role of fruit aromas in mediating interactions with insect herbivores and pathogens. Plant Physiol 156:793–802

148. Rodríguez A, Shimada T, Cervera M, Alquézar B, Gadea J, Gómez-Cadenas A, De Ollas CJ, Rodrigo MJ, Zacarías L, Peña L (2014) Terpene down-regulation triggers defense responses in transgenic orange leading to resistance against fungal pathogens. Plant Physiol 164:321–339
149. Rogers ED, Jackson T, Moussaieff A, Aharoni A, Benfey PN (2012) Cell type-specific transcriptional profiling: implications for metabolite profiling. Plant J 70:5–17
150. Rohmer M (1999) The discovery of a mevalonate-independent pathway for isoprenoid biosynthesis in bacteria, algae and higher plants. Nat Prod Rep 16:565–574
151. Rohmer M, Knani M, Simonin P, Sutter B, Sahm H (1993) Isoprenoid biosynthesis in bacteria: a novel pathway for the early steps leading to isopentenyl diphosphate. Biochem J 295(2):517–524
152. Rolletschek H, Melkus G, Grafahrend-Belau E, Fuchs J, Heinzel N, Schreiber F, Jakob PM, Borisjuk L (2011) Combined noninvasive imaging and modeling approaches reveal metabolic compartmentation in the barley endosperm. Plant Cell 23:3041–3054
153. Roston RL, Gao J, Murcha MW, Whelan J, Benning C (2012) TGD1, -2, and -3 proteins involved in lipid trafficking form ATP-binding cassette (ABC) transporter with multiple substrate-binding proteins. J Biol Chem 287:21406–21415
154. Römer S, Fraser PD, Kiano JW, Shipton CA, Misawa N, Schuch W, Bramley PM (2000) Elevation of the provitamin a content of transgenic tomato plants. Nat Biotechnol 18:666–669
155. Saha R, Suthers PF, Maranas CD (2011) Zea mays iRS1563: a comprehensive genome-scale metabolic reconstruction of maize metabolism. PLoS One 6:e21784
156. Sahoo KK, Tripathi AK, Pareek A, Sopory SK, Singla-pareek SL (2011) An improved protocol for efficient transformation and regeneration of diverse indica rice cultivars. Plant Methods 7:49
157. Salim V, Yu F, Altarejos J, De Luca V (2013) Virus-induced gene silencing identifies *Catharanthus roseus* 7-deoxyloganic acid-7-hydroxylase, a step in iridoid and monoterpene indole alkaloid biosynthesis. Plant J 76:754–765
158. Samuels L, Kunst L, Jetter R (2008) Sealing plant surfaces: cuticular wax formation by epidermal cells. Annu Rev Plant Biol 59:683–707
159. Savidge B, Weiss JD, Wong YH, Lassner MW, Mitsky TA, Shewmaker CK, Post-Beittenmiller D, Valentin HE (2002) Isolation and characterization of homogentisate phytyltransferase genes from Synechocystis sp. PCC 6803 and Arabidopsis. Plant Physiol 129:321–332
160. Sawai S, Saito K (2011) Triterpenoid biosynthesis and engineering in plants. Front Plant Sci 2:25
161. Schilmiller AL, Miner DP, Larson M, McDowell E, Gang DR, Wilkerson C, Last RL (2010) Studies of a biochemical factory: tomato trichome deep expressed sequence tag sequencing and proteomics. Plant Physiol 153:1212–1223
162. Seaver SMD, Henry CS, Hanson AD (2012) Frontiers in metabolic reconstruction and modeling of plant genomes. J Exp Bot 63:2247–2258
163. Shintani D, DellaPenna D (1998) Elevating the vitamin E content of plants through metabolic engineering. Science 282(80):2098–2100
164. Sitbon F, Jonsson L (2001) Sterol composition and growth of transgenic tobacco plants expressing type-1 and type-2 sterol methyltransferases. Planta 212:568–572
165. Steinbrenner J, Sandmann G (2006) Transformation of the green alga *Haematococcus pluvialis* with a phytoene desaturase for accelerated astaxanthin biosynthesis. Appl Environ Microbiol 72:7477–7484
166. Strenkert D, Schmollinger S, Schroda M (2013) Heat shock factor 1 counteracts epigenetic silencing of nuclear transgenes in *Chlamydomonas reinhardtii*. Nucleic Acids Res 41:5273–5289
167. Suzuki R, Ito N, Uno Y, Nishii I, Kagiwada S, Okada S, Noguchi T (2013) Transformation of lipid bodies related to hydrocarbon accumulation in a green alga, *Botryococcus braunii* (Race B). PLoS One 8:e81626
168. Tissier A (2012) Trichome specific expression: promoters and their applications. In Çiftçi YÖ (ed) Transgenic plants-advances and limitations, March 2012. INTECH, pp 353–378

169. Unsicker SB, Kunert G, Gershenzon J (2009) Protective perfumes: the role of vegetative volatiles in plant defense against herbivores. Curr Opin Plant Biol 12:479–485
170. Vickers CE, Gershenzon J, Lerdau MT, Loreto F (2009) A unified mechanism of action for volatile isoprenoids in plant abiotic stress. Nat Chem Biol 5:283–291
171. Vranová E, Coman D, Gruissem W (2012) Structure and dynamics of the isoprenoid pathway network. Mol Plant 5:318–333
172. Wallaart TE, Bouwmeester HJ, Hille J, Poppinga L, Maijers NC (2001) Amorpha-4,11-diene synthase: cloning and functional expression of a key enzyme in the biosynthetic pathway of the novel antimalarial drug artemisinin. Planta 212:460–465
173. Wang C-T, Liu H, Gao X-S, Zhang H-X (2010) Overexpression of G10H and ORCA3 in the hairy roots of *Catharanthus roseus* improves catharanthine production. Plant Cell Rep 29:887–894
174. Wang G, Tang W, Bidigare RR (2005) Terpenoids as therapeutic drugs and pharmaceutical agents. In: Zhang L, Demain AL (eds) Natural products. Humana Press, Totowa, pp 197–227
175. Wang B, Wang J, Zhang W, Meldrum DR (2012) Application of synthetic biology in cyanobacteria and algae. Front Microbiol 3:344
176. Weiss TL, Roth R, Goodson C, Vitha S, Black I, Azadi P, Rusch J, Holzenburg A, Devarenne TP, Goodenough U (2012) Colony organization in the green alga *Botryococcus braunii* (Race B) is specified by a complex extracellular matrix. Eukaryot Cell 11:1424–1440
177. Williams TCR, Poolman MG, Howden AJM, Schwarzlander M, Fell DA, Ratcliffe RG, Sweetlove LJ (2010) A genome-scale metabolic model accurately predicts fluxes in central carbon metabolism under stress conditions. Plant Physiol 154:311–323
178. Wink M (2003) Evolution of secondary metabolites from an ecological and molecular phylogenetic perspective. Phytochemistry 64:3–19
179. Wright LP, Rohwer JM, Ghirardo A, Hammerbacher A, Ortiz-Alcaide M, Raguschke B, Schnitzler J-P, Gershenzon J, Phillips MA (2014) Deoxyxylulose 5-phosphate synthase controls flux through the methylerythritol 4-phosphate pathway in Arabidopsis. Plant Physiol 165:1488–1504
180. Wu S, Jiang Z, Kempinski C, Eric Nybo S, Husodo S, Williams R, Chappell J (2012) Engineering triterpene metabolism in tobacco. Planta 236:867–877
181. Wu S, Schalk M, Clark A, Miles RB, Coates R, Chappell J (2006) Redirection of cytosolic or plastidic isoprenoid precursors elevates terpene production in plants. Nat Biotechnol 24:1441–1447
182. Wurtzel ET, Cuttriss A, Vallabhaneni R (2012) Maize provitamin a carotenoids, current resources, and future metabolic engineering challenges. Frontiers in Plant Science 3:29
183. Ye X, Al-Babili S, Klöti A, Zhang J, Lucca P, Beyer P, Potrykus I (2000) Engineering the provitamin A (beta-carotene) biosynthetic pathway into (carotenoid-free) rice endosperm. Science 287:303–305
184. Yu Z-X, Li J-X, Yang C-Q, Hu W-L, Wang L-J, Chen X-Y (2012) The jasmonate-responsive AP2/ERF transcription factors AaERF1 and AaERF2 positively regulate artemisinin biosynthesis in *Artemisia annua* L. Mol Plant 5:353–365
185. Yu B, Lydiate DJ, Young LW, Schäfer UA, Hannoufa A (2008) Enhancing the carotenoid content of *Brassica napus* seeds by downregulating lycopene epsilon cyclase. Transgenic Res 17:573–585
186. Zhang W-J, Dewey RE, Boss W, Phillippy BQ, Qu R (2013) Enhanced Agrobacterium-mediated transformation efficiencies in monocot cells is associated with attenuated defense responses. Plant Mol Biol 81:273–286
187. Zhang L, Jing F, Li F, Li M, Wang Y, Wang G, Sun X, Tang K (2009) Development of transgenic *Artemisia annua* (Chinese wormwood) plants with an enhanced content of artemisinin, an effective anti-malarial drug, by hairpin-RNA-mediated gene silencing. Biotechnol Appl Biochem 52:199–207
188. Zhang Y, Li Y, Guo Y, Jiang H, Shen X (2009) A sesquiterpene quinone, dysidine, from the sponge *Dysidea villosa*, activates the insulin pathway through inhibition of PTPases. Acta Pharmacol Sin 30:333–345

189. Zhao L, Chang W, Xiao Y, Liu H, Liu P (2013) Methylerythritol phosphate pathway of isoprenoid biosynthesis. Annu Rev Biochem 82:497–530
190. Zook M, Hohn T, Bonnen A, Tsuji J, Hammerschmidt R (1996) Characterization of novel sesquiterpenoid biosynthesis in tobacco expressing a fungal sesquiterpene synthase. Plant Physiol 112:311–318
191. Zou H-W, Tian X-H, Ma G-H, Li Z-X (2013) Isolation and functional analysis of ZmLTP3, a homologue to Arabidopsis LTP3. Int J Mol Sci 14:5025–5035

Adv Biochem Eng Biotechnol (2015) 148: 201–214
DOI 10.1007/10_2014_278

Published Online: 12 October 2014

Advances in the Analysis of Volatile Isoprenoid Metabolites

Matthias Wüst

Abstract The advances in the analysis of volatile isoprenoid metabolites from sample preparation to separation and detection over the past 3 years are discussed in this overview. Novel headspace sampling techniques like the so-called high concentration capacity techniques are compared with the classical liquid extraction and distillation methods. The advantages of multidimensional separation techniques in the analysis of complex samples are outlined and commercially available dedicated heart-cutting and comprehensive GC systems are described.

Keywords Headspace sampling · Heart-cutting MDGC · Comprehensive GC · SPME

Contents

M. Wüst (✉)
Institute of Nutritional and Food Sciences (IEL), University of Bonn,
Endenicher Allee 11–13, 53115 Bonn, Germany
e-mail: matthias.wuest@uni-bonn.de

1 Introduction

This chapter gives an overview of the latest developments in the field of volatile isoprenoid metabolite analysis, from sample preparation to separation and detection. The aim is not to provide a comprehensive review of this topic but rather to outline the most important technical advances. Therefore, special attention has been paid to methods that have attained technical maturity in recent years, are available on the market, and have enlarged the toolbox of scientists working on volatile isoprenoid metabolites, such as plant scientists, natural products chemists, bioengineers, and food scientists (just to mention a few).

2 Sample Preparation Techniques

Sample preparation is an often underestimated part of the analytical process, but it has a decisive influence on the accuracy of the results. A representative sample from an often heterogeneous material has to be selected, and some degree of sample preparation to remove substances that interfere in the analysis of the desired analyte (the so-called matrix) and pre-concentration is necessary. The sample as such can be homogeneous and almost free of interfering matrix (e.g. volatiles in a defined headspace emitted from a plant) or heterogeneous and extremely matrix laden (e.g. volatiles in a food sample, volatiles produced by recombinant microorganisms in a culture broth). Thus, the sample type and purpose of the analysis (qualitative profiling *versus* relative and absolute quantification) determine the method to be chosen for sample preparation.

2.1 Classical Methods: Liquid Extraction, Sorptive Extraction, and Distillation

Liquid extraction using an organic solvent is the classic method for isolating volatile isoprenoids from solid or liquid samples [16]. This method has been compared with more recent methods, such as stir bar sorptive extraction (SBSE), headspace solid-phase microextraction (HS-SPME), and solid-phase extraction (SPE) [13, 19]. These studies have shown that liquid extraction and SPE have still some advantages to offer, including a better recovery of more polar analytes when compared to HS-SPME and SBSE. Moreover, all volatile compounds (low, medium, and high volatility) can be isolated in one extraction step. However, liquid extraction and SPE are relatively tedious and time consuming. The extracts have to be concentrated, and interference with nonvolatile matrix compounds, such as lipids and solvent impurities, is frequently observed. Combining liquid extraction with solvent-assisted flavor evaporation has been shown to efficiently remove these matrix compounds [11], but it adds another tedious operation to the sample preparation procedure. Simultaneous distillation/solvent extraction combines both operations; however, due to the heating of the sample, the risk of artifact formation is quite high [7]. Therefore, these classic methods have been replaced in recent years by a large variety of headspace sampling methods, which will be discussed in the next section. However, lipid-rich matrices still present an analytical challenge, even when using headspace methods. Thin-layer high vacuum distillation and extraction with medium-chain triglyceride and subsequent thermal desorption gas chromatography have been demonstrated to be alternatives for lipid-rich or pure fat samples [20].

2.2 Headspace Sampling

Headspace sampling is a unique technique that allows sample preparation, preconcentration, and cleanup in parallel; thus, it is exceptionally well suited for the isolation of volatile isoprenoids. This has led to the development of numerous headspace methods and techniques with unprecedented performance. Numerous excellent reviews on the headspace sampling of plant volatiles [10, 26, 30] and of flavors and fragrances [3] have been published. Besides the classical static and dynamic headspace methods [23], the so-called high concentration capacity headspace techniques (HCC-HS) have emerged and act as a bridge between static and dynamic methods [3]. HCC-HS techniques are based on either a static or dynamic accumulation of volatiles on very small amounts of polymers or solvents, operating in sorption and/or adsorption modes. Table 1 gives an overview of the most important HCC-HS techniques.

Among the methods listed in Table 1, HS-SPME is a simple, fast, sensitive, and convenient technique that has gained the largest acceptance during the last two decades [6, 22]. If precise quantification of analytes is needed, one has to keep in mind that their accumulation in HS-SPME depends on their vapor pressure in the gas phase. Thus, it depends in an incalculable way on temperature, release from the matrix, and the vapor pressure of other volatiles present. In this case, a stable isotope dilution

Table 1 Overview of common high-concentration capacity headspace techniques

Technique	Common abbreviation	Solvent based	Sorbent based	Recent review(s)
Headspace solid-phase microextraction	HS-SPME		X	[22, 34]
Headspace liquid-phase microextraction[a]	HS-LPME	X		[3, 30]
In-tube sorptive extraction	ITEX		X	[3]
Headspace sorptive extraction	HSSE		X	[3, 30]

[a] Also called headspace single drop microextraction (HS-SDME)

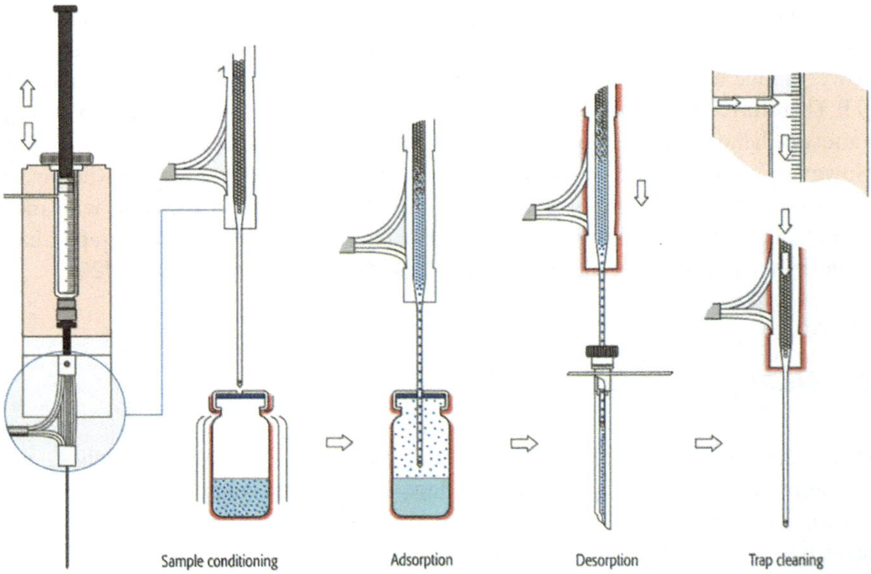

Fig. 1 ITEX-2 sample extraction procedure (copyright by CTC Analytics AG; reproduced with permission). The sample is heated and/or agitated in a sealed sample vial until equilibrium is achieved (*sample conditioning*). The ITEX trap needle pierces the sealed sample vial and the heated gas tight syringe pumps the gas through the ITEX trap (*adsorption*). The loaded ITEX trap is flash heated up to 350 °C and desorbed into the hot injector (*desorption*). After thermal desorption, the hot ITEX trap is cleaned with inert flash gas (*trap cleaning*)

assay is the method of choice to obtain accurate and precise results [1]. Alternatively, two kinetic calibration methods, on-fiber standardization and measurement using a predetermined sampling rate, were used as quantification methods [2].

Recent research has shown that sulfonated graphene/polystyrol coatings for SPME are exceptionally well suited for the analysis of trace terpenes from fennel and star anise samples [33]. However, HS-SPME has a limited concentration capability due to the sampling being in the static headspace mode and due to the low amounts of polymer (typically 0.6 μl on a 100-μm polydimethyl siloxane fiber). To overcome these limits the in-tube extraction (ITEX) device was introduced in 2006 by CTC Analytics AG (Zwingen, Switzerland). The fully automatable device features a sorbent volume of 160 μl in a packed large diameter needle of a headspace syringe [12]. In 2009, a new ITEX-2 system, which can be mounted on any widely used PAL-type autosampler without modification, was released [15]. Using the headspace syringe as a pump, a part of the headspace of the preconditioned sample vial is pumped repeatedly through this microtrap. To gain sensitivity, the number of pumping strokes can be increased or several vials can be sequentially extracted. Figure 1 illustrates the ITEX-2 sampling procedure. Method detection limits for 2-methylisoborneol and geosmin of 0.03 and 0.06 μg/l, respectively, could be achieved for aqueous samples [15]. Several applications of ITEX for the analysis of volatiles from plants and food have been published [14, 24, 25, 31, 32].

3 Separation Techniques

Since its invention in the 1950s, gas chromatography (GC) has been the method of choice for the separation, identification, and quantification of volatile isoprenoids [16]. Despite the high chromatographic efficiency, overlapping peaks are still a problem in the analysis of complex samples, such as essential oils and food aroma extracts. To tackle this problem, multidimensional techniques have been developed. Here, two columns (the first dimension column ^{1}D and the second dimension column ^{2}D) with orthogonal separation characteristics (polar vs. nonpolar) are connected to improve the selectivity of the chromatographic system. During the last decade, two multidimensional techniques—comprehensive GC (GCxGC) and heart-cutting multidimensional GC (MDGC)—have considerably evolved and attained technical maturity. An excellent review comparing both techniques has been published by Marriott et al. [17] and the following sections outline some recent technical advances.

3.1 Heart-Cutting Multidimensional Gas Chromatography

The first common commercial systems for MDGC were developed in the 1980s and 1990s (e.g. SICHROMAT-2 by Siemens, Germany), but they are no longer available. However, advances in GCxGC may have spurred the re-evaluation of MDGC technologies [17]. Today, dedicated MDGC systems are available from various companies, as listed in Table 2.

Table 2 Currently available dedicated MDGC systems

Switching device	Manufacturer	Resource for technical details (accessed 15.9.2014)
Multi-Deans switching technology	Shimadzu Corporation	http://www.shimadzu.com/an/gc/multidimgc/multidimmdgc.html
Multi-column switching	Gerstel	http://www.gerstel.com/en/multidimensional-gc-mcs.htm
Capillary-flow technology	Agilent technologies	http://www.chem.agilent.com/Library/brochures/5989-9384EN.pdf
Swafer microchannel flow technology	Perkin-Elmer	http://www.perkinelmer.com/Catalog/Product/ID/N6520273
SilFlow microchannel device	SGE	http://www.sge.com/uploads/02/38/02387fae8947b8cc3d46be977b901aa9/BR-0364-A.pdf

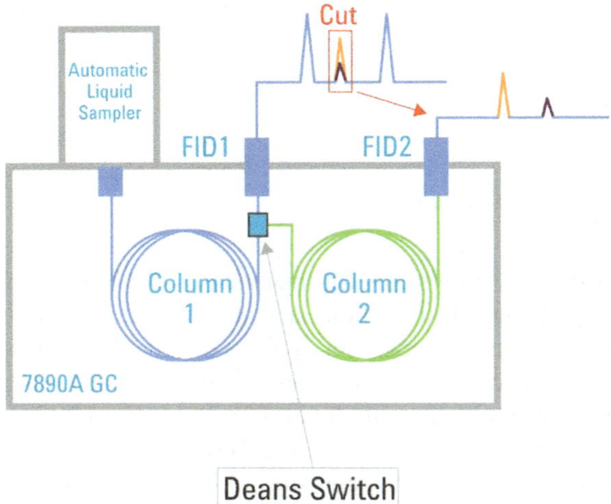

Fig. 2 The basic principle of heart-cutting MDGC (copyright by Agilent Technologies, reproduced with permission). Column 1 and column 2 are connected by a switching device (e.g. a Deans switch). Co-eluting or insufficiently resolved peaks on column 1 are selectively transferred to column 2 and separated due to the orthogonal separation characteristics

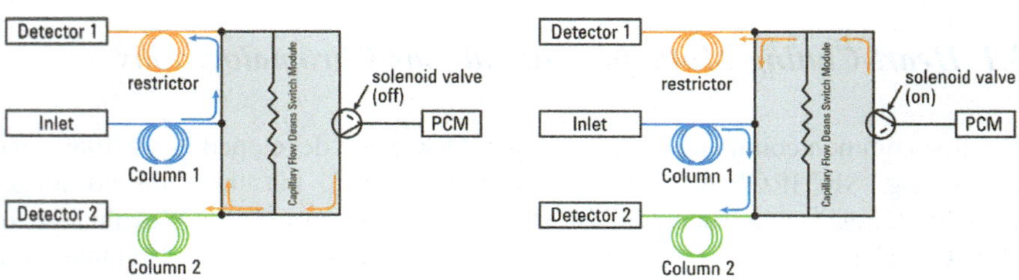

Fig. 3 Deans switch module (copyright by Agilent Technologies, reproduced with permission). *Left* The effluent from column 1 goes directly to detector 1 through the short restrictor when the Deans Switch is in the "no cut" position. *Right* The effluent from column 1 is diverted to column 2 and detector 2 when the Deans Switch is in the "cut" position

The basic principle of heart-cutting MDGC is illustrated in Fig. 2. Column 1 and column 2 are connected by a switching device (e.g. a Deans switch). Co-eluting or insufficiently resolved peaks on column 1 are selectively transferred to column 2 and separated due to the orthogonal separation characteristics. Figure 3 demonstrates how the Deans switch directs effluent to the selected column to separate and quantify the compounds of interest. Comprehensive theories of a Deans switch and a multicolumn switching system have been described [4, 5].

3.2 Comprehensive GC

In GCxGC, the switching device typical for the MDGC system, which connects the first dimension column ^{1}D and the second dimension column ^{2}D, is replaced by a so-called modulator enabling continuous heart-cutting and reinjection into the ^{2}D column [28]. Thus, small segments of 2–10 s are focused and periodically transferred into the ^{2}D column. Two types of modulators are frequently used in commercially available GCxGC systems: the thermal and the flow modulator. A Zoex ZX1 modulator is used in GCxGC systems offered by Shimadzu and Markes (Table 3). It is a two-stage thermal modulator using cold and hot jet gas for modulation. The principle and the technical design of this double loop modulator are shown in Figs. 4 and 5, respectively.

Short and narrow ^{2}D columns are used because the analysis in the second dimension has to be finished before the next injection from the ^{1}D column is made. Peak width in the ^{2}D column is between 50 and 200 ms and requires a spectral collection frequency of 80–10 Hz to give five data points across the full peak width for peak reconstruction (the narrower the peak, the larger the required spectral

Table 3 Available dedicated GCxGC systems from various companies

Manufacturer	Modulation type	Detector	Resource for technical details (accessed 15.9.2014)
LECO corporation	Cooled thermal dual-stage modulator or consumable-free thermal modulator[a]	TOF-MS; up to 500 spectra/sec (500 Hz) in full scan mode	http://www.leco.com/products/separation-science/gcxgc-tofms/pegasus-4d-gcxgc-tofms
Shimadzu corporation	Zeox ZX1 thermal modulator or consumable-free ZX2 Thermal Modulator[a]	qMS; 50 spectra/sec using a 290-amu mass range (m/z = 40–330; 50 Hz)	http://www.shimadzu.co.uk/gas-chromatograph-mass-spectrometry
Markes international	Zoex ZX1 cooled thermal two-stage loop modulator or consumable-free ZX2 thermal modulator[a]	TOF-MS; 50 spectra/sec using a 560-amu mass range (m/z = 40–600; 50 Hz) with 200 spectra per data point	http://www.markes.com/Products/Mass-spectrometry/GCxGC.aspx
Agilent technologies	Flow modulator	FID with 200-Hz data collection rate; MSD can be used with a splitter over a limited scan range	http://www.chem.agilent.com/en-US/products-services/Instruments-Systems/Gas-Chromatography/GC-x-GC-Flow-Modulator/Pages/default.aspx

[a] Modulates volatile and semivolatile compounds over the C7+ range

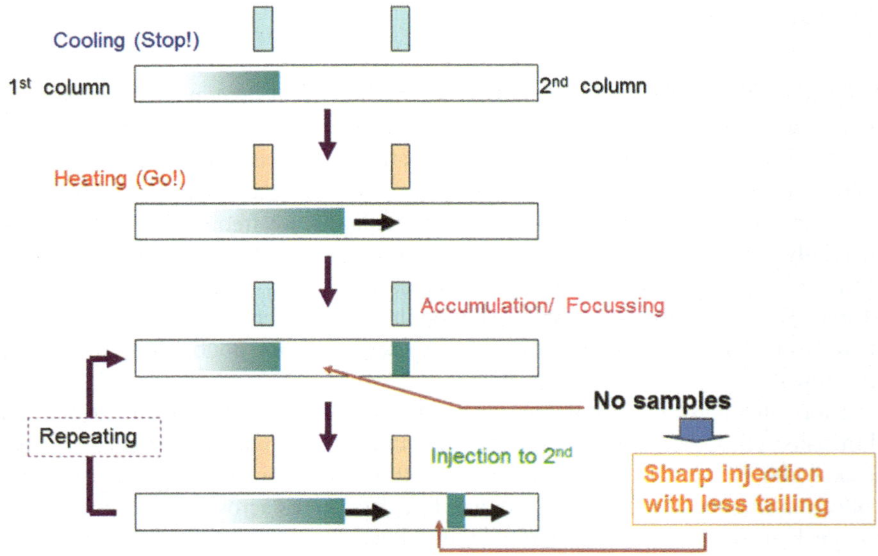

Fig. 4 Principle of a two-stage modulator (copyright by Shimadzu Corporation, reproduced with permission). Two-stage modulation allows sharp injection with less tailing to the second dimension column. Generally, two cold jet nozzles and two hot jet nozzles are required for two-stage modulation. A column loop reduces the number of nozzles in the Zoex design (see also Fig. 5)

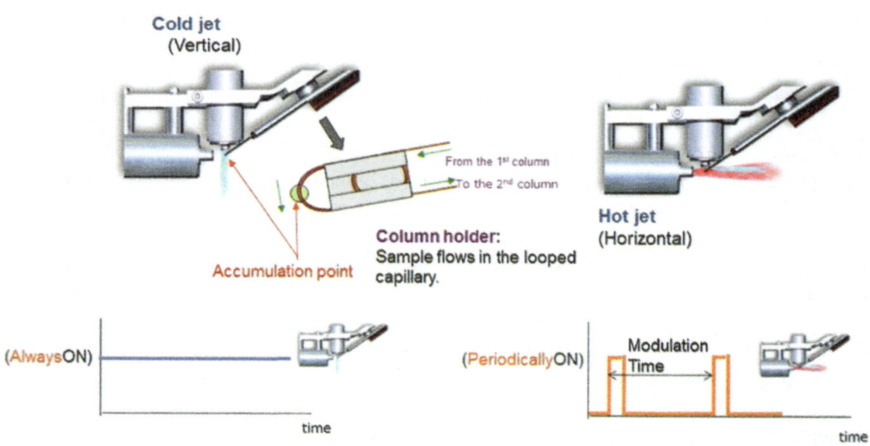

Fig. 5 Zoex ZX1 modulator operation (copyright by Shimadzu Corporation, reproduced with permission). Cold jet gas is continuously blowing on the point of the looped capillary and cryofocuses sample peaks. Hot jet gas periodically blowing on the point of the looped capillary and heats up the spot to inject the sample into the ^{2}D column

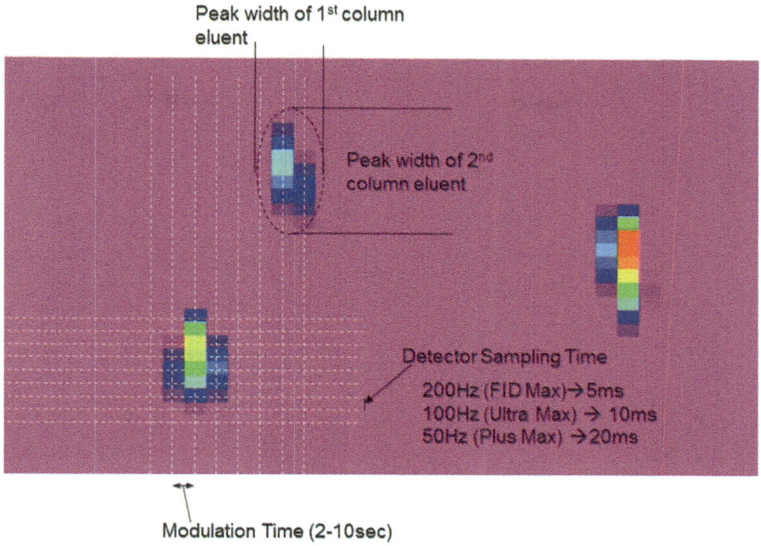

Fig. 6 Sampling intervals of a two-dimensional chromatogram (copyright by Shimadzu Corporation, reproduced with permission). Sampling times are determined by the modulation time for the ^{1}D column and the sampling time of the detector for the ^{2}D column

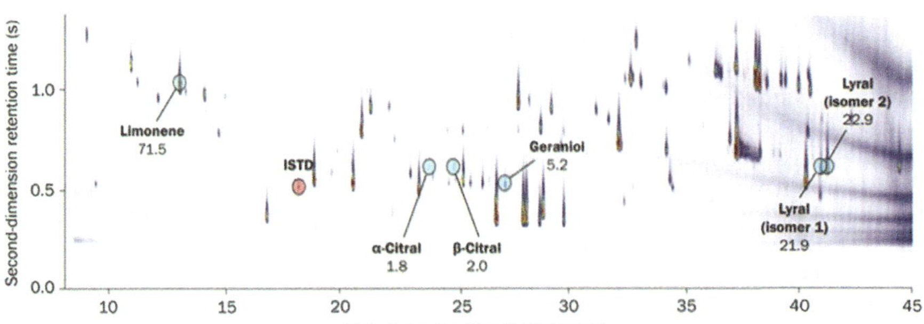

Fig. 7 Quantified terpenes in a perfume sample analyzed by GCxGC time-of-flight mass spectrometry using thermal modulation. Values are given in ppm. (Copyright by Markes International; reproduced with permission.)

collection frequency). The chromatogram is usually visualized as a contour plot, and the z-axis can be cut at a given intensity to yield a two-dimensional contour plot (see Figs. 6 and 7).

Because the cold gas is cooled by liquid nitrogen, also highly volatile compounds can be modulated. However, large amounts of liquid nitrogen are needed, which represents a large increase in the cost of operation. Therefore, flow modulation is a feasible alternative for analysts seeking a simple, low-cost GCxGC approach [21].

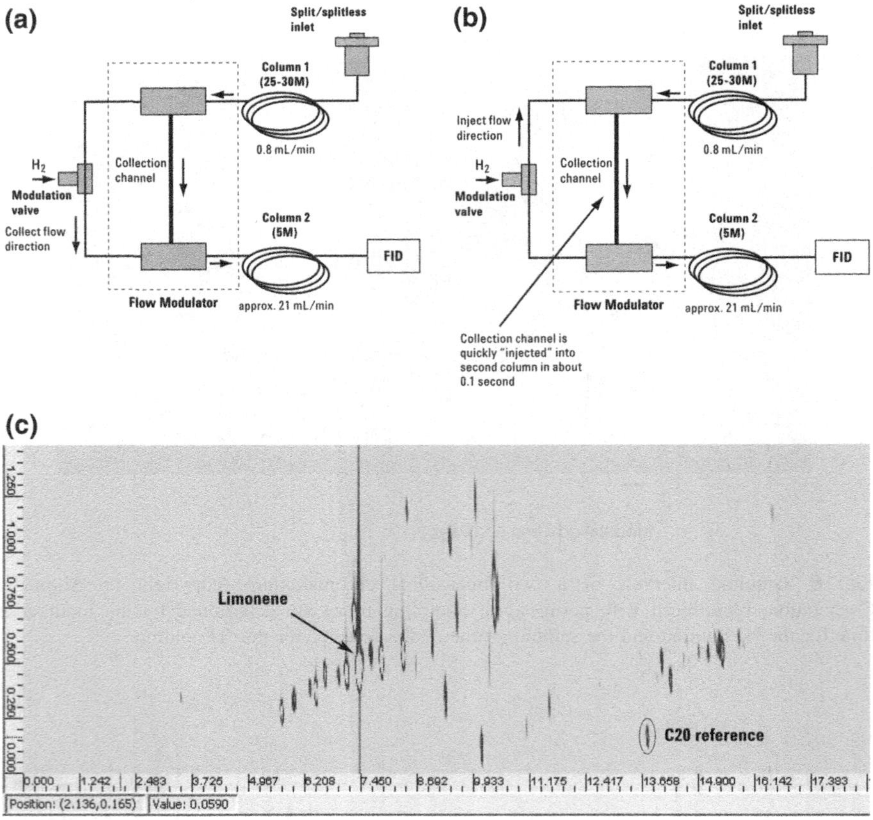

Fig. 8 Principle of flow modulation (copyright by Agilent Technologies; reproduced with permission). **a** The collection channel is filled with hydrogen gas from a previous injection cycle flush. **b** Hydrogen gas flow from the three-way solenoid valve is directed to the top tee; a high flow of typically 20 mL/min for approximately 0.1 s rapidly flushes the collection channel, transferring material in a very narrow band onto the second column, where any analytes collected in the channel undergo rapid separation. **c** Two-dimensional chromatogram of a lime oil sample analyzed by GCxGC FID using flow modulation

A flow modulator based on a Deans switch design has been introduced by Agilent Technologies and is shown in Fig. 8. It has to be kept in mind that a mass spectrometer can only be used with a splitter over limited scan range due to the high flow rate. Moreover, thermal modulation provides potentially narrower peaks in the second dimension when compared to flow modulation; this is still the best choice if performance is the paramount concern and cost is not a determining factor [21].

Due to the narrow peaks eluting from the ^{2}D column, fast-scanning time-of-flight mass spectrometry (TOF-MS) is the preferred detector in GCxGC systems and has gained a firm position in the market [27]. However, in 2010, Shimadzu released a very fast scanning quadrupol mass spectrometer (qMS) that can operate at a 20,000 amu/s scan speed and generate 50 spectra/sec using a 290-amu mass range (m/z = 40–330) [18]. This GCxGC-qMS system has been used for the analysis of isoprenoid plant volatiles and essential oils [8, 9, 29]. Available dedicated GCxGC systems from various companies are summarized in Table 3.

4 Outlook

The toolbox for the scientists working on volatile isoprenoid metabolites has considerably grown in recent years. The most exciting new tools are certainly GCxGC-MS systems, with fully integrated software for instrument control and entirely automated data processing. Previously, GCxGC-MS system setup and software implementation had proven to be inadequate in dealing with the highly complex data and inhibited use by the less specialized scientist. Today, sample preparation, separation, and detection of volatile isoprenoid metabolites can be efficiently managed by taking advantage of numerous systems on the market. This will spur the analysis of complex mixtures of volatile isoprenoid metabolites in the pharmaceutical, biological, environmental, biotechnological, and clinical fields.

References

1. Alonso J, González P (2013) Isotope Dilution Mass Spectrometry. Royal Society of Chemistry, Cambridge, p 472
2. Bai Z, Pilote A, Sarker PK, Vandenberg G, Pawliszyn J (2013) In vivo solid-phase microextraction with in vitro calibration: determination of off-flavor components in live fish. Anal Chem 85:2328–2332. doi:10.1021/ac3033245
3. Bicchi C, Cordero C, Liberto E, Sgorbini B, Rubiolo P (2012) 4.01—Headspace sampling in flavor and fragrance field. In: Pawliszyn J (ed) Comprehensive sampling and sample preparation, Academic Press, Oxford, pp 1–25
4. Boeker P, Haas T, Lammers PS (2013) Theory and practice of a variable dome splitter for gas chromatography-olfactometry. J Chromatogr A 1286:200–207. doi:10.1016/j.chroma.2013.02.057
5. Boeker P, Leppert J, Mysliwietz B, Lammers PS (2013) Comprehensive theory of the deans' switch as a variable flow splitter: fluid mechanics, mass balance, and system behavior. Anal Chem 85:9021–9030. doi:10.1021/ac401419j
6. Bojko B, Cudjoe E, Gomez-Rios GA, Gorynski K, Jiang R, Reyes-Garces N, Risticevic S, Silva EAS, Togunde O, Vuckovic D, Pawliszyn J (2012) SPME—Quo vadis? Anal Chim Acta 750:132–151. doi:10.1016/j.aca.2012.06.052
7. Chaintreau A (2001) Simultaneous distillation-extraction: from birth to maturity—review. Flavour Fragr J 16:136–148. doi:10.1002/ffj.967
8. Cordero C, Liberto E, Bicchi C, Rubiolo P, Reichenbach SE, Tian X, Tao Q (2010) Targeted and non-targeted approaches for complex natural sample profiling by GCxGC-qMS. J Chromatogr Sci 48:251–261
9. Cordero C, Zebelo SA, Gnavi G, Griglione A, Bicchi C, Maffei ME, Rubiolo P (2012) HS-SPME-GCxGC-qMS volatile metabolite profiling of Chrysolina herbacea frass and Mentha spp. leaves. Anal Bioanal Chem 402:1941–1952. doi:10.1007/s00216-011-5600-4

10. D'Alessandro M, Turlings TCJ (2006) Advances and challenges in the identification of volatiles that mediate interactions among plants and arthropods. Analyst 131:24–32. doi:10.1039/b507589k
11. Engel W, Bahr W, Schieberle P (1999) Solvent assisted flavour evaporation—a new and versatile technique for the careful and direct isolation of aroma compounds from complex food matrices. Eur Food Res Technol 209:237–241. doi:10.1007/s002170050486
12. Jochmann MA, Yuan X, Schilling B, Schmidt TC (2008) In-tube extraction for enrichment of volatile organic hydrocarbons from aqueous samples. J Chromatogr A 1179:96–105. doi:10.1016/j.chroma.2007.11.100
13. Komes D, Ulrich D, Lovric T, Schippel K (2005) Isolation of white wine volatiles using different sample preparation methods. Vitis 44:187–193
14. Laaks J, Jochmann MA, Schilling B, Molt K, Schmidt TC (2014) In-tube extraction-GC-MS as a high-capacity enrichment technique for the analysis of alcoholic beverages. J Agric Food Chem 62:3081–3091. doi:10.1021/jf405832u
15. Laaks J, Jochmann MA, Schilling B, Schmidt TC (2010) In-tube extraction of volatile organic compounds from aqueous samples: an economical alternative to purge and trap enrichment. Anal Chem 82:7641–7648. doi:10.1021/ac101414t
16. Lockwood GB (2001) Techniques for gas chromatography of volatile terpenoids from a range of matrices. J Chromatogr A 936:23–31. doi:10.1016/S0021-9673(01)01151-7
17. Marriott PJ, Chin S-T, Maikhunthod B, Schmarr H-G, Bieri S (2012) Multidimensional gas chromatography. Trac-Trends Anal Chem 34:1–21. doi:10.1016/j.trac.2011.10.013
18. Purcaro G, Tranchida PQ, Ragonese C, Conte L, Dugo P, Dugo G, Mondello L (2010) Evaluation of a rapid-scanning quadrupole mass spectrometer in an apolar × ionic-liquid comprehensive two-dimensional gas chromatography system. Anal Chem 82:8583–8590. doi:10.1021/ac101678r
19. Sanchez-Palomo E, Alanon ME, Diaz-Maroto MC, Gonzalez-Vinas MA, Perez-Coello MS (2009) Comparison of extraction methods for volatile compounds of Muscat grape juice. Talanta 79:871–876. doi:10.1016/j.talanta.2009.05.019
20. Schimanski S, Krings U, Berger RG (2013) Rapid analysis of volatiles in fat-containing matrices for monitoring bioprocesses. Eur Food Res Technol 237:739–746. doi:10.1007/s00217-013-2047-1
21. Seeley JV (2011) Flow-modulated comprehensive two-dimensional gas chromatography. In: Mondello L, (ed) Comprehensive chromatography in combination with mass spectrometry, Wiley, Hoboken, p 145–170
22. Silva EAS, Risticevic S, Pawliszyn J (2013) Recent trends in SPME concerning sorbent materials, configurations and in vivo applications. Trac-Trends Anal Chem 43:24–36. doi:10.1016/j.trac.2012.10.006
23. Snow NH, Bullock GP (2010) Novel techniques for enhancing sensitivity in static headspace extraction-gas chromatography. J Chromatogr A 1217:2726–2735. doi:10.1016/j.chroma.2010.01.005
24. Socaci SA, Socaciu C, Muresan C, Farcas A, Tofana M, Vicas S, Pintea A (2014) Chemometric discrimination of different tomato cultivars based on their volatile fingerprint in relation to lycopene and total phenolics content. Phytochem Anal 25:161–169. doi:10.1002/pca.2483
25. Socaci SA, Socaciu C, Tofana M, Rati IV, Pintea A (2013) In-tube Extraction and GC-MS analysis of volatile components from wild and cultivated sea buckthorn (Hippophae rhamnoides L. ssp Carpatica) berry varieties and juice. Phytochem Anal 24:319–328. doi:10.1002/pca.2413
26. Tholl D, Boland W, Hansel A, Loreto F, Rose USR, Schnitzler JP (2006) Practical approaches to plant volatile analysis. Plant J 45:540–560. doi:10.1111/j.1365-313X.2005.02612.x
27. Tranchida PQ, Mondello L, Poynter SDH, Shellie R.A (2011a) Comprehensive two-dimensional gas chromatography combined with mass spectrometry. In: Mondello L (ed) Comprehensive chromatography in combination with mass spectrometry. Wiley, Hoboken, p 171–242

28. Tranchida PQ, Purcaro G, Dugo P, Mondello L, Purcaro G (2011) Modulators for comprehensive two-dimensional gas chromatography. TrAC Trends Anal Chem 30:1437–1461. doi:10.1016/j.trac.2011.06.010
29. Tranchida PQ, Shellie RA, Purcaro G, Conte LS, Dugo P, Dugo G, Mondello L (2010) Analysis of fresh and aged tea tree essential oils by using GCxGC-qMS. J Chromatogr Sci 48:262–266
30. Yang C, Wang J, Li D (2013) Microextraction techniques for the determination of volatile and semivolatile organic compounds from plants: a review. Anal Chim Acta 799:8–22. doi:10.1016/j.aca.2013.07.069
31. Zapata J, Lopez R, Herrero P, Ferreira V (2012) Multiple automated headspace in-tube extraction for the accurate analysis of relevant wine aroma compounds and for the estimation of their relative liquid-gas transfer rates. J Chromatogr A 1266:1–9. doi:10.1016/j.chroma.2012.10.015
32. Zapata J, Mateo-Vivaracho L, Lopez R, Ferreira V (2012) Automated and quantitative headspace in-tube extraction for the accurate determination of highly volatile compounds from wines and beers. J Chromatogr A 1230:1–7. doi:10.1016/j.chroma.2012.01.037
33. Zhang C, Zhang Z, Li G (2014) Preparation of sulfonated graphene/polypyrrole solid-phase microextraction coating by in situ electrochemical polymerization for analysis of trace terpenes. J Chromatogr A 1346:8–15. doi:10.1016/j.chroma.2014.04.043
34. Zhu F, Xu J, Ke Y, Huang S, Zeng F, Luan T, Ouyang G (2013) Applications of in vivo and in vitro solid-phase microextraction techniques in plant analysis: a review. Anal Chim Acta 794:1–14. doi:10.1016/j.aca.2013.05.016

Adv Biochem Eng Biotechnol (2015) 148: 215–250
DOI 10.1007/10_2014_296

Published Online: 15 February 2015

Terpene Hydroxylation with Microbial Cytochrome P450 Monooxygenases

Simon Janocha, Daniela Schmitz and Rita Bernhardt

Abstract Terpenoids comprise a highly diverse group of natural products. In addition to their basic carbon skeleton, they differ from one another in their functional groups. Functional groups attached to the carbon skeleton are the basis of the terpenoids' diverse properties. Further modifications of terpene olefins include the introduction of acyl-, aryl-, or sugar moieties and usually start with oxidations catalyzed by cytochrome P450 monooxygenases (P450s, CYPs). P450s are ubiquitously distributed throughout nature, involved in essential biological pathways such as terpenoid biosynthesis as well as the tailoring of terpenoids and other natural products. Their ability to introduce oxygen into nonactivated C–H bonds is unique and makes P450s very attractive for applications in biotechnology. Especially in the field of terpene oxidation, biotransformation methods emerge as an attractive alternative to classical chemical synthesis. For this reason, microbial P450s depict a highly interesting target for protein engineering approaches in order to increase selectivity and activity, respectively. Microbial P450s have been described to convert industrial and pharmaceutically interesting terpenoids such as ionones, limone, valencene, resin acids, and triterpenes (including steroids) as well as vitamin D_3. Highly selective and active mutants have been evolved by applying classical site-directed mutagenesis as well as directed evolution of proteins. As P450s usually depend on electron transfer proteins, mutagenesis has also been applied to improve the interactions between P450s and their respective redox partners. This chapter provides an overview of terpenoid hydroxylation reactions catalyzed by bacterial P450s and highlights the achievements made by protein engineering to establish productive hydroxylation processes.

Keywords Terpenes · Terpenoids · P450 · Monooxygenase · CYP106A1 · CYP106A2 · Adrenodoxin · Adrenodoxin reductase · Abietic acid · Boswellic acid · Dammarane · Steroid · Vitamin D_3 · Ionone · Valencene · Ferredoxin · Ferredoxin reductase · Protein engineering

S. Janocha · D. Schmitz · R. Bernhardt (✉)
Department of Biochemistry, Saarland University, Campus B2 2,
66123 Saarbruecken, Germany
e-mail: ritabern@mx.uni-saarland.de

Abbreviations

Adx	Adrenodoxin
P450s	Cytochrome P450 enzymes
BM3	Cytochrome P450 102A1
$P450_{cam}$	Cytochrome P450 101A1
CPR	Cytochrome P450 reductase
DMSO	Dimethylsulfoxide
epPCR	Error-prone polymerase chain reaction
Fdx	Ferredoxin
FdR	Ferredoxin reductase
FMN	Flavine mononucleotide
FAD	Flavine adenine dinucleotide
NAD(P)H	Nicotine amide adenine dinucleotide (phosphate)
Pdx	Putidaredoxin
THF	Tetrahydrofuran
VD_3	Vitamin D_3
$1\alpha,25(OH)_2VD_3$	1α,25-dihydroxyvitamin D_3
CYP	cytochrome P450
KBA	11-keto-β-boswellic acid
PDB	Protein Data Bank
VDR	Vitamin D receptor

Contents

1 Introduction

Although the biosynthesis of terpenoids[1] in plants is broadly studied, biosynthesis and modifications of the terpenoid skeleton in microorganisms has been much less investigated. Therefore, microorganism reaction pathways of terpenoid biosynthesis as well as involved enzymes are much less identified and characterized as compared with the plant systems. An essential step in the biosynthesis of terpenes is the cyclization of isoprene subunit chains by different enzymes belonging to the family of terpene synthases. Depending on the number of isoprene units, the resulting structures can be classified into hemiterpenoids (one isoprene unit), monoterpenoids (two isoprene units), sesquiterpenoids (three isoprene units), diterpenoids (four isoprene units), sesterterpenoids (five isoprene units), triterpenoids (six isoprene units), tetraterpenoids (eight isoprene units), and polyterpenoids with a larger number of isoprene units. Terpene synthases have also been found in microorganisms, indicating the pivotal role of terpenoids in this kingdom of life [1]. In addition to their basic carbon skeleton, terpenoids also differ from one another in their functional groups, which form the basis for their diverse properties. Further modifications of terpene olefins include the introduction of acyl-, aryl-, or sugar moieties and usually start with oxidations catalyzed by cytochrome P450 monooxygenases (P450s, CYPs) [2].

P450s are heme-containing enzymes that are ubiquitously distributed in all biological kingdoms and exhibit activity towards a diverse range of substrates [3]. Members of this enzyme family are not only involved in the biosynthesis of physiologically important compounds such as terpenes, steroids, vitamins, and bile acids, but also play a major role in the detoxification of xenobiotics, herbicides, and insecticides. The central step in the P450-catalyzed oxidation is the activation of oxygen by the reduced heme iron of the enzyme and the incorporation of one atom of molecular oxygen into the substrate while the other oxygen atom is reduced to water [4]. The reaction is usually a hydroxylation reaction, but P450s are also able to catalyze reactions such as heteroatom oxygenation, dealkylation, epoxidation, aromatic hydroxylation, reduction, dehalogenation, and even C–C bond formation or cleavage [5, 6]. P450 enzymes are categorized into families and subfamilies depending on the sequence identity, with members of the same family being defined as usually having ≥40 % sequence identity. Representatives of the same subfamily are always >55 % identical [7]. As external monooxygenases, P450s depend on NAD(P)H and corresponding electron transfer proteins in order to perform the enzymatic reaction. According to the composition of their electron transport chain, P450 enzymes can be grouped into different classes [8]. The most common classes

---

[1] The difference between terpenes and terpenoids is that compounds belonging to the latter category contain additional functional groups whereas terpenes are solely composed of carbon and hydrogen. Cytochromes P450 are often involved in the further functionalization of terpenes into terpenoids, but terpenoids themselves can also be substrates for P450 enzymes. For this reason the terms terpenes and terpenoids are used synonymously throughout the chapter.

are shown in Fig. 1. For example, plant P450s are localized in the endoplasmic reticulum, belong to class II of the P450 group, and need a NADPH-dependent FAD and FMN-containing cytochrome P450 reductase (CPR) for their activity, whereas many of the bacterial P450s get the necessary electrons via a FAD-containing reductase and an iron–sulfur protein (ferredoxin). A special case is CYP102A1,

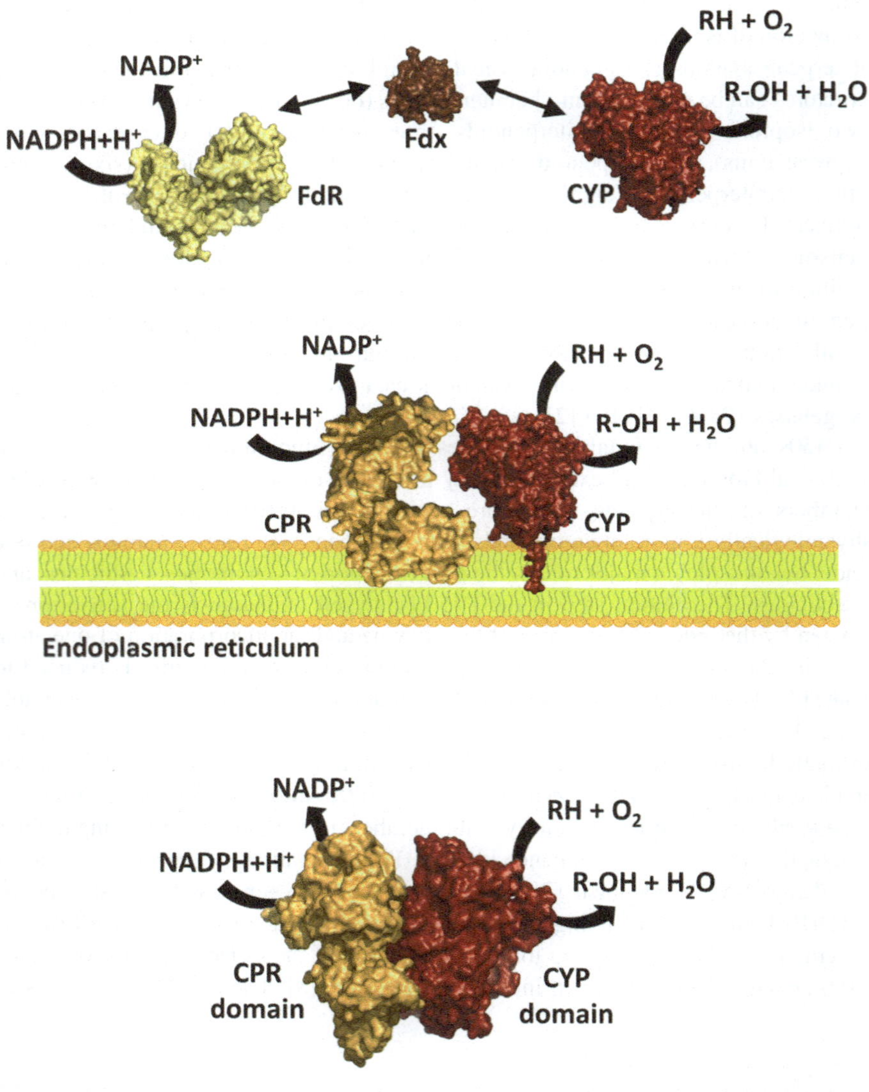

Fig. 1 Schematic organization of different cytochrome P450 systems and their classification according to [8]. *Top* class I, soluble bacterial system (e.g., CYP101A1). *Middle* class II, microsomal system (e.g., plant P450s). *Bottom* class VIII, bacterial CPR-P450 fusion system (e.g., CYP102A1)

where a FAD- and FMN-containing reductase domain is fused within one gene with a corresponding heme domain (Fig. 1).

In addition to their central role in essential biological pathways such as terpenoid biosynthesis, their ability to introduce oxygen into nonactivated C–H bonds under normal pressure and at room temperature makes P450s very attractive for applications in the field of biotechnology. Their tremendous potential is supported by the diverse spectrum of catalyzed reactions on a vast range of complex molecules and the possibility of applying protein engineering techniques to improve the characteristics of P450s further for biotechnological purposes [3, 9]. Although at present only a few microbial P450s have been described as being involved in the modification of terpenoids, their importance will certainly increase in the following years. In this review we summarize present knowledge on the identification, characterization, and engineering of microbial P450s for terpene modification.

2 Terpene Hydroxylation and Cytochromes P450

Terpenoids are the largest group of natural products and conservative estimates suggest that there are tens of thousands of terpenoids present in plants and fungi. Additionally, it has become increasingly clear during the last decade that many bacteria also produce terpenes. A major boost in this research area was given by newly developed techniques of genome mining, in which bioinformatic tools are used to screen bacterial genome sequences for conserved functional domains typical of terpene synthases [1]. The combination of genome mining and biochemical characterization leads to the simultaneous discovery of new natural products and the responsible biosynthetic enzymes. In fact, this approach has revealed a plethora of previously unknown enzymes, among them many cytochromes P450. In many cases, however, the physiological role of the biosynthetic pathways and their products, respectively, still need to be elucidated. Because the involved enzymes catalyze a variety of unusual biochemical reactions, they provide a wide range of possible applications, not least for sustainable production processes in the field of biotechnology.

Terpenes and their derivatives possess a broad spectrum of applications in varying areas and are therefore very interesting for different sectors of industry. Prominent examples include the anticancer agent taxol and the antimalaria drug artemisinin for the pharmaceutical industry or volatile monoterpenoids such as geraniol, citral, and menthol for the flavor and fragrance industries. In general terms, terpenoids possess a vast number of biological properties that can often be influenced by modification of their base structure. Especially the introduction of a hydroxyl group usually shows an influence on the volatility, efficacy, solubility, and toxicity of the substance. However, regio- and enantioselective hydroxylation through chemical synthesis is difficult to achieve due to side reactions, low yields, and the need of expensive catalysts. In contrast, the application of P450 enzymes for bioconversion processes allows the regio- and stereoselective introduction of oxygen under mild reaction conditions [10]. As a result, P450s are very attractive for the production and modification of terpenoids. Despite their impressive synthetic potential, practical application of P450s in industrial bioconversion processes is still limited [9]. Expression of typically membrane-bound plant P450s in microbial hosts tends to pose problems and results in low solubility and protein yields. Further challenges including their cofactor dependency and their overall low activity make these enzymes a difficult system with which to work. In view of these problems, soluble bacterial P450 systems are often considered as a substitute [11]. They usually exhibit higher catalytic turnover rates and their solubility and high expression level in heterologous hosts make them easier to handle in the laboratory. Due to these properties, bacterial P450s constitute ideal targets for protein engineering and are, therefore, not limited to the reactions that they naturally catalyze.

3 Origin and Identification of Microbial Terpene-Hydroxylating P450s

The number of described cytochrome P450 sequences is rapidly growing, with more than 21,000 representatives being named roughly about 50 years after the discovery of this enzyme family (http://drnelson.uthsc.edu/CytochromeP450.html) [7, 12]. Nonetheless, the examples for microbial P450 enzymes capable of converting substrates of terpene origin are still limited. Table 1 provides an overview of the best investigated microbial P450s that are known to act on terpenoids. It can be expected that this number will grow very quickly in the coming years due to industrial interest in developing processes that are biomimics of natural pathways and more sustainable, for example, using less energy, water, and hazardous compounds. Especially in the field of terpene oxidation, biotransformation methods (i.e., methods that utilize microorganisms or enzyme systems) emerge as an attractive alternative [13]. Some of their most prominent advantages are the ability to proceed under mild conditions and the lack of toxic waste production compared with many traditional chemical methods [14]. The use of microorganisms in terpene biotransformation started in the 1950s and was originally based on the discovery of microbial terpene metabolism. In pioneering studies, a pseudomonad isolated from sewage sludge was found to decompose (+)-camphor [15]. The responsible enzyme for the initial hydroxylation of camphor was later identified as CYP101A1 ($P450_{cam}$) [16, 17], the first described P450s of bacterial origin. Similarly, the oxidation of terpenes by fungi was reported in the same period with the discovery of α-pinene conversion into several oxygenated products by *Aspergillus niger* [18]. Further monoterpenoid substrates of different *A. niger* strains include, for instance, linalool and limonene [19, 20]. Although there are indications that P450s are not involved in the transformation of limonene by *A. niger*, most microorganism-mediated conversions of terpenes seem to proceed via cytochrome P450 monooxygenases.

Table 1 Microbial Cytochromes P450 with activities towards terpenoids

Cytochrome P450	Organism	Catalyzed reaction	Reference
P450cam (CYP101A1)	*Pseudomonas putida*	Camphor to 5-hydroxycamphor Diverse terpenoid hydroxylations by engineered enzyme variants	[117]
P450 BM-3 (CYP102A1)	*Bacillus megaterium* ATCC 14581	Diverse terpenoid hydroxylations by engineered enzyme variants	[118–120]
P450moxA (CYP105)	*Nonomuraea recticatena* IFO 14525	Oleanoic acid to queretaroic acid	[121]
CYP105A1	*Streptomyces griseolus*	Vitamin D_3 to 1α,25-dihydroxyvitamin D_3 Hydroxylation of abietane-type diterpene resin acids Epoxidation of pimarane-type diterpene resin acids	[11, 76]
CYP105A2	*Pseudonocardia autotrophica*	Vitamin D_3 to 25-hydroxyvitamin D_3	[122]
CYP105A3	*Streptomyces carbophilus*	Compacting to pravastatin	[123]
CYP106A1	*Bacillus megaterium* DSM319	11-Keto-β-boswellic acid (KBA) to 7β-hydroxy-KBA, 15α-hydroxy-KBA and 7β,15α-dihydroxy-KBA	[81]
CYP106A2	*Bacillus megaterium* ATCC 13368	Hydroxylation of several di- and triterpenoids	[73, 80, 83]
CYP107 (Vdh)	*Pseudonocardia autotrophica*	Vitamin D_3 to 1α-hydroxy- and 25-hydroxyvitamin D_3	[124]
CYP107 family member	*Sebekia benihana*	Vitamin D_3 1α- and 25-hydroxylation	[125]
P450terp (CYP108)	*Pseudomonas* sp.	α-Terpinol to 7-hydroxyterpineol	[126]
CYP109B1	*Bacillus subtilis*	(+)-Valencene to (+)-nootkatone via cis- or trans-nootkatol α-Ionone to 3-hydroxy-α-ionone β-Ionone to 4-hydroxy-β-ionone	[69]
CYP109D1	*Sorangium cellulosum* So ce56	α-Ionone to 3-hydroxy-α-ionone β-Ionone to 4-hydroxy-β-ionone	[45]
CYP110C1	*Nostoc* sp. PCC 7120	Germacrene A hydroxylase	[127, 128]
P450lin (CYP111)	*Pseudomonas incognita*	Linalool to 8-hydroxylinalool	[129]
CYP120A1	*Synechocystis* sp. PCC 6803	Various hydroxylations of retinoic acid	[130]
CYP153A6	*Mycobacterium* sp. HXN-1500	(−)-Limonene to perrilyl alcohol	[58]

(continued)

Table 1 (continued)

Cytochrome P450	Organism	Catalyzed reaction	Reference
CYP170A1	*Streptomyces coelicolor* A3(2) and *Streptomyces avermitilis*	Epi-isozizaene to (4S)- and (4R)-albaflavenols to albaflavenone Moonlighting terpene synthase	[131, 132]
CYP175A1	*Thermus thermophilus* HB27	β-Carotene to β-cryptoxanthin	[133]
P450cin (CYP176A1)	*Citrobacter braakii*	Cineole to (S)-6β-hydroxycineole	[107]
P450-cin	*Bacillus cereus* UI-1477	1,4-Cineole hydroxylations	[134]
CYP226A family members	*Burkholderia xenovorans* LB400	Abietane diterpenoid hydroxylation	[71]
CYP238A1	*Pseudomonas putida* KT2440	*cis*- and *trans*-Nerolidol to the 9-hydroxy product Binding of acyclic and cyclic terpene alcohols such as farnesol, nerolidol, linalool, and terpineol	[135]
CYP264B1	*Sorangium cellulosum* So ce56	α-Ionone to 3-hydroxy-α-ionone β-Ionone to 3-hydroxy-β-ionone	[53]
P450(camr)	*Rhodococcus* sp. NCIMB 9784	(+)-Camphor to 6-endo-hydroxycamphor	[136]
Putative CYPs CotB3 +CotB4	*Streptomyces melanosporofaciens* MI614-43F2	Cyclooctat-9-en-7-ol to cyclooctat-9-en-5,7-diol to cyclooctatin	[137]

The number of microorganisms described to convert terpenes is continuously growing. A major driving force behind this is the extensive research for the discovery of novel microbial production routes towards flavors and fragrances and the rising interest and demand of bioactive terpenoid compounds for pharmaceutical applications. A straightforward approach in identifying an adequate biocatalyst is the screening of microorganisms that can use the substrate as the sole carbon source, thereby indicating the existence of a substrate-degrading metabolic pathway. The first step is the isolation and characterization of organisms that come into close contact with terpenoids in their natural habitat. Examples include the isolation of the α- and β-pinene degrading *Bacillus pallidus* BR425 from pine trees [21] and the identification of limonene-degrading microorganisms from orange peel [22]. Other interesting targets for screening approaches are organisms that are known to have a well-developed xenobiotic metabolism, which in turn indicates the presence of interesting P450s. Both approaches are extensively applied and contribute to the description of newly identified terpene biotransformations by microorganisms, whereas in many cases the P450 enzymes that catalyze the reactions are not yet identified. The identification and isolation of the genes and enzymes, however, opens up the possibility of applying protein engineering techniques to overcome the problem of low transformation rates, which are still the main obstacle for large-scale applications [9]. Furthermore, the cloning of P450s allows their expression in heterologous whole-cell hosts that can enable a more effective terpene oxidation process. For several examples given in Table 1, the terpene-hydroxylating activity of the cytochrome P450 was not described until years after the first isolation and characterization of the enzyme. This is often the case with so-called orphan P450s, for which the natural substrate is not known, or xenobiotic-degrading P450s with a broad substrate range [23]. In such circumstances, screening of compound libraries is a common method to examine the substrate range and often results in the identification of new terpenoid substrates. On the other hand, the examples of CYP102A1 (P450 BM-3) and CYP101A1 ($P450_{cam}$) have proven that protein engineering is a powerful method to develop tailored enzymes for the oxidation of a given terpene substrate.

Taken together, the rapidly growing number of P450 structures, the continuous identification of microorganisms capable of converting terpenes, triggered by the trend of classical chemical oxidation methods for terpenes gradually being replaced by biotechnological methods, as well as recent progress in the development of effective protein engineering methods, will surely lead to a steady increase in the identification and importance of microbial cytochromes P450 for terpene hydroxylation.

4 Engineering of Microbial Cytochromes P450 for Terpene Hydroxylation

Numerous protein engineering approaches have been applied to improve the performance of P450s further, and most work has been focused on CYP101A1 from *Pseudomonas putida* and CYP102A1 from *Bacillus megaterium*. Both are well-characterized, representative enzymes that can be easily expressed in *E. coli*. They are also the first two P450 enzymes for which crystal structures could be solved [24, 25]. CYP101A1 naturally oxidizes the terpenoid (1R)-camphor to 5-exo-(1R)-hydroxycamphor, the first step in the camphor metabolism pathway of *P. putida*. For the completion of the electron transfer reaction, the enzyme depends on two additional electron transfer proteins: putidaredoxin, a ferredoxin of the [2Fe–2S] type, and putidaredoxin reductase, a FAD-containing enzyme. It can therefore be assigned to class I of the P450 group. In contrast, the CYP102A1 is a self-sufficient fusion enzyme that contains both a P450 and a reductase domain and therefore only requires NADPH and oxygen to function (class VIII) [8]. As a result, the catalytic activity of this fatty acid hydroxylase is comparatively high, with a k_{cat} value of more than 5,000 min^{-1} for the hydroxylation of arachidonic acid [26]. In comparison, most other P450 reactions are very slow, with k_{cat} values of less than 100 min^{-1}.

The different protein engineering methods that have been applied to alter P450 enzymes can be grouped into different categories and are illustrated in the following sections using CYP101A1 and CYP102A1 as examples.

4.1 Engineering by Directed Evolution

Directed evolution is a method that mimics natural selection in order to generate proteins with new or improved activities. This approach starts with the generation of a library of mutated P450 enzymes using techniques that randomly introduce mutations into the protein sequence, for example, error-prone polymerase chain reaction (*ep*PCR). After screening for new or improved activities, the respective mutants are isolated and then used to generate new libraries by random mutagenesis or recombination technologies. These iterative cycles continue until the desired mutant is obtained. The possibilities of this technique are manifold and span not only the improvement of catalytic activity and change of regio- or stereoselectivity but also the conversion of nonnatural substrates. Many studies have been performed to engineer the heme domain-coding region of CYP102A1 using directed evolution. Examples include the improvement of activity: the wild-type CYP102A1 typically hydroxylates saturated and nonsaturated fatty acids, but also has a low hydroxylation activity for β-ionone of less than 1 min^{-1}. Two rounds of *ep*PCR resulted in the triple mutant A74E F87V P386S, which exhibited an increase in activity of up to 300-fold compared to the wild-type enzyme [27]. Furthermore, there are many examples demonstrating that directed evolution is a powerful tool for the generation of P450 libraries with completely new activities. However, because these activities are often accompanied by low enzyme activity and stability, the most promising candidates usually need additional optimization by rational protein design methods. Other strategies to enhance the performance of CYP102A1 as catalyst in chemical synthesis using laboratory evolution methods aim towards an improvement of enzyme stability [28]. Among the broad range of substrates that are hydroxylated by CYP102A1 are also water-insoluble compounds, for example, the sesquiterpene valencene. In these cases, polar organic cosolvents are added to increase substrate solubility and to achieve high catalytic efficiency. Applying directed evolution, the tolerance of CYP102A1 towards the cosolvents dimethylsulfoxide (DMSO), tetrahydrofuran (THF), acetone, acetonitrile, dimethylformamide, and ethanol could be significantly increased [29].

4.2 Engineering by Rational Protein Design

In contrast to directed evolution, rational protein design relies on methods such as site-directed mutagenesis with the aim of rationally altering the properties of an enzyme. A fundamental prerequisite for this method is a solid structural basis in terms of profound knowledge of the key amino acid residues that are involved in the characteristics to be influenced. The analysis of protein structural models, especially in their substrate-bound state, provides us with some insight into the mechanism of substrate binding and furnishes ideas for rational engineering. The role of CYP101A1 and CYP102A1 as model enzymes for the rational protein design of cytochromes P450 is highlighted by the fact that there are crystal structure data available for several dozens of mutants of both enzymes (see Protein Data Bank (PDB) at: http://www.rcsb.org/pdb/). Most studies of these enzymes focus on altering the specificity and regioselectivity of the catalyzed hydroxylation reactions. Analysis of the active site of CYP101A1 and comparisons of the structure of target substrates with that of camphor led to the design of active-site mutants with greatly enhanced activity for the oxidation of (+)-α-pinene [30]. Based on the enzyme/substrate contacts revealed by the structure of the triple mutant F87W/Y96F/V247L with (+)-α-pinene bound within the active site, additional mutations were carried out and resulted in an improved selectivity for the formation of the natural fragrances and flavors (+)-*cis*-verbenol and (+)-verbenone [31]. In addition, results from studies of other mutants suggest that CYP101A1 can be engineered to oxidize compounds substantially larger than camphor by manipulating the active site volume and topology to maneuver the substrate binding orientation [32].

A central starting point for the rational protein design of CYP102A1 was the identification of a hydrophobic channel between the surface of the enzyme and the proximal side of the heme. The hydrophobic residues in this channel, especially amino acid F87, have been shown to affect both the substrate specificity and regioselectivity [33]. Replacements of this residue were used in many studies as the starting point for the creation of new site-directed mutants with specificities towards nonnatural substrates. Another CYP102A1 mutant created by rational mutagenesis for the hydroxylation of shorter-chain fatty acids was unexpectedly capable of hydroxylating various alkanes and cycloalkanes, indicating some promise for engineering the enzyme for isoprenoid substrates [34].

4.3 Engineering by Chimeragenesis

Chimeragenesis imitates the natural principle of gene recombination by using techniques such as DNA shuffling and SteP recombination to produce a variety of chimeric genes [35, 36]. This method is often more efficient than random mutagenesis for generating mutants with new properties, inasmuch as the amino acid substitutions have already proven to be successful in one of the parents. However, genes normally need to exhibit a sequence identity of 70 % or higher in order to be successfully recombined. Problems may arise due to the generally low sequence identity of P450s of less than 20 % and recombination at crossover points that produce unstable and inactive mutants. An effective tool to overcome this shortcoming in designing P450 chimeras is the structure-based computation algorithm SCHEMA, which was designed to identify optimal crossover points for generating protein fusions [37]. Chimeragenesis can be applied using P450 enzymes with a high sequence homology to change the substrate selectivity and activity. Additionally, the construction of chimeras between soluble and membrane-bound P450s may provide a means of generating soluble P450s that are functionally active in microbial hosts. The creation of diverse libraries of chimeric P450s therefore opens exciting opportunities for the engineering of P450s.

In the case of CYP102A1, a "scanning chimeragenesis" approach was used to study the roles of individual structural elements that determine the substrate specificity towards terpenoids [38]. For this, several amino acid residues of the CYP102A1 substrate binding site were replaced with a homologous fragment from CYP4C7, an insect P450 catalyzing the regio- and stereospecific hydroxylation of the sesquiterpenoid farnesol to (10E)-12-hydroxyfarnesol [39]. The regiospecificity of monooxygenation was shifted and resulted in the formation of new products not seen with wild-type CYP102A1. Further studies eventually led to a chimeric enzyme that produces 12-hydroxyfarnesol as the major product, with an approximately twofold increase in turnover number towards farnesol as compared with wild-type CYP102A1 [40]. Another promising application for chimeragenesis is the production of chimeric oxygenases in which the P450 domain is fused to the reductase domain of a self-sufficient P450. A good example for this is the fusion of CYP101A1 to the reductase domain of P450RhF, comprising flavin mononucleotide- and NADH-binding motifs and a [2Fe-2S] ferredoxin-like center [41]. The resulting catalytically self-sufficient chimera can achieve high conversion rates and has a high potential for applications in biocatalysis, especially of the natural terpene substrates. Additionally, the production of a library of self-sufficient chimeric P450 enzymes and mutants can also be a useful tool for high-throughput screening protocols for the identification of novel biooxidation activities [42].

4.4 Engineering Improved Redox Partner Interaction

The above-described generation of self-sufficient cytochrome P450 variants addresses an important issue in P450-catalyzed reaction. As external monooxygenases, most P450s have to interact with partner proteins to receive electrons for oxygen activation and substrate conversion, and in most cases, the rate-limiting step is the transfer of the second electron to the hydroxylase component. Improving redox partner interaction and electron transfer is, therefore, an important tool to enhance product formation and thereby improves the potential of P450 systems for biotechnological applications [3, 9, 43].

Investigations of the native CYP101A1 system, consisting of the monooxygenase, the electron transport protein putidaredoxin, and putidaredoxin reductase expressed in *E. coli*, showed a marked enhancement in the conversion of camphor when utilizing C73S/C85S-Pdx, a stable putidaredoxin mutant instead of the wild-type protein [44]. Aside from homologous redox systems, choosing heterologous redox partners and their further tuning by protein design can also increase the activities of CYP systems. A good example of this is the mammalian electron transfer protein adrenodoxin (Adx) that turned out to be three- to fivefold more efficient in supporting the substrate conversion by a myxobacterial CYP than the autologous redox systems [45]. Because the bacterial Pdx transfers electrons much faster to its redox partner CYP101A1, an "evolutionary" approach was used to produce a truncated Adx-mutant with an additional amino acid exchange (S112W) that exhibits higher similarity to Pdx than the wild-type. This mutant showed a 75-fold increase in the homologous CYP11A1-dependent cholesterol conversion system [46]. Its usefulness for the conversion of terpenoids by bacterial CYPs needs to be investigated. However, the truncated Adx variant proved to be a very efficient heterologous redox partner in many bacterial P450 bioconversion systems, also in those converting terpenoids [47].

These results demonstrate that protein engineering methods are not limited to the hydroxylase component of P450 systems and further improvement of redox partner interaction can surely lead to increased substrate conversion.

5 Examples for Terpenoid Conversions Using Microbial Cytochromes P450

The following sections provide selected examples for the conversion of important terpenoids by microbial cytochromes P450 and thereby show different approaches in exploiting the biotechnological potential of these enzymes. Aside from monoterpenes (limonene), sesquiterpenes (valencene), diterpenes (resin acids), and triterpenes, additional examples are given for terpene derivatives such as norisoprenoids (ionone), steroids, and vitamin D_3.

5.1 Ionones

Ionones (α-, β-, and γ-) are norisoprenoids that are derived from the degradation of carotenoids. They are aroma components of floral scents and can be found in a variety of essential oils. The organoleptic properties of ionones and especially their oxygenated derivatives have attracted the attention of the fragrance and flavor industries. Additionally, α-ionone and its derivatives form important components of insect lures and can favor insect pollination [48], whereas hydroxy-β-ionone is an important intermediate for the synthesis of carotenoids and deoxyabscisic acid, a synthetic analogue of the phytohormone abscisic acid [49]. Because of the potential industrial applications, several microbial transformation studies of α- and β-ionone were performed using various fungal strains including *Aspergillus* sp. [50] and bacterial strains including *Streptomyces* sp. [51].

Cytochrome P450 monooxygenases are very useful for the oxygenation of ionones and several P450s from actinomycetes that are able to hydroxylate α- and β-ionone have been identified. Among those are CYP105A1 and CYP105B1 are from *Streptomyces griseolus* and CYP105D1 is from *Streptomyces griseus*. Using a recombinant expression system in *E. coli*, all three enzymes were shown to convert both α- and β-ionone into hydroxylated products. Although the activity of CYP105A1 and CYP105D1 towards α-ionone was rather low, CYP105B1 was able to produce 3-hydroxy-α-ionone with a conversion rate of more than 50 % [52]. Comparable results could be obtained for the conversion of β-ionone, with 4-hydroxy-β-ionone being the main product. Regarding the moderate enantioselectivity of CYP105B1 towards β-ionone, this enzyme provides an ideal target for biocatalyst improvement, either by rational protein design or directed evolution.

In contrast to the hydroxylation of α- and β-ionone with members of the CYP105 family where a mixture of different hydroxylated products has been observed in addition to the main product, two P450s from the soil-dwelling myxobacterium *Sorangium cellulosum* So ce 56 were found to hydroxylate both α- and β-ionone in a very regioselective manner. For CYP109D1, the exclusive production of the main products of the CYP105 catalyzed reactions, namely 3-hydroxy-α-ionone from α-ionone and 4-hydroxy-β-ionone from β-ionone, could be shown using a reconstituted P450 system [45]. The reason for the high regioselectivity of the enzymatic hydroxylation lies in the stereochemistry of ionones. The presence of two methyl groups at C-1 directs any oxidative attack towards C-3. Likewise, the electronic activation of the allylic hydrogens by the double bond of the cyclohexane ring governs the regioselective hydroxylation to position C-4 of β-ionone. Nevertheless, another P450 from this organism, namely CYP264B1, was shown to hydroxylate both α- and β-ionone in a highly regioselective manner at the C-3 position, giving 3-hydroxy-α-ionone and 3-hydroxy-β-ionone, respectively [53]. For the elucidation of the structural basis for the selectivity of ionone hydroxylation, both α- and β-ionone were docked into the active sites of CYP109D1 and CYP264B1 (see Fig. 2). The computational data justified the experimental observations and form the structural basis for advanced studies using rational protein design.

Concerning the activity, the most promising P450 for the hydroxylation of ionones is still CYP102A1. Based on earlier studies on the expansion of its substrate range [34], several P450 CYP102A1 mutants have been constructed using rational protein design as well as directed evolution that are able to hydroxylate β-ionone at C-4 regioselectively with high activity and rather high coupling efficiency [27].

As for most terpenoid substrates, the number of identified P450s that are able to accept ionones as substrates is constantly increasing. With CYP101C1 from *Novosphingobium aromaticivorans* DSM12444, a homologue of CYP101A1 has been described very recently that does not bind camphor but is capable of binding and hydroxylating ionone derivatives including α- and β-ionone and β-damascone [54]. Scheme 1 gives an overview of P450 catalyzed ionone conversions.

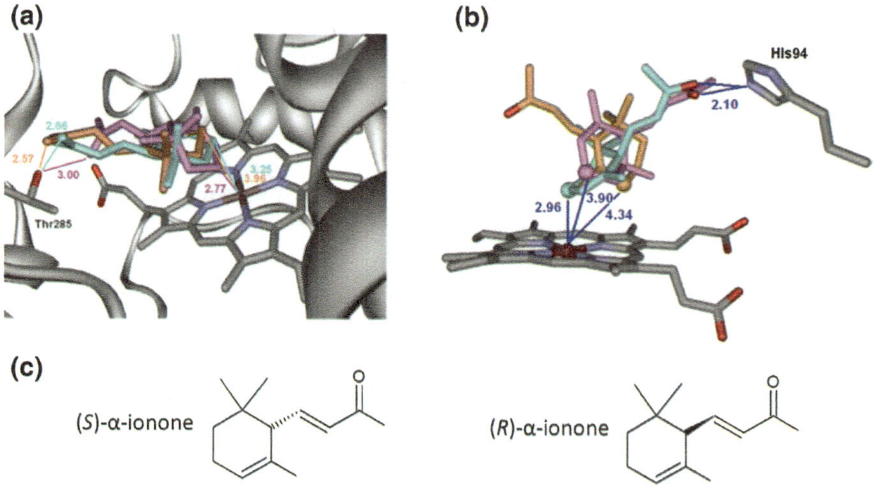

Fig. 2 Docking positions of ionone in the homology model of CYP264B1 (**a**) and CYP109D1 (**b**). The *R*-isomer of α-ionone is shown in pink, the *S*-isomer in cyan, and β-ionone in orange. The distances between the position where hydroxylation occurs and the heme iron are given in Ångstroms. (**c**) *R*- and *S*-isomer of α-ionone. Figure (**a**) taken from [53] and Figure (**b**) taken from [45]

5.2 Limonene

Limonene is a chiral cyclic monoterpene, the D-enantiomer of which constitutes the main component of many citrus peel oils, such as that from oranges. D-Limonene is usually obtained via centrifugal separation or steam distillation of citrus fruits and areas of application include cosmetics and perfumery as well as food manufacturing where it is used as a flavoring. It is also added to cleaning products due to its lemon/orange odor. In contrast, the enantiomer L-limonene (e.g., found in noble fir and peppermint plants) has a piney/turpentine-like odor. The main metabolites of limonene in the human body are (+)- and (−)-perillyl alcohol, which are products of 7-hydroxylation by CYP2C9 and CYP2C19 in the liver [55]. The CYP-catalyzed reaction of (−)-limonene to (−)-perillyl alcohol is shown in Scheme 2. Aside from its antibacterial and antifungal effects, (−)-perillyl alcohol is of considerable pharmaceutical interest due to its anticarcinogenic properties [56]. As with many other terpenoids, the extraction of perillyl alcohol from plant tissues yields only small amounts of this compound and the chemical oxidation of readily accessible limonene results in a complex mixture of products. The production of significant quantities of side products, namely carveol, carvone, and terpineol, also occurred in the case of the first microbial enzyme system described to transform limonene to perillyl alcohol in *Bacillus stearothermophilus* BR388 [57]. In a study with the aim of identifying bacteria capable of regiospecific hydroxylation of limonene, 1,800 bacterial strains were grown on a range of relatively reduced substrates such as

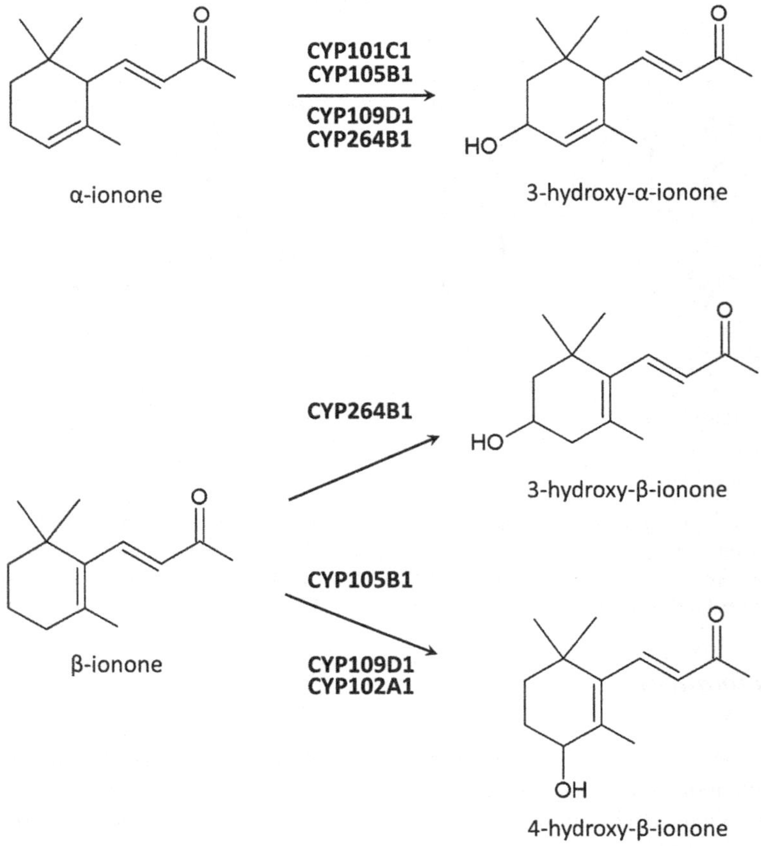

Scheme 1 CYP-catalyzed reactions of α- and β-ionone

toluene, naphthalene, and various alkanes. Because previous results demonstrated that many catabolic oxygenases accept a wide range of unnatural substrates, this procedure was chosen to identify oxygenases involved in catabolic pathways that would probably also accept L-limonene as a substrate. Using this approach, CYP153A6 from *Mycobacterium* sp. strain HXN-1500 was found to oxidize limonene exclusively at the carbon atom in position 7 (see Scheme 2) [58]. In recent studies, metabolic engineering was applied with the aim of utilizing this enzyme for limonene and perillyl alcohol production from simple carbon sources in *E. coli* [59]. To this end, the genes of a heterologous mevalonate pathway, a GPP synthase, a limonene synthase, and CYP153A6 were expressed in the bacterial host. After several gene modifications to improve the enzyme availability and activity, L-limonene titers reached 435 mg/l and perillyl alcohol was produced with 105 mg/l from 1 % glucose. These results mark an important step in the production of valuable terpenoids from simple sugars using microbial enzymes and hosts.

Scheme 2 CYP-catalyzed reaction of (−)-limonene to (−)-perillyl alcohol

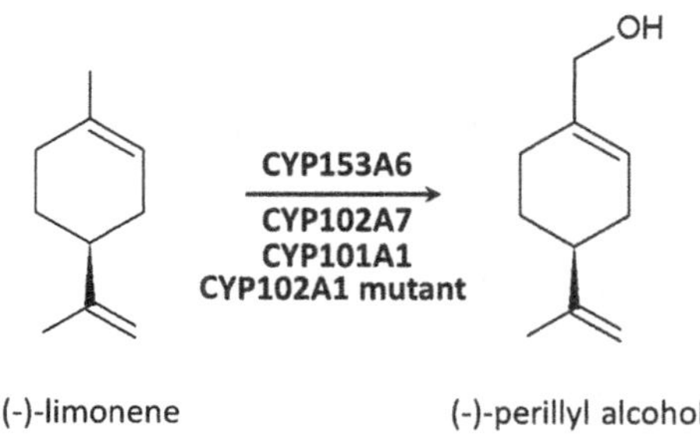

Another P450 that has been tested for perillyl alcohol production in *E. coli* is CYP101A1. After coexpression with its homologous redox partners from *P. putida*, the whole-cell 7-hydroxylation of L-limonene to (−)-perillyl alcohol could be confirmed [60]. CYP102A1, the second major model P450, is also able to produce perillyl alcohol from limonene. In an approach to optimize the regiospecificity through rational protein design, this enzyme has been engineered to increase the selectivity of limonene hydroxylation at the terminal allylic C-7 position by up to 97 % [61]. The applied methods could also be useful in the engineering of other limonene hydroxylases such as CYP102A7 from *Bacillus licheniformis* [62] and CYP147F1 from *Streptomyces peucetius* [63].

5.3 Valencene

The bicyclic sesquiterpenoid (+)-nootkatone was first isolated from the Nootka cypress tree *Callitropsis nootkatensis*, but it can also be found in the fruits of citrus plants including grapefruit and pomelo. It exhibits unique odor characteristics and is, therefore, a highly valued and highly demanded aroma compound for the food and flavor industry. In addition, (+)-nootkatone is also an effective and environmentally friendly repellent of insects such as mosquitos, ticks, and bedbugs [64, 65]. Limitations in application are dictated by the costly extraction of this compound from natural sources in connection with inadequate yields. Access to (+)-nootkatone in industrial production processes is, therefore, often provided by oxidation of the cheap and abundantly available sesquiterpene (+)-valencene. However, chemical oxidation methods often rely on unsafe agents such as *tert*-butyl peracetate or *tert*-butyl hydroperoxide in combination with surface-functionalized silica-supported metal catalysts [66]. As an alternative to chemical oxidation methods, a huge number of biotechnological processes using whole-cell biotransformations in bacteria and fungi, as well as plants have been described [67]. In addition to laccases and peroxidases, many of these biotransformation reactions are catalyzed by enzymes of the cytochrome P450 superfamily. The involvement of a P450 enzyme is often detected indirectly as in the case of (+)-valencene bioconversion in the ascomycete *Chaetomium globosum*, where a decrease in product concentrations could be observed

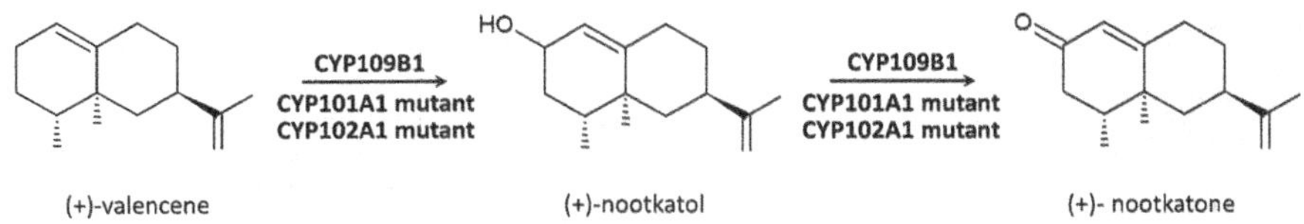

Scheme 3 CYP-catalyzed reaction of (+)-valencene to (+)-nootkantone via (+)-nootkatol

after addition of the P450 inhibitor proadifen [68]. The direct identification of efficient and regioselective P450 enzymes matching commercial requirements still remains a challenge. In a study testing 125 bacterial P450 enzymes, CYP109B1 from *Bacillus subtilis* was found to catalyze the oxidation of (+)-valencene at position C-2 to nootkatol and (+)-nootkatone with the production of 120 mg l^{-1} of the desired products after optimizing the reaction [69] (see Scheme 3). However, the biooxidation suffered from the formation of undesired multioxygenated products that can appear when the active site of the enzyme allows different binding orientations for an unnatural substrate. A similar observation could be made in the case of (+)-valencene conversion with CYP102A1, where the selectivity pattern of the reaction suggests multiple orientations of the substrate within the active site [70]. In contrast, different mutants of CYP101A1 could oxidize (+)-valencene with a regioselectivity for C-2 oxidation of more than 85 % yield, whereas the wild-type enzyme is not able to accept (+)-valencene as a substrate. The protein engineering challenges therefore differ for these two P450s: both substrate specificity and selectivity of the CYP102A1 require improvement whereas the CYP101A1 exhibits the desired selectivity but the activity has to be increased.

5.4 Resin Acid Diterpenoids

Diterpenes are derived from geranylgeranyl pyrophosphate and usually have the molecular formula $C_{20}H_{32}$. They form the basis for important compounds such as the vitamin precursors retinol and phytol. In addition, many diterpenes and their derivatives exhibit antibacterial, antifungal, and anti-inflammatory actions. Prominent representatives are the tricyclic diterpenoids of the abietane- and pimarane-type which are gathered under the name resin acids and can be found in tree resins. Other examples include labdane, a bicyclic diterpene that forms the core for a wide range of natural products, and taxadiene, an important intermediate in the synthesis of the mitotic inhibitor taxol used in cancer chemotherapy. The number of described microbial P450 enzymes that can be used for the biotransformation of diterpenoid substrates is still limited. Recently, two members of the CYP226 family able to hydroxylate different abietane-type diterpenoids could be identified in *Burkholderia xenovorans* LB400 [71]. This proteobacterium is well known for its noncatabolic physiological adaptation to organic compounds and both CYP226 enzymes have distinct roles in the abietane catabolism. Many other organisms that are capable of abietane diterpenoid biodegradation have been described and the involvement of P450s has been proven in several cases, such as the *Pseudomonas abietaniphila* BKME-9 [72]. It can, therefore, be speculated that the number of cytochromes P450 described as diterpenoid hydroxylases will sharply grow in the future. In addition to those enzymes involved in catabolic pathways, orphan P450s also play an important role when it comes to the description of hydroxylating activity towards new substrates. For example, the CYP106A2 from *B. megaterium* ATCC 13368, whose natural substrate is still unknown but which is known to hydroxylate different steroids, was recently also described as a regioselective allylic bacterial diterpene hydroxylase [73]. By screening of a compound library consisting of more than 16,000 potential substrates, the diterpene derivative dihydrochinopimaric acid could be identified as binding in the CYP106A2 active site. Further studies with structurally related compounds showed the conversion of abietic acid to 12α- and 12β-hydroxyabietic acid, respectively (see Scheme 4). The existence of many mutants of CYP106A2, created by directed evolution as well as rational protein design [74, 75], provides an excellent source to exploit the potential of this P450 for further biotechnological applications. Due to their diverse biological properties, the production of hydroxylated diterpene derivatives could be especially useful for the pharmaceutical industry. A vivid example is the conversion of abietic acid into 15-hydroxyabietic acid catalyzed by CYP105A1 from *S. griseolus* [14, 76] (see Scheme 4). This substance is the precursor of the strong contact allergen 15-hydroperoxyabietic acid and only hard to obtain from natural sources or by chemical means. The availability via the recently established biotechnological process gives a new and promising basis for the production of this compound for test developments against allergic compounds.

Much effort has been undertaken to engineer microbial hosts for the production of functionalized diterpenoids, for example, in the production of important drugs

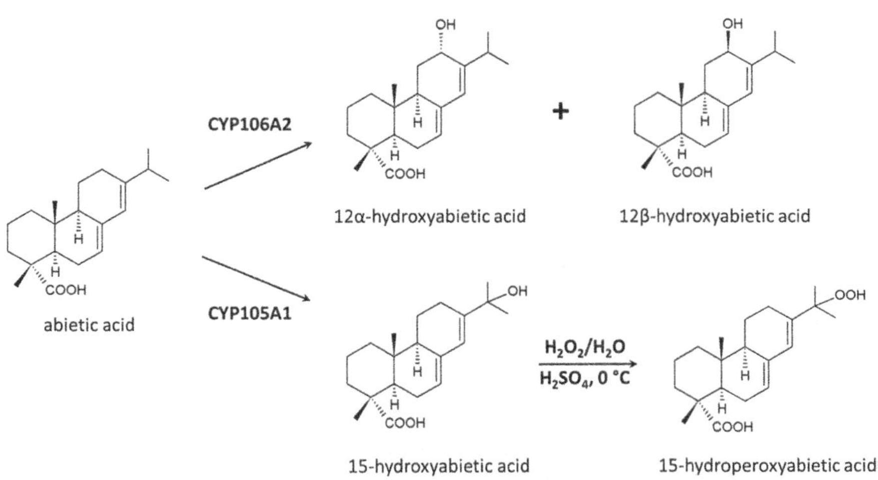

Scheme 4 CYP-catalyzed hydroxylations of abietic acid

such as taxol. In most cases, researchers focus on the plant P450s of the homologous biosynthetic pathways. However, because their use is accompanied with problems and challenges of the membrane localization of these enzymes, the identification and engineering of bacterial P450s for these reactions could prove to be very beneficial.

5.5 Triterpenoids

Triterpenoids consist of six isoprene units and the simplest triterpene regarding the complexity of the carbon skeleton is squalene. This compound is produced via head-to-head condensation of two C15 units of farnesyl diphosphate and constitutes the precursor of polycyclic triterpenes. From a biological perspective, the most important triterpenoid structures are the oleanane, ursane, lupane, and dammarane/euphane carbon skeletons [77]. In contrast to mono- and sesquiterpenes, which are mostly found in essential oils of plants, triterpenes are predominantly found in resin and balsams. Their biological effects are diverse and comprise anti-inflammatory, hepatoprotective, analgesic, antimicrobial, antimycotic, virostatic, immunomodulatory, and tonic effects [77]. However, the use of many triterpenoids for pharmaceutical applications is restricted due to hemolytic and cytostatic properties. In order to overcome these limitations and to expand the range of usable triterpenes, a modification of these compounds by means of chemical or biotechnological techniques is possible [78]. For this purpose, microbial transformation is particularly suitable because it often results in a modification with a high regio- and enantioselectivity and many microorganisms that are able to transform triterpenoids have been described [78]. An example is *B. megaterium* ATCC 13368, which was shown to perform an hydroxylation at the C-1 or C-11 site of the antimelanoma agent betulinic acid [79]. The aforementioned CYP106A2 from this organism belongs to the few described microbial P450s that can transform triterpenes. Recent studies showed that this P450 is able to hydroxylate the pentacyclic triterpene 11-keto-β-boswellic acid (KBA) to 15α-hydroxy-KBA, depicting an interesting reaction because this position is not accessible via synthetic chemical methods [80].

Comprehensive studies with CYP106A1, another member of the CYP106 family, revealed that this P450 mainly hydroxylates KBA at position 7β. Additional products, namely 15α-hydroxy-KBA and 7β,15α-dihydroxy-KBA, were formed in a lower amount [81]. These novel KBA derivatives can possibly show an enhanced bioavailability or improved pharmaceutical activities compared to the substrate KBA or can be the basis for additional modifications by chemical methods [81]. This is especially interesting inasmuch as pentacyclic triterpenes have gained significant importance because their chemical structures resemble those of steroids and there is an increasing demand especially for boswellic acids for pharmaceutical and medicinal studies [82]. Scheme 5 shows the hydroxylation of 11-keto-β-boswellic acid by CYP106 family members. Moreover, the substrate range of CYP106A2 is not limited to KBA. The screening of a natural product library revealed the binding

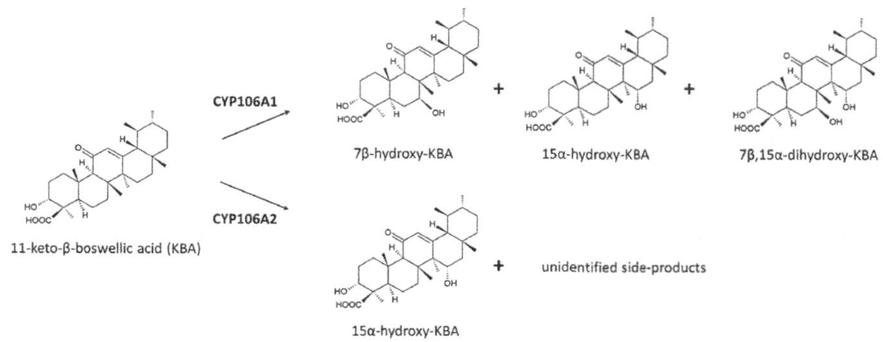

Scheme 5 Hydroxylation of 11-keto-β-boswellic acid by CYP106 family members

of additional triterpenes and conversion of the tetracyclic dammarane-type triterpenoid dipterocarpol to 7β,11α-dihydroxydipterocarpol could be shown [83]. These are very recent research results, therefore thus far no protein engineering studies have been performed to exploit fully the potential of P450s for triterpenoid functionalization. Considering the background of two closely related P450 enzymes, CYP106A1 and CYP106A2, that are both capable of hydroxylating the same substrate with a different regioselectivity, the starting point for the elucidation of the structural basis that determines this selectivity, and subsequent approaches to influence it, seems to be very promising.

5.6 Steroids

Like triterpenoids, steroids are derived from squalene and constitute a class of very important molecules that are involved in the regulation of essential processes in mammals. They are synthesized from cholesterol, which is biologically produced from lanosterol in animals. The fact that steroidal drugs belong to the second most marketed drugs next to antibiotics [84] highlights the important role of steroids in biotechnology [3, 85]. Because the chemical synthesis of steroid compounds requires multiple reaction steps, harsh reaction conditions, and cost-intense catalysts that are often toxic to organisms and harmful to the environment, microbial biotransformations have been used for the generation of novel steroidal drugs for decades [86]. This carries the advantage that low-cost precursors such as cholesterol or diosgenin can be exploited for further hydroxylations in positions that are interesting for the production of novel hydroxysteroids (7α, 9α, 11α, 11β, 16α, and 17α). Although the history of microbial steroid production started in the early 1950s, the steroid-modifying enzymes have not yet been characterized in most cases. An exception is CYP154C5 from *Nocardia farcinica*, which was identified as potent 16α-hydroxylase for testosterone and similar steroids [87]. Moreover, CYP106A2, which was identified in the 1970s to be responsible for the ability of *B. megaterium*

ATCC 13368 to convert progesterone into 15β-hydroxyprogesterone [88–90], is also able of catalyzing a 15β-hydroxylation of other 3-oxo-Δ4-steroids such as 11-deoxycorticosterone, testosterone, and 11-deoxycortisol [74, 75, 80, 91]. Later on it was found that di- and triterpenes can also be highly selectively hydroxylated by CYP106A2 [73, 80, 83, see above].

As a soluble prokaryotic P450, CYP106A2 depicts an excellent candidate to establish a steroid bioconversion process. It can be expressed with high yields in *E. coli* and *B. megaterium* and has already been applied for the preparative hydroxylation of several steroids, including dehydroepiandrosterone, pregnenolone, testosterone, deoxycorticosterone, deoxycortisol, and progesterone [92–94]. Recent studies showed that the main hydroxylation position of CYP106A2 is influenced by the character of groups attached to C-3 in the steroid core: 3-oxo-Δ4-steroids are preferentially hydroxylated at the 15β position whereas the 7β-position is favored when converting 3-hydroxy-Δ5-steroids [92]. However, the hydroxylation reaction is not strictly regioselective and the formation of different side products occurs in many CYP106A2-dependent bioconversions as, for instance, in the case of progesterone where 11α-, 6β-, and 9α-hydroxyprogesterone are formed. Albeit these hydroxylations occur to a much lower extent, their products are of higher interest from an industrial point of view. Rational protein design and directed evolution approaches have, therefore, been performed to shift the selectivity of CYP106A2 in favor of progesterone 11α-hydroxylation [28, 75, 95]. To this end, the primary structures of CYP106A2 and the mammalian steroid-11β-hydroxylase CYP11B1 were aligned in order to identify amino acids that could be involved in the determination of stereoselectivity in the steroid hydroxylation. The greatest differences were found in the substrate recognition sites of both enzymes, and using site-directed mutagenesis, five amino acid substitutions (S394I, A395L, T396R, G397P, and Q398S) were performed to change the respective amino acids in CYP106A2 into those of CYP11B1. Two mutants, A395L and G397P, increased the 11α-hydroxylation efficiency of CYP106A2 by a factor of four [75]. To improve the 11α-hydroxylation further, saturation mutagenesis at positions A395 and G397 was performed and more than 13,000 transformants were screened. The mutants A395I and A395I/G397K have 8.9-fold and 11.5-fold higher 11α-hydroxylation activity compared with the wild-type. However, the total activity of the mutants was reduced by approximately 50 % compared with the wild-type enzyme. To increase the activity of these mutants, site-directed mutagenesis was applied to introduce previously identified mutations that result in higher active enzyme variants. This approach eventually led to the identification of three mutants (A106T/A395I/R409L, M155I/F165L/A395I, and D217V/A395I) with significantly increased k_{cat}/K_m values and 41–55 % 11α-hydroxylating selectivity. These mutants as well as the highly selective T89N/A395I and A106T/A395I mutants were tested for in vivo 11α-hydroxylation of progesterone using a whole-cell bioconversion system established for CYP106A2 in *E. coli.* The highest in vivo selectivity (80.9 % 11α-hydroxyprogesterone) was achieved with the double mutant T89N/A395I, demonstrating an impressive alteration in regioselectivity [28].

Due to the high interest in hydroxylated steroids, numerous studies also examined the potential of the well-characterized CYP102A1 as steroid hydroxylase and many mutants of this enzyme that catalyze the conversion of steroids were engineered over the past years. The first CYP102A1 variant described as steroid hydroxylase is the triple mutant R47L/F87V/L88Q. This mutant converts testosterone into 2β-hydroxy-, 15β-hydroxy-, and 16β-hydroxytestosterone. To identify amino acids crucial for testosterone biotransformation, the respective single and double mutants were also investigated. However, none of the single mutants was able to hydroxylate testosterone, whereas the double mutants F87V/L188Q and R47L/F87V converted testosterone into the same products as the triple mutant, although with a four- to tenfold lower conversion rate. Furthermore, positive homotropic cooperativity was observed with testosterone as substrate, suggesting that multiple substrate molecules can bind simultaneously into the binding pocket of CYP102A1 R47L/F87V/L188Q. Due to the lack of activity of the mutant R47L/L188Q, it was assumed that the mutation F87V is crucial for the testosterone hydroxylation activity of CYP102A1 [96]. Analysis of the ligand-free crystal structure revealed that phenylalanine 87 lies perpendicular to the heme at the end of the substrate-access channel. The central role of this residue in substrate selectivity has been studied before by replacing the large nonpolar residue with smaller ones such as alanine, resulting in the acceptance of a number of nonnatural substrates, due to the reduction of steric hindrance in the active site. The role of F87 in substrate selectivity and regioselectivity during testosterone conversion was studied using the highly active testosterone-hydroxylating mutant CYP102A1 M11, containing eight mutations in total [97]. Saturation mutagenesis was performed at position 87 to introduce all 20 canonical amino acids and testosterone was used to characterize the regioselectivity of the mutants. It was found that all mutants were able to catalyze testosterone hydroxylation, although the activities and regioselectivities differ [56]. It is noteworthy that these CYP102A1 mutants are thus far the only known microbial enzymes that catalyze the 16β-hydroxylation of steroids with F87I being the mutation with the highest 16β-hydroxylating activity [56].

Another recent study showed that CYP102A1 F87A hydroxylates testosterone to a mixture of 2β- and 15β-hydroxytestosterone (51 and 46 %), and 3 % 6β- and 16β-hydroxytestosterone. Applying laboratory-controlled evolution with the iterative combinatorial active-site saturation test (CAST) approach, a mutant with >95 % regioselectivity in favor of 2β-hydroxytestosterone as well as a mutant with opposite regioselectivity leading to >95 % 15β-hydroxytestosterone could be evolved [98]. The investigation of the ability of these mutants to hydroxylate other steroids in addition to testosterone showed that mutant F87A was also active towards progesterone. Interestingly, the substrate change shifted the selectivity from C-15 to C-16, with a low selectivity resulting in a mixture of 2β- and 16β-products of 18:82. By testing the mutants created with testosterone as substrate, a CYP102A1 variant with 100 % 2β-hydroxylating activity and a highly selective 16β-hydroxylating mutant producing 91 % 16β-hydroxyprogesterone were identified [98]. Among other amino acids that play important roles in the selectivity and activity of substrate conversion is alanine 82. Mutations at this position altered the selectivity towards 16β-hydroxylation for both testosterone and norethisterone and led to a 42-fold increase in V_{max} for 16β-hydroxylation [98]. Mutations in this protein region can also increase the coupling efficiency of steroid hydroxylation, most probably because of a more efficient water exclusion from the active site due to a stronger substrate–enzyme interaction. Other studies concerning the hydroxylation of testosterone by different CYP102A1 mutants revealed a shift in the stereoselectivity from the 16β- to the 16α product resulting from a single amino acid substitution from serine to isoleucine at position 72 [99]. These examples highlight the power of protein engineering in the generation of highly specific steroid-hydroxylases based on the knowledge about important regions that determine substrate specificity and reaction selectivity. Scheme 6 highlights selected steroid hydroxylations catalyzed by bacterial cytochrome P450s.

5.7 Vitamin D_3

Similar to steroids, Vitamin D_3 (VD_3) is derived from lanosterol, a tetracyclic triterpenoid compound. VD_3 is an important circulating secosteroid that regulates bone and calcium metabolism, cellular proliferation, differentiation, and immune response. In addition to these classical roles, a variety of nonclassical activities is described including effects on muscle function, cardiovascular homeostasis, and neuronal function [100]. In animals, it is produced in the skin by UV-light activation of 7-dehydrocholesterol and then transported into the liver by a vitamin D binding protein where it is hydroxylated at C-25 by CYP2R1. The following 1α-hydroxylation of 25(OH)D_3 into the active form 1α,25-dihydroxyvitamin D_3 is catalyzed by CYP27B1. Although this mitochondrial P450 is expressed in many tissues, only the kidneys are considered as a production site for circulating 1α,25$(OH)_2VD_3$. The active VD_3 mediates a variety of endocrine effects via binding to the vitamin D receptor (VDR), and regulation of VDR-dependent gene

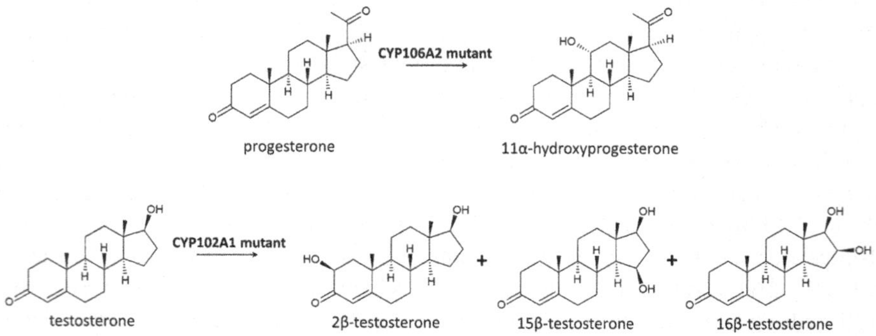

Scheme 6 Exemplary steroid hydroxylations catalyzed by P450s

expression [101, 102]. VD_3 is essential for the human body and deficiency is associated with a variety of diseases and overall mortality. Pharmaceutical applications of VD_3, for instance, in the treatment of hypothyroidism, osteoporosis, and chronic renal failure, rely on the use of the active compound and result in a high demand for $1\alpha,25(OH)_2D_3$. Chemical synthesis of hydroxylated vitamin D requires approximately 20 steps and achieves only low yields, therefore there is great interest in a simple biotechnological route towards the dihydroxylated VD_3. One approach that has been realized for the biotechnological production of $1\alpha,25(OH)_2D_3$ is the expression of CYP27B1 for the hydroxylation of $25(OH)D_3$ at position 1α in recombinant *E. coli* cells. However, the 25-hydroxylated derivative is an expensive precursor for biotechnological approaches and a system capable of hydroxylating vitamin D_3 at both the 1α- and 25-positions would be much more efficient from an economical point of view. The first prokaryotic production of $1\alpha,25(OH)_2D_3$ was described in 1992 after screening several hundred *Streptomyces* strains towards their ability to hydroxylate VD_3. In this way, *Streptomyces sclerotialus* FERM BP-1370 and *Streptomyces roseosporus* FERM BP-1574 were found to convert $25(OH)D_3$ and $1\alpha(OH)D_3$, respectively, to $1\alpha,25(OH)_2D_3$ [103]. Given that none of the strains converted VD_3 directly into $1\alpha,25(OH)_2D_3$, an additional 500 bacterial and 400 fungal strains were investigated in further studies, and eventually led to the identification of 12 strains converting vitamin D_3 directly into the dihydroxylated derivative. Among these 12 strains, the fungus *Amycolata autotrophica* FERM BP-1573 exhibited the highest activity. Bioconversions with *A. autotrophica* (now renamed *Pseudonocardia autotrophica*) yielded 8.3 mg $25(OH)D_3$/l culture and 0.17 mg $1\alpha,25(OH)_2D_3$/l culture from cultivations in a 200-l fermenter after 120 h. Mapping the chromosome of *A. autotrophica* led to the identification of P450VD25, which was classified as CYP105A2 and cloned into *Streptomyces lividans*, where it provided 25-hydroxylation activity towards vitamin D_3. The VD_3 1α-hydroxylase (Vdh) from *P. autotrophica* could be identified as a member of the CYP107 family and conversion of VD_3 to $1\alpha,25(OH)_2D_3$ via $25(OH)D_3$ (see Scheme 7) was shown with this enzyme after purification from recombinant *E. coli*. It can, therefore, be assumed that CYP107 has both VD_3 25- and $25(OH)D_3$ 1α-hydroxylation activity [104]. This P450 is now applied in the bioconversion of vitamin D_3 into $1\alpha,25(OH)_2D_3$ [105]. To increase the hydroxylation efficiency of CYP107, random mutagenesis was performed and 1,000

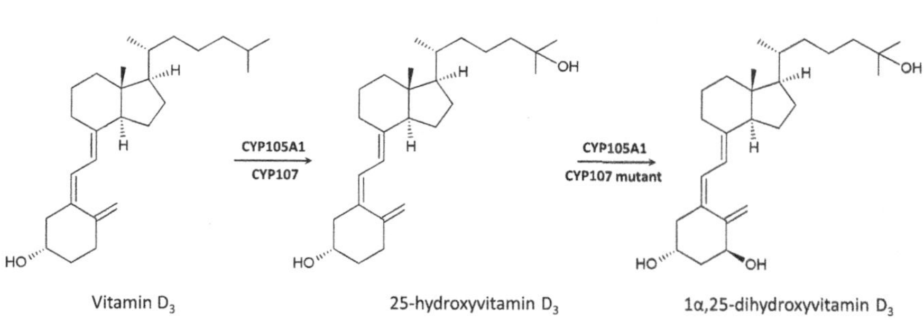

Scheme 7 CYP-catalyzed hydroxylations of vitamin D_3

transformants were tested towards increased vitamin D_3 biotransformation. The results indicated eight hotspots associated with higher activity. Based on these findings, site-directed mutagenesis studies revealed four mutations with high vitamin D hydroxylating activity: T70R, V156S, E216A, and E384R. Each of these mutants showed a two- to threefold elevated biotransformation activity compared with the wild-type enzyme expressed in *E. coli*. Furthermore, the combination of the mutants showed a synergistic effect on enzymatic activity and the mutant combining all four amino acid substitutions showed a 7.9 times higher $1\alpha,25(OH)_2D_3$ product formation rate. Because the described amino acid residues are all located in the loop regions of the protein that undergo dramatic conformational changes upon substrate binding, the mutations may increase the flexibility of substrate binding sites or other domains involved in catalysis [104]. These results show that the hotspots for protein engineering of cytochromes P450 in order to achieve higher activities are not restricted to the active site of the enzyme. Another example of this is the highly active CYP107 mutant T107A. This residue is located in the ferredoxin binding site and crystallographic and kinetic analysis revealed that the mutation changes the protein conformation from an open to a closed state which increases the binding affinity with ferredoxin [106].

Another P450 that is capable of hydroxylating VD_3 is CYP105A1 from *S. griseolus*, an enzyme with a broad substrate range that was first identified in connection with its characteristic of metabolizing sulfonylurea herbicides. Sharing a sequence similarity of >55 % with CYP105A2, its ability to accept VD_3 as a substrate was further investigated. Cloning and functional expression in *E. coli* confirmed that CYP105A1 hydroxylates VD_3 [21] (see Scheme 7). In contrast to CYP105A2, which lacks the 1α-hydroxylase activity, CYP105A1 even converts VD_3 into $1\alpha,25(OH)_2D_3$ via $25(OH)D_3$. However, the dihydroxylated VD_3 is formed in much lower amounts than the main metabolite $25(OH)D_3$. Further drawbacks are the additional production of a trihydroxylated VD_3 derivative and the overall low activity towards these substrates. With the objective of increasing enzyme activity by structure-based protein design, the crystal structure of CYP105A1 was solved and the amino acid residues Arg73 and Arg84, located within the active site, were identified as key residues in vitamin D_3 catalysis [107]. Based on this finding, a series of single and double mutants was produced by site-directed mutagenesis leading to the identification of the double mutant R73V/R84A with a 435-fold higher specificity constant compared with the wild-type CYP105A1. The vitamin D_3 25-hydroxylation activity of the optimized CYP105A1 is comparable to the wild-type CYP107, whereas its 1α-hydroxylation activity towards $25(OH)D_3$ is approximately one order of magnitude higher. Co-crystallization of the mutants with the reaction product revealed that the mutation R84A most probably increases the coupling efficiency of CYP105A1 by stabilizing the oxygenated form of the enzyme, whereas R73V (or R73A) alters the hydrogen-bond network resulting in a rearrangement in the water surrounding of the 1α-hydroxygroup [108]. The double mutant was used to establish a whole-cell VD_3 conversion system with *S. lividans* TK23 as host strain and VD_3 was hydroxylated into $1\alpha,25(OH)_2D_3$ to a notable amount [109]. These studies depict an excellent example for the straightforward route from identification of an enzyme that is capable of biotransforming a given substrate, its subsequent optimization using structure-guided protein design, and the following development of a biotechnological process.

6 Use of Whole-Cell Systems

Although cytochromes P450 might seem the enzyme species of choice for biooxidation, their application faces more challenges compared to enzymes such as hydrolases, lyases, or transferases. As mentioned before, the low activity and stability of P450s as well as their need for auxiliary electron transfer proteins and expensive cofactors pose challenges in the application of these enzymes [9]. Certain progress has been made regarding the use of isolated P450 enzymes, among them different techniques of cofactor regeneration and immobilization of the protein in order to achieve higher stability. Nevertheless, the isolation of proteins is a time-consuming and expensive task and the applications of P450s in industrial processes are mostly restricted to fermentation. This carries the advantage of increased protein stability and efficient cofactor supply by the host cell metabolism. A major hurdle that remains, however, is the low productivity and yield of these biotransformations. Most P450-catalyzed processes are close to the minimum requirements and almost all examples of implemented industrial P450 processes concentrate on the production of compounds with high pharmaceutical interest for which the final product concentrations do not need to be as high as for the production of fine chemicals [110]. The use of whole cells also delivers new adjusting screws for optimization. The most efficient process can only be ensured when all necessary components are available at the right concentration, time, and location in the cell for optimal reaction conditions. The first question that needs to be addressed is, therefore, the selection of the appropriate host for the biotransformation. In the majority of cases, *E. coli* is the host organism of choice as the techniques of molecular biology are well developed and most P450s can be expressed as stable proteins displaying high activity. There are, however, cases where other organisms lead to better results. An example is the CYP153A6-catalyzed bioconversion of (−)-limonene to (−)-perillyl alcohol which reached 3.0 U/g in *P. putida* Gpo12 compared with 0.1 U/g in *E. coli* based on dry cell weight [58]. Because the transfer of large molecules across the cell membrane can become a limiting factor, the substrate availability within the host cell is an important point to be considered. Whole-cell bioconversion of abietic acid using different host organisms revealed that the transport of the diterpene resin acid into the gram-negative *E. coli* cell is much lower compared to the gram-positive *B. megaterium* [14]. In such cases, the use of permeabilizing agents or host engineering, for example, through cloning of additional transporter proteins, might be a promising approach to circumvent this problem. Another important issue in bioconversion processes is the formation of unwanted side products. In the case of the CYP153A6-catalyzed (−)-limonene hydroxylation in *P. putida* KT2440, the side products perillyl aldehyde and perillic acid constituted up to 26 % of the total amount of oxidized terpenes. In contrast, no perillic acid and only minor amounts of perillyl aldehyde were formed when *E. coli* W3110 was used as a host [111]. A good way to achieve a higher concentration of the desired product is the setup of aqueous-organic two-liquid–phase systems. Using this approach, by-product formation in the biooxidation of (+)-valencene by recombinant *E. coli* cells expressing CYP109B1 could be significantly reduced [69].

A particularly exciting and challenging task in the use of whole cells is the construction of complete biosynthetic pathways. The metabolic engineering of several enzymatic steps allows the biosynthesis of complex products from simple carbon sources. An impressive example is the production of the antimalaria drug artemisinin in yeast and *E. coli*, where the overexpression of genes from the mevalonate biosynthetic pathway from *S. cerevisiae* resulted in an increased availability of endogenous precursors for terpenoid biosynthesis and thus artemisinic acid levels that are suitable for an industrial production process [112, 113]. In further studies that propose a novel semibiosynthetic route, the native CYP71AV1 from *Artemisia annua* that catalyzes the conversion of amorpha-4,11-diene to artemisinic acid was replaced by the bacterial CYP102A1. Saturation mutagenesis of key residues in the active site of this substrate-promiscuous P450 led to a mutant that was able to perform an epoxidation of amorpha-4,11-diene which could then be transformed by high-yielding chemistry to artemisinin [114]. This example illustrates the high potential of whole-cell systems that express terpenoid synthases together with P450 genes as interesting platforms for the easy and cheap production of valuable terpenoid compounds from simple carbon sources.

7 Concluding Remarks

Members of the cytochrome P450 superfamily significantly contribute to the diversification of natural compounds [115]. They usually depict the first enzymes to modify an initial terpene synthase product in plants. Although major advances concerning the engineering of plant P450s towards easier use for isoprenoid functionalization have been made during the last two decades, these enzymes still pose great challenges, especially in terms of microbial expression. Therefore, microbial, especially soluble bacterial, P450s catalyzing terpene hydroxylation became the focus of interest for biotechnological production of terpenoid compounds because they are easier to express in bacteria. As shown in this review, various bacterial P450s have been described catalyzing an efficient conversion of terpenoids with significant yields of up to several hundred mg/l/d, thus demonstrating the potential of these families of P450 enzymes for the production of terpenoids for the fragrance, flavor, and pharmaceutical industries. The discovery and characterization of further genome sequences as well as focused screening for terpenoid conversion of known bacteria will further expand the source of P450s displaying activity towards terpenoids.

As P450s depend on redox partners, the identification, cloning, and coexpression of these electron transfer proteins might, however, comprise a significant barrier. While plant P450s use only one kind of electron transfer protein (CPR, Fig. 1), the redox partners of bacterial P450s might differ significantly [8]. However, it was shown that heterologous reconstitution of enzymatic activity using redox partners from other organisms, especially using bovine adrenodoxin can be very efficient [47]. In addition, strategies using fusion proteins with known redox domains may be a promising tool to overcome this obstacle [9]. The applicability of P450s for biotechnological purposes (e.g., terpenoid production) is further underlined by the major advantage of P450s in terms of their robustness against mutations, particularly in their substrate binding pocket. Although the basis for this ability is not yet fully elucidated, the high conformational variability and nonpolar nature of the active site seem to make an important contribution [116]. This high evolvability in terms of maintaining a folded structure and the catalytic activity, therefore, provides an excellent starting point for both rational and evolutionary approaches to improve substrate scope, selectivity, and activity. Such engineering methods can be used especially effectively with bacterial P450s, which are soluble and under normal conditions can be easily expressed with very high yields in bacteria. Successful examples have been described demonstrating the change in the selectivities of investigated P450s as well as improvements concerning activities and stabilities (see Sect. 5). With the decreasing costs associated with the redesign and engineering of these proteins in conjunction with higher chances of a successful outcome, P450-catalyzed syntheses become increasingly attractive for biotechnological applications. This is particularly true for the industrial production of terpenoid compounds in fields such as pharmaceuticals or flavors and fragrances.

Taken together, the examples described in this chapter show that microbial P450s are promising tools for the production of modified terpenes and illustrate how enzyme engineering can be used to redesign P450s in order to solve challenging chemical problems. The vastly growing number of P450 crystal structures deposited in the protein databank creates a valuable foundation for future studies using rational protein design to create enzymes that accommodate a wider range of unnatural substrates and exhibit improved or altered activity. Moreover, coexpression of engineered plant and bacterial P450s and combination of their different selectivities for the production of completely novel terpenoid structures will open new horizons for the production of flavor and fragrance as well as pharmaceutical compounds.

Acknowledgments: The work is supported by a grant of the Deutsche Forschungsgemeinschaft to R.B. (Be 1343/23-1). S.J. is supported by the INTERREG program of the European Union, and D.S. by a grant from the Saarland Ministry of Economics and Science.

References

1. Cane DE, Ikeda H (2012) Exploration and mining of the bacterial terpenome. Acc Chem Res 45(3):463–472
2. Kirby J, Keasling JD (2009) Biosynthesis of plant isoprenoids: perspectives for microbial engineering. Annu Rev Plant Biol 60:335–355
3. Bernhardt R (2006) Cytochromes P450 as versatile biocatalysts. J Biotechnol 124(1):128–145
4. Ortiz de Monetellano PR (2010) Hydrocarbon hydroxylation by cytochrome P450 enzymes. Chem Rev 110(2):932–948
5. Sono M, Roach MP, Coulter ED, Dawson JH (1996) Heme-containing oxygenases. Chem Rev 96(7):2841–2888
6. Guengerich FP, Munro AW (2013) Unusual cytochrome p450 enzymes and reactions. J Biol Chem 288(24):17065–17073
7. Nelson DR (2009) The cytochrome p450 homepage. Hum Genomics 4(1):59–65
8. Hannemann F, Bichet A, Ewen KM, Bernhardt R (2007) Cytochrome P450 systems–biological variations of electron transport chains. Biochim Biophys Acta 1770(3):330–344
9. Bernhardt R, Urlacher VB (2014) Cytochromes P450 as promising catalysts for biotechnological application: chances and limitations. Appl Microbiol Biotechnol 1–19
10. Urlacher VB, Girhard M (2012) Cytochrome P450 monooxygenases: an update on perspectives for synthetic application. Trends Biotechnol 30(1):26–36
11. Urlacher VB, Lutz-Wahl S, Schmid RD (2004) Microbial P450 enzymes in biotechnology. Appl Microbial Biotechnol 64(3):317–325
12. Nelson DR (2013) A world of cytochrome P450s. Philos Trans R Soc Lond B: Biol Sci 368:1–4
13. Wohlgemuth R (2009) The locks and keys to industrial biotechnology. N Biotechnol 25(4):204–213
14. Janocha S, Bernhardt R (2013) Design and characterization of an efficient CYP105A1-based whole-cell biocatalyst for the conversion of resin acid diterpenoids in permeabilized *Escherichia coli*. Appl Microbiol Biotechnol 97(17):7639–7649
15. Bradshaw WH, Conrad HE, Corey EJ, Gunsalus IC, Lednicer DJ (1959) Microbiological degradation of (+)-camphor. J Am Chem Soc 81:5507
16. Gunsalus IC, Ganguli BN, Katagiri M, Tsibris JC, Debrunner P, Frauenfelder H (1968) Oxygenation: a specific soluble cytochrome p-450 coupled enzyme complex. Science 160(3826):438–439
17. Katagiri M, Ganguli BN, Gunsalus IC (1968) A soluble cytochrome P-450 functional in methylene hydroxylation. J Biol Chem 243(12):3543–3546
18. Bhattacharyya PK, Prema BR, Kulkarni BD, Pradhan SK (1960) Microbiological transformation of terpenes: hydroxylation of alpha-pinene. Nature 187:689–690
19. Demyttenaere JC, Willemen HM (1998) Biotransformation of linalool to furanoid and pyranoid linalool oxides by *Aspergillus niger*. Phytochemistry 47(6):1029–1036
20. Menendez P, Garcia C, Rodriguez P, Moyna P, Heinzen H (2002) Isolation and screening of d-limonene-resistant microorganisms. Braz Arch Biol Techn 45(2):111–114
21. Sawada N, Sakaki T, Yoneda S, Kusudo T, Shinkyo R, Ohta M, Inouye K (2004) Conversion of vitamin D_3 to 1α,25-dihydroxyvitamin D_3 by *Streptomyces griseolus* cytochrome P450SU-1. Biochem Biophys Res Commun 320(1):156–164
22. Chang HC, Oriel P (1994) Bioproduction of perillyl alcohol and related monoterpenes by isolates of *Bacillus stearothermophilus*. J Food Sci 59(3):660–662
23. Guengerich FP, Tang Z, Salamanca-Pinzón SG, Cheng Q (2010) Characterizing proteins of unknown function: orphan cytochrome p450 enzymes as a paradigm. Mol Interv 10(3):153–163
24. Poulos TL, Finzel BC, Gunsalus IC, Wagner GC, Kraut J (1985) The 2.6-A crystal structure of *Pseudomonas putida* cytochrome P-450. J Biol Chem 260(30):16122–16130

25. Ravichandran KG, Boddupalli SS, Hasermann CA, Peterson JA, Deisenhofer J (1993) Crystal structure of hemoprotein domain of P450BM-3, a prototype for microsomal P450's. Science 261(5122):731–736
26. Capdevila JH, Wei S, Helvig C, Falck JR, Belosludtsev Y, Truan G, Graham-Lorence SE, Peterson JA (1996) The highly stereoselective oxidation of polyunsaturated fatty acids by cytochrome P450BM-3. J Biol Chem 271(37):22663–22671
27. Urlacher VB, Makhsumkhanov A, Schmid RD (2006) Biotransformation of beta-ionone by engineered cytochrome P450 BM-3. Appl Microbiol Biotechnol 70(1):53–59
28. Nguyen KT, Virus C, Günnewich N, Hannemann F, Bernhardt R (2012) Changing the regioselectivity of a P450 from C15 to C11 hydroxylation of progesterone. Chembiochem. 13(8):1161–1166
29. Wong TS, Arnold FH, Schwaneberg U (2004) Laboratory evolution of cytochrome p450 BM-3 monooxygenase for organic cosolvents. Biotechnol Bioeng 85(3):351–358
30. Bell SG, Chen X, Xu F, Rao Z, Wong LL (2003) Engineering substrate recognition in catalysis by cytochrome P450cam. Biochem Soc Trans 31(Pt 3):558–562
31. Bell SG, Chen X, Sowden RJ, Xu F, Williams JN, Wong LL, Rao Z (2003) Molecular recognition in (+)-alpha-pinene oxidation by cytochrome P450cam. J Am Chem Soc 125 (3):705–714
32. Bell SG, Harford-Cross CF, Wong LL (2001) Engineering the CYP101 system for in vivo oxidation of unnatural substrates. Protein Eng 14(10):797–802
33. Graham-Lorence S, Truan G, Peterson JA, Falck JR, Wei S, Helvig C, Capdevila JH (1997) An active site substitution, F87V, converts cytochrome P450 BM-3 into a regio- and stereoselective (14S,15R)-arachidonic acid epoxygenase. J Biol Chem 272(2):1127–1135
34. Appel D, Lutz-Wahl S, Fischer P, Schwaneberg U, Schmid RD (2001) A P450 BM-3 mutant hydroxylates alkanes, cycloalkanes, arenes and heteroarenes. J Biotechnol 88(2):167–171
35. Stemmer WP (1994) DNA shuffling by random fragmentation and reassembly: in vitro recombination for molecular evolution. Proc Natl Acad Sci USA 91(22):10747–10751
36. Zhao H, Giver L, Shao Z, Affholter JA, Arnold FH (1998) Molecular evolution by staggered extension process (StEP) in vitro recombination. Nat Biotechnol 16(3):258–261
37. Otey CR, Silberg JJ, Voigt CA, Endelman JB, Bandara G, Arnold FH (2004) Functional evolution and structural conservation in chimeric cytochromes p450: calibrating a structure-guided approach. Chem Biol 11(3):309–318
38. Murataliev MB, Trinh LN, Moser LV, Bates RB, Feyereisen R, Walker FA (2004) Chimeragenesis of the fatty acid binding site of cytochrome P450BM3. Replacement of residues 73-84 with the homologous residues from the insect cytochrome P450 CYP4C7. Biochemistry 43(7):1771–1780
39. Sutherland TD, Unnithan GC, Andersen JF, Evans PH, Murataliev MB, Szabo LZ, Mash EA, Bowers WS, Feyereisen R (1998) A cytochrome P450 terpenoid hydroxylase linked to the suppression of insect juvenile hormone synthesis. Proc Natl Acad Sci USA 95 (22):12884–12889
40. Chen CK, Berry RE, Shokhireva TKh, Murataliev MB, Zhang H, Walker FA (2010) Scanning chimeragenesis: the approach used to change the substrate selectivity of fatty acid monooxygenase CYP102A1 to that of terpene omega-hydroxylase CYP4C7. J Biol Inorg Chem 15(2):159–174
41. Robin A, Roberts GA, Kisch J, Sabbadin F, Grogan G, Bruce N, Turner NJ, Flitsch SL (2009) Engineering and improvement of the efficiency of a chimeric [P450cam-RhFRed reductase domain] enzyme. Chem Commun (Camb) 18:2478–2480
42. Robin A, Köhler V, Jones A, Ali A, Kelly PP, O'Reilly E, Turner NJ, Flitsch SL (2011) Chimeric self-sufficient P450cam-RhFRed biocatalysts with broad substrate scope. Beilstein J Org Chem 7:1494–1498
43. Ewen KM, Kleser M, Bernhardt R (2011) Adrenodoxin: the archetype of vertebrate-type [2Fe–2S] cluster ferredoxins. Biochim Biophys Acta 1814(1):111–125

44. Mouri T, Kamiya N, Goto M (2006) Increasing the catalytic performance of a whole cell biocatalyst harboring a cytochrome p450cam system by stabilization of an electron transfer component. Biotechnol Lett 28(18):1509–1513
45. Khatri Y, Girhard M, Romankiewicz A, Ringle M, Hannemann F, Urlacher VB, Hutter MC, Bernhardt R (2010) Regioselective hydroxylation of norisoprenoids by CYP109D1 from *Sorangium cellulosum* So ce56. Appl Microbiol Biotechnol 88(2):485–495
46. Schiffler B, Kiefer M, Wilken A, Hannemann F, Adolph HW, Bernhardt R (2001) The interaction of bovine adrenodoxin with CYP11A1 (cytochrome P450scc) and CYP11B1 (cytochrome P45011beta). Acceleration of reduction and substrate conversion by site-directed mutagenesis of adrenodoxin. J Biol Chem 276(39):36225–36232
47. Ewen KM, Ringle M, Bernhardt R (2012) Adrenodoxin—a versatile ferredoxin. IUBMB Life 64(6):506–512
48. Ishida T, Enomoto H, Nishida R (2008) New attractants for males of the solanaceous fruit fly *Bactrocera latifrons*. J Chem Ecol 34(12):1532–1535
49. Tamura S, Nagao M (1970) Syntheses and biological activities of new plant growth inhibitors structurally related to abscisic acid. Agric Biol Chem 34(9):1393–1401
50. Larroche C, Creuly C, Gros J-B (1995) Fed-batch biotransformation of β-ionone by *Aspergillus niger*. Appl Microbiol Biotechnol 43:222–227
51. Lutz-Wahl S, Fischer P, Schmidt-Dannert C, Wohlleben W, Hauer B, Schmid RD (1998) Stereo- and regioselective hydroxylation of alpha-ionone by *Streptomyces strains*. Appl Environ Microbiol 64(10):3878–3881
52. Celik A, Flitsch SL, Turner NJ (2005) Efficient terpene hydroxylation catalysts based upon P450 enzymes derived from actinomycetes. Org Biomol Chem 3(16):2930–2934
53. Ly TT, Khatri Y, Zapp J, Hutter MC, Bernhardt R (2012) CYP264B1 from *Sorangium cellulosum* So ce56: a fascinating norisoprenoid and sesquiterpene hydroxylase. Appl Microbiol Biotechnol 95(1):123–133
54. Ma M, Bell SG, Yang W, Hao Y, Rees NH, Bartlam M, Zhou W, Wong LL, Rao Z (2011) Structural analysis of CYP101C1 from *Novosphingobium aromaticivorans* DSM12444. Chembiochem 12(1):88–99
55. Miyazawa M, Shindo M, Shimada T (2002) Metabolism of (+)- and (−)-limonenes to respective carveols and perillyl alcohols by CYP2C9 and CYP2C19 in human liver microsomes. Drug Metab Dispos 30(5):602–607
56. Vottero E, Rea V, Lastdrager J, Honing M, Vermeulen NP, Commandeur JN (2011) Role of residue 87 in substrate selectivity and regioselectivity of drug-metabolizing cytochrome P450 CYP102A1 M11. J Biol Inorg Chem 16(6):899–912
57. Oriel PJ, Savithiry S, Chang HC (1997) Process for the preparation of monoterpenes using bacterium containing recombinant DNA. US patent 5688673
58. Van Beilen JB, Holtackers R, Lüscher D, Bauer U, Witholt B, Duetz WA (2005) Biocatalytic production of perillyl alcohol from limonene by using a novel *Mycobacterium* sp. cytochrome P450 alkane hydroxylase expressed in *Pseudomonas putida*. Appl Environ Microbiol 71(4):1737–1744
59. Alonso-Gutierrez J, Chan R, Batth TS, Adams PD, Keasling JD, Petzold CJ, Lee TS (2013) Metabolic engineering of *Escherichia coli* for limonene and perillyl alcohol production. Metab Eng 19:33–41
60. Kim D, de Montellano PRO (2009) Tricistronic overexpression of cytochrome P450cam, putidaredoxin, and putidaredoxin reductase provides a useful cell-based catalytic system. Biotechnol Lett 31(9):1427–1431
61. Seifert A, Antonovici M, Hauer B, Pleiss J (2011) An efficient route to selective bio-oxidation catalysts: an iterative approach comprising modeling, diversification, and screening, based on CYP102A1. Chembiochem 12(9):1346–1351
62. Dietrich M, Eiben S, Asta C, Do TA, Pleiss J, Urlacher VB (2008) Cloning, expression and characterisation of CYP102A7, a self-sufficient P450 monooxygenase from *Bacillus licheniformis*. Appl Microbiol Biotechnol 79(6):931–940

63. Bhattarai S, Liou K, Oh TJ (2012) Homology modeling and docking studies of *Streptomyces peucetius* CYP147F1 as limonene hydroxylase. J Microbiol Biotechnol 22(7):917–922
64. Dietrich G, Dolan MC, Peralta-Cruz J, Schmidt J, Piesman J, Eisen RJ, Karchesy JJ (2006) Repellent activity of fractioned compounds from *Chamaecyparis nootkatensis* essential oil against nymphal Ixodes scapularis (Acari: Ixodidae). J Med Entomol 43(5):957–961
65. Jordan RA, Schulze TL, Dolan MC (2012) Efficacy of plant-derived and synthetic compounds on clothing as repellents against Ixodes scapularis and *Amblyomma americanum* (Acari: Ixodidae). J Med Entomol 49(1):101–106
66. Gavira C, Höfer R, Lesot A, Lambert F, Zucca J, Werck-Reichhart D (2013) Challenges and pitfalls of P450-dependent (+)-valencene bioconversion by *Saccharomyces cerevisiae.* Metab Eng 18:25–35
67. Fraatz MA, Berger RG, Zorn H (2009) Nootkatone—a biotechnological challenge. Appl Microbiol Biotechnol 83(1):35–41
68. Kaspera R, Krings U, Nanzad T, Berger RG (2005) Bioconversion of (+)-valencene in submerged cultures of the ascomycete *Chaetomium globosum.* Appl Microbiol Biotechnol 67(4):477–483
69. Girhard M, Machida K, Itoh M, Schmid RD, Arisawa A, Urlacher VB (2009) Regioselective biooxidation of (+)-valencene by recombinant *E. coli* expressing CYP109B1 from *Bacillus subtilis* in a two-liquid-phase system. Microb Cell Fact 8:36
70. Sowden RJ, Yasmin S, Rees NH, Bell SG, Wong LL (2005) Biotransformation of the sesquiterpene (+)-valencene by cytochrome P450cam and P450BM-3. Org Biomol Chem 3 (1):57–64
71. Smith DJ, Patrauchan MA, Florizone C, Eltis LD, Mohn WW (2008) Distinct roles for two CYP226 family cytochromes P450 in abietane diterpenoid catabolism by *Burkholderia xenovorans* LB400. J Bacteriol 190(5):1575–1583
72. Smith DJ, Martin VJ, Mohn WW (2004) A cytochrome P450 involved in the metabolism of abietane diterpenoids by *Pseudomonas abietaniphila* BKME-9. J Bacteriol. 186(11):3631–3639
73. Bleif S, Hannemann F, Lisurek M, von Kries JP, Zapp J, Dietzen M, Antes I, Bernhardt R (2011) Identification of CYP106A2 as a regioselective allylic bacterial diterpene hydroxylase. Chembiochem 12(4):576–582
74. Virus C, Lisurek M, Simgen B, Hannemann F, Bernhardt R (2006) Function and engineering of the 15beta-hydroxylase CYP106A2. Biochem Soc Trans 34(Pt 6):1215–1218
75. Lisurek M, Simgen B, Antes I, Bernhardt R (2008) Theoretical and experimental evaluation of a CYP106A2 low homology model and production of mutants with changed activity and selectivity of hydroxylation. Chembiochem 9(9):1439–1449
76. Janocha S, Zapp J, Hutter M, Kleser M, Bohlmann J, Bernhardt R (2013) Resin acid conversion with CYP105A1: an enzyme with potential for the production of pharmaceutically relevant diterpenoids. Chembiochem 14(4):467–473
77. Dzubak P, Hajduch M, Vydra D, Hustova A, Kvasnica M, Biedermann D, Markova L, Urban M, Sarek J (2006) Pharmacological activities of natural triterpenoids and their therapeutic implications. Nat Prod Rep 23(3):394–411
78. Muffler K, Leipold D, Scheller M-C, Haas C, Steingroewer J, Bley T, Neuhaus HE, Mirata MA, Schrader J, Ulber R (2011) Biotransformation of Triterpenes. Proc Biochem 46:1–15
79. Chatterjee P, Kouzi SA, Pezzuto JM, Hamann MT (2000) Biotransformation of the antimelanoma agent betulinic acid by *Bacillus megaterium* ATCC 13368. Appl Environ Microbiol. 66(9):3850–3855
80. Bleif S, Hannemann F, Zapp J, Hartmann D, Jauch J, Bernhardt R (2012) A new *Bacillus megaterium* whole-cell catalyst for the hydroxylation of the pentacyclic triterpene 11-keto-β-boswellic acid (KBA) based on a recombinant cytochrome P450 system. Appl Microbiol Biotechnol 93(3):1135–1146
81. Brill E, Hannemann F, Zapp J, Brüning G, Jauch J, Bernhardt R (2014) A new cytochrome P450 system from *Bacillus megaterium* DSM319 for the hydroxylation of 11-keto-β-boswellic acid (KBA). Appl Microbiol Biotechnol 98(4):1701–1717

82. Jauch J, Bergmann J (2003) An efficient method for large-scale preparation of 3-O-Acetyl-11-oxo-β-boswellic acid and other boswellic acids. Eur J Org Chem 24:4752–4756
83. Schmitz D, Zapp J, Bernhardt R (2012) Hydroxylation of the triterpenoid dipterocarpol with CYP106A2 from *Bacillus megaterium*. FEBS J. 279(9):1663–1674
84. Tong WY, Dong X (2009) Microbial biotransformation: recent developments on steroid drugs. Recent Pat Biotechnol 3(2):141–153
85. Bernhardt R (2014) Mammalian and bacterial cytochromes P450 involved in steroid hydroxylation: regulation of catalysis and selectivity, and potential applications. Yamazaki H (ed) Fifty years of cytochrome P450 research. Springer, Berlin
86. Donova MV, Egorova OV (2012) Microbial steroid transformations: current state and prospects. Appl Microbiol Biotechnol 94(6):1423–1447
87. Bracco P, Janssen DB, Schallmey A (2013) Selective steroid oxyfunctionalisation by CYP154C5, a bacterial cytochrome P450. Microb Cell Fact. 12:95
88. Berg A, Carlstrom K, Gustafsson JA, Ingelman-Sundberg M (1975) Demonstration of a cytochrome P-450-dependent steroid 15beta-hydroxylase in *Bacillus megaterium*. Biochem Biophys Res Commun 66(4):1414–1423
89. Berg A, Gustafsson JA, Ingelman-Sundberg M (1976) Characterization of a cytochrome P-450-dependent steroid hydroxylase system present in *Bacillus megaterium*. J Biol Chem 251(9):2831–2838
90. Berg A, Ingelman-Sundberg M, Gustafsson JA (1979) Isolation and characterization of cytochrome P-450meg. Acta Biol Med Ger 38(2–3):333–344
91. Simgen B, Contzen J, Schwarzer R, Bernhardt R, Jung C (2000) Substrate binding to 15beta-hydroxylase (CYP106A2) probed by FT infrared spectroscopic studies of the iron ligand CO stretch vibration. Biochem Biophys Res Commun 269(3):737–742
92. Schmitz D, Zapp J, Bernhardt R (2014) Steroid conversion with CYP106A2—production of pharmaceutically interesting DHEA metabolites. Microb Cell Fact 13(1):81
93. Zehentgruber D, Hannemann F, Bleif S, Bernhardt R, Lütz S (2010) Towards preparative scale steroid hydroxylation with cytochrome P450 monooxygenase CYP106A2. Chembiochem 11(5):713–721
94. McAleer WJ, Jacob TA, Turnbull LB, Schoenewaldt EF, Stoudt TH (1958) Hydroxylation of progesterone by *Bacillus cereus* and *Bacillus megaterium*. Arch Biochem Biophys 73(1):127–130
95. Virus C, Bernhardt R (2008) Molecular evolution of a steroid hydroxylating cytochrome P450 using a versatile steroid detection system for screening. Lipids 43(12):1133–1141
96. Van Vugt-Lussenburg BM, Damsten MC, Maasdijk DM, Vermeulen NP, Commandeur JN (2006) Heterotropic and homotropic cooperativity by a drug-metabolising mutant of cytochrome P450 BM3. Biochem Biophys Res Commun 346(3):810–818
97. Van Vugt-Lussenburg BM, Stjernschantz E, Lastdrager J, Oostenbrink C, Vermeulen NP, Commandeur JN (2007) Identification of critical residues in novel drug metabolizing mutants of cytochrome P450 BM3 using random mutagenesis. J Med Chem 50(3):455–461
98. Kille S, Zilly FE, Acevedo JP, Reetz MT (2011) Regio- and stereoselectivity of P450-catalysed hydroxylation of steroids controlled by laboratory evolution. Nat Chem 3(9):738–743
99. Venkataraman H, Beer SB, Bergen LA, van Essen N, Geerke DP, Vermeulen NP, Commandeur JN (2012) A single active site mutation inverts stereoselectivity of 16-hydroxylation of testosterone catalyzed by engineered cytochrome P450 BM3. Chembiochem 13(4):520–523
100. Zittermann A, Gummert JF (2010) Nonclassical vitamin D action. Nutrients 2(4):408–425
101. Bikle DD (2014) Vitamin D metabolism, mechanism of action, and clinical applications. Chem Biol 21(3):319–329
102. Schuster I (2011) Cytochromes P450 are essential players in the vitamin D signaling system. Biochim Biophys Acta 1814(1):186–199
103. Sasaki J, Miyazaki A, Saito M, Adachi T, Mizoue K, Hanada K, Omura S (1992) Transformation of vitamin D_3 to 1α,25-dihydroxyvitamin D_3 via 25-hydroxyvitamin D_3 using *Amycolata* sp. strains. Appl Microbiol Biotechnol 38(2):152–157

104. Fujii Y, Kabumoto H, Nishimura K, Fujii T, Yanai S, Takeda K, Tamura N, Arisawa A, Tamura T (2009) Purification, characterization, and directed evolution study of a vitamin D_3 hydroxylase from *Pseudonocardia autotrophica*. Biochem Biophys Res Commun 385 (2):170–175
105. Sakaki T, Sugimoto H, Hayashi K, Yasuda K, Munetsuna E, Kamakura M, Ikushiro S, Shiro Y (2011) Bioconversion of vitamin D to its active form by bacterial or mammalian cytochrome P450. Biochim Biophys Acta 1814(1):249–256
106. Yasutake Y, Nishioka T, Imoto N, Tamura T (2013) A single mutation at the ferredoxin binding site of P450 Vdh enables efficient biocatalytic production of 25-hydroxyvitamin D (3). Chembiochem 14(17):2284–2291
107. Sugimoto H, Shinkyo R, Hayashi K, Yoneda S, Yamada M, Kamakura M, Ikushiro S, Shiro Y, Sakaki T (2008) Crystal structure of CYP105A1 (P450SU-1) in complex with 1alpha,25-dihydroxyvitamin D_3. Biochemistry. 47(13):4017–4027
108. Hawkes DB, Adams GW, Burlingame AL, de Montellano PRO, De Voss JJ (2002) Cytochrome P450(cin) (CYP176A), isolation, expression, and characterization. J Biol Chem 277(31):27725–27732
109. Hayashi K, Yasuda K, Sugimoto H, Ikushiro S, Kamakura M, Kittaka A, Horst RL, Chen TC, Ohta M, Shiro Y, Sakaki T (2010) Three-step hydroxylation of vitamin D_3 by a genetically engineered CYP105A1: enzymes and catalysis. FEBS J 277(19):3999–4009
110. Julsing MK, Cornelissen S, Bühler B, Schmid A (2008) Heme-iron oxygenases: powerful industrial biocatalysts? Curr Opin Chem Biol 12(2):177–186
111. Cornelissen S, Julsing MK, Volmer J, Riechert O, Schmid A, Bühler B (2013) Whole-cell-based CYP153A6-catalyzed (S)-limonene hydroxylation efficiency depends on host background and profits from monoterpene uptake via AlkL. Biotechnol Bioeng 110 (5):1282–1292
112. Ro DK, Paradise EM, Ouellet M, Fisher KJ, Newman KL, Ndungu JM, Ho KA, Eachus RA, Ham TS, Kirby J, Chang MC, Withers ST, Shiba Y, Sarpong R, Keasling JD (2006) Production of the antimalarial drug precursor artemisinic acid in engineered yeast. Nature 440(7086):940–943
113. Chang MC, Eachus RA, Trieu W, Ro DK, Keasling JD (2007) Engineering *Escherichia coli* for production of functionalized terpenoids using plant P450s. Nat Chem Biol 3(5):274–277
114. Dietrich JA, Yoshikuni Y, Fisher KJ, Woolard FX, Ockey D, McPhee DJ, Renninger NS, Chang MC, Baker D, Keasling JD (2009) A novel semi-biosynthetic route for artemisinin production using engineered substrate-promiscuous P450(BM3). ACS Chem Biol 4(4):261–267
115. Omura T (2013) Contribution of cytochrome P450 to the diversification of eukaryotic organisms. Biotechnol Appl Biochem 60(1):4–8
116. Jung ST, Lauchli R, Arnold FH (2011) Cytochrome P450: taming a wild type enzyme. Curr Opin Biotechnol 22(6):809–817
117. Zhang Z, Sibbesen O, Johnson RA, de Montellano PRO (1998) The substrate specificity of cytochrome P450cam. Bioorg Med Chem 6(9):1501–1508
118. Warman AJ, Roitel O, Neeli R, Girvan HM, Seward HE, Murray SA, McLean KJ, Joyce MG, Toogood H, Holt RA, Leys D, Scrutton NS, Munro AW (2005) Flavocytochrome P450 BM3: an update on structure and mechanism of a biotechnologically important enzyme. Biochem Soc Trans 33(Pt 4):747–753
119. Whitehouse CJ, Bell SG, Tufton HG, Kenny RJ, Ogilvie LC, Wong LL (2008) Evolved CYP102A1 (P450BM3) variants oxidise a range of non-natural substrates and offer new selectivity options. Chem Commun (Camb) 8:966–968
120. Whitehouse CJ, Bell SG, Wong LL (2012) P450(BM3) (CYP102A1): connecting the dots. Chem Soc Rev 41(3):1218–1260
121. Fujii Y, Hirosue S, Fujii T, Matsumoto N, Agematu H, Arisawa A (2006) Hydroxylation of oleanolic acid to queretaroic acid by cytochrome P450 from *Nonomuraea recticatena*. Biosci Biotechnol Biochem 70(9):2299–2302

122. Kawauchi H, Sasaki J, Adachi T, Hanada K, Beppu T, Horinouchi S (1994) Cloning and nucleotide sequence of a bacterial cytochrome P-450VD25 gene encoding vitamin D-3 25-hydroxylase. Biochim Biophys Acta 1219(1):179–183
123. Watanabe I, Nara F, Serizawa N (1995) Cloning, characterization and expression of the gene encoding cytochrome P-450sca-2 from *Streptomyces carbophilus* involved in production of pravastatin, a specific HMG-CoA reductase inhibitor. Gene 163(1):81–85
124. Yasutake Y, Fujii Y, Nishioka T, Cheon WK, Arisawa A, Tamura T (2010) Structural evidence for enhancement of sequential vitamin D_3 hydroxylation activities by directed evolution of cytochrome P450 vitamin D_3 hydroxylase. J Biol Chem 285(41):31193–31201
125. Ban JG, Kim HB, Lee MJ, Anbu P, Kim ES (2014) Identification of a vitamin D_3-specific hydroxylase genes through actinomycetes genome mining. J Ind Microbiol Biotechnol 41(2):265–273
126. Peterson JA, Lu JY, Geisselsoder J, Graham-Lorence S, Carmona C, Witney F, Lorence MC (1992) Cytochrome P-450terp. Isolation and purification of the protein and cloning and sequencing of its operon. J Biol Chem 267(20):14193–14203
127. Agger SA, Lopez-Gallego F, Hoye TR, Schmidt-Dannert C (2008) Identification of sesquiterpene synthases from *Nostoc punctiforme* PCC 73102 and Nostoc sp. strain PCC 7120. J Bacteriol 190(18):6084–6096
128. Harada H, Shindo K, Iki K, Teraoka A, Okamoto S, Yu F, Hattan J-I, Utsumi R, Misawa N (2011) Efficient functional analysis system for cyanobacterial or plant cytochromes P450 involved in sesquiterpene biosynthesis. Appl Microbiol Biotechnol 90(2):467–476
129. Ropp JD, Gunsalus IC, Sligar SG (1993) Cloning and expression of a member of a new cytochrome P-450 family: cytochrome P-450lin (CYP111) from *Pseudomonas incognita.* J Bacteriol 175(18):6028–6037
130. Kühnel K, Ke N, Cryle MJ, Sligar SG, Schuler MA, Schlichting I (2008) Crystal structures of substrate-free and retinoic acid-bound cyanobacterial cytochrome P450 CYP120A1. Biochemistry 47(25):6552–6559
131. Takamatsu S, Lin X, Nara A, Komatsu M, Cane DE, Ikeda H (2011) Characterization of a silent sesquiterpenoid biosynthetic pathway in *Streptomyces avermitilis* controlling epi-isozizaene albaflavenone biosynthesis and isolation of a new oxidized epi-isozizaene metabolite. Microb Biotechnol. 4(2):184–191
132. Zhao B, Lin X, Lei L, Lamb DC, Kelly SL, Waterman MR, Cane DE (2008) Biosynthesis of the sesquiterpene antibiotic albaflavenone in *Streptomyces coelicolor* A3(2). J Biol Chem 283(13):8183–8189
133. Blasco F, Kauffmann I, Schmid RD (2004) CYP175A1 from *Thermus thermophilus* HB27, the first beta-carotene hydroxylase of the P450 superfamily. Appl Microbiol Biotechnol 64(5):671–674
134. Liu W, Rosazza JP (1993) A soluble *Bacillus cereus* cytochrome P-450cin system catalyzes 1,4-cineole hydroxylations. Appl Environ Microbiol 59(11):3889–3893
135. Bell SG, French L, Rees NH, Cheng SS, Preston G, Wong LL (2013) A phthalate family oxygenase reductase supports terpene alcohol oxidation by CYP238A1 from *Pseudomonas putida* KT2440. Biotechnol Appl Biochem 60(1):9–17
136. Grogan G, Roberts GA, Parsons S, Turner NJ, Flitsch SL (2002) P450(camr), a cytochrome P450 catalysing the stereospecific 6- endo-hydroxylation of (1 R)-(+)-camphor. Appl Microbiol Biotechnol 59(4–5):449–454
137. Kim SY, Zhao P, Igarashi M, Sawa R, Tomita T, Nishiyama M, Kuzuyama T (2009) Cloning and heterologous expression of the cyclooctatin biosynthetic gene cluster afford a diterpene cyclase and two p450 hydroxylases. Chem Biol 16(7):736–743

Adv Biochem Eng Biotechnol (2015) 148: 251–286
DOI 10.1007/10_2015_321

Published Online: 18 April 2015

Bioprocess Engineering for Microbial Synthesis and Conversion of Isoprenoids

Hendrik Schewe, Marco Antonio Mirata and Jens Schrader

Abstract Isoprenoids represent a natural product class essential to living organisms. Moreover, industrially relevant isoprenoid molecules cover a wide range of products such as pharmaceuticals, flavors and fragrances, or even biofuels. Their often complex structure makes chemical synthesis a difficult and expensive task and extraction from natural sources is typically low yielding. This has led to intense research for biotechnological production of isoprenoids by microbial de novo synthesis or biotransformation. Here, metabolic engineering, including synthetic biology approaches, is the key technology to develop efficient production strains in the first place. Bioprocess engineering, particularly in situ product removal (ISPR), is the second essential technology for the development of industrial-scale bioprocesses. A number of elaborate bioreactor and ISPR designs have been published to target the problems of isoprenoid synthesis and conversion, such as toxicity and product inhibition. However, despite the many exciting applications of isoprenoids, research on isoprenoid-specific bioprocesses has mostly been, and still is, limited to small-scale proof-of-concept approaches. This review presents and categorizes different ISPR solutions for biotechnological isoprenoid production and also addresses the main challenges en route towards industrial application.

Keywords Bioprocess · In situ product removal · Isoprenoids · Solvent tolerance · Terpenoids

---

H. Schewe · J. Schrader (✉)
DECHEMA Research Institute, Biochemical Engineering, Frankfurt, Germany
e-mail: schrader@dechema.de

M.A. Mirata
Speciality Ingredients Research & Technology (RSI), Lonza Ltd, Basel, Switzerland

Contents

1 Introduction

Isoprenoids (the terms "isoprenoid", "terpene", and "terpenoid" are used synonymously throughout this work) are a class of natural products present in all living organisms. The great diversity of structures accompanied by many varied biological functions lets isoprenoids play a vital role in basic intra- and intercellular processes, for example, as components of cell membranes, respiratory chains, photosynthetic light harvesting systems, or as hormones. All isoprenoids are derived from prenyl diphosphate precursors and consist of one or more C_5 isoprene units which are condensed to different isoprenoid classes (C_5: hemiterpenoids, C_{10}: monoterpenoids, C_{15}: sesquiterpenoids, C_{20}: diterpenoids, C_{25}: sesterterpenoids, C_{30}: triterpenoids, C_{40}: tetraterpenoids and $> C_{45}$: polyterpenoids). Terpenoids are major components of plant essential oils and extracts whose antiseptic and medicinal effects and in particular their sensory properties and value for food preservation have been used since the Middle Ages [8]. Nowadays, industrially relevant isoprenoid molecules cover a wide range of products such as flavors and fragrances [83], pharmaceuticals [20], and biofuels [66] corresponding to multibillion US$ markets.

The prime examples illustrating the tremendous progress made in isoprenoid biotechnology are the microbial de novo synthesis of farnesene as both fragrance and biofuel precursor [102], of amorphadiene and artemisinic acid as antimalaria drug artemisinin precursors [65, 103], and of isoprene as a chemical commodity and monomer for synthetic rubber synthesis [105]. Also, the plant cell culture based biosynthesis of precursors of the anticancer blockbuster drug Taxol [51] represents an important industrially applied isoprenoid biotechnology. Finally, in the area of microbial isoprenoid conversion, a number of high-yielding approaches have been published as well, for example, for the biotransformations of limonene to perillyl alcohol [21, 22], to perillic acid [54], and to α-terpineol [9] or of α-pinene oxide to isonovalal [34], to name only a few. The reader is kindly referred to the respective chapters in the same volume series addressing the biotechnology of specific isoprenoid products.

Due to their often complex structure, chemical synthesis of isoprenoids is usually difficult and expensive, and isolation from their natural sources is also typically low yielding. In contrast, biotechnological strategies take advantage of the pronounced regio- and stereo selectivities of enzymatic reactions (usually by use of microbial cells) and the possibility of overproducing isoprenoids in tailored bioprocesses. One strategy is whole-cell biocatalysis, which is usually favored if structurally closely related precursor molecules are abundantly available, for example, the monoterpenes limonene or pinene for carvone or verbenone production, respectively [44, 46], and therefore represent cheap starting materials. The target products are often obtained by selective oxyfunctionalization reactions catalyzed by only one or a few enzymes, such as P450 monooxygenases and alcohol dehydrogenases. The other strategy follows the cell factory notion, that is, the de novo biosynthesis with microorganisms, whose metabolism has to be comprehensively re-engineered for isoprenoid production starting from inexpensive carbon sources, such as sugar [20]. Metabolic engineering typically involves optimization of the flux through isoprenoid biosynthetic pathways, that is, the methylerythritol phosphate or MEP pathway, also referred to as the 1-deoxy-D-xylulose 5-phosphate or DXP pathway, and the mevalonate (MVA) pathway, depending on the host strain or strategy used. Both pathways lead to the key C_5 isoprenoid intermediate isopentenyl diphosphate, from which isoprenoid synthesis is usually continued towards the final product via a cascade comprising prenyltransferase(s), a terpene synthase, and often further terpene decorating enzymes [19].

The question remains why, despite the many economically interesting applications of isoprenoids and the intrinsic advantages of enzymes as highly selective biological catalysts for isoprenoid formation, only few industrial-scale bioprocesses have been established to date. In addition to the obvious challenges associated with a comprehensive genetic reprogramming of cellular metabolism, research is also consistently hampered by a couple of physicochemical properties of isoprenoids. A recurring phenomenon in bioconversion or de novo synthesis of isoprenoids is the hydrophobicity and volatility of the substrate and/or product, and the fact that these properties are often accompanied by a pronounced cytotoxicity. However, hydrophobicity and volatility of isoprenoids can be exploited for developing selective in situ product removal (ISPR) techniques which are usually also mandatory if product toxicity is a major issue. Alternatives might be the improvement of microbial resilience by genetic engineering or the use of solvent-tolerant microbes. For this review chapter we compiled bioprocess engineering approaches for microbial (bacterial and fungal) synthesis and conversion of isoprenoids with special focus on technical solutions for ISPR. We present laboratory- and industrial-scale bioprocesses and discuss the major hurdles towards industrial implementation.

Microbial processes for the production of carotenoids and steroids as well as isoprenoid production by plant cell cultures and by engineering in vivo plant isoprenoid biosynthesis are outside the scope of this review.

1.1 In situ Product Removal (ISPR)

Considering the physicochemical and biochemical properties of isoprenoids (hydrophobicity, volatility, and toxicity, all of which are discussed in later chapters of this review), a successful biotechnological production of isoprenoids has to rely on both metabolic engineering and process engineering in order to overcome inhibitory (reversible loss of activity) or toxic (irreversible loss of activity) effects of isoprenoid precursors or products. Due to the characteristics of isoprenoids, bioprocess development needs to focus on the integration of ISPR techniques. ISPR techniques have already been thoroughly investigated since the early 1980s due to a recurring problem of bioprocessing: the productivity of biocatalytic processes is frequently limited by the need to operate the reaction under conditions unsuited to the biocatalyst [49, 89]. Although chemical synthesis can be relatively flexible in matters of reaction conditions, biological systems have evolved over hundreds of millions of years to function under relatively limited conditions, that is, in a comparably narrow "process window". Applying the same economical necessities that guide optimization of industrial chemical production processes to biological systems and force biocatalysts to function outside the environment for which they have evolved leads to compromises over bioreactor design and operation. Because inhibition of enzymatic reactions or microbial growth is usually observed at elevated product concentrations, dilute product streams and low productivities are characteristics of biotechnological processes [84, 89]. A technical approach to overcome the problem of low productivities is the application of ISPR for the immediate separation of a product from the producing biocatalyst [35]. The separation is achieved by compartmentalization of biocatalyst and substrate and/or product thereby increasing the productivity or yield of a biotechnological process by

- Overcoming inhibitory toxic effects of the product or the substrate
- Minimizing product loss due to degradation or uncontrolled release
- Reduction of liquid volume by concentration of the product in a separate phase
- Reducing the total number of downstream processing steps.

Although the compartmentalization displays the aforementioned beneficial effects it also introduces mass transfer fluxes of the product and/or substrate over interfaces thereby creating new possible bottlenecks influencing process productivity and yield. In order to minimize adverse effects by limited mass transfer the choice of the appropriate ISPR technique is as important as it is challenging.

Several techniques are available to separate inhibitory or toxic substances from biotechnological processes depending on their physicochemical properties. Molecular weight, hydrophobicity, volatility, charge, and specific binding properties of a compound can help to decide the appropriate ISPR technique [35]. Five main techniques are available to remove a product from the vicinity of the biocatalyst [89]. Extraction into another phase makes use of differences in hydrophobicity of substrate or product and biocatalyst. The receiving second phase may be an organic solvent, a supercritical fluid, an ionic liquid, or another aqueous phase. ISPR by perstraction involves the separation of the extractive phase from the biocatalyst-containing phase by a membrane. High volatility favors evaporation as a separation technique. Evaporation is accomplished by stripping or vacuum distillation. Pervaporation and transmembrane distillation are applied as membrane-supported evaporation techniques. Separation by size is achieved by dialysis, electrodialysis, reverse osmosis, or nanofiltration. Depending on the charge of the target molecules, adsorption to hydrophobic carriers, ion-exchange resins and affinity adsorption can be used. Finally, a charged product may be precipitated by the addition of a counter-ion.

Figure 1 depicts possible configurations of bioreactor and separation units and different modes of operation for the five principal ISPR techniques described above.

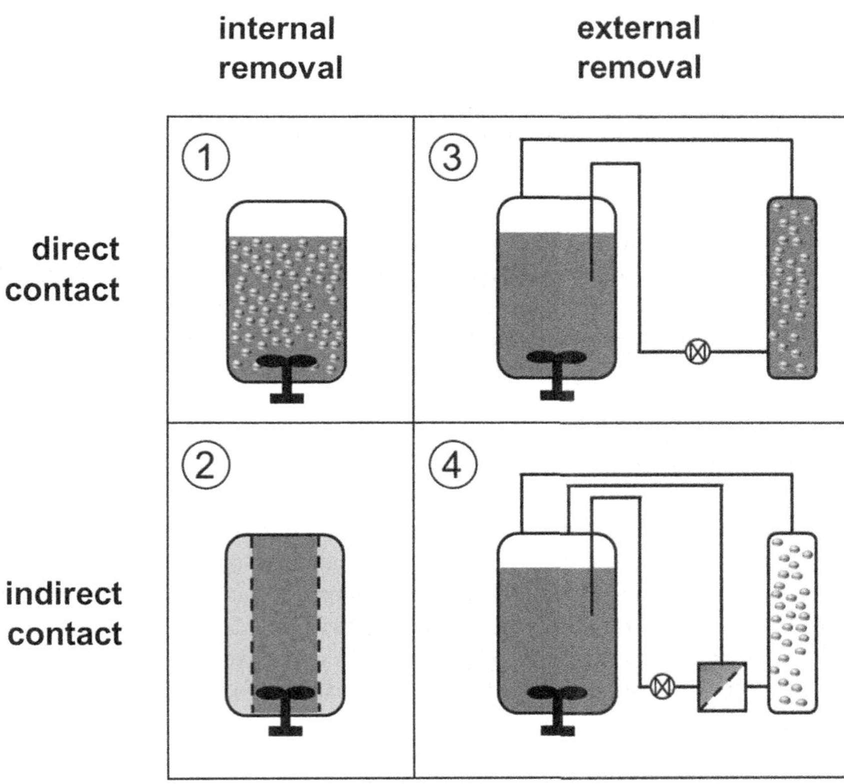

Fig. 1 Options of in situ product removal (ISPR) from a bioreactor. The ISPR can take place inside the reactor (internal) or in an external loop. The biocatalyst can be in direct contact with the extractive phase or may be separated by various techniques, such as a membrane (shown as *dotted lines*) or a gel matrix. The biocatalyst-containing phase is shown in *dark gray*, the extractive phase is shown in *light gray*. The complexity of the setup increases from 1 to 4. Modified after Woodley et al. [108]

The physicochemical properties of isoprenoid compounds are reflected in the chosen ISPR techniques listed in Tables 1 and 2. The hydrophobicity and volatility of isoprenoid compounds favor extraction, hydrophobic adsorption, or evaporation. Water solubilities of terpenes are generally low, with those for terpene hydrocarbons far lower than those for oxyfunctionalized terpenes. As an example, the saturation concentration of limonene in water at 25 °C is around 0.15 mM, that of carveol 19 mM [33]. The low water solubility of many terpenes and their high activity coefficients in dilute aqueous solution results in their excessive accumulation in the gas phase over the liquid phase, as compared to an ideal system according to Raoult's law. This leads to relatively quick stripping from an aqueous phase under typical bioreactor conditions. For instance, limonene is completely removed from an aqueous solution of 7.5 mg/L in a 3-L reactor, which is stirred at 1000 rpm and aerated with 0.9 L/min, within only 30 min [33]. This makes it comprehensible, that even when applying certain ISPR methods substrate and/or product loss may occur in significant amounts [15, 21, 64]. The following chapters give examples of application of ISPR techniques in bioprocess engineering for synthesis and conversion of isoprenoids using the categories depicted in Fig. 1.

2 Internal and External Removal of Isoprenoid Products by Direct Contact

Bioprocesses with a direct contact ISPR technique according to numbers 1 and 3 in Fig. 1 usually consist of a biocatalyst-containing aqueous phase and an immiscible second phase. The second phase may be located inside the bioreactor (internal removal) or in an external loop (external removal). The second phase functions typically as either a product sink removing potentially inhibitory products from the aqueous phase or a substrate reservoir delivering substrates at low concentrations to the aqueous phase. This combination of substrate release and product removal is often referred to as substrate feeding and product removal (SFPR) and can be seen as a method to avoid the toxic effect of substrate and product. The second phase has a larger affinity for the target compounds than the aqueous phase, allowing in the case of substrate delivery the loading of large amounts of poorly water-soluble substrate into the second phase. This extractive bioprocess concept is often referred to as a two-phase partitioning bioreactor (TPPB) [24, 52]. In two-liquid phase partitioning bioreactors for microbial processes the extractive second phase consists of a hydrophobic liquid phase [28]. As an alternative to a liquid second phase a solid extractive phase (e.g., a solid adsorbent material) can be used. Whether a liquid or solid phase is used, the fundamental intention is to enhance the mass transfer rate of apolar toxic compounds with low bioavailability and to control their delivery or removal. The considerations involved in the selection of an appropriate extractive phase, and the advantages and drawbacks of the application of a liquid or a solid phase are discussed in the next two sections.

Table 1 Bioprocesses for synthesis and conversion of isoprenoids with internal and external in situ product removal by direct contact using aqueous–organic or aqueous–solid two-phase systems. Only literature describing cultivations in bioreactors is cited

ISPR details	Organism	Substrate	Product	t (h)	V_{tot} (L)	Cell condition	X (g L^{-1})	c (g L^{-1})	$Y_{P/X}$ (g g^{-1})	STY (mg L^{-1} h^{-1})	Reference
Internal SFPR STR extractive phase hexadecane (log P_{Oct} = 8.9) V_{Aq} = 0.5 L V_{Org} = 0.5 L	*Pseudomonas rhodesiae CIP* 107491	α-pinene oxide	Isonovalal	20	1	Permeabilized	–	266	–	13,300	Fontanille and Larroche [34]
Internal SFPR STR extractive phase sunflower oil V_{Aq} = 0.15 L V_{Org} = 0.15 L	*Sphingobium* sp.	Limonene	α-terpineol	144	0.3	Resting	4.6	23	5	160	Bicas et al. [9]
Internal SFPR STR extractive phase bis (2-ethylhexyl) phthalate (log P_{Oct} = 7.5) V_{Aq} = 0.5 L V_{Org} = 1 L	Recombinant *Pseudomonas putida* KT2440	Limonene	Perillyl alcohol	24	1.5	Growing	20.1	4.37	0.22	182	Cornelissen et al. [22]
Internal SFPR STR extractive phase bis (2-ethylhexyl) phthalate (log P_{Oct} = 7.5) V_{Aq} = 0.5 L V_{Org} = 1 L	Recombinant *Escherichia coli*	Limonene	Perillyl alcohol	26	1.5	Growing	29.3	5.96	0.20	229	Cornelissen et al. [21]

(continued)

Table 1 (continued)

ISPR details	Organism	Substrate	Product	t (h)	V_{tot} (L)	Cell condition	X (g L^{-1})	c (g L^{-1})	$Y_{P/X}$ (g g^{-1})	STY (mg L^{-1} h^{-1})	Reference
Internal SFPR STR extractive phase bis (2-ethylhexyl) phthalate (log P_{Oct} = 7.5) V_{Aq} = 1 L V_{Org} = 0.55 L	Recombinant *Pseudomonas putida*	Limonene	Perillyl alcohol	75	1.55	Growing	15	2.3	0.15	31	van Beilen et al. [97]
Internal SFPR STR extractive phase diisononylphthalate (log P_{Oct} = 9.4) V_{Aq} = 0.15 L V_{Org} = 0.1 L	Recombinant *Escherichia coli*	α-pinene	α-pinene oxide verbenol myrtenol	5	0.25	Growing	10	0.61	0.061	122	Schewe et al. [80]
Internal SFPR STR extractive phase silicone oil V_{Aq} = 3.1 L V_{Org} = 0.5 L	*Rhodococcus erythropolis* DCL14	*Trans*-carveol	Carvone	28.75	3.6	Growing	0.5	0.823	1.65	29	Morrish et al. [56]
Internal SFPR STR extractive phase hexadecane (log P_{Oct} = 8.9) V_{Aq} = 1. 5 L V_{Org} = 0.15 L	*Saccharomyces cerevisiae*	Geraniol	Citronellol	60	1.65	Resting	–	4.9	–	82	Valadez-Blanco et al. [96]

(continued)

Table 1 (continued)

ISPR details	Organism	Substrate	Product	t (h)	V_{tot} (L)	Cell condition	X (g L^{-1})	c (g L^{-1})	$Y_{P/X}$ (g g^{-1})	STY (mg L^{-1} h^{-1})	Reference
Internal SFPR STR extractive phase hexadecane (log P_{Oct} = 8.9) V_{Aq} = 0.4 L V_{Org} = 0.4 L	*Pseudomonas rhodesiae* CIP107491	α-pinene oxide	Novalic acid	8	0.8	Resting	7	8	1.14	1,000	Linares et al. [48]
Internal SFPR STR extractive phase hexadecane (log P_{Oct} = 8.9) V_{Aq} = 1.6 L V_{Org} = 0.1 L	Recombinant *Saccharomyces cerevisiae*	Geraniol	Citronellol	11.5	1.7	Resting	6.25	0.176	0.028	15	Doig et al. [29]
Internal SFPR air-driven column reactor extractive phase dodecane (log P_{Oct} = 6.8) V_{Aq} = 0.07 L V_{Org} = 0.04 L	*Rhodococcus erythropolis* DCL14	*Trans*-carveol	Carvone	460	0.11	Growing	–	38.9	–	85	de Carvalho and da Fonseca [27]
Internal extraction by dodecane (log P_{Oct} = 6.8) STR V_{Aq} = 1.2 L V_{Org} = 0.12 L	Recombinant *Pichia pastoris*	De novo synthesis; glucose, methanol	Nootkatone	108	1.32	Growing	100	0.187	0.002	1.7	Wriessnegger et al. [109]

(continued)

Table 1 (continued)

ISPR details	Organism	Substrate	Product	t (h)	V_{tot} (L)	Cell condition	X (g L^{-1})	c (g L^{-1})	$Y_{P/X}$ (g g^{-1})	STY (mg L^{-1} h^{-1})	Reference
Internal extraction by dodecane (log P_{Oct} = 6.8) STR V_{Aq} = 1.2 L V_{Org} = 0.3 L	Recombinant *Escherichia coli*	De novo synthesis; LB-medium	Taxadiene	120	1.5	Growing	19	0.095	0.005	0.8	Boghigian et al. [12]
Internal extraction by dodecane (log P_{Oct} = 6.8) STR V_{Aq} = 1 L V_{Org} = 0.1 L	Recombinant *Escherichia coli*	De novo synthesis; glycerol, yeast extract	Sclareol	48	1.1	Growing	60	1.46	0.024	30	Schalk et al. [79]
Internal extraction by Diisononylphthalate (log P_{Oct} = 9.4) STR V_{Aq} = 1 L V_{Org} = 0.5 L	Recombinant *Escherichia coli*	De novo synthesis; glucose	Limonene	45	1.5	Growing	61.8	0.9	0.015	20	Willrodt et al. [107]
Internal extraction by dodecane (log P_{Oct} = 6.8) STR V_{Aq} = 6 L V_{Org} = 0.6 L	Recombinant *Escherichia coli*	De novo synthesis; TB-medium	Amorpha-4,11-diene	60	6.6	Growing	5.4	0.48	0.09	8	Newman et al. [60]
Internal extraction by dodecane (log P_{Oct} = 6.8) STR V_{Aq} = 1 L V_{Org} = 0.1 L	Recombinant *Saccharomyces cerevisiae*	De novo synthesis; glucose	α-santalene	62	1.1	Growing	–	0.0072	–	1.16×10^{-4}	Scalcinati et al. [77]

(continued)

Table 1 (continued)

ISPR details	Organism	Substrate	Product	t (h)	V_{tot} (L)	Cell condition	X (g L^{-1})	c (g L^{-1})	$Y_{P/X}$ (g g^{-1})	STY (mg L^{-1} h^{-1})	Reference
Internal extraction by dodecane (log P_{Oct} = 6.8) STR V_{Aq} = 1 L V_{Org} = 0.2 L	Recombinant *Escherichia coli*	De novo synthesis; complex medium	Taxadiene	120	1.2	Growing	10	0.17	0.017	1.4	Ajikumar et al. [1]
Internal extraction by dodecane (log P_{Oct} = 6.8) STR V_{Aq} = 1 L V_{Org} = 0.1 L	Recombinant *Escherichia coli*	De novo synthesis; glucose	Amorpha-4,11-diene	140	1.1	Growing	90	27.4	0.3	196	Tsuruta et al. [94]
Internal extraction by dodecane (log P_{Oct} = 6.8) STR V_{Aq} = 4 L V_{Org} = 0.3 L	Recombinant *Saccharomyces cerevisiae*	De novo synthesis; galactose	Cubebol	76	4.3	Growing	10	5.86×10^{-4}	5.86×10^{-5}	7.7×10^{-6}	Asadollahi et al. [7]
Internal extraction by methyl oleate (log P_{Oct} = 7.6) STR V_{Aq} = 1.8 L V_{Org} = 0.2 L	Recombinant *Saccharomyces cerevisiae*	De novo synthesis; ethanol	Amorpha-4,11-diene	116	2	Growing	–	41	–	353	Westfall et al. [103]
Internal extraction by isopropyl myristate (log P_{Oct} = 7.4) STR V_{Aq} = 0.8 L V_{Org} = 0.4 L	Recombinant *Saccharomyces cerevisiae*	De novo synthesis; ethanol	Artemisinic acid	140	1.2	Growing	–	25	–	179	Paddon et al. [65]

(continued)

Table 1 (continued)

ISPR details	Organism	Substrate	Product	*t* (h)	V_{tot} (L)	Cell condition	*X* ($g\ L^{-1}$)	*c* ($g\ L^{-1}$)	$Y_{P/X}$ ($g\ g^{-1}$)	STY ($mg\ L^{-1}\ h^{-1}$)	Reference
Internal SFPR, STR adsorption to hydrophobic polymer	*Corynespora cassiicola* DSM 62475	Racemic linanool	Linalool oxides	120	0.5	Growing	n.d.	4.6	–	38	Bormann et al. [15]
Internal SFPR, STR adsorption to hydrophobic polymer	*Rhodococcus erythropolis* DCL14	*Trans*-carveol	Carvone	48.75	3.83	Growing	0.7	4.83	6.9	99	Morrish and Daugulis [57]
External SFPR, STR adsorption to anion exchange polymer	*Pseudomonas putida* DSM 12264	Limonene	Perillic acid	168	0.16	Growing	9	31	3.44	185	Mirata et al. [54]
Internal SFPR, Pad-packed Interface bioreactor extractive phase decane (log P_{Oct} = 5.8)	*Hansenula saturnus* IF0 0809	Racemic citronellol	Citronellic acid	288	1.1	Biofilm	–	18	–	62.5	Oda et al. [61]
Internal SFPR, Multistory interface bioreactor extractive phase decane (log P_{Oct} = 5.8)	*Pichia kluyveri* IF0 1165	Racemic citronellol	Citronellyl acetate	432	7	Biofilm	–	14.5	–	33.6	Oda et al. [62]

t reaction time; V_{tot} reaction volume (organic and aqueous phase); *X* cell dry weight concentration; *c* product concentration in relation to total volume; $Y_{P/X}$ product yield coefficient; *STY* space time yield; *ISPR* in situ product removal; *SFPR* substrate feeding and product removal; *STR* Stirred tank reactor

Table 2 Bioprocesses for synthesis and conversion of isoprenoids with internal and external in situ product removal by indirect contact. Only literature describing cultivations in bioreactors is cited

ISPR details	Organism	Substrate	Product	t (h)	V_{tot} (L)	Cell condition	X (g L^{-1})	c (g L^{-1})	$Y_{P/x}$ (g g^{-1})	STY (mg L^{-1} h^{-1})	Reference
Stripping, absorption into hexadecane STR	*Saccharomyces cerevisiae*	Geraniol	Citronellol	45	5.2	Resting	10	0.09	0.009	2	Arifin et al. [4]
Stripping, absorption into hexadecane STR	*Saccharomyces cerevisiae*	Geraniol	Citronellol	36	5.2	Growing	3	0.045	0.015	1	Arifin et al. [5]
Stripping, adsorption to hydrophobic polymer STR	*Pleurotus ostreatus* DSMZ 1020	β-myrcene	Perillene	240	1.8	Growing	210 (wet biomass)	0.08		0.3	Krings and Berger [45]
Stripping, adsorption to hydrophobic polymer STR	*Streptomyces citreus* CBS 109.60	De novo synthesis; complex medium	Geosmin	4	2.5	Growing	2	2.6×10^{-3}	1.3×10^{-3}	0.7	Pollak and Berger [68]
Stripping, adsorption to hydrophobic polymer STR	*Cystoderma carcharias* CBS 157.47	Citronellol	3,7-dimethyl-1,6,7-octanetriol	240	2	Growing	15	0.866	0.058	3.6	Onken and Berger [64]
Pervaporation STR	*Ceratocystis moniliformis* ATCC 12861	De novo synthesis; glucose, potato–dextrose broth	Geraniol citronellol	260	3.5	Growing	2	0.04 0.02	0.02 0.01	0.2 0.1	Bluemke and Schrader [11]

(continued)

Table 2 (continued)

ISPR details	Organism	Substrate	Product	t (h)	V_{tot} (L)	Cell condition	X (g L^{-1})	c (g L^{-1})	$Y_{P/x}$ (g g^{-1})	STY (mg L^{-1} h^{-1})	Reference
Perstraction in membrane reactor, Extractive phase hexadecane STR	*Pseudomonas fluorescens* NCIMB 11671	α-pinene oxide	Isonovalal	380	2.4	Resting	60	45	0.75	118	Boontawan and Stuckey [14]
Perstraction in membrane reactor, Extractive phase hexadecane STR	*Saccharomyces cerevisiae*	Geraniol	Citronellol	170	0.66	Resting	–	4.25	–	25	Valadez-Blanco et al. [96]
Perstraction in membrane reactor, Extractive phase hexadecane STR	*Saccharomyces cerevisiae*	Geraniol	Citronellol	11.5	1.7	Resting	6.25	0.141	0.022	12	Doig et al. [29]

2.1 Two-Liquid Phase Bioprocesses

In two-liquid phase bioprocesses a biphasic reaction medium is created by the addition of a solvent with preferably low water-solubility to an aqueous phase containing the biocatalyst. Biphasic systems assure compartmentalization, thus avoiding inhibitory (reversible loss of activity) or toxic effects (irreversible loss of activity) on the biocatalyst and minimizing substrate or product loss by volatilization. The stirred-tank reactor is by far the most popular reactor type used in isoprenoid bioprocesses. Out of the 36 isoprenoid bioprocesses with ISPR presented in this review 33 used stirred-tank reactors (c.f. Tables 1 and 2). Another reactor type is the airlift or bubble column reactor where interface mixing is accomplished by gas bubbles [27]. Bioconversion or de novo synthesis of isoprenoids in liquid-impelled loop reactors [98] have not yet been described for isoprenoid bioprocesses. In a stirred-tank reactor the interfacial area available for mass transfer between the two immiscible phases is easily regulated via agitator speed [50]. The two phases form a fine emulsion under operational conditions. The average droplet diameter depends, among other parameters, on agitator speed, resulting in a direct relation of total specific surface area and agitator speed [99]. However, because the extractive phase needs to be hydrophobic in nature in order to be able to extract a hydrophobic product or deliver a hydrophobic substrate, it may convey toxic effects to the biocatalyst. Hence, biocompatibility is one of the criteria for the selection of a suitable organic phase. The toxic effect of an organic solvent or its biocompatibility can be correlated with its log P_{Oct} [47]. The log P_{Oct} value describes the logarithm of the partition coefficient of an organic solvent in a standard octanol–water system. Organic solvents with log P_{Oct} values greater than 4 are usually suitable for direct contact with biocatalysts whereas lower log P_{Oct} values generally result in loss of activity [18]. The log P_{Oct} value helps to limit the screening effort by reducing the amount of possible organic solvents to a manageable number. Sikkema and colleagues described why relatively low aqueous concentrations of hydrocarbons can have a devastating effect on cell membranes [87]. In their studies with liposomes prepared from *Escherichia coli* phospholipids a correlation between the membrane–buffer partition coefficients and the octanol–water partition coefficients (log P_{Oct}) of several hydrocarbons was found. According to this correlation, the sparingly water-soluble monoterpene limonene (log $P_{\mathrm{Oct}} = 4.46$, water solubility 0.15 mmol/L) would partition into the cell membrane and accumulate to a membrane concentration of up to 728 mmol/L (99 g/L). Consequences of hydrocarbon accumulation into and interaction with the cell envelope, the entity of cell membranes and cell wall, are cell membrane swelling, destruction of proton motive force, alterations in enzyme activities, and loss of cell membrane or cell wall integrity or function [17, 28, 88].

In addition to biocompatibility, other desirable criteria for an organic solvent as extractive phase have been identified and phase volume ratio (organic volume/aqueous volume) effects have to be considered (Table 3). The phase volume ratio affects the amounts of water-soluble and poorly water-soluble organic components present in the bioreactor, the degree of reduction or elimination of substrate or product

Table 3 Desirable characteristics, selection criteria, and phase volume ratio considerations of an organic phase in two-liquid phase bioprocesses [18, 50]

Desirable solvent characteristics
Favorable distribution coefficient for product
High selectivity
Low emulsion-forming tendency
Low aqueous solubility
Chemical and thermal stability
Favorable properties for product recovery
Nonbiodegradability
Nonhazardous
Inexpensive
Available in bulk quantities
Biocompatibility
Factors to consider
Solubility limit of S and P in organic phase
Solvent recycling options
Environmental impact
Phase volume ratio effects
Ease of pH control
Nature of dispersed phase
Possibility of phase inversion
Ease of phase separation
Interfacial area available for mass transfer
Inter-droplet mixing

inhibition, and the absolute and specific interfacial areas available for mass transport [50]. The most common approach is to operate with the aqueous phase as the continuous phase with a phase volume ratio typically less than 0.3–0.4 (c.f. Table 1).

A couple of publications illustrate the challenges involved with solvent screening following the desired properties listed in Table 3. Usually a number of organic solvents with appropriate log P_{Oct} values are compiled and experiments conducted to determine biocompatibility, partition coefficients for substrate, and/or product and biodegradability [6, 22, 37, 56, 80]. To test for biocompatibility an activity profile can be created by plotting an organic solvent log P_{Oct} against biocatalyst activity. However, this activity profile is specific to a particular biocatalyst and the extent of its contact with the organic phase [50]. The extent of contact depends on many process-specific factors such as available interfacial area, power input by mixing or shaking, concentration of surface-active components, concentration of biocatalyst, and the amount of organic solvent, to name a few. These factors may change during the bioprocess or with bioprocess scale. Therefore, the selection of a suitable organic solvent remains empirical and is often limited to the investigated setup. For example, using recombinant *E. coli* for the biotransformation of α-pinene (log $P_{Oct} = 4.5$) presented as a separate organic phase in shaking flasks resulted in product formation [81]. In contrast, the application of stirred-tank bioreactors using the same biocatalyst, aqueous medium, and phase volume ratio led to a complete loss of biocatalyst activity due to extended contact of the biocatalyst with α-pinene. Activity could be regained by the use of a different organic phase with a higher log P_{Oct} value [80] as a reservoir for α-pinene, "masking" its detrimental effects.

Searching for a suitable organic phase for the biotransformation of carveol to carvone with *Rhodococcus erythropolis* DCL14, Morrish and colleagues determined the critical log P_{Oct} of *R. erythropolis* DCL14 by exposing the cells to a selection of 12 organic solvents with log P_{Oct} ranging from 1.25 to 9.04 [56]. The critical log P_{Oct} of *R. erythropolis* DCL14 was found to be 5 in shake flask experiments. After further shake flask studies on biodegradability and determination of partition coefficients three organic solvents remained for bioreactor testing. Bioconversions were now tested in a stirred-tank reactor at a phase volume ratio of 0.16. In the end silicone oil proved to be the most suitable solvent although 1-dodecene showed superior affinity for the product. In contrast to experimental data from shaking flask experiments strong emulsion formation was observed when 1-dodecene was used as the organic phase in the stirred-tank reactor. The aforementioned examples illustrate the setup-specific nature of solvent screening experiments. Screening at smaller scales is less laborious and time consuming, however, the results are often not transferable to larger scale, especially if shake flask experiments are intended to optimize aqueous–organic two-liquid phase bioprocesses. Here, parallel fermentation systems are far superior as they are run under realistic process conditions at small scale.

The formation of emulsions, although in some cases advantageous for creation of a large interfacial area, may create significant problems in separation of the organic phase for further product removal; furthermore, organic phase or cell recycling represents an important practical issue in downstream processing of industrial-scale bioprocesses [39]. Surface-active components, often produced by microorganisms themselves as a protective mechanism or to improve substrate availability, impede phase separation and promote the formation of stable emulsions. Phenomena associated with stable emulsions in isoprene-producing bioprocesses were described as difficulties in determination of cell, substrate, or product concentration [56]; formation of a third phase partitioning of cell mass in the organic phase [26]; or problems in recovery of organic solvent and product [96]. The tendency for emulsion formation can change with phase volume ratio. A phase volume ratio of 0.2 led to a complete emulsification of dodecane in a stirred-tank reactor bioconversion of carveol to carvone by *R. erythropolis* whereas at a phase volume ratio of 0.4 half of the solvent was emulsified [26]. Although cell migration into the organic phase and problems in phase separation were reported, the bioprocess with complete emulsification of the organic phase yielded the highest carvone production rates compared to other configurations such as a membrane reactor.

Although strong emulsion formation of hexadecane with the aqueous phase during the reduction of geraniol to citronellol by *Saccharomyces cerevisiae* caused the authors to eventually employ a membrane reactor (indirect contact, c.f. Fig. 1) which actually eliminated emulsification, the space–time yield and product concentration was higher when the direct contact ISPR technique in a stirred-tank reactor was used [29, 96]. Furthermore, the spontaneous formation of an oil-in-water emulsion was found advantageous in an aqueous–organic two-phase bioprocess for isonovalal production from α-pinene oxide by *Pseudomonas rhodesiae*

[34]. Considering that emulsion droplet sizes in aqueous–organic two-phase systems in stirred-tank reactors can be as small as 10 μm resulting in organic–aqueous interfacial areas of 130 m^2 dm^{-3} [82], mass transfer limited reactions can profit from emulsion formation. However, droplet-stabilizing surface-active components such as fermentation broth, proteins, biomass, and biosurfactants can also considerably reduce the overall mass transfer coefficient by adsorbing to the aqueous–organic interface thereby damping turbulence transfer across the interface or, in the case of biomass, probably by blockage of the available interfacial area by multi-layers of biomass [69]. These examples demonstrate the mixed blessing of emulsion formation in aqueous–organic two-phase processes: that is, increased mass transfer rates versus problems in phase separation and product or cell recovery.

Additional aspects of solvent selection may involve the physicochemical nature of the organic phase. Vegetable oil was favored over hexadecane in the biotransformation of limonene to α-terpineol by *Sphingobium* sp. because a vegetable oil phase helped to achieve good anaerobic conditions that increased α-terpineol synthesis [9]. Other effects of the organic phase on biocatalyst or reaction can be: an influence on the enantiomeric specificity of the reaction accompanied by reduction of the e.e. value [96] or degradation of the organic phase by the microorganism [27].

In addition to instant removal from the vicinity of the biocatalyst the organic phase can serve as a protective environment for the product. Direct contact with an aqueous phase can cause autoxidation of terpenoid products and immediate extraction into an organic phase makes product accumulation possible in the first place such as in the case of nootkatol and nootkatone [37] or pinene oxide [80].

Long-chain alkanes (dodecane, hexadecane), silicone or vegetable oil, or diesters of phthalic acid [bis (2-ethylhexyl) phthalate, diisononyl phthalate] are examples for commonly used biocompatible organic solvents with high log P_{Oct} values in isoprenoid-producing bioprocesses (c.f. Table 1). Alternatively, a neat phase of substrate may be used for both substrate reservoir and extractive phase for the product, forming a SFPR system as published by Savithiry and colleagues who used undiluted limonene as the extractive phase during the bioconversion of limonene to α-terpineol by recombinant *E. coli* [76]. During the de novo production of biofuel sesquiterpenes farnesene or bisabolene by recombinant *S. cerevisiae* and *E. coli*, the insoluble product readily separates from the aqueous phase forming an aqueous–organic two-phase system in the course of the bioprocess which then acts as a product sink for further product extraction [67, 74]. Of course application of undiluted terpenoid substrate or product in an aqueous–organic two-phase system implies insignificant toxic and inhibitory effects on the biocatalyst.

Biofilm formation has been described as disadvantageous in terpene bioconversion studies when they interfered with substrate or product transport between the extractive phase and the biocatalyst [27, 57], however, biofilms can be used deliberately in biocatalytic approaches. Using catalytic biofilms allows for the exploitation of the advantages of a continuous reactor operation such as cell reuse, steady-state conditions with constant product qualities, and avoidance of downtimes, while at the same time the disadvantages of immobilization including reduced absolute activity and viability of the biocatalyst, additional costs, and lack of universally applicable immobilization concepts are avoided [38]. For isoprenoids, the only example described thus far followed an alternative approach using biofilms in direct contact with an extractive phase. The microbes are grown on nutrient-soaked carrier materials and then brought into direct contact with the substrate via a biocompatible organic carrier solution [61, 62]. Oxygen is supplied through the organic phase which may exhibit higher oxygen solubility compared to water [63]. A major disadvantage, however, is that the nutrient and water content and accumulation of metabolic by-products in the hydrophilic carrier material cannot be controlled during the process.

Considering the isoprenoid bioprocesses with direct contact and internal removal listed in Table 1 it is notable that only laboratory-scale systems with total reaction volumes typically below 10 L have been published and that most publications describe a proof-of-concept without further investigating product purification from the receiving phase. Exceptions are the distillation of limonene produced by recombinant *E. coli* from diisononyl phthalate [107], the purification of artemisinic acid from isopropyl myristate [65], and a stirred-tank reactor operated in continuous cultivation mode with a continuous recovery of the product α-santalene and regeneration of the extractive solvent dodecane [78]. On the shaking flask scale the recovery of bisabolene from the organic phase and its purification and subsequent chemical treatment have been described [67]. The only example of an industrial-scale isoprenoid bioprocess with direct internal removal of the product remains the production of the sesquiterpene farnesene by Amyris Inc. [102]. However, technical details of this process were not published.

2.2 Solid Phase Extraction

An extractive liquid organic phase of a TPPB can be replaced by a solid phase. It was found that some thermoplastics have strong affinities for hydrophobic organic molecules and show partitioning behavior similar to organic solvents and can therefore function as second phase and product sink or substrate source [3]. Polymers display several advantages and disadvantages in comparison to a liquid phase (Table 4). Desirable polymer characteristics in solid–liquid TPPBs are also listed in Table 4.

Few researchers employed the adsorption of isoprenoid products to polymers directly added to the aqueous phase (c.f. Table 1). Usually dispersed organic solvent droplets of an oil in water emulsion in a two-liquid phase-partitioning bioreactor form a larger available surface area and allow higher mass transfer rates because the droplet size is significantly smaller (μm scale) compared to the typical size of solid polymer materials (mm scale). For example, with an estimated droplet size of 30 μm, 50 mL of organic solvent in 1 L of aqueous phase form a total specific interfacial area of 8.7 $m^2\ dm^{-3}$ [99]. In contrast, 50 g of solid spherical or cylindrical beads with diameter of 2–4 mm result in a total specific interfacial area of 0.07 $m^2\ dm^{-3}$ in 1 L of aqueous phase [72]. Furthermore, equilibrium in a

Table 4 Advantages, disadvantages, and desirable properties of polymer beads for direct contact ISPR [73]

Advantage
Biocompatible
No emulsion formation
Resistance to microbial degradation
Reusable
Ease of separation from aqueous phase
Polymer structure can be tailored to enhance selective absorption of the desired target molecule(s)
Disadvantage
Low capacity determined by surface area
Mass transfer limitations due to biofilm formation
Delayed equilibrium between solid and liquid phase due to low diffusion rates
Desirable properties
Commercially available at a low cost
Nonhazardous
Nontoxic to the employed organisms
Not consumed as carbon and energy source or otherwise biodegradable
Not promoting biofilm formation under operational conditions
Possessing desirable affinity for the target molecule(s)
Thermally stable for sterilization purposes
Stable in aqueous medium at the pH and electrolyte concentration of the employed culture medium
Stable in medium employed to load polymer with target compounds

liquid–liquid system is established almost instantaneously whereas equilibrium between a solid and liquid phase can be delayed due to diffusion into the solid polymer matrix [72]. Nevertheless, application of solid polymer materials as the extractive phase can lead to improved volumetric productivities compared to two-liquid phase-partitioning bioreactors. Utilizing two important advantages of solid polymers, the biocompatibility and low emulsion formation tendency, the volumetric productivity of the biotransformation of carveol to carvone by *R. erythropolis* DCL14 was improved by a factor of 3.5 in comparison to a two-liquid phase-partitioning bioreactor where strong emulsion formation was observed [57]. The fact that, in contrast to experiments in a two-liquid phase-partitioning bioreactor, no dramatic morphological changes to the cells were observed provided proof for superior biocompatibility of polymer beads. Selection criteria for a suitable polymer material were bioavailability, partition coefficients, thermal stability, and biofilm formation tendency. The ease of separation of solid polymer material from the aqueous phase and the straightforward approach for product recovery led to high product concentrations and purities in eluates [15, 57] or even a crystallized product of 98 % purity [54].

On a laboratory scale product extraction from polymer beads may be much easier than the removal of products from a liquid–organic extracting phase where distillation of the solvent or back extraction using another immiscible solvent is required. However, on an industrial scale the additional step of polymer bead separation and subsequent product recovery may disproportionately increase process costs. Another problem arises when polymer beads are used in bioconversions where substrate and product share similar physicochemical properties. In these cases the polymer material may not be selective enough which can lead to decrease of conversion yields. Polymer beads were used to feed the substrate linalool and simultaneously remove the product linalool oxide during the bioconversion catalyzed by *Corynespora cassiicola* DSM 62475 [15]. Inasmuch as linalool and linalool oxide share similar adsorption characteristics to the polymer material used, 35 % of the substrate linalool was still eluted from polymer beads at the end of the process, thereby reducing the conversion yield. If the substrate and the product possess significantly different physicochemical properties polymer material can be successfully used in a SFPR approach. Anion exchange polymer beads (Amberlite IRA 410 Cl) were used to remove perillic acid efficiently from limonene bioconversion catalyzed by *Pseudomonas putida* GS1 [54]. The same adsorber was used by Alonso-Gutierrez et al. [2] to remove the product perillyl alcohol selectively from a de novo biosynthesis with a recombinant *E. coli* shake flask culture while avoiding detracting the immediate precursor in the isoprenoid pathway, limonene. The latter problem occurred when dodecane was used as an overlying organic phase instead, disabling perillyl alcohol formation.

3 Internal and External Removal of Isoprenoid Products by Indirect Contact

Bioprocesses with an indirect contact ISPR technique according to numbers 2 and 4 in Fig. 1 usually separate the biocatalyst from the extractive phase. The extractive phase may be located inside the bioreactor (internal removal) or in an external loop (external removal).

3.1 Stripping and Pervaporation

The high vapor pressure of some isoprenoids favors their transition from the liquid phase to the vapor phase at bioprocess temperatures. If the vapor phase is mobile as in the case of bioreactor off-gas the isoprenoid compound is stripped from the liquid phase. Stripping can effectively reduce the concentration of the isoprenoid product in the aqueous phase and avoid toxic effects on the biocatalyst. A straightforward way to remove isoprenoid volatile products from bioprocesses is their separation from bioreactor off-gas streams by adsorption to a solid-phase material such as activated carbon [105] or thermoplastics [45, 64, 68], or absorption into a hydrophobic liquid [4, 5]. The receiving phase has a certain affinity and capacity for the product: that is, at certain product loadings breakthrough of product leading to product loss occurs if no technical countermeasures (gas sensor, change of reservoir) have been installed [64, 105]. In addition to affinity and capacity of the receiving phase variables determining the efficiency of stripping are temperature and abundance of other gases such as CO_2, low molecular weight contaminants, or water vapor [105]. Desorption from adsorber materials can be achieved with organic solvents or, in the case of isoprene, by steam or nitrogen followed by condensation of the product vapor into the concentrated liquid product. Purification from absorbing liquids is usually achieved by distillation or back extraction using another immiscible solvent. Provided that the physicochemical properties allow the application of this ISPR technique stripping can be seen as a flexible and relatively easily scalable method.

If the product first diffuses through a membrane before it is evaporated into the vapor phase this is called pervaporation or in case of a hydrophobic membrane, organophilic pervaporation. The membrane forms a selective barrier between the aqueous phase (feed) and the vapor phase (permeate). Because pervaporation selectivity is primarily governed by differences in transport rate of components through the membrane and not by vapor pressure it can also be applied to less volatile compounds. Although organophilic pervaporation has always been considered a membrane-based technique that has great potential for mild, natural aroma recovery, industrial application remains at pilot-scale operations at the most [106] in contrast to hydrophilic pervaporation, which is industrially used for dewatering of organic solvents. Advantages of organophilic pervaporation are that available organophilic pervaporation membranes act as both sterile barrier and separation medium, rendering cell separation unnecessary. The membranes are biocompatible, incur negligible fouling tendency, and are steam sterilizable. Despite the many advantages of the combination of organophilic pervaporation with biosynthesis or conversion of isoprenoids in a bioreactor only a few such processes have been described in the literature thus far. During cultivation of *Ceratocystis moniliformis* ATCC 12861 ethyl acetate, propyl acetate, isobutyl acetate, isoamyl acetate, citronellol, and geraniol were isolated by organophilic pervaporation [11]. The choice of membrane material influenced product composition. Using a poly-octylmethylsiloxane membrane the process yielded higher concentrations of esters; using a polyetheramide-block–copolymer membrane the permeate contained higher concentrations of the terpenoids citronellol and geraniol. Another approach also used a flat-sheet organophilic pervaporation membrane to demonstrate proof-of-principle for in situ separation of the C13-norisoprenoid β-ionone from enzymatic β-carotene cleavage [59]. The biggest challenge still hampering a broader application of organophilic pervaporation for in situ volatiles separation is the limited transmembrane flux of the product. Here, membrane modules arranged as tube bundles with much larger interfacial membrane surface areas would most probably help in significantly improving overall performance.

3.2 Perstraction in Membrane Reactors

In a two-liquid phase bioreactor setup the biocatalyst-containing phase and the extractive phase can be separated by a membrane. The aim of such a membrane reactor is to avoid the consequences of direct contact between the aqueous and the organic phases such as emulsion formation and toxic effects on the biocatalyst. The biocatalyst and organic phases are separated by the membrane, therefore the range of solvents that can be used to assist biosynthesis or bioconversions can be expanded to solvents with high product affinity but low biocompatibility [96]. On the downside, a membrane may exhibit a considerable resistance to mass transfer between the aqueous and the organic phase [13, 14, 96]. For instance, according to the transport mechanism of a solute from the bulk aqueous phase to the bulk organic phase through a flat-sheet, hydrophobic, asymmetric, polyimide, organic solvent nanofiltration membrane there are five possible bottlenecks for solute transfer [85, 96]: transfer from the aqueous solution bulk to the aqueous/membrane interface through the aqueous liquid film, partition into the membrane, diffusion across the membrane, partition from the membrane into the organic liquid, transfer from the membrane/organic–liquid interface to the bulk of the organic liquid through the organic–liquid film. Because of these mass transfer constraints membrane bioreactors often show inferior performance when compared to an aqueous–organic two-phase system with direct contact [29, 96]. An integral part of a membrane reactor layout is the determination of the overall mass transfer coefficient of the solute of interest [13]. Together with the specific activity of the biocatalyst a suitable membrane surface area to biomass ratio can be calculated [29, 104]. By selecting a suitable membrane surface area to biomass ratio mass transfer limitations and the associated reduction of biocatalyst specific yields can be avoided.

Three types of membranes have been used in membrane reactors for isoprenoid synthesis and conversion: nonporous, microporous, and nanoporous. Silicone rubber is a nonporous membrane material displaying high permeability and selectivity to small hydrophobic molecules and prevents aqueous breakthrough and is impermeable to ionic species and macromolecules. The high permeability of silicone rubber is due to extensive swelling in the presence of organic solvents [30]. The silicone membrane swelling must be considered in membrane bioreactor design to prevent membrane bursting or solvent breakthrough. Silicone rubber tubing submerged into a liquid phase of a stirred-tank reactor is a practical and straightforward implementation of a membrane reactor [14, 27]. The organic phase may be pumped through the silicone rubber tubing which is submerged in the aqueous phase such as during the bioconversion of geraniol to citronellol by *S. cerevisiae* [29] or the bioconversion of α-pinene oxide to isonovalal by *Pseudomonas fluorescens* NCIMB 11671 [14]. Alternatively, the aqueous phase containing the biocatalyst *R. erythropolis* DCL14 was circulated inside a silicone tube that was immersed in dodecane containing the substrate carveol and the accumulated product carvone [27]. However, this mode led to biofilm formation inside the silicone tubing along with a reduction in mass transfer of substrate and product across the membrane, a problem often observed in membrane bioreactor setups.

A relatively hydrophilic microporous nylon membrane was used in a membrane reactor for the kinetic resolution of racemic menthyl acetate by *Bacillus subtilis* NCIMB 11705 [104]. A crucial point in process operation of microporous membranes is the careful control of the transmembrane pressure to avoid phase breakthrough, which leads to mixing of the two phases and to the same emulsification problems experienced in direct contact systems [29]. Surface-active components of the aqueous phase such as biosurfactants, proteins, or the cells themselves can cause a considerable reduction in the membrane breakthrough pressure [95]. However, in the case of kinetic resolution of racemic menthyl acetate the cells were placed on the same side of the membrane as the organic phase, pure racemic menthyl acetate, and the aqueous phase removed acetate from the vicinity of the cells. No breakthrough of either phase was observed provided the right pressure was maintained. Although cell stability was lower in the membrane reactor setup compared to an aqueous–organic two-phase system, probably due to better contact with the phase interface, the specific activity obtained in the membrane reactor was about fivefold that in the emulsion system with direct contact [104].

A nanofiltration membrane was used in a membrane bioreactor with the bioconversion of geraniol to citronellol by *S. cerevisiae* as the model reaction [96]. Measurements of partition coefficients and overall mass transfer coefficients identified toluene (log P_{Oct} = 2.4) as the superior extractive phase in comparison to hexadecane (log P_{Oct} = 8.8). As expected, bioconversion experiments with direct internal contact in an aqueous–organic two-phase system yielded no product when toluene was used whereas hexadecane proved to be biocompatible. Applying the nanofiltration membrane to separate the aqueous and organic phases showed higher space–time yields but at the same time lower e.e. values when toluene was used instead of hexadecane. Due to mass transfer limitations even the membrane reactor with toluene was outperformed by the direct contact ISPR system using the aqueous–organic two-phase system. The mass transfer resistance could be located in the membrane for the hexadecane system whereas for the toluene system the contribution of the aqueous liquid film mass transfer resistance became predominant [96].

The immobilization of cells in carrier materials can be seen as a compartmentalization of the biocatalyst and product. Fungal mycelia of *Penicillium digitatum* immobilized in calcium alginate were used to convert limonene to α-terpineol [92]. Alginate beads could be reused in batch conversions or applied in a continuous bioreactor setup. However, the alginate matrix formed a diffusional barrier for oxygen and limonene and decreased the substrate affinity by increasing the apparent K_M of the bioconversion.

4 Towards Industrial Isoprenoid Processes

Few microbial bioprocesses involving isoprenoids have been developed to industrial scale. Notable examples are the de novo productions of isoprene realized by the joint research of Genencor and Goodyear [105], of farnesene by Amyris Inc. [102], and of artemisinic acid for artemisinin by Amyris/Sanofi [65]. The basis of an industrial bioprocess is always a suitable biocatalyst able to accumulate high product titers in a short time. For this, whole-cell biocatalysts or microbial cell factories have to be optimized in iterative cycles by the tools of systems biology and metabolic engineering, but they also have to be integrated successfully into the setting of an industrial process. A simple constraint might be that this process requires integration of biotechnology and chemical engineering at a skill level not usually present in a single company [105].

4.1 Challenges of ISPR Implementation

Focusing on the technical side of bioprocess development, realization of ISPR techniques in industrial processes generally faces a number of challenges. A respective ISPR technique usually involves high investment costs for research and development and the construction of rather complex facilities. These measures have been usually dedicated to a single process, thereby reducing the flexibility of plant setup and increasing the pressure of efficient capital utilization [89]. However, nowadays with the concept of microbial platform strains, an array of different terpenoid products may be produced by the same microbe under almost identical fermentation conditions, which makes the use of process settings with ISPR more affordable ("one process—many products"). ISPR research and development are further hampered by the fact that ISPR techniques are either not easily scalable from laboratory conditions to production scale or that the suitable ISPR technique itself may change with process scale. Further consideration has to be put into the required product purity for subsequent process steps. Depending on the final application of the product of a bioprocess, legal limitations, such as regulations of product, process, work, and environmental safety, may further affect the choice of the ISPR technique. Adequately addressing all these points may lead to prolonged process development times and an extended time-to-market for a new product. In consequence, a less elaborated but faster process may be preferred [91].

As one of the few publications on industrial-scale isoprenoid bioprocesses the biological production of isoprene illustrates the aforementioned challenges [105]. It also highlights a central key to successful implementation of a large-scale bioprocess: an integrated approach that, from the choice of host strain, to the design and engineering of an isoprene cell factory, to its cultivation/fermentation, and finally to product recovery and purification, never loses the focus on large-scale applicability.

4.2 Industrial Production Strain Prerequisites

Most proof-of-concept research projects focus on only one or a couple of aspects of a bioprocess, thereby separating the microorganism and the isoprenoid synthesis or the conversion from the rest of the process design and neglecting demands of large-scale operations for certain organism properties [23]. In addition to those micro-organism properties generally desirable from a bioprocess-oriented viewpoint, such as the ability to use low-cost feedstocks, tolerance to extreme conditions, or resistance to inhibitors, isoprenoid bioprocesses would benefit from biocatalysts with low-emulsion–forming tendency and tolerance against organic compounds. This would dramatically increase the ease of product recovery in aqueous–organic two-phase systems simply by gravity separation without the need for centrifugation or de-emulsification techniques which further raise process costs [39]. For price competitive "low-cost high-volume" products such as biofuels the possibility of operating a bioprocess in a continuous mode including the possibility of cell reuse is a valuable measure to increase productivity and lower capital cost [23]. A closed bioreactor infrastructure to prevent contaminations operating in batch or fed-batch mode, as is usually the case in laboratory-scale experiments, can significantly increase infrastructure costs and will most likely not be cost-competitive enough at the required scale [102]. Continuous operation demands a robust biocatalyst able to tolerate conditions unfavorable to competing contaminating organisms, such as extreme pH, salt concentrations, or temperatures in order to maintain monoseptic conditions. Continuous cultures require a genetically stable strain with constantly high productivity. Unfortunately, production strains may change their phenotype during production conditions. The overproduction of a molecule in a fermentation process is usually an energy burden for the microorganism. Any mutation that leads to a decrease in that metabolic burden yields faster-growing mutants with decreased productivity that eventually overgrow the initial production host. Therefore, production organisms must be maintained under strong selection so that they do not lose their beneficial feature. Production processes are situations of prolonged cultivation where selection pressure can hardly be maintained. Once this selection process is interrupted by prolonged cultivation or repeated inoculation of new media, productivity may be lost [111]. However, commonly used techniques to maintain selection pressure such as antibiotic resistance markers (e.g., ampicillin or kanamycin) may not be applicable at large scale because of concerns of drug-resistant isolates being released into the environment [102]. The commonly used laboratory genetic tool to transfer genes, the plasmid, establishes a further potential route for horizontal gene transfer [93] and introduces a certain degree of genetic instability [90]. For the de novo production of isoprenoids from sugar-based raw materials cells need to reduce the carbohydrate and get rid of the carbohydrate oxygen. In biological systems this is achieved by the use of reducing equivalents [$NAD(P)H_2$] and rejection of excess oxygen as water. Redox imbalances, i.e., the excess formation of NAD(P)H2, which is often observed in production hosts, need to be removed aerobically (by oxidation to water). Therefore, bioreactor aeration must be carefully adjusted to avoid excessive NAD(P)H_2 oxidation and, thus, a decrease of carbohydrate-to-hydrocarbon yields below theoretical maximum values corresponding to reduced overall process productivity [10]. Another way to improve productivity is to divide a bioprocess into a growth phase, where nutrients are employed to build up biomass rapidly, and a production phase, where nutrients are exclusively used for product formation. In laboratory environments this is accomplished by the application of induction systems including IPTG in *E. coli* or galactose in *S. cerevisiae*. On an industrial scale these induction systems would not be cost-competitive and alternative routes to limit growth while allowing production are necessary [102]. An easy and well-scalable way to achieve such a "metabolic switch," that is, separation of growth and production phases, can be carbon or nitrogen restriction [94]. These examples of industrial-scale concerns on the one side and laboratory routine on the other side demonstrate a certain discrepancy caused by the different driving forces of scientific research and economic feasibility.

4.3 Terpene Toxicity

It is obvious that the choice of the host organism may ultimately determine the success or failure of an isoprenoid production process [102]. Each organism utilized in isoprenoid bioprocesses has its advantages and disadvantages that need to be weighed up against each other. Isoprenoids are essential components of living cells. However, their biotechnological overproduction in bioreactor systems may cause significant stress to exposed organisms. The toxicity of isoprenoids towards single cells varies with their structure. Due to their presumed natural function as defense molecules, toxins, growth inhibitors, or repellents, monoterpenoids are generally highly toxic to microorganisms as are some sesqui- and diterpenoids as well [36]; however, prominent examples of biotechnologically synthesized sesquiterpenes, such as farnesene or bisabolene are tolerated by the producing microbe in high concentrations [16, 67, 74]. Toxicity of hydrophobic compounds to microbial cells is often explained by accumulation of hydrophobic molecules in the lipid bilayer of the cell membrane ultimately resulting in a loss of its function as a permeability barrier, as a protein and reaction matrix, and as an energy transducer [41]; additionally, effects on other components of the cell envelope, for example, the cell wall have been described [17].

4.4 Microbial Solvent Tolerance

Microbial cells have developed different mechanisms to adapt to the presence of toxic solvents. In short, these mechanisms involve morphological adaptions, changes of the energy status, modification of the cell membrane fluidity, changes in the cell wall and outer membrane, modifications of surface properties such as charge and hydrophobicity, transformation, and degradation of the solvent and active transport of solvents from the membrane into the environment by an energy-dependent efflux system [40]. The reader is kindly referred to one of the many detailed reviews on this subject [25, 41, 70, 71, 75, 86]. The tolerance of micro-organisms towards organic and toxic compounds varies from organism to organism. Two different strategies to benefit from microbial solvent tolerance in bioprocesses with hydrophobic substrates or products are the use of well-known solvent-tolerant strains as production hosts or, alternatively, to transfer solvent-tolerance–conveying elements to well-established production hosts.

4.5 Solvent-Tolerant Hosts

Considering the usually hydrophobic and often toxic nature of isoprenoids, the choice of highly solvent-tolerant microbes, such as the gram-negative *P. putida* or the gram-positive *R. erythropolis* (c.f. Table 1) as whole-cell biocatalysts for the biotransformation of monoterpenes seems obvious. However, solvent-tolerance mechanisms are a consequence of natural selection on stress conveyed by hydrophobic substances. Some of the protective mechanisms of solvent tolerance may prevent substrate access to or product release from the biocatalyst thereby reducing product yields. For example, the outer membrane of gram-negative bacteria, in particular the lipopolysaccharide layer, was found to be an effective permeability barrier for monoterpenes [34]. Permeabilization by freeze drying or chemicals enhanced reaction rates. However, permeabilization is only applicable when the respective catalytic reaction is cofactor independent and will most likely prove to be too expensive to be applied in industrial-scale processes. Its metabolic versatility and its solvent tolerance make *P. putida* a promising biocatalyst for the bioconversion of terpenes [55]. A convincing example for the robustness of *P. putida* in the presence of monoterpenes is the conversion of (+)-limonene to (+)-perillic acid, a natural monoterpenoic acid with high potential as an antimicrobial cosmetic ingredient [54]. The microbe efficiently produces the target terpenoid without by-product formation in a fed-batch process where the precursor limonene can reach levels in the gram per liter range forming a separate organic phase in the bioreactor. However, the metabolic versatility of pseudomonads may also be a hindrance for establishing a bioprocess as it can result in unwanted side reactions and decrease product yields. Recombinant *P. putida* GPo12 was first described as a biocatalyst for the bioconversion of limonene to perillyl alcohol, a potential anticancer compound, catalyzed by a P450 monooxygenase from *Mycobacterium* sp. strain HXN-1500 in an aqueous–organic two-phase system [97]. Further research increased productivity sixfold by switching to *P. putida* KT2440 and conducting a careful carbon source selection and optimization of cell physiology [22].

However, additional research revealed that unknown enzymatic activities of *P. putida* KT2440 oxidized the product perillyl alcohol to the unwanted side products perillyl aldehyde and perillic acid. These side products constituted up to

26 % of the total amount of oxidized terpenes [21]. Another example for host-mediated unwanted side reactions is the reduction of geraniol to citronellol by unspecific dehydrogenases of the whole-cell biocatalyst *S. cerevisiae* [96]. Concerning tolerance to hydrophobic and toxic molecules such as the monoterpene limonene, which, for example, may serve as a precursor for jet biofuels in the future [10, 66], *P. putida* should have a clear advantage over monoterpene-sensitive *S. cerevisiae* [16, 17, 54]. However, the first de novo production of a monoterpene with *P. putida* has been described only very recently [53]. By introducing a heterologous bacterial mevalonate pathway from *Myxococcus xanthus* together with a plant geraniol synthase, *P. putida* GS1 produced about 200 mg/L geranic acid in a bioreactor. The monoterpenoic acid showed high efficiencies as a fungicide for corn protection [110]. The acid is produced due to endogenous oxidative enzymes converting geraniol, the direct product of the terpene synthase reaction, to the corresponding acid without accumulation of intermediates. If it were possible to narrow down the pronounced oxidative capacity of *P. putida* (e.g., by selective gene deletions or by running an appropriate process regime), this solvent-tolerant bacterium could be an interesting host for de novo production of monoterpene alcohols or hydrocarbons as well.

In an industrial process population heterogeneity is an unwanted side-effect that leads to incalculable productivities. Population heterogeneity can often be observed in *P. putida* cultivations [42]. Population heterogeneity functions as a safeguard of the respective population against fast-changing environmental conditions such as hydrophobicity. For example, direct contact of *P. putida* S12 with a toluene phase initiates a mutator element leading to the formation of a subpopulation with increased survival rates [101]. In this case population heterogeneity can be seen as another mechanism of solvent tolerance. Other sources of population heterogeneity are epigenetic modifications, asymmetric cell division, different growth and cell cycle states, microenvironmental conditions, and permanently changing levels of transcripts and proteins [58]. Inasmuch as transformation and degradation are mechanisms of solvent tolerance (see above), and solvent tolerance is often achieved by a combination of several protective mechanisms, metabolic versatility, population heterogeneity, possible side reactions, and an increased tendency to create stable emulsions are the flipside of solvent-tolerant strains.

4.6 Transfer of Solvent Tolerance

Taking a look at renewable diesel platforms, *S. cerevisiae* or other similar fungal systems are, in addition to *E. coli*, another popular choice for a development platform. Its genetic accessibility, robust and well-studied genetics, tolerance to low pH, resistance to osmotic stress, and immunity to viral contaminations make *S. cerevisiae* a production host with many of the requested properties beneficial in industrial bioprocesses [43]. In addition, problems with certification of *S. cerevisiae* genetically modified organisms on the government level are unlikely due to its well-established reputation as a nonhazardous organism. However, biofuel toxicity still limits production and it will be necessary to improve tolerance to further increase yields [31]. An approach to benefit from increased solvent tolerance without the drawback of unwanted side reactions can be the expression of efflux pumps, heat shock proteins, membrane modifying proteins, and activation of general stress response genes in suitable expression hosts [31]. Recently, heterologous expression of a specialized ABC efflux transporter from the fungal pathogen *Grosmannia clavigera* enhanced survival of *S. cerevisiae* in the presence of a mixture of the monoterpenes (+)-limonene, (+)-3-carene, racemic α-pinene, and (−)-β-pinene [100]. A couple of energy-dependent efflux systems have been described to increase tolerance against terpenoids in well-established production hosts and thereby sometimes improving process parameters. By first screening a library of primarily uncharacterized hydrophobe/amphiphile efflux pumps of the family of resistance-nodulation–division (RND) pumps from gram-negative bacteria and then heterologously expressing an efflux pump from *Alcanivorax borkumensis* limonene tolerance of *E. coli* was increased resulting in a 64 % improvement in limonene yield of a limonene production strain [32].

The outer-membrane protein alkL has been shown to increase transmembrane transport of monoterpene substrate limonene and product perillyl alcohol [21], thereby decreasing the reduction of mass transfer across the outer membrane of gram-negative *E. coli* and increasing reaction rates [21].

5 Conclusions

Despite the many exciting applications of isoprenoids as biofuels, pharmaceuticals, flavors and fragrances, and the advantages of microbial isoprenoid production, research on isoprenoid bioprocesses has mostly been and still is limited to proof-of-concept approaches with a focus on enhancing product concentrations and yields. Often no further considerations are given on critical targets for process technology and process scale implementation such as suitability of the strain employed, ease of product separation, purification, or comparison of different ISPR techniques. This can be attributed to the fact that most of the published research is conducted in nonprofit environments under idealized laboratory conditions where the industrial relevant considerations such as time-to-market, GMO certification, strain robustness, efficient capital utilization, or lot-to-lot variations of feedstock sources are not comparably relevant. In light of the physicochemical and biochemical properties of isoprenoids an industrial-scale isoprenoid bioprocess without ISPR, that is, compartmentalization of biocatalyst and substrate and/or product to avoid inhibitory or toxic effects on the biocatalyst or to avoid product loss by evaporation or degradation, is hardly imaginable. Because the implementation of a suitable ISPR technique is a laborious, time-consuming, expensive, and often an empirical process that needs to be verified on every new process scale, only few industrial-size microbial isoprenoid production processes have been published to date.

Experiences from the implementation of large-scale biofuel processes suggest that large-scale microbial isoprenoid bioprocesses would also benefit from an integrated approach of systems biology, metabolic and process engineering. The focus on process design and large-scale applicability during the whole development of a production strain has proven to be a challenging but ultimately rewarding necessity for a successful upscaling of an isoprenoid bioprocess. Considering the final process, the choice of a solvent-tolerant microbe such as *P. putida* instead of baker's yeast or *E. coli* as the host may be an alternative if the production of toxic compounds such as monoterpenoids is aimed at and/or technical substrates containing inhibitors have to be used. However, the genetic tools available still lag behind those developed for yeast and *E. coli* and genetic instability and metabolic diversity might be further disadvantages of *P. putida* as the host strain. Nevertheless, for the time being it cannot be foreseen whether the transfer of solvent-tolerance mechanisms to conventional hosts by metabolic and cell engineering will be a successful endeavor; thus both strategies are worth being followed in this segment of isoprenoid products.

References

1. Ajikumar PK, Xiao W-H, Tyo KEJ, Wang Y, Simeon F, Leonard E, Mucha O, Phon TH, Pfeifer B, Stephanopoulos G (2010) Isoprenoid pathway optimization for taxol precursor overproduction in *Escherichia coli*. Science 330(6000):70–74
2. Alonso-Gutierrez J, Chan R, Batth TS, Adams PD, Keasling JD, Petzold CJ, Lee TS (2013) Metabolic engineering of *Escherichia coli* for limonene and perillyl alcohol production. Metab Eng 19:33–41
3. Amsden BG, Bochanysz J, Daugulis AJ (2003) Degradation of xenobiotics in a partitioning bioreactor in which the partitioning phase is a polymer. Biotechnol Bioeng 84(4):399–405
4. Arifin AA, Don MM, Uzir MH (2011) Baker's yeast mediated biotransformation of geraniol into citronellol using a continuous-closed-gas-loop bioreactor (CCGLB) system. Biochem Eng J 56(3):219–224
5. Arifin AA, Don MM, Uzir MH (2011) The feasibility of growing cells of Saccharomyces cerevisiae for citronellol production in a continuous-closed-gas-loop bioreactor (CCGLB). Bioresour Technol 102(19):9318–9320
6. Asadollahi MA, Maury J, Møller K, Nielsen KF, Schalk M, Clark A, Nielsen J (2008) Production of plant sesquiterpenes in Saccharomyces cerevisiae: effect of ERG9 repression on sesquiterpene biosynthesis. Biotechnol Bioeng 99(3):666–677
7. Asadollahi MA, Maury J, Patil KR, Schalk M, Clark A, Nielsen J (2009) Enhancing sesquiterpene production in Saccharomyces cerevisiae through in silico driven metabolic engineering. Metab Eng 11(6):328–334
8. Bakkali F, Averbeck S, Averbeck D, Idaomar M (2008) Biological effects of essential oils—a review. Food Chem Toxicol 46(2):446–475
9. Bicas JL, Fontanille P, Pastore GM, Larroche C (2010) A bioprocess for the production of high concentrations of R-(+)-α-terpineol from R-(+)-limonene. Process Biochem 45(4):481–486
10. Blanch HW (2012) Bioprocessing for biofuels. Curr Opin Biotechnol 23(3):390–395
11. Bluemke W, Schrader J (2001) Integrated bioprocess for enhanced production of natural flavors and fragrances by *Ceratocystis moniliformis*. Biomol Eng 17(4–5):137–142

12. Boghigian B, Myint M, Wu J, Pfeifer B (2011) Simultaneous production and partitioning of heterologous polyketide and isoprenoid natural products in an *Escherichia coli* two-phase bioprocess. J Ind Microbiol Biotechnol 38(11):1809–1820
13. Boontawan A, Stuckey DC (2005) Mass Transfer of Terpenes through a Silicone Rubber Membrane in a Liquid-Liquid Contacting System. Biotechnol Prog 21(6):1680–1687
14. Boontawan A, Stuckey DC (2006) A membrane bioreactor for the biotransformation of α-pinene oxide to isonovalal by *Pseudomonas fluorescens* NCIMB 11671. Appl Microbiol Biotechnol 69(6):643–649
15. Bormann S, Etschmann M, Mirata M-A, Schrader J (2012) Integrated bioprocess for the stereospecific production of linalool oxides from linalool with *Corynespora cassiicola* DSM 62475. J Ind Microbiol Biotechnol 39(12):1761–1769
16. Brennan TC, Turner CD, Kromer JO, Nielsen LK (2012) Alleviating monoterpene toxicity using a two-phase extractive fermentation for the bioproduction of jet fuel mixtures in *Saccharomyces cerevisiae*. Biotechnol Bioeng 109(10):2513–2522
17. Brennan TCR, Krömer JO, Nielsen LK (2013) Physiological and transcriptional responses of *Saccharomyces cerevisiae* to d-Limonene show changes to the cell wall but not to the plasma membrane. Appl Environ Microbiol 79(12):3590–3600
18. Bruce LJ, Daugulis AJ (1991) Solvent selection strategies for extractive biocatalysis. Biotechnol Prog 7(2):116–124
19. Chandran SS, Kealey JT, Reeves CD (2011) Microbial production of isoprenoids. Process Biochem 46(9):1703–1710
20. Chang MCY, Keasling JD (2006) Production of isoprenoid pharmaceuticals by engineered microbes. Nat Chem Biol 2(12):674–681
21. Cornelissen S, Julsing MK, Volmer J, Riechert O, Schmid A, Bühler B (2013) Whole-cell-based CYP153A6-catalyzed (*S*)-limonene hydroxylation efficiency depends on host background and profits from monoterpene uptake via AlkL. Biotechnol Bioeng 110 (5):1282–1292
22. Cornelissen S, Liu S, Deshmukh AT, Schmid A, Buhler B (2011) Cell physiology rather than enzyme kinetics can determine the efficiency of cytochrome P450-catalyzed C-H-oxyfunctionalization. J Ind Microbiol Biotechnol 38(9):1359–1370
23. Cuellar MC, Heijnen JJ, van der Wielen LAM (2013) Large-scale production of diesel-like biofuels—process design as an inherent part of microorganism development. Biotechnol J 8 (6):682–689
24. Daugulis AJ (1997) Partitioning bioreactors. Curr Opin Biotechnol 8(2):169–174
25. de Bont JAM (1998) Solvent-tolerant bacteria in biocatalysis. Trends Biotechnol 16 (12):493–499
26. de Carvalho CCCR, da Fonseca MMR (2002) Maintenance of cell viability in the biotransformation of (−)-carveol with whole cells of *Rhodococcus erythropolis*. J Mol Catal B Enzym 19–20:389–398
27. de Carvalho CCR, da Fonseca MMR (2002) Influence of reactor configuration on the production of carvone from carveol by whole cells of *Rhodococcus erythropolis* DCL14. J Mol Catal B: Enzym 19–20:377–387
28. Déziel E, Comeau Y, Villemur R (1999) Two-liquid-phase bioreactors for enhanced degradation of hydrophobic/toxic compounds. Biodegradation 10(3):219–233
29. Doig SD, Boam AT, Leak DI, Livingston AG, Stuckey DC (1998) A membrane bioreactor for biotransformations of hydrophobic molecules. Biotechnol Bioeng 58(6):587–594
30. Doig SD, Boam AT, Livingston AG, Stuckey DC (1999) Mass transfer of hydrophobic solutes in solvent swollen silicone rubber membranes. J Membr Sci 154(1):127–140
31. Dunlop M (2011) Engineering microbes for tolerance to next-generation biofuels. Biotechnol Biofuels 4(1):32
32. Dunlop MJ, Dossani ZY, Szmidt HL, Chu HC, Lee TS, Keasling JD, Hadi MZ, Mukhopadhyay A (2011) Engineering microbial biofuel tolerance and export using efflux pumps. Mol Sys Biol 7(1):n/a–n/a

33. Fichan I, Larroche C, Gros JB (1999) Water solubility, vapor pressure, and activity coefficients of terpenes and terpenoids. J Chem Eng Data 44(1):56–62
34. Fontanille P, Larroche C (2003) Optimization of isonovalal production from α-pinene oxide using permeabilized cells of Pseudomonas rhodesiae CIP 107491. Appl Microbiol Biotechnol 60(5):534–540
35. Freeman A, Woodley JM, Lilly MD (1993) In situ product removal as a tool for bioprocessing. Biotechnology (N Y) 11(9):1007–1012
36. Gershenzon J, Dudareva N (2007) The function of terpene natural products in the natural world. Nat Chem Biol 3(7):408–414
37. Girhard M, Machida K, Itoh M, Schmid RD, Arisawa A, Urlacher VB (2009) Regioselective biooxidation of (+)-valencene by recombinant E. coli expressing CYP109B1 from *Bacillus subtilis* in a two-liquid-phase system. Microb Cell Fact 8:36
38. Halan B, Buehler K, Schmid A (2012) Biofilms as living catalysts in continuous chemical syntheses. Trends Biotechnol 30(9):453–465
39. Heeres AS, Picone CSF, van der Wielen LAM, Cunha RL, Cuellar MC (2014) Microbial advanced biofuels production: overcoming emulsification challenges for large-scale operation. Trends Biotechnol 32(4):221–229
40. Heipieper H, Neumann G, Cornelissen S, Meinhardt F (2007) Solvent-tolerant bacteria for biotransformations in two-phase fermentation systems. Appl Microbiol Biotechnol 74 (5):961–973
41. Isken S, Bont de JAM (1998) Bacteria tolerant to organic solvents. Extremophiles 2:229–238
42. Jahn M, Seifert J, von Bergen M, Schmid A, Bühler B, Müller S (2013) Subpopulation-proteomics in prokaryotic populations. Curr Opin Biotechnol 24(1):79–87
43. Kampranis SC, Makris AM (2012) Developing a yeast cell factory for the production of terpenoids. Comput Struct Biotechnol J 3:e201210006
44. Kaspera R, Krings U, Pescheck M, Sell D, Schrader J, Berger RG (2005) Regio- and stereoselective fungal oxyfunctionalisation of limonenes. Z Naturforsch C 60(5–6):459–466
45. Krings U, Berger R (2008) In situ recovery of the aroma compound perillene from stirred-tank cultured *Pleurotus ostreatus* using gas stripping and adsorption on polystyrene. Biotechnol Lett 30(8):1347–1351
46. Krings U, Lehnert N, Fraatz MA, Hardebusch B, Zorn H, Berger RG (2009) Autoxidation versus Biotransformation of α-Pinene to Flavors with *Pleurotus sapidus*: Regioselective Hydroperoxidation of α-Pinene and Stereoselective Dehydrogenation of Verbenol. J Agric Food Chem 57(21):9944–9950
47. Laane C, Boeren S, Vos K, Veeger C (1987) Rules for optimization of biocatalysis in organic solvents. Biotechnol Bioeng 30(1):81–87
48. Linares D, Fontanille P, Larroche C (2009) Exploration of α-pinene degradation pathway of *Pseudomonas rhodesiae* CIP 107491. Application to novalic acid production in a bioreactor. Food Res Int 42(4):461–469
49. Lye GJ, Woodley JM (1999) Application of in situ product-removal techniques to biocatalytic processes. Trends Biotechnol 17(10):395–402
50. Lye GJ, Woodley JM (2001) Advances in the selection and design of two-liquid phase biocatalytic reactors. In: Mota M, Tramper J (eds) Cabral JMS. Taylor and Francis, London, pp 115–134
51. Malik S, Cusidó RM, Mirjalili MH, Moyano E, Palazón J, Bonfill M (2011) Production of the anticancer drug taxol in *Taxus baccata* suspension cultures: A review. Process Biochem 46(1):23–34
52. Malinowski JJ (2001) Two-phase partitioning bioreactors in fermentation technology. Biotechnol Adv 19(7):525–538
53. Mi J, Becher D, Lubuta P, Dany S, Tusch K, Schewe H, Buchhaupt M, Schrader J (2014) De novo production of the monoterpenoid geranic acid by metabolically engineered *Pseudomonas putida*. Microb Cell Fact 13(1):170
54. Mirata MA, Heerd D, Schrader J (2009) Integrated bioprocess for the oxidation of limonene to perillic acid with *Pseudomonas putida* DSM 12264. Process Biochem 44(7):764–771

55. Molina G, Pimentel M, Pastore G (2013) Pseudomonas: a promising biocatalyst for the bioconversion of terpenes. Appl Microbiol Biotechnol 97(5):1851–1864
56. Morrish JL, Brennan ET, Dry HC, Daugulis AJ (2008) Enhanced bioproduction of carvone in a two-liquid-phase partitioning bioreactor with a highly hydrophobic biocatalyst. Biotechnol Bioeng 101(4):768–775
57. Morrish JL, Daugulis AJ (2008) Improved reactor performance and operability in the biotransformation of carveol to carvone using a solid-liquid two-phase partitioning bioreactor. Biotechnol Bioeng 101(5):946–956
58. Müller S, Harms H, Bley T (2010) Origin and analysis of microbial population heterogeneity in bioprocesses. Curr Opin Biotechnol 21(1):100–113
59. Nacke C, Hüttmann S, Etschmann MW, Schrader J (2012) Enzymatic production and in situ separation of natural β-ionone from β-carotene. J Ind Microbiol Biotechnol 39(12):1771–1778
60. Newman JD, Marshall J, Chang M, Nowroozi F, Paradise E, Pitera D, Newman KL, Keasling JD (2006) High-level production of amorpha-4,11-diene in a two-phase partitioning bioreactor of metabolically engineered *Escherichia coli*. Biotechnol Bioeng 95(4):684–691
61. Oda S, Inada Y, Kato A, Matsudomi N, Ohta H (1995) Production of (S)-citronellic acid and (R)-citronellol with an interface bioreactor. J Ferment Bioeng 80(6):559–564
62. Oda S, Sugai T, Ohta H (1999) Optical resolution of racemic citronellol via a double coupling system in an interface bioreactor. J Biosci Bioeng 87(4):473–480
63. Oda S, Sugai T, Ohta H (2001) Interface bioreactor. In: Vulfson E, Halling P, Holland H (eds) Enzymes in nonaqueous solvents. Humana Press, New York, p 401–416
64. Onken J, Berger RG (1999) Biotransformation of citronellol by the basidiomycete *Cystoderma carcharias* in an aerated-membrane bioreactor. Appl Microbiol Biotechnol 51 (2):158–163
65. Paddon CJ, Westfall PJ, Pitera DJ, Benjamin K, Fisher K, McPhee D, Leavell MD, Tai A, Main A, Eng D et al (2013) High-level semi-synthetic production of the potent antimalarial artemisinin. Nature 496(7446):528–532
66. Peralta-Yahya PP, Keasling JD (2010) Advanced biofuel production in microbes. Biotechnol J 5(2):147–162
67. Peralta-Yahya PP, Ouellet M, Chan R, Mukhopadhyay A, Keasling JD, Lee TS (2011) Identification and microbial production of a terpene-based advanced biofuel. Nat Commun 2:483
68. Pollak FC, Berger RG (1996) Geosmin and related volatiles in bioreactor-cultured *Streptomyces citreus* CBS 109.60. Appl Environ Microbiol 62(4):1295–1299
69. Pursell MR, Mendes-Tatsis MA, Stuckey DC (2004) Effect of fermentation broth and biosurfactants on mass transfer during liquid–liquid extraction. Biotechnol Bioeng 85 (2):155–165
70. Ramos JL, Duque E, Gallegos M-T, Godoy P, Ramos-González MI, Rojas A, Terán W, Segura A (2002) Mechanisms of solvent tolerance in gram-negative bacteria. Ann Rev Microbiol 56(1):743–768
71. Ramos JL, Duque E, Rodriguez-Herva JJ, Godoy P, Haidour A, Reyes F, Fernandez-Barrero A (1997) Mechanisms for solvent tolerance in bacteria. The J biol chem 272(7):3887–3890
72. Rehmann L, Daugulis AJ (2007) Biodegradation of biphenyl in a solid–liquid two-phase partitioning bioreactor. Biochem Eng J 36(3):195–201
73. Rehmann L, Sun B, Daugulis AJ (2007) Polymer selection for biphenyl degradation in a solid-liquid two-phase partitioning bioreactor. Biotechnol Prog 23(4):814–819
74. Renninger NS, Newman J, Reiling KK, Regentin R, Paddon CJ, Amyris Biotechnologies, Inc., assignee (2010). Production of isoprenoids. U.S.A. patent US 7659097 B2
75. Sardessai Y, Bhosle S (2002) Tolerance of bacteria to organic solvents. Res Microbiol 153 (5):263–268
76. Savithiry N, Cheong T, Oriel P (1997) Production of α-Terpineol from escherichia coli cells expressing thermostable limonene hydratase. In: Davison B, Wyman C, Finkelstein M (eds) Biotechnology for fuels and chemicals. Humana Press, New York, p 213–220

77. Scalcinati G, Knuf C, Partow S, Chen Y, Maury J, Schalk M, Daviet L, Nielsen J, Siewers V (2012) Dynamic control of gene expression in *Saccharomyces cerevisiae* engineered for the production of plant sesquitepene α-santalene in a fed-batch mode. Metab Eng 14(2):91–103
78. Scalcinati G, Partow S, Siewers V, Schalk M, Daviet L, Nielsen J (2012) Combined metabolic engineering of precursor and co-factor supply to increase alpha-santalene production by *Saccharomyces cerevisiae*. Microb Cell Fact 11:117
79. Schalk M, Pastore L, Mirata MA, Khim S, Schouwey M, Deguerry F, Pineda V, Rocci L, Daviet L (2012) Toward a biosynthetic route to sclareol and amber odorants. J Am Chem Soc 134(46):18900–18903
80. Schewe H, Holtmann D, Schrader J (2009) P450(BM-3)-catalyzed whole-cell biotransformation of a-pinene with recombinant *Escherichia coli* in an aqueous-organic two-phase system. Appl Microbiol Biotechnol 83(5):849–857
81. Schewe H, Kaup BA, Schrader J (2008) Improvement of P450(BM-3) whole-cell biocatalysis by integrating heterologous cofactor regeneration combining glucose facilitator and dehydrogenase in *E. coli*. Appl microbiol biotechnol 78(1):55–65
82. Schmid A, Kollmer A, Witholt B (1998) Effects of biosurfactant and emulsification on two-liquid phase pseudomonas oleovorans cultures and cell-free emulsions containing n-Decane. Enzyme Microb Technol 22(6):487–493
83. Schrader J (2007) Microbial flavour production. In: Berger RG (ed) Flavours and fragrances. Springer, Berlin, pp 507–574
84. Schügerl K, Hubbuch J (2005) Integrated bioprocesses. Curr Opin Microbiol 8(3):294–300
85. See-Toh YH, Ferreira FC, Livingston AG (2007) The influence of membrane formation parameters on the functional performance of organic solvent nanofiltration membranes. J Membr Sci 299(1–2):236–250
86. Segura A, Molina L, Fillet S, Krell T, Bernal P, Muñoz-Rojas J, Ramos J-L (2012) Solvent tolerance in Gram-negative bacteria. Curr Opin Biotechnol 23(3):415–421
87. Sikkema J, de Bont JA, Poolman B (1994) Interactions of cyclic hydrocarbons with biological membranes. J Biol Chem 269(11):8022–8028
88. Sikkema J, de Bont JA, Poolman B (1995) Mechanisms of membrane toxicity of hydrocarbons. Microbiol Rev 59(2):201–222
89. Stark D, von Stockar U (2003) In situ product removal (ISPR) in whole cell biotechnology during the last twenty years. Adv Biochem Eng/Biotechnol 80:149–175
90. Summers DK, Sherratt DJ (1984) Multimerization of high copy number plasmids causes instability: cole 1 encodes a determinant essential for plasmid monomerization and stability. Cell 36(4):1097–1103
91. Takors R (2004) Ganzzell-ISPR-Prozessentwicklung: chancen und risiken. Chemie Ingenieur Technik 76(12):1857–1864
92. Tan Q, Day DF (1998) Bioconversion of limonene to α-terpineol by immobilized *Penicillium digitatum*. Appl Microbiol Biotechnol 49(1):96–101
93. Thomas CM, Nielsen KM (2005) Mechanisms of, and barriers to, horizontal gene transfer between bacteria. Nat Rev Micro 3(9):711–721
94. Tsuruta H, Paddon CJ, Eng D, Lenihan JR, Horning T, Anthony LC, Regentin R, Keasling JD, Renninger NS, Newman JD (2009) High-level production of amorpha-4,11-diene, a precursor of the antimalarial agent artemisinin, in *Escherichia coli*. PLoS One 4(2):e4489
95. Vaidya AM, Halling PJ, Bell G (1994) Surfactant-induced breakthrough effects during the operation of two-phase biocatalytic membrane reactors. Biotechnol Bioeng 44(6):765–771
96. Valadez-Blanco R, Ferreira FC, Jorge RF, Livingston AG (2008) A membrane bioreactor for biotransformations of hydrophobic molecules using organic solvent nanofiltration (OSN) membranes. J Membr Sci 317(1–2):50–64
97. van Beilen JB, Holtackers R, Lüscher D, Bauer U, Witholt B, Duetz WA (2005) Biocatalytic production of perillyl alcohol from limonene by using a novel *Mycobacterium* sp. Cytochrome P450 alkane hydroxylase expressed in *Pseudomonas putida*. Appl Environ Microbiol 71(4):1737–1744

98. Van Sonsbeek HM, Beeftink HH, Tramper J (1993) Two-liquid-phase bioreactors. Enzyme Microb Technol 15(9):722–729
99. Wang DIC, Ochoa A (1972) Measurements on the interfacial areas of hydrocarbon in yeast fermentations and relationships to specific growth rates. Biotechnol Bioeng 14(3):345–360
100. Wang Y, Lim L, DiGuistini S, Robertson G, Bohlmann J, Breuil C (2013) A specialized ABC efflux transporter GcABC-G1 confers monoterpene resistance to Grosmannia clavigera, a bark beetle-associated fungal pathogen of pine trees. New Phytol 197(3):886–898
101. Wery J, Hidayat B, Kieboom J, de Bont JAM (2001) An insertion sequence prepares *Pseudomonas putida* S12 for severe solvent stress. J Biol Chem 276(8):5700–5706
102. Westfall PJ, Gardner TS (2011) Industrial fermentation of renewable diesel fuels. Curr Opin Biotechnol 22(3):344–350
103. Westfall PJ, Pitera DJ, Lenihan JR, Eng D, Woolard FX, Regentin R, Horning T, Tsuruta H, Melis DJ, Owens A et al (2012) Production of amorphadiene in yeast, and its conversion to dihydroartemisinic acid, precursor to the antimalarial agent artemisinin. PNAS 109(3):E111–E118
104. Westgate S, Vaidya AM, Bell G, Halling PJ (1998) High specific activity of whole cells in an aqueous-organic two-phase membrane bioreactor. Enzyme Microb Technol 22(7):575–577
105. Whited GM, Feher FJ, Benko DA, Cervin MA, Chotani GK, McAuliffe JC, LaDuca RJ, Ben-Shoshan EA, Sanford KJ (2010) Development of a gas-phase bioprocess for isoprene-monomer production using metabolic pathway engineering. Ind Biotechnol 6(3):152–163
106. Willemsen JHA, Dijkink BH, Togtema A (2004) Organophilic pervaporation for aroma isolation—industrial and commercial prospects. Membr Technol 2004(2):5–10
107. Willrodt C, David C, Cornelissen S, Bühler B, Julsing MK, Schmid A (2014) Engineering the productivity of recombinant *Escherichia coli* for limonene formation from glycerol in minimal media. Biotechnol J. doi:10.1002/biot.201400023
108. Woodley JM, Bisschops M, Straathof AJJ, Ottens M (2008) Future directions for in situ product removal (ISPR). J Chem Technol Biotechnol 83(2):121–123
109. Wriessnegger T, Augustin P, Engleder M, Leitner E, Müller M, Kaluzna I, Schürmann M, Mink D, Zellnig G, Schwab H et al (2014) Production of the sesquiterpenoid (+)-nootkatone by metabolic engineering of *Pichia pastoris*. Metab Eng 24:18–29
110. Yang T, Stoopen G, Yalpani N, Vervoort J, de Vos R, Voster A, Verstappen FWA, Bouwmeester HJ, Jongsma MA (2011) Metabolic engineering of geranic acid in maize to achieve fungal resistance is compromised by novel glycosylation patterns. Metab Eng 13 (4):414–425
111. Zelder O, Hauer B (2000) Environmentally directed mutations and their impact on industrial biotransformation and fermentation processes. Curr Opin Microbiol 3(3):248–251

Part III
From Biosynthesis to Biotechnological Production of Selected Isoprenoids

Adv Biochem Eng Biotechnol (2015) 148: 289–318
DOI 10.1007/10_2014_303

Published Online: 28 January 2015

Isoprene

Claudia E. Vickers and Suriana Sabri

Abstract Isoprene is a volatile C_5 hydrocarbon. It is produced by a wide variety of organisms and has been shown to play a role in protection of plants under abiotic stress conditions. It also has many different uses as an industrial chemical: most notably as a precursor for synthetic rubbers, but also for production of elastomers, copolymers, adhesives, and specialised chemicals. Modifying and/or engineering isoprene production in plants has the potential to contribute to engineered stress resistance. Moreover, as petrochemical sources of isoprene increase in price and become more scarce, bioproduction routes through microbial processes are becoming more attractive. Here we examine biotechnological aspects of isoprene production and review the current state of the art for both microbial-based industrial bioprocesses and plant engineering.

Keywords Isoprene · Metabolic engineering · Abiotic stress · Rubber · Industrial biotechnology · Microbial fermentation · Synthetic biology

Abbreviations

Ac	Acetyl
DMAPP	Dimethylallyl pyrophosphate
DX	Deoxyxylulose
DXP	Deoxyxylulose 5-phosphate
DXR	Deoxyxylulose reductase
DXS	Deoxyxylulose synthase
FPP	Farnesyl pyrophosphate

C.E. Vickers (✉)
Australian Institute for Bioengineering and Nanotechnology, The University of Queensland, St Lucia, QLD 4072, Australia
e-mail: c.vickers@uq.edu.au

S. Sabri
Enzyme and Microbial Technology Research Center, Department of Microbiology, Faculty of Biotechnology and Biomolecular Sciences, Universiti Putra Malaysia, 43400 UPM Serdang, Selangor, Malaysia

GC	Gas chromotography
GA3P	Glyceraldehyde-3-phosphate
HDR	Hydroxymethylbutenyl diphosphate reductase
HDS	Hydroxymethylbutenyl diphosphate synthase
HMBPP	Hydroxymethylbutenyl diphosphate
HMGR	Hydroxymethylglutaryl-CoA reductase
MK	Mevalonate kinase
IDI	Isopentenyl diphosphate isomerase
IPP	Isopentenyl pyrophosphate
IspS	Isoprene synthase
MEP	Methylerythritol phosphate
MVA	Mevalonate
MVD	Diphosphomevalonate decarboxylase
PGL	Phosphogluconolactonase
PMK	Phosphomevalonate kinase
s	Second(s)
ROS	Reactive oxygen species
Tg	Teragram

Contents

1 Introduction

Williams [146] first identified the volatile C_5 hydrocarbon hemiterpene isoprene (2-methyl-1,3-butadiene) as a product of destructive distillation (thermal decomposition) of gutta percha (natural latex) and caoutchouc (natural rubber). It was recognised that a five-carbon ‘isoprene unit’ forms a building block for the vast array of different natural products in the terpene (isoprenoid) family; this is known as the ‘isoprene rule’ [102]. Synthesis of these products does not proceed via isoprene itself, but through the universal C_5 prenyl phosphate precursors, dimethylallyl pyrophosphate (DMAPP) and its isomer isopentenyl pyrophosphate (IPP). These are both produced by the mevalonate (MVA) and methylerythritol phosphate (MEP) isoprenoid pathways (Fig. 1). Isoprene was identified as a leaf emission in the mid-1900s [63, 95, 106, 107]. Subsequently, many isoprene-producing plants,

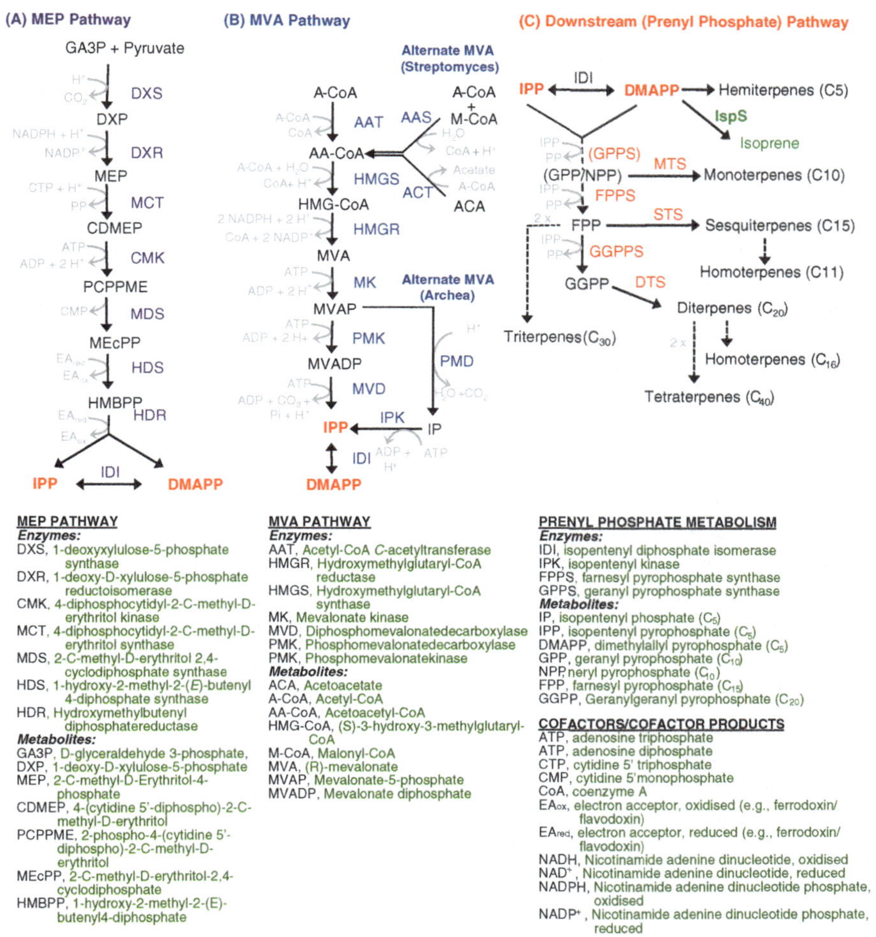

Fig. 1 Isoprenoid pathways **a** MEP pathway, found in plant chloroplasts, most bacteria, and some eukaryotic parasites [45, 98, 100, 154]. **b** Mevalonate (MVA) pathway, with various modifications, found in most eukaryotes, archaea, and some bacteria [41, 79, 90]. Both pathways produce the five-carbon intermediates isopentenyl pyrophosphate (IPP) and dimethylallyl pyrophosphate (DMAPP). **c** downstream prenyl phosphate metabolism continues from DMAPP and IPP, with the availability of prenyltransferase enzymes catalysing condensation of prenyl phosphate precursors determining which prenyl phosphates are available to the organism. Isoprene, a hemiterpene, is produced by the action of isoprene synthase (IspS) on DMAPP. Figure modified from [133, 135]

primarily in the C3 group, have been identified [60, 94]. Isoprene is produced in large amounts by phytoplankton in the oceans [84], and animals and bacteria can also emit isoprene [70, 114].

Isoprene produced by plants contributes an enormous amount of carbon—some 440–660 Tg (million tons) per year—to the atmosphere [53]. This isoprene has a significant impact on atmospheric chemistry. It is highly reactive and affects the oxidative capacity of the troposphere [92, 129]. Models indicate that it extends the residence time of greenhouse gases [113], thereby having a climate forcing effect. Isoprene also contributes to local pollution: it reacts with free radicals and ozone, the products of which contribute to formation of secondary organic aerosol particles, ozone, and carbon monoxide; and in the presence of high levels of nitric oxides (typically from industrial activities) it also reacts to produce tropospheric ozone [35, 39, 132]. These pollutants can have severe negative effects both on human health and on plant/animal agriculture. Moreover, because isoprene production is a function of temperature and light, it plays a role in biosphere–climate chemistry feedback [113]. The biological role of isoprene production in plants appears to be the protection of photosynthetic and other processes under stress conditions [80, 118, 136].

Williams [146] observed that the liquid isoprene he distilled reacted with atmospheric ozone over time to become ozonised, and that on distilling the ozonised liquid, a 'violent reaction' took place, whereby all unaltered hydrocarbon volatilised and the remaining product solidified suddenly to produce a 'pure, white, amorphous mass', which was unique and had the chemical formulation $C_{10}H_8O$, the first observation of (presumably) naphthol. Isoprene polymerises with itself and various different partners to form polymers and copolymers with many different properties. It can therefore be used to make a wide variety of useful products. These include synthetic rubbers, elastomers, thermoplastic elastomer block copolymers (e.g., styrene–isoprene–styrene, SIS), adhesives, and specialized chemicals (e.g., vitamins, pharmaceuticals, flavourings and perfumes, and epoxy hardeners) [16, 25, 105, 126, 142]. The primary isoprene product is synthetic rubber (*cis*-1,4-polyisoprene); this is used to make vehicle tires, surgical gloves, golf balls, and adhesives among others. Approximately 1 million tons of petrochemical-derived isoprene are made per annum [142]. Market growth per annum is estimated at 1–2 %, and there is a potential market of 5 billion kg/year through product expansion [83].

Industrial isoprene is currently produced mainly from petroleum as a by-product of thermal cracking of naphtha or gas oil [105, 142]. This petroleum-derived isoprene is available in limited supply, occasionally precipitating global shortages; it therefore suffers from price volatility due to fluctuating petrochemical prices. In addition, it has a high carbon footprint [126, 142]. The production process is relatively expensive and energy-intensive, and yields may be insufficient for future demand [16]. As fossil resources become more expensive to extract and purify, the price of isoprene (and other petrochemical by-products) will also increase. A cost-effective and renewable alternative source of isoprene is therefore attractive.

Although some plants produce enormous amounts of isoprene (Hewitt et al. 2004), harvesting it from this source is not feasible. However, microbes which normally cannot produce isoprene can be engineered to produce it at significant levels. Production of biochemicals in microbial systems as alternatives to petrochemical routes is a rapidly growing industry, and microbial bioprocesses can be significantly less expensive and more straightforward compared with other methods [125]. The California company Genencor (now owned by DuPont), in collaboration with The Goodyear Tire & Rubber Company, has pioneered development of microbe-based isoprene bioprocesses [54, 142], and several other research groups/companies are also working in the field. Moreover, the biological role of isoprene lends itself to the possibility of engineering for improved stress resistance in crop plants. Herein, we examine the biotechnological aspects of isoprene and review the current state of the art for both microbial-based industrial bioprocesses and plant engineering.

2 Isoprene Synthase

The enzyme isoprene synthase (IspS; EC 4.2.3.27; systematic name dimethylallyl-diphosphate diphosphate-lyase [isoprene-forming]) is responsible for the prodigious global production of isoprene. Isoprene synthase catalyses the removal of pyrophosphate from DMAPP to produce isoprene and pyrophosphate (Fig. 2). Mg^{2+} or Mn^{2+} are required as cofactors, with Mg^{2+} being the preferred cofactor [120, 121, 144].

IspS genes and/or proteins have been characterised from several *Populus* sp. (poplar/aspen; [13, 49, 85, 109, 119, 121, 138, 139]), *Quercus ilex* (oak; [13, 73]), *Salix discolour* (pussy willow; [144, 145]), *Salix babylonica* (weeping willow; [57]), *Liquidambar styraciflua* (sweetgum; [57]), *Arachis hypogaea* (peanut; [13]), *Myrtus communis* (myrtle; [57]), *Pueraria montana/Pueraria lobata* (kudzu; [13, 57, 119]), and *Mucuna bracteata* [57]. Although many prokaryotes also produce isoprene [46, 70], a prokaryotic *IspS*/IspS has yet to be identified.

Isoprene synthase belongs to the Tps-b group of angiosperm isoprenoid synthases [52, 115, 119]. The genes have six introns and seven exons [49, 85, 109, 115, 119, 138, 139]. In poplar (and presumably also in other plants), isoprene synthase is found in three to four copies; these genes are differentially regulated [139], but it is thus far

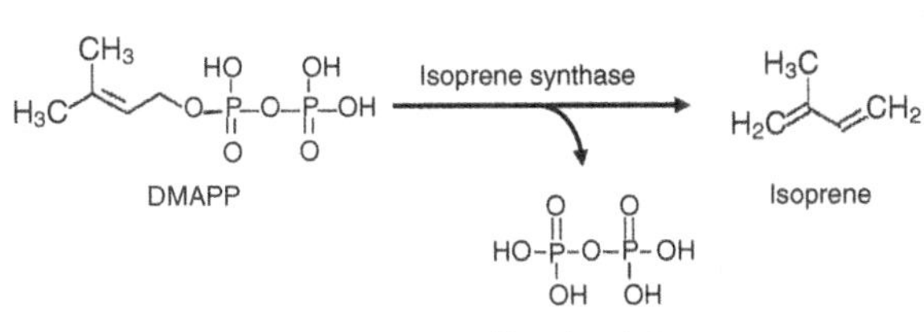

Fig. 2 Synthesis of isoprene from DMAPP as catalysed by isoprene synthase

unclear what the biological significance of this is. In in vitro analyses, isoprene synthase enzymes have a broad pH optimum (usually peaking at ~7–8) and temperature optima between 40–60 °C [47, 69, 73, 120]. They are relatively inefficient enzymes, with a high K_m (~2–20 mM) and variable but generally low k_{cat} values [13, 73, 111, 119, 121, 145, 151]. Substrate availability and thermal influence on enzyme kinetics have a strong controlling effect on isoprene emission [117]. IspS enzymes from different source organisms also have different susceptibilities to substrate-mediated inhibition [13]. The crystal structure of several IspS enzymes has been solved [14, 25, 68].

3 Engineering Isoprene Production in Microbes

Although plants produce large quantities of isoprene, approaches for engineering industrial production (for use as an industrial chemical) have focused on the use of microbial fermentation. Although plants represent an attractive starting point due to the naturally high emission levels of some species and the potential for direct conversion of carbon dioxide into isoprene via photosynthesis, bioprocess requirements are far cheaper and simpler in microorganisms. The challenges of capturing isoprene from a plant-emitting system would add significant cost and complication to the process. Microbial fermentation represents a sustainable alternative to produce isoprene directly from abundant and cost-effective renewable resources, for example, simple sugars sourced from plants, and ultimately, lignocellulosic feedstocks [134].

Isoprene production from bacteria was first identified by Kuzma et al. [70]. Of the bacteria surveyed, *Bacillus subtilis* was the highest producer [46, 70]. However, isoprene production in *B. subtilis* is only 50 μg/L/h^{-1} [64], insufficient for commercial processes. Common industrial microorganisms such as *Escherichia coli* and *Saccharomyces cerevisiae* (yeast) do not naturally produce isoprene. In order to achieve sufficient titres from microbial bioprocesses, substantial engineering is required, starting with introducing an appropriately engineered isoprene synthase gene. Due to the low native flux through isoprenoid pathways in most microbes, it is also necessary to engineer upstream pathways to provide sufficient precursors for industrial-scale isoprenoid production. This includes engineering central carbon metabolism to improve precursor availability or balance subcellular precursor concentrations for isoprenoid pathways, engineering through the core isoprenoid pathways themselves (MEP/MVA), and engineering availability of prenyl phosphate precursors (including controlling competition from downstream prenyltransferases). In the case of isoprene, the synthase uses DMAPP so the latter engineering steps are not required. With respect to engineering isoprene production in microbes, much of the experimental data is found in the patent literature, and we refer to patents frequently in the sections below.

3.1 Engineering Isoprene Synthase for Microbial Expression

Despite isoprene being widely produced by different microbes, and despite intensive search efforts [64, 123, 150], no microbial isoprene synthase has been reported to date. Indeed, only plant isoprene synthases have been discovered (see above). Consequently, it is necessary to transfer a plant isoprene synthase into microbes for engineering purposes. To optimise expression and activity, significant engineering is required prior to transfer of plant-encoded enzymes into microbes. Briefly, this involves (1) removal of intron sequences (i.e., use of a cDNA sequence), (2) removal of the chloroplast targeting sequence (if present), (3) codon optimisation of the sequence, and (4) selection of appropriate transcription and translation control sequences (see [135] for details). Further engineering, such as introduction of mutations that improve catalytic activity, thermostability, or other industrially useful traits, might also be performed.

Prior to any engineering, an appropriate *IspS* gene must be selected. Isoprene synthases from different source organisms vary significantly in their performance with respect to isoprene production in microbes [13, 25, 33, 57, 142].

Removal of introns (or selection of a cDNA sequence) and truncation to remove the chloroplast sequence are usually performed first (see below). Modifying isoprene synthase codon usage from plant to *E. coli* improves isoprene production in *E. coli* [25, 31]. Peptide tags, which are often included in cloning constructs and used for isolation of resulting proteins, can interfere with catalytic properties of isoprene synthase [25, 85] and should be avoided for industrial applications when designing constructs.

Isoprene synthase proteins include a chloroplast targeting peptide [85, 108, 119, 139, 140]; removal of this peptide increases isoprene production in microbes [25, 57, 85, 140]. As for other isoprenoid synthase genes [30, 147], selection of the truncation site for removal of the chloroplast leader sequence is critical in isoprene synthase [25]. Mapping of several truncation sites in a *Populus alba* isoprene synthase indicated that, of those tested, removal of 156 nucleotides (a 52-amino–acid deletion) with addition of an ATG (yielding a protein starting with MEA-RRS…) provided the highest levels of isoprene [25]. This truncation improved the k_{cat} 2-fold, the K_m 1.3-fold, the K_i 1.6-fold, and the specific activity 2-fold.

Somewhat surprisingly, decreasing the copy number using a low copy-number expression vector can improve isoprene production; isoprene synthase apparently forms inclusion bodies when overexpressed [31, 33]. Titrating expression from different promoters, albeit not yet directly examined, will most likely have combinatorial effects with copy number. This should be considered during construct design.

Although there is significant variation in catalytic performance, in general the poor catalytic properties of IspS enzymes (in particular the high $K_{m(DMAPP)}$ and low k_{cat} values) make the wild-type isoprene synthases unattractive as industrial enzymes. IspS proteins also exhibit substrate inhibition at varying levels for different enzymes [13, 14]. An *A. hypogaea* IspS is relatively resistant to substrate inhibition compared to a *P. alba* IspS (which is commonly used for engineering), as long as high concentrations of the cofactor Mg^{2+} (100 mM) are available. The *A. hypogaea* IspS produced more isoprene in in vivo studies than a *P. alba* IspS, but also has an in vitro K_m of almost an order of magnitude more than the *P. alba* IspS.

A variety of targeted and random approaches was carried out by Bott et al. [25] to identify IspS mutants with improved catalytic activities. A number of mutations that conferred improved activities were identified. Bott et al. [25] also solved the crystal structure of IspS using a number of IspS enzymes and constructs. The structures suggested that several crucial loops forming the active site of isoprene synthase are flexible, and the authors noted the potential to modify metal binding, diphosphate recognition, DMAPP chain binding, and active site; a large number of potentially useful mutations were suggested [25]. A thorough mutation analysis of many different residues was then performed using a truncated (MEARR…) *P. alba* IspS [14]. This revealed a number of mutations that improved activities when nonpolar residues were converted to polar residues, presumably improving solubility (which had been noted to be poor). Many other mutations that improved specific activity in vitro and/or isoprene production in vivo were also identified. A selection of high-performing mutations (numbering based on MEAA… truncation) includes: A453N in a 'flexible loop' of the substrate binding site; L494P/L494C, a hydrophobic residue in a surface-exposed loop, where mutations to hydrophilic residues are thought to affect protein folding, solubility, or activity; and T536I/T536F/T536Y (a 'miscellaneous' mutation that yielded significant increases in activity with several substitutions). Crystallisation confirmed a change in conformation of the loop containing residues 490–497 where the L494P mutation is situated; this mutant provided a twofold higher specific activity, a decreased K_m and an increased k_{cat} in in vivo studies. Both L494P and T536F proved to confer thermostability, with the latter being more effective. It was suggested that the thermostability of T536F was due to replacement of a polar amino acid residue with a large hydrophobic residue (thereby increasing local hydrophobicity). Multiple sequence analysis demonstrated that both proline at 494 and I/V/F (hydrophobic residues) are common in isoprene synthases/terpene synthases (respectively). These data probably provide some clues as to why different IspS proteins from different species perform significantly differently in terms of isoprene production in microbes.

Another mutagenesis/selection approach was later used to further improve solubility (and improve the in vivo activity) of IspS [97]. To identify improved mutants, challenges with increasing concentrations of DMAPP in both in vitro and in vivo systems were used. The in vivo method involved titration of flux through an engineered MVA pathway, thereby taking advantage of the toxicity of prenyl phosphate precursors (see below) to link improved catalytic properties with growth rate and allow facile selection of desirable mutants. An S288C mutation (numbering based on MEA… truncation) repeatedly emerged from different screens. Other mutations that conferred either improved catalytic properties (K_m, k_{cat}, V_{max}) or improved expression/solubility were also identified. Combining the S288C mutation with an A3T mutation provided a three- to fivefold enhancement.

3.2 Engineering Precursor Availability

One of the largest obstacles to efficient microbial biosynthesis of isoprene is the production of its precursor, DMAPP. In the absence of temperature variation, DMAPP availability is the primary driver for isoprene production in plants [117, 139, 140], and precursor availability in microbes is a key limiting factor for engineering isoprenoid production in microbes [135]. Thus, engineering of the upstream isoprenoid pathway (either MEP or MVA, depending on the production organism) is required. This kind of pathway engineering is necessary for production of any isoprenoid compound in industrial microbes, because isoprenoids share common metabolic precursors. Genetic platforms resulting from work with isoprene can therefore be applied to the biosynthesis of other valuable products (and vice versa).

A problem encountered in isoprenoid pathway engineering is that high levels of prenyl phosphate precursors (DMAPP/IPP/FPP) are apparently toxic to cells [74, 82, 122, 150]. Furthermore, it is known that several prenyl phosphates act as feedback regulators to control flux through isoprenoid pathways [11, 12, 50, 58]. In order to avoid toxicity/feedback from buildup of prenyl phosphates, it is necessary to have a metabolic sink for these precursors in place prior to engineering increased flux. Isoprene is an ideal sink, because it is produced directly from one of the C_5 precursors (DMAPP), thereby providing an immediate drain for the prenyl phosphate pool. The high $K_{m(DMAPP)}$ of IspS also means that it is unlikely to compete effectively with downstream prenyltransferases that use DMAPP as a substrate. This means that potential growth defects caused by depleting the pool of prenyl phosphates required for synthesis of essential isoprenoids (primarily FPP) are avoided. Moreover, being volatile (b.p. = 34 °C), it partitions into the gas phase under normal cultivation conditions and is unlikely to cause toxicity problems in itself.

Broad approaches for engineering improved carbon flux in microbes through both the MVA and the MEP pathways have been reviewed recently [135]; here we focus on published approaches specific to isoprene production. Readers should note that additional approaches used for other isoprenoid products (in particular, for controlling flux upstream of the DMAPP/IPP node and for controlling downstream competition) will most likely also be effective for isoprene production.

3.3 Engineering Isoprene Production in E. coli

E. coli was the first microorganism to be engineered for isoprene production, and the best results by far have been achieved in this organism. The first recombinant isoprene-producing *E. coli* was generated as part of a study to identify the first *IspS* gene [85]. A hybrid poplar (*P. alba* × *P. tremula*, aka *P.* × *canescens*) isoprene synthase cDNA was used; truncation to remove the putative chloroplast targeting peptide resulted in a 1.85-fold increase in isoprene production. However, as later studies also found, simply introducing an isoprene synthase into *E. coli* results in only very low yields; in this case, only 1.85 nmol isoprene/mg cellular protein.

It should be noted that the choice of strain has a significant effect on initial production of isoprenoids in both *E. coli* and yeast [24, 34, 99, 127]. Strain also affects isoprene production in *E. coli*; for example, *E. coli* BL21(λDE3) produces 10 times higher titres than *E. coli* FM5, and *E. coli* ATCC11303 (an *E. coli* B derivative) performs even better than BL21(λDE3) in fed-batch culture [31]. We have also observed significant interstrain differences, both in isoprene production capability upon introduction of an isoprene synthase and in 'engineerability' (the effect of upstream pathway engineering) in our laboratory (Bongers, Behrendorff, Nielsen and Vickers, unpublished data).

Initial attempts at improving isoprene production focused on engineering through the native MEP pathway. Introduction of an heterologous MVA pathway significantly improves titres. Both pathways can be used in parallel for production, or engineering can be focused on the MVA pathway, which to date has delivered the best yields [17, 37] as discussed below.

3.3.1 Engineering Core MEP Pathway Flux

E. coli uses a MEP pathway for native isoprenoid production. Condensation of the central carbon intermediates pyruvate and glyceraldehyde-3-phosphate (GA3P) by deoxyxylulose synthase (DXS; see Fig. 1) to generate deoxyxylulose 5-phosphate (DXP) represents the primary rate-limiting step in the MEP pathway [3, 55]. Coexpression of an homologous or heterologous deoxyxylulose synthase (*dxs*; see Fig. 1) provides a two- to fourfold increase in isoprene [31, 57, 85].

Conversion of DXP to 2-C-methyl-D-erythritol 4-phosphate (MEP) by deoxylulose reductase (DXR) can also be rate-limiting [3]. A titre of 94 mg/L isoprene with a rate of 2.8 mg/L/h^{-1} was achieved by Zhao et al. [155] in *E. coli* fed-batch cultures when *Populus nigra* (black poplar) *IspS* was overexpressed together with the *E. coli dxs* and *dxr* genes. When heterologous *B. subtilis dxs* and *dxr* were used for overexpression, isoprene production was further enhanced 3.3-fold to 314 mg/L. This suggests that use of heterologous enzymes can provide more efficient catalysis, either from improved native catalytic properties or avoidance of native posttranslational regulatory control. Significant differences between the primary sequences of the homologous and heterologous genes were observed, supporting these possibilities.

In addition to *dxs* and *dxr*, overexpression of isopentenyl diphosphate isomerase (*idi*), which converts IPP to DMAPP and vice versa, improves isoprene production [25, 31, 33]. This presumably allows rapid balancing of the intracellular IPP and DMAPP pools, and may be particularly important given the heavy relative drain on the DMAPP pool exerted by IspS. In the absence of release at the DXS bottleneck by overexpression of *dxs*, overexpression of *idi* has minimal effect [33]. Coexpression of *dxs* and *idi* with engineered kudzu isoprene synthase provides a five- to 12-fold increase in isoprene compared to the IspS alone, allowing titres of up to 300 mg/L broth using *E. coli* strain BL21 [31, 33]. Strains overexpressing *E. coli dxs* and *idi* along with a *P. alba IspS* produced 35–70 mg/L isoprene on rich media in batch culture [76, 93].

The last two steps of the pathway are catalysed by hydroxymethylbutenyl diphosphate synthase (HDS/IspG/GcpE) and hydroxymethylbutenyl diphosphate reductase (HDR/IspH/LytB; see Fig. 1), both of which are $[4Fe\text{–}4S]^{2+}$ (iron–sulfur cluster) enzymes [112, 152]. Following upstream release at the DXS node and overexpression of IDI to balance the C_5 prenyl phosphate pools, flux through these two enzymes becomes rate-limiting. Under these conditions, 2-C-methyl-D-erythritol-2,4-cyclodiphosphate (MEcPP, aka cMEPP), the substrate for IspG (GcpE/HDS), accumulates in the cell and is even exported into the medium [156]. It was also observed that a strain expressing kudzu *IspS*, yeast *idi*, and *E. coli dxs* produced more isoprene under noninducing conditions than under inducing conditions, and that the strain also accumulated (*E*)-4-hydroxy-3-methylbut-2-enyl diphosphate (HMBPP), the substrate for HDR [37]. This suggested that HDR (IspH) activity was limiting.

Overexpression may help in some cases; for example, inclusion of an HDS overexpression, which has been shown to increase isoprenoid production when other bottlenecks are released [156], can further improve isoprene titres twofold in a strain overexpressing *dxs* and *idi* [76]. However, the complex cofactor requirements and biochemistry of the enzymes themselves must also be addressed.

The 4Fe–4S cluster is very sensitive to oxidative damage. Oxidative stress causes accumulation of MEcPP in mycobacteria and corynebacteria, most likely due to the interference at the 4Fe–4S group [10]. In *E. coli*, the iron–sulfur cluster (isc) operon enzymes are involved in formation of the 4Fe–4S cluster in Fe–S enzymes [131]. The *isc* enzymes also assist in repair of the 4Fe–4S clusters after oxidative damage [43]. In the presence of appropriate concentrations of flavodoxin and flavodoxin reductase, HDR activity in extracts from *E. coli* overexpressing both *IspH* and the *isc* operon increased 200-fold in vitro [51]. Removal of the *isc* operon repressor (*iscR*) was successfully used to increase specific productivity of isoprene by up to threefold in vivo (though a reduction in growth rate was also observed [37].

Another approach to improve activities of 4Fe–4S enzymes is to improve electron transfer to support redox reactions. Accordingly, overexpression of the *E. coli* electron transferase flavodoxin I (*fldA*) increased isoprene production one- to twofold. Increased expression of IspS did not improve the isoprene titre, suggesting that in this case, IspS activity was not limiting.

After overexpression of *fldA*, the HDS and HDR nodes were further targeted by overexpression of *Thermosynechococcus elongates IspG* (*GcpE*) and *IspH* (*LytB*; [37]). *T. elongates* does not encode an *fldA*; instead it apparently uses a reducing shuttle system consisting of ferredoxin (*petF*) and ferredoxin-NADP(+) oxidoreductase (*petH*; [89]). Expressing *GcpE*, *petF*, and *petH* in strains with a *P. alba* IspS (MEARR… truncation), yeast *idi* and variable levels of *dxs* overexpression increased isoprene production by 10–80 %, with a stronger response observed with increasing promoter *dxr* strength. Both the *dxs*-only and the *dxs/GcpE/petF/petH* strains exhibited accumulation of MEcPP. However, surprisingly, MEcPP was just as high —and in some cases, higher—in the *GcpE/petF/petH* strains than in the strains without *GcpE* overexpressed. Despite this, the HDR (GcpE/IspH) product HMBPP was observed to accumulate in *GcpE* strains, indicating that the overexpression successfully relieved at least a proportion of the flux constraint at that node. Its accumulation suggested that HDS (IspG) activity was still limiting. Heterologous overexpression of a second *IspG* from a different species and an *IspH* gene from *Anabaena* further improved titres [88]. A strain with several modifications, including overexpressions of *dxs*, *dxr*, *idi*, *fldA-ispG*, *Anabaena IspH-petF-petH*, a lower MVA pathway, a truncated *P. alba IspS*, and a restored chromosomal *pgl* (see the following section) provided titres of 8.4 g/L isoprene in a fed-batch fermentation [88].

3.3.2 Balancing MEP Pathway Precursors

Balancing of GA3P and pyruvate pools to prevent accumulation of either precursor has been shown previously to be important for isoprenoid production via the MEP pathway [4, 5, 48]. Two studies have demonstrated, unsurprisingly, that this is also important for isoprene production. The first examined different pathways for glucose utilisation, namely the Embden–Meyerhof pathway (EMP), Entner–Doudoroff pathway (EDP), pentose phosphate pathway (PPP), and Dahms pathway [76]. Carbon redirection to specific pathways was achieved through various strategies: glucose-6-phosphate isomerase (*pgi*) was disrupted to divert carbon away from the preferred EMP and towards PPP (in this case flux also occurs to a lesser extent via the EDP); a double *pgi* and gluconate-6-phosphate dehydrogenase (*gnd*) knock-out blocks both EMP and PPP, channelling flux towards the EDP; feeding D-xylulose restricts glycolytic carbon flux to the PPP; and finally, carbon from xylose was channelled into a Dahms pathway constructed by introducing a xylonate catabolic pathway, knocking out the PPP by disrupting *xylA*, and introducing D-xylose dehydrogenase to convert D-xylose to D-xylonate. The EDP, which produces equimolar GA3P: pyruvate, provided the highest titres; combining this with an engineered MEP pathway (overexpression of *dxs*, *idi*, and *IspG*) and a modified *P. alba IspS* provided titres of ~225 mg/L isoprene (about threefold higher than the engineered MEP pathway alone).

The second study utilised a novel approach to balance precursor levels on galactose by forcing galactose utilisation via the De Ley–Doudoroff pathway [93]. This involved blocking the Leloir pathway for galactose utilisation and engineering a complete De Ley–Doudoroff pathway by introducing a missing dehydrogenase gene from *Pseudomonas syringae* to catalyse conversion of galactose to galactonase. Equimolar ratios of GA3P and pyruvate were achieved in the resulting strain. Coupling this pathway with overexpression of *E. coli dxs* + *idi* and a *P. alba IspS* resulted in a 3.7-fold improvement over the MEP pathway engineering alone. The final titre was 260 mg/L isoprene, very similar to the first study. This result confirmed the importance of balancing precursor concentrations; however, it would not be feasible in an industrial setting due to the cost of galactose.

Other approaches for balancing GA3P and pyruvate that might increase isoprene production include overexpressing glucose-6-phosphate dehydrogenase (G6PDH, encoded by *zwf*) alone or in combination with the modifications described above, that is, limiting expression of *pgi* to redirect flux to PPP and EDP, and/or limiting expression of *gnd* to limit flux to the PPP and increase flux to the EDP [37].

3.3.3 Addressing Remaining MEP Flux Limitations

The fact that the MEP pathway delivers higher theoretical yields for isoprenoids from simple sugars than the MVA pathway [101, 133] has driven extensive research efforts towards improving MEP pathway flux. However, despite the significant engineering that has thus far been performed, the MEP pathway still delivers lower yields/titres than heterologously expressed MVA pathways (see below).

Flux is still clearly limited at the 4Fe–4S enzymes, and resolving the problems at these nodes will take further research. In addition, the MEP pathway relies heavily on reducing equivalents (required at both the 4Fe–4S nodes and at DXR), and it was anticipated early that increasing their availability would improve MEP pathway flux [4]. Ultimately, as flux bottlenecks are relieved, improving carbon channelling into MEP pathway precursors (in addition to keeping central carbon intermediates balanced, as described above) will then become limiting.

Avoiding loss of carbon through side pathway reactions may also help. In *E. coli*, DXP (the product of DXS) is used as a precursor for the B-group vitamins thiamine and pyroxidine. Deletion/downregulation of thiamine and pyridoxine pathways to minimise carbon loss/competition/feedback from those pathways might also improve MEP pathway flux for isoprene production [37]. However, given that these vitamins are essential, and that thiamine (as thiamine diphosphate, TPP) is a cofactor for DXS, care should be taken with this approach.

Finally, it has recently been shown that both IPP and DMAPP compete with TPP at the TPP binding site in a *Populus trichocarpa* DXS, thereby causing competitive inhibition of DXS [12]. Presumably, this is also the case for other DXS enzymes. Moreover, given that TPP is a cofactor for many different enzymes, and that IPP/DMAPP have very similar binding mechanisms to TPP, increased DMAPP/IPP pools might cause significant metabolic perturbation (this may indeed be the mechanism by which IPP/DMAPP exert toxicity). This feedback inhibition and the potential side-effects underline the importance of ensuring that prenyl phosphate pools do not accumulate in engineered strains. Further improvement of the IspS enzyme catalytic properties, in parallel with flux modification to ensure that isoprene production is balanced with essential pathway requirements downstream with DMAPP, may be required.

3.3.4 Engineering the MVA Pathway in *E. coli*

Although substantial advances have been achieved by engineering through the native MEP pathway, the best yields to date are significantly below calculated theoretical maxima for the pathway [101, 137]. It appears that MEP carbon flux is heavily regulated, and as yet we do not understand enough about this regulation to overcome it [135]. In order to circumvent this problem, either a partial MVA pathway (with mevalonate supplementation; [32]) or a complete MVA pathway [82] have been introduced into *E. coli* strains engineered for isoprenoid production. This approach was also successful for improving isoprene production [15, 31, 33,

57, 142, 157]. An MVA pathway was constructed by using *mvaE* (which encodes a dual function enzyme that has both acetyl-CoA acetyltransferase activity and HMG-CoA reductase activity; see Fig. 1) and *mvaS* (mevalonate synthase) from *Enterococcus faecalis* (referred to generically as the 'upper' MVA pathway) combined with mevalonate kinase (MK), phosphomevalonate kinase (PMK), diphosphomevalonate decarboxylase (MVD), and IDI from *Saccharomyces cerevisiae* (referred to generically as the 'lower' MVA pathway). In the presence of an engineered kudzu isoprene synthase, this engineered MVA pathway provided a 3-fold increase in isoprene production compared to the engineered MEP pathway, and was capable of producing 20 g/L in 40-h fermentation [33]. A next-generation strain that included a feedback-resistant version of MK from *Methanosarcina mazei* and an engineered *P. alba* isoprene synthase produced 35 g/L isoprene in fed-batch culture (see further comments regarding bioprocess considerations below; [15, 33]). The isoprene yield from glucose was 5.2 %. This strain provides a 73-fold improvement over IspS alone; compared to an engineered MEP pathway, the MVA pathway provides a 10–20-fold improvement in isoprene production [31]. Judicious selection of MVA pathway genes, including screening genes from different source organisms or improved catalytic activity and selection of feedback-resistant versions of hydroxymethylglutaryl-CoA reductase (HMGR) and MK (the enzymes catalysing the primary and secondary rate-limiting steps, respectively) can also increase MVA-produced isoprene [13, 15, 31, 142]. Examining MVA production from the 'upper MVA pathway' (steps up to MVA synthesis) and titrating copy number have demonstrated that the upper pathway flux is quite good when using *E. faecalis* enzymes, however, flux is limited by steps downstream of MVA when it is integrated onto the chromosome [31]. Varying the source of the upper pathway genes can have a significant effect on isoprene production, however, the effect is dependent on the catalytic properties of the isoprene synthase, as different IspS enzymes exhibit different levels of substrate inhibition [13].

3.3.5 Improving MVA Pathway Precursor Availability

Modifications to the central carbon metabolism in MVA pathway engineered strains can also improve isoprene production. The MVA pathway initiates at the central carbon metabolite acetyl-CoA, and the first two steps involve condensation of three acetyl-CoA molecules to produce HMG-CoA (see Fig. 1b). There are multiple competing pathways that use acetyl-CoA, and decreasing this competition where possible increases production of desirable biochemicals. Moreover, decreasing other indirect carbon sinks and increasing availability of reducing equivalents can also improve flux through desired pathways.

Overexpressing phosphogluconolactonase (*pgl*) to improve pentose phosphate pathway flux (and possibly also suppress posttranslational glucosylation of heterologously expressed proteins) results in a two- to threefold increase in specific productivity of isoprene over the parent strain, which contained a *P. alba IspS,* engineered MVA pathway, and an *M. masei* MK [16, 33, 142]. Titres of 80 g/L were reported using a strain with a truncated, codon-optimised *P. alba* isoprene synthase engineered for improved catalytic properties in combination with an MVA pathway that included feedback-resistant enzymes, overexpression of *pgl*, and an optimised fed-batch process (see the section on bioprocess considerations below; [16]).

Further investigations into carbon and cofactor competition started with construction of a base strain with downregulated citrate synthase (*gltA*, a TCA-cycle enzyme that competes for acetyl-CoA; [17]). Combining this with knock-out of *pgl* and overexpressing phosphoketolase (*pkl*) to scavenge carbon from xylulose-5-phosphate into isoprenoid pathway precursors further increased isoprene titres [17]. This approach is thought to improve the balance or availability of pathway precursors, however, gene dose and the type of *pkl* are important. Combining overexpression of a plasmid-borne *Enterococcus gallinarum pkl* with other MVA pathway engineering steps and a truncated *P. alba IspS* provided titres of up to 123.6 g/L, overall volumetric productivities up to 2.21 g/L/h and 16.3–17.4 % yields of isoprene on glucose [17]. These are the highest isoprene titres thus far reported.

3.3.6 Controlling Downstream Prenyl Phosphate Feedback

As discussed above (see the section, 'Addressing Remaining MEP Pathway Flux Constraints') MEP pathway-specific feedback issues are particularly problematic for DMAPP/IPP. However, as also discussed above, prenyl phosphates have been shown to drive feedback regulation of pathway flux in both the MVA and MEP pathways. Because of the high $K_{m(DMAPP)}$ of IspS relative to downstream prenyl transferases, in a strain with increased upstream flux to IPP and DMAPP there is likely to be a buildup of downstream prenyl phosphates. This can in turn feed back to control pathway flux. Approaches to minimise prenyl phosphate accumulation by coexpression of monoterpene and sesquiterpene synthases with IspS have been suggested to improve isoprene accumulation [37] and are integrated in the highest-producing MVA-based strains [17]. Appropriate balancing of IspS activity with upstream flux and downstream requirements for essential pathway products will be required for both MEP and MVA engineering going into the future. Titration of these complex factors requires high-throughput systems to analyse isoprenoid products [20].

3.4 Novel Isoprene Production Pathways

An alternative route for conversion of DMAPP to isoprene, presumably to circumvent the poor catalytic properties of isoprene synthase, has been proposed [27]. 3-Methyl-2-buten-l-ol has a similar carbon backbone to isoprene, and can be biotransformed into isoprene by *E. coli* engineered with a strawberry (*Fragaria* sp.) acyl-CoA transferase (SAAT) and a *Castellaniella defragrans* strain 65Phen linalool dehydratase-isomerase (LDI; [27]). The SAAT isomerises 3-methyl-2-buten-l-ol to the aldehyde 2-methyl-3-buten-l-ol (prenal), and the LDI dehydrates

2-methyl-3-buten-l-ol into isoprene. Although not directly demonstrated, it was proposed that DMAPP can be converted into 3-methyl-2-buten-l-ol through the action of a *Pinus sabiniana* methylbutenol synthase.

Both the MEP and MVA pathways operate optimally under aerobic conditions due to their energetic requirements. Moreover, their theoretical maxima are sub-optimal compared to other pathways. To address these issues, novel theoretical isoprene production pathways that operate effectively under anaerobic conditions have recently been proposed [40]. Four pathways that proceed via 2,3-dihydroxyisovalerate (DHIV), an amino acid pathway intermediate which has an isoprene carbon skeleton and is produced from pyruvate via just two enzymatic steps, are described. Each pathway involves a series of reduction and dehydration reactions to produce isoprene and would require introduction of five to seven additional enzymes (many of which are theoretical activities). All four pathways are redox balanced (unlike the MVA pathway); one is ATP neutral and three of them are net ATP-positive (unlike the MEP pathway, which has a net cost of 3 ATP). Although the work is currently theoretical, it offers an exciting potential approach, especially given that ATP-producing pathways can be coupled to growth rate through increased pathway flux, thereby allowing the option of applying adaptive evolution for pathway improvement.

3.5 Engineering Isoprene Production in Other Organisms

Outside of *E. coli*, a handful of other microorganisms have been engineered for isoprene production, but the results have been somewhat disappointing in terms of titres/yields.

S. cerevisiae, which uses the MVA pathway, was engineered with truncated wild-type or codon-optimised *P. montana IspS* cDNA sequences [61]. Codon optimisation did improve isoprene production about twofold; however, although isoprene was detected in both strains, levels were very low, and the dominant products were hydroylated isoprene derivatives. Combined isoprene + derivatives in headspace measurements were estimated at ~500 μg/L culture (this excludes derivatives that were retained in the aqueous phase). A truncated *Mucuna bracteata IspS* was also tested in *S. cerevisiae*, but titres were only 16.1 μg/L under the conditions tested [57]. These results cannot be compared directly to the previous experiment, but they are not very encouraging. Overcoming both the low titres (relative to *E. coli*) and propensity for bioconversion will be necessary before yeast can become an attractive isoprene production organism.

Isoprene production by overexpression of an engineered kudzu IspS (truncated, codon-optimised for *E. coli*) in a variety of different constructs (different promoters and different copy numbers) was also tested in *Pantoea citra* [33]. The best production observed was about 10 μg/L. The truncated *M. bracteata IspS* yielded slightly better results in *Pantoea ananatis* (63 μg/L; [57]). The *M. bracteata IspS* was also tested in *Corynebacterium glutamicum* (24.2 μg/L) and *Enterobacter aerogenes* (316 μg/L; [57]).

Upon expression of the kudzu construct in *B. subtilis*, isoprene titres were increased threefold over native production, from 400 μg/L to 1.2 mg/L in batch culture; in a fed-batch culture titres peaked at 30 mg/L. Similar titres were achieved using an engineered MVA pathway. Truncated kudzu and poplar IspS genes codon-optimised and expressed in *Yarrowia lipolytica* using a number of different approaches; titres ranged from 0.5–1.0 μg/L in the culture headspace [33]. The engineered kudzu IspS was also expressed in *Trichoderma reesei*; titres peaked at 0.5 μg/L. A codon-optimised version was also expressed in *Streptomyces alba*, which naturally produces low levels of isoprene; isoprene production was increased tenfold but levels were still extremely low (0.75 ppm in headspace samples; [33]).

Engineering photosynthetic organisms for isoprene production is an attractive proposition, as it provides the potential to circumvent feedstock production for bioprocesses. With this aim, a truncated (*sans* chloroplast targeting sequence), tagged *P. montana IspS* cDNA was introduced into a glucose-sensitive version of the photosynthetic bacterium *Synechocystis* sp. PC6803 [75]. Only very low levels of IspS protein were detected. Codon optimisation significantly enhanced protein production, and this latter strain produced ~50 μg isoprene per gram dry cell weight per day. This was noted to be equivalent to ~4 μg isoprene/L/h^{-1} [61], compared to production of 50 μg/L/h^{-1} in wild-type *B. subtilis* [64]. The same *IspS* construct was later used with a glucose-sensitive version of *Synechocystis* sp. PC6803 [21]. Isoprene production peaked at ~100–130 μg/L culture [21, 22]. Introducing a heterologous MVA pathway improved production to slightly over 300 μg/L culture [22].

In an alternative approach to engineer photosynthetic isoprene production, the hybrid poplar IspS was tagged and codon-optimised for expression in the cyanobacterium *Thermococcus elongatus* [6]. This thermophilic organism has an optimal growth temperature of 55 °C, potentially rendering it a very attractive isoprene production organism inasmuch as (a) volatilisation is favoured, and (b) IspS enzymes have high temperature optima (see above). Mutations to improve thermostability were included in the gene design. However, isoprene production was low (25 μg/L per 30 min in the off-gas). As for yeast, and other production organisms, significant effort will be required to produce photosynthetic cyanobacterial strains that can compete with *E. coli* for isoprene production.

3.6 Bioprocess Considerations

A number of general and more specific considerations apply to microbial isoprene bioprocesses. First, as isoprene is a gas under typical bioprocess conditions, it volatilises into the headspace of the culture. This provides both advantages and disadvantages: purification is far more straightforward from the gas phase; however, it must be trapped from this phase, ideally in a continuous process.

Accumulation of isoprenoids is usually done in *E. coli* at decreased temperature (30 °C), as this minimises formation of inclusion bodies from overexpressed proteins. This may have a significant effect on IspS activity, depending on where the temperature optimum of the enzyme sits. As engineering to improve solubility and thermostability progresses (see above), titration of the bioprocess temperature—taking into account the cost-benefit analyses—may provide further improvements in yields.

Isoprene is flammable at high concentrations; consequently, it is necessary to engineer bioprocess conditions that maximise isoprene yields while keeping isoprene in the culture headspace at low enough levels to remain nonflammable and prevent serious hazards. A combination of computer modelling and experimental testing to determine the flammability limits of isoprene in the presence of various gases (among other considerations) was used to develop bioprocesses running outside the flammability envelope of isoprene [31].

Methods for recovering, purifying, and polymerising isoprene for industrial applications have also been developed [25, 31]. Isoprene can be recovered by adsorption to a solid phase, partition into a liquid phase, or direct compression/condensation. For off-gas methods, excess H_2O and CO_2 can be removed by passing the off-gas through sodium hydroxide pellets and isoprene condensed by cryotrapping in a continuous process. Both the nature and level of contaminants is important as polymerisation catalysts are sensitive to particular classes of compounds [142]. Isoprene produced via this biological route is purer than petrochemical-derived isoprene and therefore offers significant downstream processing advantages [31].

In bioprocesses it is often desirable to decouple the growth (biomass accumulation) phase from the production phase. When isoprene production genes (synthases and engineered isoprenoid pathway genes) are placed under inducible promoters, expression (and thus induction of isoprene production) can be titrated for optimal effect [31]. In this way, production of isoprene precursors (in this case, through the mevalonate pathway) and isoprene can be separated from cell growth. Different induction times can significantly affect titres.

Fed-batch processes produce significantly higher titres and yields than batch processes with the amount of isoprene being proportional to the amount of carbon source fed [33]. Optimising the feed times is important to maximise yields [31]. Isoprene yields are higher on rich media than on minimal media, with glucose/yeast extract fed-batch cultures providing the best results [31, 33].

For bulk biochemicals, the feedstock price is the primary bioprocess cost driver [101, 148]. This has driven research into alternative feedstocks, including sucrose [8, 9, 29, 104], which has a number of advantages over traditional glucose [7, 134, 137]; glycerol, a highly reduced industrial biofuel by-product [44]; and of course lignocellulosic feedstocks, which are challenging but have high potential [23, 42, 96]. A variety of different carbon sources, including glucose, lignocellulosic sources, glycerol, and the like, has also been tested for isoprene production [27, 31, 33, 38]. The advantages of glycerol are particularly applicable to isoprenoids, as they themselves are highly reduced. Use of glycerol provides improved yields of isoprenoids [27, 66, 82, 149], most likely due to improved energy and redox balances [44]. Improved isoprene yields from glycerol were achieved by improving glycerol catabolism through introduction of either a glycerol kinase and a glycerol-3-phosphate dehydrogenase or a glycerol dehydrogenase and a dihydroxyacetone kinase [27].

For photosynthetic production of isoprene from cyanobacteria (and potentially, microalgae), a diffusion-based process allowing both uptake of carbon dioxide and emission of isoprene has been developed [21]. This involves use of a custom-designed two-phase (gas/aqueous) photobioreactor. The reactor is sealed, and the headspace is periodically flushed with carbon dioxide in parallel with recovery of the gas-phase products (isoprene). As discussed above, substantial engineering will be required to make such a process competitive with current *E. coli*-based processes; however, the benefit of photosynthetic conversion would be significant. The photobioreactor process can potentially be applied to other photosynthetically produced volatiles.

4 Engineering Isoprene Production in Plants

Although the exact mechanism remains a topic of research, it is now apparent that isoprene production provides protection to plants against a variety of abiotic stresses [80, 118, 136]. Naturally nonemitting model species (tobacco and *Arabidopsis*) emit isoprene and show resistance to oxidative and thermal stress when isoprene synthase is introduced [78, 110, 138]. This has opened the discussion on the possibility of exploiting the biological role of isoprene by engineering agricultural species (which do not typically produce isoprene) for improved stress resistance through introducing isoprene synthase [135].

However, although the complex emission patterns observed in naturally emitting plants can be reproduced in engineered plants [138, 140], the genetics and molecular regulation of isoprene synthase are relatively complex [139]. This may present unforeseen problems when trying to reproduce the protective effects observed in nature [135]. For example, protection of photosynthetic processes is observed in isoprene-emitting transgenic tobacco exposed to drought [103, 128], however, broad changes in isoprenoid, nonstructural carbohydrate, and phenylpropanoid metabolism [128], and reductions in plant productivity [103], may be observed. Moreover, although engineering tobacco plants with isoprene synthase can deter herbivores [71, 72], isoprene emission in engineered *Arabidopsis* can interfere with recruitment of herbivore parasites, thus upsetting complex tritrophic interactions [77]. This would obviously be undesirable in an agricultural setting.

Isoprene production appears to mediate in the cellular reactive oxygen balance/response system, particularly under stress conditions; but differences are also observed under nonstress conditions, suggesting that the response system is essentially ‘primed’ for a rapid effective response to oxidative stress [136, 138]. Moreover, loss of carbon in the form of isoprene represents a cost to the plant; this may be only a few percent of recently fixed carbon, or may form up to 10–20 % under increased temperature [56, 86, 116]. When plants are suffering from abiotic stress, emissions are often increased/maintained even under photosynthetic limitation [28, 91, 130], thereby further increasing the relative proportion of lost carbon. Both of these elements could affect the observed metabolic and productivity perturbations [103, 128]. A final consideration is the possible effect of perturbed reactive oxygen response systems under pathogen attack, where carefully controlled and specific ROS responses are required for protection against different pathogens.

The opposite proposition—that is, engineering plants to decrease/remove isoprene emission—has also been considered [19]. Tree species, including many that are grown as forestry crops (e.g., pine, eucalyptus, and poplar; [65]), are contributors to the global atmospheric isoprene budget. Isoprene emission in plantation forests can affect local climate and air quality [59, 143]. Moreover, minimising carbon loss to isoprene might increase productivity in forestry species. When transgenic hybrid poplar plants that do not emit isoprene (emission suppressed) were compared to empty vector controls over two growing seasons, no differences in growth or biomass accumulation were observed under the growth conditions used. Modelling suggested a 2.2 % reduction in carbon loss in the absence of isoprene emission; this may contribute to improved biomass accumulation over the lifecycle of a tree. However, although the trees exhibited reduced susceptibility to the fungal pathogen *Pollaccia radiosa* (poplar shoot blight), they also became more attractive to herbivores. Moreover, given the demonstrated protective effects of isoprene under abiotic stress conditions (see above), this approach should be treated with caution, especially as it has been indicated that suppression of isoprene emission in these transgenic plants also has a negative effect on thermotolerance [18].

5 Summary, Conclusions, Outlook

The biology, atmospheric chemistry, and industrial applications of isoprene have proved rich grounds for discovery over a period of scientific investigation spanning over one and a half centuries from its discovery. As our society begins its shift from a petrochemical society to a biochemical society, biological engineering of isoprene production has emerged at the forefront of the new age of biotechnology. To date, engineering in the industrial workhorse *E. coli* has proved to be by far the most successful. Other organisms, including the lab/industry favourite, yeast, will require significant work (both in bioengineering and in bioprocess engineering) to approach the titres, rates, and yields observed in *E. coli*.

As reviewed above, many different engineering steps and approaches have been applied to isoprene production in *E. coli*. A wide variety of other gene targets and engineering approaches that have been shown to influence production of related isoprenoid compounds might also increase flux towards isoprene. These include targets that increase flux to pathway precursors, balance cofactor/precursor concentrations, regulate overall isoprenoid accumulation, balance expression of pathway enzymes, and avoid accumulation of toxic intermediates [1, 2, 5, 17, 26, 36, 62, 67, 81, 124, 141, 153, 155]. As of 2010, the tires produced using isoprene from the Genencor process cost $150,000 each [54], obviously an unattractive proposition for the average motorist. No current estimates are available, but significant advances in bioengineering and bioprocess conditions have been made since then, and current yields are much higher.

With respect to engineering isoprene production in plants, the cost and benefit of the pleiotropic effects discussed above would need to be carefully evaluated in agricultural/silvicultural settings. It was recently suggested that the patchy taxonomic distribution of isoprene emission might be due to isoprene emission only being advantageous under a narrow range of environmental and phenotypical conditions [87]; if this is the case, a better understanding of those conditions is required. Changing climate conditions in the future, which are now accepted as a scientific reality regardless of the political debate on the cause, may have a significant effect on this cost-benefit analysis. All of these considerations should be taken into account, and potential mitigating strategies should be investigated, when designing potential agronomic engineering strategies involving isoprene emission [135].

References

1. Ajikumar PK, Tyo K, Carlsen S, Mucha O, Phon TH, Stephanopoulos G (2008) Terpenoids: opportunities for biosynthesis of natural product drugs using engineered microorganisms. Mol Pharm 5(2):167–190. doi:10.1021/mp700151b
2. Ajikumar PK, Xiao W-H, Tyo KEJ, Wang Y, Simeon F, Leonard E, Mucha O, Phon TH, Pfeifer B, Stephanopoulos G (2010) Isoprenoid pathway optimization for taxol precursor overproduction in *Escherichia coli*. Science 330(6000):70–74. doi:10.1126/science.1191652
3. Albrecht M, Misawa N, Sandmann G (1999) Metabolic engineering of the terpenoid biosynthetic pathway of *Escherichia coli* for production of the carotenoids β-carotene and zeaxanthin. Biotechnol Lett 21(9):791–795
4. Alper H, Jin YS, Moxley JF, Stephanopoulos G (2005) Identifying gene targets for the metabolic engineering of lycopene biosynthesis in *Escherichia coli*. Metab Eng 7 (3):155–164. doi:10.1016/j.ymben.2004.12.003
5. Alper H, Miyaoku K, Stephanopoulos G (2005) Construction of lycopene-overproducing *E. coli* strains by combining systematic and combinatorial gene knockout targets. Nat Biotechnol 23(5):612–616
6. Anderson S (2013) Methods for the direct conversion of carbon dioxide into a hydrocarbon using a metabolically engineering photosynthetic organism. USA Patent US8609385, 17 Dec 2013
7. Archer C, Kim J, Jeong H, Park J, Vickers C, Lee S, Nielsen L (2011) The genome sequence of *E. coli* W ATCC 9637: comparative genome analysis and an improved genome-scale model of *E. coli*. BMC Genomics 12(9):9. doi:10.1186/1471-2164-1112-1189
8. Arifin Y, Archer C, Lim S, Quek L-E, Sugiarto H, Marcellin E, Vickers C, Krömer J, Nielsen L (2014) *Escherichia coli* W shows fast, highly oxidative sucrose metabolism and low acetate formation. Appl Microbiol Biotechnol 98(21):9033–9044. doi:10.1007/s00253-014-5956-4

9. Arifin Y, Sabri S, Sugiarto H, Krömer JO, Vickers CE, Nielsen LK (2011) Deletion of *cscR* in *Escherichia coli* W improves growth and poly-3-hydroxyburyrate (PHB) production from sucrose in fed batch culture. J Biotechnol 156:275. doi:10.1016/j.jbiotec.2011.07.003
10. Artsatbanov VY, Vostroknutova GN, Shleeva MO, Goncharenko AV, Zinin AI, Ostrovsky DN, Kapreliants AS (2012) Influence of oxidative and nitrosative stress on accumulation of diphosphate intermediates of the non-mevalonate pathway of isoprenoid biosynthesis in corynebacteria and mycobacteria. Biochemistry 77(4):362–371. doi:10.1134/S0006297912040074
11. Banerjee A, Sharkey TD (2014) Methylerythritol 4-phosphate (MEP) pathway metabolic regulation. Nat Prod Rep. doi:10.1039/C3NP70124G
12. Banerjee A, Wu Y, Banerjee R, Li Y, Yan H, Sharkey TD (2013) Feedback inhibition of deoxy-D-xylulose 5-phosphate synthase regulates the methyl erythritol 4-phosphate pathway. J Biol Chem 288(23):16926–16936. doi:10.1074/jbc.M113.464636
13. Beatty MK, Hayes K, Hou Z, Meyer DJ, Nannapaneni K, Rife CL, Wells DH, Zastrow-Hayes GM (2013) Legume isoprene synthase for production of isoprene. USA Patent US 20130330709, 12 Dec 2013
14. Beck ZQ, Bott RR, Rife CL, Wells D, Miller JV (2013) Three-dimensional structure of isoprene synthase and its use thereof for generating variants. USA Patent US8518686, 27 Aug 2013
15. Beck ZQ, Calabria AR, Miller MC, Vaviline DV, Nielsen AT (2013) Increased isoprene production using the archaeal lower mevalonate pathway. USA Patent US8361762, 29 Jan 2013
16. Beck ZQ, Cervin MA, Nielsen AT, Peres CM (2013) Compositions and methods of PGL for the increased production of isoprene. USA Patent US8455236B2, 4 Jun 2013
17. Beck ZQ, Eliot AC, Peres CM, Vaviline DV (2013) Utilization of phosphoketolase in the production of mevalonate, isoprenoid precursors, and isoprene. Application US20130089906, 11 April 2013
18. Behnke K, Ehlting B, Teuber M, Bauerfeind M, Louis S, Hansch R, Polle A, Bohlmann J, Schnitzler JP (2007) Transgenic, non-isoprene emitting poplars don't like it hot. Plant J 51:485–499
19. Behnke K, Grote R, Brüggemann N, Zimmer I, Zhou G, Elobeid M, Janz D, Polle A, Schnitzler J-P (2012) Isoprene emission-free poplars—a chance to reduce the impact from poplar plantations on the atmosphere. New Phytol 194(1):70–82. doi:10.1111/j.1469-8137.2011.03979.x
20. Behrendorff JBYH, Vickers CE, Chrysanthopoulos P, Nielsen LK (2013) 2,2-Diphenyl-1-picrylhydrazyl as a screening tool for recombinant monoterpene biosynthesis. Microb Cell Fact 12:76. doi:10.1186/1475-2859-12-76
21. Bentley FK, Melis A (2012) Diffusion-based process for carbon dioxide uptake and isoprene emission in gaseous/aqueous two-phase photobioreactors by photosynthetic microorganisms. Biotechnol Bioeng 109(1):100–109. doi:10.1002/bit.23298
22. Bentley FK, Zurbriggen A, Melis A (2014) Heterologous expression of the mevalonic acid pathway in cyanobacteria enhances endogenous carbon partitioning to isoprene. Mol Plant 7 (1):71–86. doi:10.1093/mp/sst134
23. Blanch HW, Simmons BA, Klein-Marcuschamer D (2011) Biomass deconstruction to sugars. Biotechnol J 6(9):1086–1102. doi:10.1002/biot.201000180
24. Boghigian BA, Salas D, Ajikumar PK, Stephanopoulos G, Pfeifer B (2012) Analysis of heterologous taxadiene production in K- and B-derived *Escherichia coli*. Appl Microbiol Biotechnol 93(4):1651–1661. doi:10.1007/s00253-011-3528-4
25. Bott RR, Cervin MA, Kellis JT, Jr, McAuliffe JC, Miasnikov A, Peres CM, Rife CL, Wells DH, Weyler W, Whited GM (2012) Isoprene synthase variants for improved microbial production of isoprene. USA Patent US8173410, 8 May 2012
26. Bramucci MG, Brzostowicz PC, Cheng Q, Kostichka KN, Rouviere PE, Nagarajan V, Tao L, Thomas SM (2006) Genes involved in isoprenoid compound production. USA Patent US7034140, 25 April 2006

27. Bresdow S, Doneske S, Li M, Zhou H, Monticello DJ, Campbell P (2014) Microorganisms and processes for the conversion of glycerol to isoprene. Application WO2014015210 A2, 23 Jan 2014
28. Brilli F, Barta C, Fortunati A, Centritto M, Loreto F (2006) The contribution of different carbon sources to isoprene synthesis during and after drought in *Populus alba* leaves. Geophys Res Abs 8:EGU06-A-04113
29. Bruschi M, Boyes S, Sugiarto H, Nielsen LK, Vickers CE (2011) A transferrable sucrose utilization approach for non-sucrose-utilizing *Escherichia coli* strains. Biotech Adv 30 (5):1001–1010. doi:10.1016/j.biotechadv.2011.08.019
30. Burke C, Croteau R (2002) Geranyl diphosphate synthase from Abies grandis: cDNA isolation, functional expression, and characterization. Arch Biochem Biophys 405(1):130–136. doi:10.1016/S0003-9861(02)00335-1
31. Calabria AR, Cervin MA, Chotani GK, La Duca R, McAuliffe JC, Miller MC, Sabo TA, Sanford KJ, Webster EL, Whited GM (2013) Compositions and methods for producing isoprene free of C5 hydrocarbons under decoupling conditions and/or safe operating ranges. USA Patent US8420360, 16 April 2013
32. Campos N, Rodríguez-Concepción M, Sauret-Gueto S, Gallego F, Lois LM, Boronat A (2001) *Escherichia coli* engineered to synthesize isopentenyl diphosphate and dimethyl allyl diphosphate from mevalonate: a novel system for the genetic analysis of the 2-C-methyl-D-erythritol 4-phosphate pathway for isoprenoid biosynthesis. Biochem J 353:59–67
33. Cervin MA, Chotani GK, Feher FJ, La Duca R, McAuliffe JC, Miasnikov A, Peres CM, Puhala AS, Sanford KJ, Valle F, Whited GM (2012) Compositions and methods for producing isoprene. USA Patent US8420360 B2, 16 Oct 2012
34. Chae HS, Kim KH, Kim SC, Lee PC (2010) Strain-dependent carotenoid productions in metabolically engineered *Escherichia coli*. Appl Biochem Biotechnol 162(8):2333–2344. doi:10.1007/s12010-010-9006-0
35. Chameides WL, Lindsay RW, Richardson J, Kiang CS (1998) The role of biogenic hydrocarbons in urban photochemical smog: atlanta as a case study. Science 241:1473–1475
36. Chandran SS, Kealey JT, Reeves CD (2011) Microbial production of isoprenoids. Process Biochem 46(9):1703–1710. doi:10.1016/j.procbio.2011.05.012
37. Chotani GK, McAuliffe JC, Miller MC, Muir RE, Vaviline DV, Weyler W (2013) Isoprene production using the DXP and MVA pathway. US8507235 B2, 13 Aug 2013
38. Chotani GK, Nielsen A, Sanford KJ (2013) Reduction of carbon dioxide emission during isoprene production by fermentation. WO/2010/031068
39. Claeys M, Graham B, Vas G, Wang W, Vermeylen R, Pashynska V, Cafmeyer J, Guyon P, Andreae MO, Artaxo P, Maenhaut W (2004) Formation of secondary organic aerosols through photooxidation of isoprene. Science 303(5661):1173–1176
40. Coelho PS, Farrow MF, Smith MA (2014) *De novo* metabolic pathways for isoprene biosynthesis. World intellectual property organisation patent application WO2014066892 A1, 1 May 2014
41. Dellas N, Thomas ST, Manning G, Noel JP, Weigel D (2013) Discovery of a metabolic alternative to the classical mevalonate pathway. eLife 2:e00672. doi:10.7554/eLife.00672
42. Dionisi D, Anderson JA, Aulenta F, McCue A, Paton G (2014) The potential of microbial processes for lignocellulosic biomass conversion to ethanol: a review. J Chem Technol Biotechnol. doi:10.1002/jctb.4544
43. Djaman O, Outten FW, Imlay JA (2004) Repair of oxidized iron-sulphur clusters in *Escherichia coli*. J Biol Chem 279(43):44590–44599. doi:10.1074/jbc.M406487200
44. Durnin G, Clomburg J, Yeates Z, Alvarez PJJ, Zygourakis K, Campbell P, Gonzalez R (2009) Understanding and harnessing the microaerobic metabolism of glycerol in *Escherichia coli*. Biotechnol Bioeng 103(1):148–161. doi:10.1002/bit.22246
45. Eisenreich W, Bacher A, Arigoni D, Rohdich F (2004) Biosynthesis of isoprenoids via the non-mevalonate pathway. Cell Mol Life Sci V61(12):1401–1426
46. Fall R, Copley SD (2000) Bacterial sources and sinks of isoprene, a reactive atmospheric hydrocarbon. Environ Microbiol 2(2):123–130

47. Fall R, Wildermuth MC (1998) Isoprene synthase: from biochemical mechanism to emission algorithm. J Geophys Res 103(D19):25599–25610
48. Farmer WR, Liao JC (2001) Precursor balancing for metabolic engineering of lycopene production in *Escherichia coli*. Biotechnol Progr 17(1):57–61
49. Fortunati A, Barta C, Brilli F, Centritto M, Zimmer I, Schnitzler JP, Loreto F (2008) Isoprene emission is not temperature-dependent during and after severe drought-stress: a physiological and biochemical analysis. Plant J 55(4):687–697
50. Goldstein JL, Brown MS (1990) Regulation of the mevalonate pathway. Nature 343 (6257):425–430
51. Gräwert T, Kaiser J, Zepeck F, Laupitz R, Hecht S, Amslinger S, Schramek N, Schleicher E, Weber S, Haslbeck M, Buchner J, Rieder C, Arigoni D, Bacher A, Eisenreich W, Rohdich F (2004) IspH protein of *Escherichia coli*: studies on iron–sulfur cluster implementation and catalysis. J Am Chem Soc 126(40):12847–12855. doi:10.1021/ja0471727
52. Gray DW, Breneman SR, Topper LA, Sharkey TD (2011) Biochemical characterization and homology modelling of methylbutenol synthase and implications for understanding hemiterpene synthase evolution in plants. J Biol Chem 286(23):20582–20590. doi:10.1074/jbc.M111.237438
53. Guenther A, Karl T, Harley P, Wiedinmyer C, Palmer PI, Geron C (2006) Estimates of global terrestrial isoprene emissions using MEGAN (model of emissions of gases and aerosols from nature). Atmos Chem Phys 6:3181–3210
54. Hansegard J (2010) Sugar plus science equals $150,000 tires. Wall Street J, 8 Dec 2010, http://online.wsj.com/news/articles/SB10001424052748703493504576007112633560914
55. Harker M, Bramley PM (1999) Expression of prokaryotic 1-deoxy-D-xylulose-5-phosphatases in *Escherichia coli* increases carotenoid and ubiquinone biosynthesis. FEBS Letters 448(1):115–119. doi:10.1016/s0014-5793(99)00360-9
56. Harley P, Guenther A, Zimmerman P (1996) Effects of light, temperature and canopy position on net photosynthesis and isoprene emission from sweetgum (*Liquidambar styraciflua*) leaves. Tree Physiol 16(1–2):25–32
57. Hayashi Y, Harada M, Takaoka S, Fukushima Y, Yokoyama K, Nishio Y, Tajima Y, Mihara Y, Nakata K (2014) Isoprene synthase and gene encoding the same, and method for producing isoprene monomer. USA Patent Application Number US2014/0113344
58. Hemmerlin A (2013) Post-translational events and modifications regulating plant enzymes involved in isoprenoid precursor biosynthesis. Plant Sci 203–204:41–54
59. Hewitt CN, MacKenzie AR, Di Carlo P, Di Marco CF, Dorsey JR, Evans M, Fowler D, Gallagher MW, Hopkins JR, Jones CE, Langford B, Lee JD, Lewis AC, Lim SF, McQuaid J, Misztal P, Moller SJ, Monks PS, Nemitz E, Oram DE, Owen SM, Phillips GJ, Pugh TAM, Pyle JA, Reeves CE, Ryder J, Siong J, Skiba U, Stewart DJ (2009) Nitrogen management is essential to prevent tropical oil palm plantations from causing ground-level ozone pollution. Proc Natl Acad Sci USA 106(44):18447–18451. doi:10.1073/pnas.0907541106
60. Hewitt CN, Street RA, Scholefield PA (2004) Biogenic VOC Emission Database. Lancaster University. http://www.es.lancs.ac.uk/cnhgroup/download.html. Accessed on 6 Oct 2014
61. Hong S-Y, Zurbriggen AS, Melis A (2012) Isoprene hydrocarbons production upon heterologous transformation of *Saccharomyces cerevisiae*. J Appl Microbiol 113(1):52–65. doi:10.1111/j.1365-2672.2012.05319.x
62. Jiang M, Stephanopoulos G, Pfeifer BA (2012) Toward biosynthetic design and implementation of *Escherichia coli*-derived paclitaxel and other heterologous polyisoprene compounds. Appl Environ Microbiol 78(8):2497–2504. doi:10.1128/aem.07391-11
63. Jones CA, Rasmussen RA (1975) Production of isoprene by leaf tissue. Plant Physiol 55 (6):982–987
64. Julsing M, Rijpkema M, Woerdenbag H, Quax W, Kayser O (2007) Functional analysis of genes involved in the biosynthesis of isoprene in *Bacillus subtilis*. Appl Microbiol Biotechnol 75:1377–1384
65. Kesselmeier J, Staudt M (1999) Biogenic volatile organic compounds (VOC): an overview on emission, physiology and ecology. J Atmos Chem V33(1):23–88

66. Kim YS, Lee JH, Kim NH, Yeom SJ, Kim SW, Oh DK (2011) Increase of lycopene production by supplementing auxiliary carbon sources in metabolically engineered *Escherichia coli*. Appl Microbiol Biotechnol 90(2):489–497. doi:10.1007/s00253-011-3091-z
67. Klein-Marcuschamer D, Ajikumar PK, Stephanopoulos G (2007) Engineering microbial cell factories for biosynthesis of isoprenoid molecules: beyond lycopene. Trends Biotechnol 25 (9):417–424
68. Köksal M, Zimmer I, Schnitzler JP, Christianson DW (2010) Structure of isoprene synthase illuminates the chemical mechanism of teragram atmospheric carbon emission. J Mol Biol 402(2):363–373
69. Kuzma J, Fall R (1993) Leaf isoprene emission rate is dependent on leaf development and the level of isoprene synthase. Plant Physiol 101(2):435–440
70. Kuzma J, Nemecek M, Pollock WH, Fall R (1995) Bacteria produce the volatile hydrocarbon isoprene. Curr Microbiol V30(2):97–103
71. Laothawornkitkul J, Paul ND, Vickers CE, Possell M, Mullineaux PM, Hewitt CN, Taylor JE (2008) The role of isoprene in insect herbivory. Plant Signal Behav 3(12):1141–1142
72. Laothawornkitkul J, Paul ND, Vickers CE, Possell M, Taylor JE, Mullineaux PM, Hewitt CN (2008) Isoprene emissions influence herbivore feeding decisions. Plant Cell Environ 31 (10):1410–1415. doi:10.1111/j.1365-3040.2008.01849.x
73. Lehning A, Zimmer I, Steinbrecher R, Brüggemann N, Schnitzler JP (1999) Isoprene synthase activity and its relation to isoprene emission in *Quercus robur* L. leaves. Plant Cell Environ 22(5):495–504. doi:10.1046/j.1365-3040.1999.00425.x
74. Leonard E, Ajikumar PK, Thayer K, Xiao W-H, Mo JD, Tidor B, Stephanopoulos G, Prather KLJ (2010) Combining metabolic and protein engineering of a terpenoid biosynthetic pathway for overproduction and selectivity control. Proc Nat Acad Sci 107(31):13654–13659. doi:10.1073/pnas.1006138107
75. Lindberg P, Park S, Melis A (2010) Engineering a platform for photosynthetic isoprene production in cyanobacteria, using *Synechocystis* as the model organism. Metab Eng 12 (1):70–79. doi:10.1016/j.ymben.2009.10.001
76. Liu H, Sun Y, Ramos KRM, Nisola GM, Valdehuesa KNG, Lee WK, Park SJ, Chung W-J (2013) Combination of entner-doudoroff pathway with MEP increases isoprene production in engineered *Escherichia coli*. PLoS ONE 8(12):e83290. doi:10.1371/journal.pone.0083290
77. Loivamäki M, Mumm R, Dicke M, Schnitzler JP (2008) Isoprene interferes with the attraction of bodyguards by herbaceous plants. Proc Natl Acad Sci USA 104 (45):17430–17435
78. Loivamäki M, Gilmer F, Fischbach RJ, Sorgel C, Bachl A, Walter A, Schnitzler JP (2007) Arabidopsis, a model to study biological functions of isoprene emission? Plant Physiol 144:1066–1078
79. Lombard J, Moreira D (2011) Origins and early evolution of the mevalonate pathway of isoprenoid biosynthesis in the three domains of life. Mol Biol Evol 28(1):87–99. doi:10.1093/molbev/msq177
80. Loreto F, Mannozzi M, Maris C, Nascetti P, Ferranti F, Pasqualini S (2001) Ozone quenching properties of isoprene and its antioxidant role in leaves. Plant Physiol 126 (3):993–1000
81. Marienhagen J, Bott M (2013) Metabolic engineering of microorganisms for the synthesis of plant natural products. J Biotechnol 163(2):166–178
82. Martin VJ, Pitera DJ, Withers ST, Newman JD, Keasling JD (2003) Engineering a mevalonate pathway in *Escherichia coli* for production of terpenoids. Nat Biotechnol 21 (7):796–802. doi:10.1038/nbt833
83. McCoy M (2008) Biobased isoprene: goodyear and genencor seek a fermentation route to important synthetic rubber raw material. Chem Eng News 86:15
84. Meskhidze N, Nenes A (2006) Phytoplankton and cloudiness in the southern ocean. Science 314(5804):1419–1423. doi:10.1126/science.1131779

85. Miller B, Oschinski C, Zimmer W (2001) First isolation of an isoprene synthase gene from poplar and successful expression of the gene in *Escherichia coli*. Planta 213(3):483–487. doi:10.1007/s004250100557
86. Monson RK, Fall R (1989) Isoprene emission from aspen leaves: influence of environment and relation to photosynthesis and photorespiration. Plant Physiol 90(1):267–274
87. Monson RK, Jones RT, Rosenstiel TN, Schnitzler JP (2013) Why only some plants emit isoprene. Plant Cell Environ 36(3):503–516
88. Muir RE, Weyler W (2014) Compositions and methods for improved isoprene production using two types of ispg enzymes. USA Patent Application US20140147902 A1
89. Okada K, Hase T (2005) Cyanobacterial non-mevalonate pathway. (*E*)-4-Hyddroxy-3-methylbut-2-enyl diphosphate synthase interacts with ferredoxin in *Thermosynechococcus elongatus* BP-1. J Biol Chem 280(21):20672–20679. doi:10.1074/jbc.M500865200
90. Okamura E, Tomita T, Sawa R, Nishiyama M, Kuzuyama T (2010) Unprecedented acetoacetyl-coenzyme a synthesizing enzyme of the thiolase superfamily involved in the mevalonate pathway. Proc Natl Acad Sci USA 107(25):11265–11270. doi:10.1073/pnas.1000532107
91. Pegoraro E, Rey A, Bobich EG, Barron-Gafford G, Grieve KA, Malhi Y, Murthy R (2004) Effect of elevated CO_2 concentration and vapour pressure deficit on isoprene emission from leaves of *Populus deltoides* during drought. Funct Plant Biol 31(12):1137–1147
92. Poisson N, Kanakidou M, Crutzen PJ (2000) Impact of non-methane hydrocarbons on tropospheric chemistry and the oxidizing power of the global troposphere: 3-dimensional modelling results. J Atmos Chem 36(2):157–230
93. Ramos K, Valdehuesa K, Liu H, Nisola G, Lee W-K, Chung W-J (2014) Combining De Ley–Doudoroff and methylerythritol phosphate pathways for enhanced isoprene biosynthesis from d-galactose. Bioproc Biosys Eng, pp 1–9. doi:10.1007/s00449-014-1228-z
94. Rasmussen R (1972) Survey of plant species that release isoprene to atmosphere. Abs Pap Am Chem Soc, 13
95. Rasmussen RA (1970) Isoprene: identified as a forest-type emission to the atmosphere. Envir Sci Tech 4(8):667–671. doi:10.1021/es60043a008
96. Reardon K (2014) Lignocellulosic biorefineries: concepts and possibilities. In: McCann MC, Buckeridge MS, Carpita NC (eds) Plants and bioenergy, Vol 4. Adv Plant Biol. Springer, New York, pp 255–265. doi:10.1007/978-1-4614-9329-7_15
97. Rife CL, Wells DH (2014) Isoprene synthase variants with improved solubility for production of isoprene. USA Patent US8735134, 27 May 2014
98. Rodríguez-Concepción M, Boronat A (2002) Elucidation of the methylerythritol phosphate pathway for isoprenoid biosynthesis in bacteria and plastids. A metabolic milestone achieved through genomics. Plant Physiol 130(3):1079–1089
99. Rodríguez-Villalón A, Perez-Gil J, Rodríguez-Concepción M (2008) Carotenoid accumulation in bacteria with enhanced supply of isoprenoid precursors by upregulation of exogenous or endogenous pathways. J Biotechnol 135(1):78–84. doi:10.1016/j.jbiotec.2008.02.023
100. Rohmer M (2003) Mevalonate-independent methylerythritol phosphate pathway for isoprenoid biosynthesis elucidation and distribution. Pure Appl Chem 75(2–3):375–387
101. Rude MA, Schirmer A (2009) New microbial fuels: a biotech perspective. Curr Opin Microbiol 12(3):274–281. doi:10.1016/j.mib.2009.04.004
102. Ruzicka L (1953) The isoprene rule and the biogenesis of terpenic compounds. Experientia 9 (10):357–367
103. Ryan AC, Hewitt CN, Possell M, Vickers CE, Purnell A, Mullineaux PM, Davies WJ, Dodd IC (2014) Isoprene emission protects photosynthesis but reduces plant productivity during drought in transgenic tobacco (*Nicotiana tabacum*) plants. New Phytol 201(1):205–216. doi:10.1111/nph.12477
104. Sabri S, Nielsen LK, Vickers CE (2013) Molecular control of sucrose utilization in *Escherichia coli W*, an efficient sucrose-utilizing strain. Appl Environ Microbiol 79 (2):478–487. doi:10.1128/aem.02544-12

105. Saltman W (ed) (1985) Isoprene Kirk-othmer concise encyclopaedia of chemical technology. Wiley, New York
106. Sanadze GA (1957) The nature of gaseous substances emitted by leaves of *Robina pseudoacacia*. Soobshch Akad Nauk Gruz SSR 19:83–86
107. Sanadze GA (1969) Light-dependent excretion of molecular isoprene by leaves. Prog Photosynth Res II:701–706
108. Sasaki K, Ohara K, Yazaki K (2005) Gene expression and characterization of isoprene synthase from *Populus alba*. FEBS Lett 579(11):2514–2518. doi:10.1016/j.febslet.2005.03.066
109. Sasaki K, Ohara K, Yazaki K (2005) Gene expression and characterization of isoprene synthase from *Populus alba*. FEBS Lett 579(11):2514–2518
110. Sasaki K, Saito T, Lamsa M, Oksman-Caldentey KM, Suzuki M, Ohyama K, Muranaka T, Ohara K, Yazaki K (2007) Plants utilize isoprene emission as a thermotolerance mechanism. Plant Cell Physiol 48(9):1254–1262
111. Schnitzler JP, Lehning A, Steinbrecher R (1997) Seasonal pattern of isoprene synthase activity in *Quercus robur* leaves and its significance for modeling isoprene emission rates. Bot Acta 110(3):240–243
112. Seemann M, Bui BT, Wolff M, Tritsch D, Campos N, Boronat A, Marquet A, Rohmer M (2002) Isoprenoid biosynthesis through the methylerythritol phosphate pathway: the (E)-4-hydroxy-3-methylbut-2-enyl diphosphate synthase (GcpE) is a [4Fe-4S] protein. Angewandte Chemie 41(22):4337–4339. doi:10.1002/1521-3773(20021115)41:22<4337:AID-ANIE4337>3.0.CO;2-K
113. Shallcross DE, Monks PS (2000) New Directions: a role for isoprene in biosphere-climate-chemistry feedbacks. Atmos Environ 34(10):1659–1660
114. Sharkey TD (1996) Isoprene synthesis by plants and animals. Endeavour 20(2):74–78
115. Sharkey TD, Gray DW, Pell HK, Breneman SR, Topper L (2013) Isoprene synthase genes form a monophyletic clade of acyclic terpene synthases in the Tps-b terpene synthase family. Evolution 67(4):1026–1040. doi:10.1111/evo.12013
116. Sharkey TD, Loreto F (1993) Water stress, temperature, and light effects on the capacity for isoprene emission and photosynthesis of kudzu leaves. Oecologia 95(3):328–333
117. Sharkey TD, Monson RK (2014) The future of isoprene emission from leaves, canopies and landscapes. Plant Cell Environ 37(8):1727–1740. doi:10.1111/pce.12289
118. Sharkey TD, Singsaas EL (1995) Why plants emit isoprene. Nature 374:769
119. Sharkey TD, Yeh S, Wiberley AE, Falbel TG, Gong D, Fernandez DE (2005) Evolution of the isoprene biosynthetic pathway in kudzu. Plant Physiol 137:700–712
120. Silver GM, Fall R (1991) Enzymatic synthesis of isoprene from dimethylallyl diphosphate in aspen leaf extracts. Plant Physiol 97(4):1588–1591
121. Silver GM, Fall R (1995) Characterization of aspen isoprene synthase, an enzyme responsible for leaf isoprene emission to the atmosphere. J Biol Chem 270(22):13010–13016
122. Sivy TL, Fall R, Rosenstiel TN (2011) Evidence of isoprenoid precursor toxicity in *Bacillus subtilis*. Biosci Biotechnol Biochem 75(12):2376–2383
123. Sivy TL, Shirk MC, Fall R (2002) Isoprene synthase activity parallels fluctuations of isoprene release during growth of *Bacillus subtilis*. Biochem Biophys Res Commun 294 (1):71–75
124. Stephanopoulos G, Alper HS, Jin YS (2006) Gene targets for enhanced carotenoid production. USA Patent US7741070, 22 June 2010
125. Stephanopoulos G (2007) Challenges in engineering microbes for biofuels production. Science 315:801–804
126. Taalman RDFM (1996) Isoprene: background and issues. Toxicology 113(1–3):242–246
127. Takahashi S, Yeo Y, Greenhagen B, McMullin T, Song L, Maurina-Brunker J, Rosson R, Noel JP, Chappell J (2007) Metabolic engineering of sesquiterpene metabolism in yeast. Biotechnol Bioeng 97(1):170–181
128. Tattini M, Velikova V, Vickers C, Brunetti C, Di Ferdinando M, Trivellini A, Fineschi S, Agati G, Ferrini F, Loreto F (2014) Isoprene production in transgenic tobacco alters

isoprenoid, non-structural carbohydrate and phenylpropanoid metabolism, and protects photosynthesis from drought stress. Plant Cell Environ 37(8):1950–1964. doi:10.1111/pce.12350

129. Thompson AM (1992) The oxidizing capacity of the earth's atmosphere—probable past and future changes. Science 256(5060):1157–1165
130. Tingey DT, Evans R, Gumpertz M (1981) Effects of environmental conditions on isoprene emission from live oak. Planta 152(6):565–570
131. Tokumoto U, Takahashi Y (2001) Genetic analysis of the *isc* operon in *Escherichia coli* involved in the biogenesis of cellular iron-sulfur proteins. J Biochem 130(1):63–71
132. Trainer M, Williams EJ, Parrish DD, Buhr MP, Allwine EJ, Westberg HH, Fehsenfeld FC, Liu SC (1987) Models and observations of the impact of natural hydrocarbons on rural ozone. Nature 329(6141):705–707
133. Vickers CE, Behrendorff JBYH, Bongers M, Brennan TCR, Bruschi M, Nielsen LK (2014) Production of industrially-relevant isoprenoid compounds in engineered microbes. In: Kamm B (ed) Microorganisms in biorefineries. Springer, Heidelberg
134. Vickers CE, Blank LM, Kromer JO (2010) Chassis cells for industrial biochemical production. Nat Chem Biol 6(12):875–877. doi:10.1038/Nchembio.484
135. Vickers CE, Bongers M, Liu Q, Delatte T, Bouwmeester H (2014) Metabolic engineering of volatile isoprenoids in plants and microbes. Plant Cell Environ Online Early. doi:10.1111/pce.12316
136. Vickers CE, Gershenzon J, Lerdau MT, Loreto F (2009) A unified mechanism of action for volatile isoprenoids in plant abiotic stress. Nat Chem Biol 5(5):283–291. doi:10.1038/Nchembio.158
137. Vickers CE, Klein-Marcuschamer D, Krömer JO (2012) Examining the feasibility of bulk commodity production in *Escherichia coli*. Biotechnol Lett 34(4):585
138. Vickers CE, Possell M, Cojocariu CI, Velikova VB, Laothawornkitkul J, Ryan A, Mullineaux PM, Hewitt CN (2009) Isoprene synthesis protects transgenic tobacco plants from oxidative stress. Plant Cell Environ 32(5):520–531. doi:10.1111/j.1365-3040.2009.01946.x
139. Vickers CE, Possell M, Hewitt CN, Mullineaux PM (2010) Genetic structure and regulation of isoprene synthase in poplar (*Populus* spp.). Plant Mole Biol 73(4–5):547–558
140. Vickers CE, Possell M, Laothawornkitkul J, Ryan AC, Hewitt CN, Mullineaux PM (2011) Isoprene synthesis in plants: lessons from a transgenic tobacco model. Plant Cell Environ 34 (6):1043–1053. doi:10.1111/j.1365-3040.2011.02303.x
141. Wang HH, Isaacs FJ, Carr PA, Sun ZZ, Xu G, Forest CR, Church GM (2009) Programming cells by multiplex genome engineering and accelerated evolution. Nature 460 (7257):894–898. doi:10.1038/nature08187
142. Whited GM, Feher FJ, Benko DA, Cervin MA, Chotani GK, McAuliffe JC, LaDuca RJ, Ben-Shoshan EA, Sanford KJ (2010) Development of a gas-phase bioprocess for isoprene-monomer production using metabolic pathway engineering. Ind Biotechnol 6(3):152–163. doi:10.1089/ind.2010.6.152
143. Wiedinmyer C, Tie X, Guenther A (2006) Future changes in biogenic isoprene emissions: how might they affect regional and global atmospheric chemistry? Earth Interact 10:1–9
144. Wildermuth MC, Fall R (1996) Light-dependent isoprene emission (characterization of a thylakoid-bound isoprene synthase in *Salix discolor* chloroplasts). Plant Physiol 112 (1):171–182
145. Wildermuth MC, Fall R (1998) Biochemical characterization of stromal and thylakoid-bound isoforms of isoprene synthase in willow leaves. Plant Physiol 116(3):1111–1123
146. Williams CG (1859) On isoprene and caoutchine. Proc R Soc London 10:516–519. doi:10.1098/rspl.1859.0101
147. Williams DC, McGarvey DJ, Katahira EJ, Croteau R (1998) Truncation of limonene synthase preprotein provides a fully active 'Pseudomature' form of this monoterpene cyclase and reveals the function of the amino-terminal arginine pair. Biochemistry 37(35):12213–12220

148. Willke T, Vorlop K (2008) Biotransformation of glycerol into 1,3-propanediol. Eur J Lipid Sci Tech 110(9):831–840. doi:10.1002/ejlt.200800057
149. Willrodt C, David C, Cornelissen S, B BH, Julsing MK, Schmid A (2014) Engineering the productivity of recombinant *Escherichia coli* for limonene formation from glycerol in minimal media. Biotechnol J. doi:10.1002/biot.201400023
150. Withers ST, Gottlieb SS, Lieu B, Newman JD, Keasling JD (2007) Identification of isopentenol biosynthetic genes from *Bacillus subtilis* by a screening method based on isoprenoid precursor toxicity. Appl Environ Microbiol 73(19):6277–6283
151. Wolfertz M, Sharkey TD, Boland W, Kühnemann F (2004) Rapid regulation of the methylerythritol 4-phosphate pathway during isoprene synthesis. Plant Physiol 135(4):1939–1945
152. Wolff M, Seemann M, Tse Sum Bui B, Frapart Y, Tritsch D, Garcia Estrabot A, Rodriguez-Concepcion M, Boronat A, Marquet A, Rohmer M (2003) Isoprenoid biosynthesis via the methylerythritol phosphate pathway: the (E)-4-hydroxy-3-methylbut-2-enyl diphosphate reductase (LytB/IspH) from *Escherichia coli* is a (4Fe-4S) protein. FEBS Lett 541 (1–3):115–120
153. Zhao J, Li Q, Sun T, Zhu X, Xu H, Tang J, Zhang X, Ma Y (2013a) Engineering central metabolic modules of *Escherichia coli* for improving β-carotene production. Metab Eng 17 (0):42–50. doi:http://dx.doi.org/10.1016/j.ymben.2013.02.002
154. Zhao L, W-c Chang, Xiao Y, H-w Liu, Liu P (2013) Methylerythritol phosphate pathway of isoprenoid biosynthesis. Ann Rev Biochem 82(1):497–530. doi:10.1146/annurev-biochem-052010-100934
155. Zhao Y, Yang J, Qin B, Li Y, Sun Y, Su S, Xian M (2011) Biosynthesis of isoprene in *Escherichia coli* via methylerythritol phosphate (MEP) pathway. Appl Microbiol Biotechnol 90(6):1915–1922. doi:10.1007/s00253-011-3199-1
156. Zhou K, Zou RY, Stephanopoulos G, Too HP (2012) Metabolite profiling identified methylerythritol cyclodiphosphate efflux as a limiting step in microbial isoprenoid production. PLOS ONE 7(11):e47513. doi:10.1371/journal.pone.0047513
157. Zurbriggen A, Kirst H, Melis A (2012) Isoprene production via the mevalonic acid pathway in *Escherichia coli* (Bacteria). BioEnergy Res 5:814–828. doi:10.1007/s12155-012-9192-4

Adv Biochem Eng Biotechnol (2015) 148: 319–354
DOI 10.1007/10_2014_289

Published Online: 25 January 2015

Biosynthesis and Biotechnology of High-Value *p*-Menthane Monoterpenes, Including Menthol, Carvone, and Limonene

Bernd Markus Lange

Abstract Monoterpenes of the *p*-menthane group are volatile secondary (or specialized) metabolites found across the plant kingdom. They are dominant constituents of commercially important essential oils obtained from members of the genera *Mentha* (Lamiaceae), *Carum* (Apiaceae), *Citrus* (Rutaceae), and *Eucalyptus* (Myrtaceae). *p*-Menthane monoterpenes have also attracted interest as chiral specialty chemicals, and the harvest from natural sources is therefore supplemented by chemical synthesis. More recently, microbial and plant-based platforms for the high-level accumulation of specific target monoterpenes have been developed. In this review chapter, I discuss the properties of the genes and enzymes involved in *p*-menthane biosynthesis and provide a critical assessment of biotechnological production approaches.

Keywords Carvone · Cineole · Essential oil · Limonene · Menthol · Metabolic engineering

List of Abbreviations

DXR	1-Deoxy-D-xylulose 5-phosphate reductoisomerase
ER	Endoplasmic reticulum
FPPS	(*E,E*)-farnesyl diphosphate synthase
ISPD	(−)-*trans*-isopiperitenol dehydrogenase
ISPR	(−)-*trans*-isopiperitenone reductase
MDR	Medium chain dehydrogenase/reductase
MEP	2*C*-methyl-D-eryhritol 4-phosphate
MFS	(+)-menthofuran synthase
MMR	(−)-menthone:(−)-menthol reductase

B.M. Lange (✉)
Institute of Biological Chemistry and M.J. Murdock Metabolomics Laboratory, Washington State University, Pullman, WA 99164-6340, USA
e-mail: lange-m@wsu.edu

MNR	(−)-menthone:(+)-neomenthol reductase
MVA	Mevalonic acid
ORF	Open reading frame
PMD	*p*-menthane-3,8-diol
PR	(+)-pulegone reductase
SDR	Short chain dehydrogenase/reductase

Contents

1 Introduction

One of the most diverse and widely distributed classes of plant monoterpenes is the *p*-menthane (1-methyl-4-isopropylcyclohexane) type (Scheme 1). Larger amounts of these metabolites are often accumulated as essential oils and oleoresins. The current methods for the commercial production of specific high-value *p*-menthane monoterpenes are chemical synthesis or isolation from natural sources. The biosynthesis of *p*-menthane monoterpenes in the genus *Mentha* occurs exclusively in specialized anatomical structures called glandular trichomes [47]. Specifically, the pathway is expressed in secretory cells, which form an eight-celled disk within glandular trichomes. Secretory cells can be isolated and have been demonstrated to be a highly enriched source of transcripts that were characterized to be directly involved in *p*-menthane monoterpene biosynthesis [19, 45]. In *Citrus* species, the fruit peel contains secretory cavities that are filled with *p*-menthane monoterpene-containing essential oil. Recently, laser-microdissection has been demonstrated to allow the isolation of epithelial cells lining these cavities, which was followed by the extraction of RNA and subsequent transcriptome analysis [90]. An integration with physiological and metabolite data allowed a first glimpse at the metabolic specialization of this highly specialized oil biosynthetic cell type. A method for the isolation of oil-accumulating subdermal secretory cavities from *Eucalyptus* leaves was also developed recently [32]. Subsequently, RNA was isolated and a cDNA fragment with high homology to a monoterpene synthase amplified.

Building on cell type-specific transcriptome sequences, functional genomics efforts have led to the cloning and characterization of many of the genes involved in *p*-menthane monoterpene biosynthesis over the last 20 years. In this review chapter,

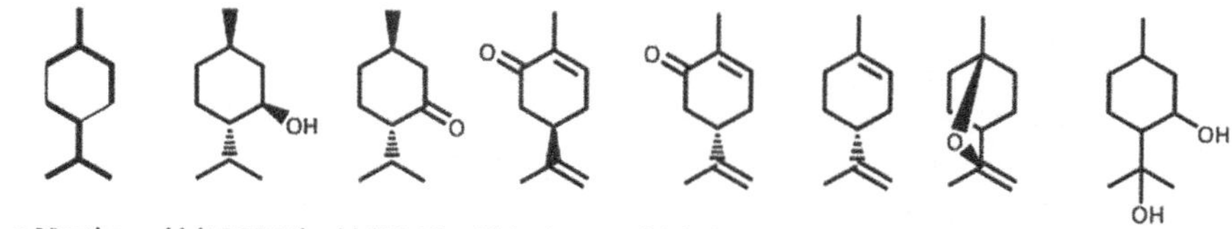

Scheme 1 Structures of selected *p*-menthane monoterpenes. The C5 units from which the parent skeleton is assembled are indicated in boldface on the left-hand side

I list the catalytic properties of the enzymes that are relevant for the discussion of *p*-menthane monoterpene formation but, for an in-depth coverage of the topic, the interested reader is referred to a comprehensive review article [19]. We also recently published an exhaustive review covering biotechnological efforts to generate volatile terpenoids in plants [48]. To avoid redundancy, this chapter focuses on attempts to produce high-value *p*-menthane monoterpenes (not biofuel components) through biotechnological methods using engineered plants and microorganisms.

2 Commercial Production and Applications for *p*-Menthane Monoterpenes

(−)-(1*R*,3*R*,4*S*)-Menthol (syn. *l*-menthol; Scheme 1) is the principal component [>60 % (w/v)] of the essential oil distilled from cornmint (*Mentha arvensis* L.). The oil of peppermint (*Mentha* x *piperita*) contains between 35 and 45 % *l*-menthol in a complex mixture with more than a dozen other volatiles [50]. *l*-Menthol has a characteristic cooling effect when ingested or applied to the skin. Its effects are mediated by TRPM8, a Na^+/Ca^{2+} ion channel that acts as a receptor for cold and cooling agents [69]. The physiological cooling effects and the minty taste underlie the widespread use of *l*-menthol in the flavor sector (e.g., chewing gum, beverages, candy, cigarettes, cosmetics), personal hygiene (e.g., toothpaste, mouthwash, skin and hair care products), and in nonprescription medical products (primarily topical analgesics, decongestants, and cough suppressants; [39]). Since the late 1980s, crystalline menthol has been registered as a pesticide for tracheal mite control in the United States. The annual production of *l*-menthol has been estimated to exceed 30,000 tons per year (t/year; [49]; Table 1), the majority of which (roughly 19,000 t/year in 2007) is obtained from cornmint essential oil by crystallization at low temperature [15]. India, the main supplier of natural *l*-menthol (roughly 80 % of the world market), has been experiencing weather-related fluctuations in production, and the compound is thus also produced synthetically on an increasingly large scale (approximately 6,300 t/year in 2009; [27]) by (in alphabetical order) BASF (from *E*/*Z*-citral), Symrise/Lanxess (from *m*-cresol), and Takasago (from myrcene; [60]).

(−)-(1*R*,4*S*)-Menthone (syn. *l*-menthone; Scheme 1) is the second most abundant monoterpene [15–20 % (v/v)] of peppermint essential oil. The taste of *l*-menthone is refreshingly cool, but sharper than that of *l*-menthol and with a bitter afternote. It is added to adjust the taste and aroma of products containing mint oils (Table 1). If stored at room temperature, an epimerization of *l*-menthone will lead to the formation of (+)-(1*R*,4*R*)-isomenthone (syn. *d*-isomenthone), the aroma of which has a greener note. Commercial *l*-menthone is therefore always sold as a mixture with up to 29 % (v/v) *d*-isomenthone. There are several commercial ant and roach killer formulations that contain menthone as an active ingredient. Natural *l*-menthone is crystallized from dementholized peppermint or cornmint essential oil at about 1,300 t/year (Table 1). The compound is also prepared synthetically from different precursors, but these routes are only of minor commercial relevance [14].

(−)-(4*R*)-*trans*-Carvone (syn. *l*-carvone; Scheme 1) occurs at high levels [50–80 % (v/v)] in the essential oil of spearmint (*Mentha spicata* L.). Spearmint oil tastes very refreshing and cooling, but is milder than peppermint oil. *l*-Carvone is used commercially in the flavor and fragrance industries, mostly for cosmetic and personal hygiene products (applications that are similar to those listed for *l*-menthol). It is also an approved insect repellent in the United States. Roughly 1,800 t of spearmint oil are produced annually [49]. However, the majority of *l*-carvone (approximately 2,000 t/year) is not obtained by fractional crystallization of natural spearmint oil but by chemical synthesis from (+)-(4*R*)-limonene, which occurs at high levels in *Citrus* rind, a by-product of juice production (particularly orange fruit). Most commercial synthesis routes are based on a strategy that involves the formation of a characteristic nitrosochloride as an intermediate and was originally developed by Reitsema [71].

(+)-(4*S*)-*trans*-Carvone (syn. *d*-carvone; Scheme 1) is the major constituent in the oils distilled from caraway fruit [*Carum carvi* L.; 50–70 % (v/v)] and dill seed [*Anethum graveolens* L.; 40–60 % (v/v)]. Its aroma is, in contrast to that of the optical antipode *l*-carvone, characteristically pungent and anise-like. Caraway fruits are used as a spice in numerous traditional dishes in India (e.g., biryani) and several European countries (e.g., sauerkraut). Caraway oil is added as a fragrance to various personal hygiene products. *d*-Carvone is also sold as a potato-sprouting inhibitor in the Netherlands. About 30–40 t/year of caraway fruit oil (from which *d*-carvone is obtained by fractional distillation) are traded on the world market (Table 1; [58]). Synthetic *d*-carvone is produced from (−)-(4*S*)-limonene, a major component of turpentine oil (a by-product of pulp and paper processing).

(+)-(4*R*)-Limonene (syn. *d*-limonene; Scheme 1) is an intermediate in the biosynthesis of *d*-carvone in caraway fruit (accumulates to 25–30 % (v/v) of the essential oil). The most characteristic source of *d*-limonene, however, is the oil obtained from the fruit peel of various *Citrus* species where it accumulates to 70–98 % (v/v) of the oil [83]. *d*-Limonene imparts the characteristic citrusy smell of orange and lemon fruit and is used widely in the flavor and fragrance industries. Orange oil is an excellent, environmentally friendly, solvent employed in adhesives,

Table 1 Commercial sources and applications of *p*-menthane monoterpenes

Monoterpene	Source		Global production	Commercial applications
	Scientific name	Common name		
(−)-(1*R*,3*R*,4*S*)-Menthol (syn. L-Menthol)	*Mentha arvensis* L. (Lamiaceae) *Mentha* x *piperita* L. (Lamiaceae)	Cornmint Japanese mint	>30,000 t/year natural from cornmint and peppermint oil (*l*-Menthol obtained by low-temperature crystallization) 6,300 t/year synthetic in 2009 Starting materials for synthetic production: Myrcene (Takasago process) m-Cresol (Symrise process) E/Z-Citral (BASF process) Price of synthetic product: $15–20/kg	Flavoring ingredient Cosmetics and personal hygiene additive Tobacco additive Pesticide (repellent of tracheal mites of honey bees) Main nonprescription medical products: Topical analgesics, decongestants, cough suppressants
(−)-(1R,4S)-Menthone (syn. L-Menthone)	*Mentha* x *piperita* L. (Lamiaceae) *Mentha arvensis* L. (Lamiaceae)	Peppermint Cornmint Japanese mint	~1,300 t/year natural from peppermint and cornmint oil (*l*-Menthone crystallized from dementholized oil) Synthetic from various precursors (only minor production)	Flavoring ingredient Pesticide ant and roach killer formulations
(−)-(4*R*)-*trans*-Carvone (syn. l-Carvone)	*Mentha spicata* L. (Lamiaceae)	Spearmint	1,800 t/year natural spearmint oil produced globally (*l*-Carvone obtained by fractional distillation) ~2,000 t/year synthetic; starting materials: *d*-Limonene from *Citrus* peel oil Price of synthetic product: $15–20/kg	Flavoring ingredient Insect repellent (approved as pesticide in United States)

(continued)

Table 1 (continued)

Monoterpene	Source		Global production	Commercial applications
	Scientific name	Common name		
(+)-(4*S*)-trans-Carvone (syn. d-Carvone)	*Carum carvi* L. (Apiaceae)	Caraway Meridian fennel Persian cumin Shahi Jeera	30–40 t/year natural caraway oil produced globally (*d*-Carvone obtained by fractional distillation) Starting materials for synthetic production: *l*-Limonene from turpentine (biproduct of pulp/paper prod.) Price of synthetic product: $40–50/kg	Flavoring ingredient Potato sprouting inhibitor (marketed under trade name Talent® in the Netherlands)
(+)-(*R*)-Limonene (syn. d-Limonene)	*Citrus* sp. (Rutaceae)	Fruit peel oil of *Citrus* species (mostly orange and lemon)	>60,000 t/year natural orange and lemon oil produced globally (*d*-Limonene obtained by cold pressing or steam distillation)	Flavoring ingredient Cosmetics and personal hygiene additive Solvent for cleaning purposes and paints Pesticide (organic weedkiller formulations)
1,8-Cineole (syn. Eucalyptol)	*Eucalyptus* sp. (Myrtaceae)	Distillates of *Eucalyptus* species	~4,000 t/year natural *Eucalyptus* oil produced globally (Eucalyptol obtained by fractional distillation)	Flavoring ingredient Cosmetics and personal hygiene additive Nonprescription medical products
p-Menthane-3,8-diols (PMD) (syn. Menthoglycols)	*Eucalyptus citriodora* (Myrtaceae)	Lemon-scented gum Lemon eucalyptus	~1,500 t/year from lemon eucalyptus oil (PMD obtained from refined oil) Starting materials for semisynthetic production: Citronellal (Takasago process)	Insect repellent (mixture of PMD isomers marketed under trade name Citriodiol®)

stain removers, household cleaners, and strippers. *d*-Limonene is also employed as an active ingredient in organic weed killer formulations. *Citrus* oils are obtained commercially by cold pressing or distillation from orange and lemon fruit peel at >60,000 t/year (Table 1; [49]). Because of the comparatively low price ($9–10/kg) of *Citrus* essential oil, which is a by-product of fruit juice production, synthetic *d*-limonene is not produced on a larger commercial scale.

1,8-Cineole (syn. eucalyptol; Scheme 1) is the principal component [50–90 % (v/v)] of the essential oil of *Eucalyptus* (various species in this genus are grown commercially; [9]). It has a camphor-like smell and a spicy cooling taste. The flavor, fragrance, and cosmetic industries employ *Eucalyptus* oil as a fragrant ingredient in numerous consumer products (e.g., baked goods, confectionery, and beverages). It is also a characteristic component of over-the-counter mouthwash formulations and cough suppressants. Technical-grade 1,8-cineole is obtained by fractional distillation from *Eucalyptus* essential oil, whereas fragrance-grade materials require chromatographic isolation from the oil (total production approximately 4,000 t/year; Table 1; [49]). Synthetic 1,8-cineole does not play a significant role in the commercial sector.

***p*-Menthane-3,8-diol** (syn. menthoglycol; Scheme 1). The essential oil of lemon eucalyptus (*Eucalyptus citriodora* Hook.; aka *Corymbia citriodora*) contains (+)-(*R*)-/(−)-(*S*)-citronellal [50–80 % (v/v)] and (+)-(*R*)-/(−)-(*S*)-citronellal [7–20 % (v/v)] as the main products (Scheme 1; [9, 49]). Commercial insect repellents derived from the oil contain diastereomeric *p*-menthane-3,8-diols (PMDs), which are obtained at about 1,500 t/year from (+)-(*R*)-/(−)-(*S*)-citronellal by acid-catalyzed cyclyzation and subsequent crystallization (Table 1; [97]).

3 Genes and Enzymes of *p*-Menthane Monoterpene Biosynthesis

Monoterpene Synthases. The cyclization of geranyl diphosphate to *l*-limonene, catalyzed by (−)-limonene synthase, constitutes the first committed step in the biosynthetic pathway specific for *p*-menthane monoterpenes in peppermint and spearmint (Fig. 1). The gene encoding (−)-limonene synthase was first cloned from spearmint and the corresponding recombinant enzyme characterized by heterologous expression in *Escherichia coli* [17]. The open reading frame (ORF) of 1,800 bp, which is 96 % identical to the peppermint orthologue [45], codes for a 600 amino acid polypeptide. After import into leucoplasts (nongreen plastids of secretory cells in glandular trichomes), the N-terminal plastidial targeting sequence of the preprotein is proteolytically cleaved to generate a mature protein of approximately 545 residues and a size of ~65 kDa [84, 91]. The catalytic cascade of the (−)-limonene synthase reaction is thought to involve (1) ionization by removal of the diphosphate from C1 of geranyl diphosphate, (2) syn-migration of diphosphate to C3 (thus forming 3*S*-linalyl diphosphate), (3) rotation of the C1–C2 ethenyl group around C3 (from a transoid to a

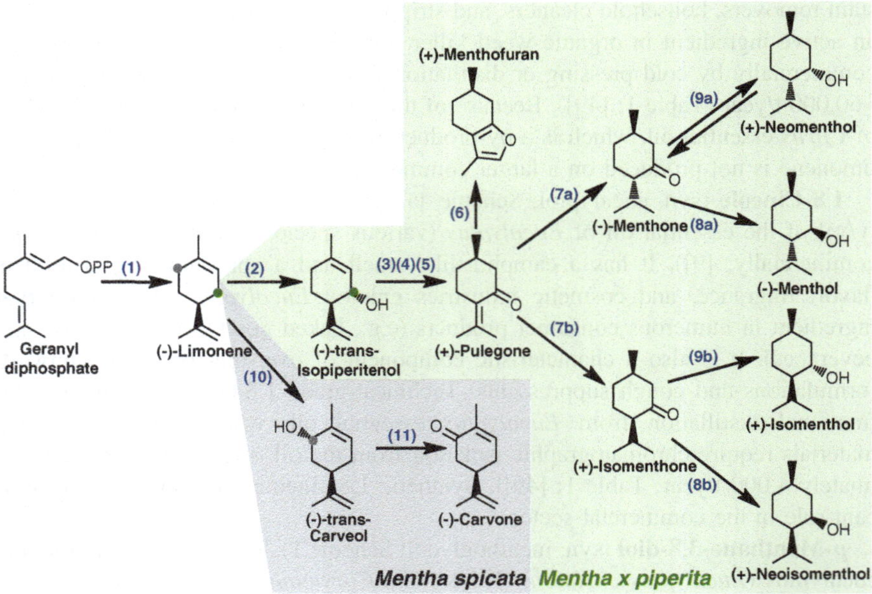

Fig. 1 *p*-Menthane monoterpene biosynthesis in peppermint (*Mentha* x *piperita* L.) and spearmint (*Mentha spicata* L.). The enzymes involved in this pathway are *1* (−)-limonene synthase; *2* (−)-limonene 3-hydroxylase; *3* (−)-*trans*-isopiperitenol dehydrogenase; *4* (−)-*trans*-isopiperitenone reductase; *5* (+)-*cis*-isopulegone isomerase; *6* (+)-menthofuran synthase; *7a* (+)-pulegone reductase ((−)-menthone-forming activity); *7b* (+)-pulegone reductase ((+)-isomenthone-forming activity); *8a* (−)-menthone:(−)-menthol reductase ((−)-menthol-forming activity); *8b* (−)-menthone:(−)-menthol reductase ((+)-neoisomenthol-forming activity); *9a* (−)-menthone:(+)-neomenthol reductase ((+)-neomenthol-forming activity); *9b* (−)-menthone:(+)-neomenthol reductase ((+)-isomenthol-forming activity); *10* (−)-limonene 6-hydroxylase; and *11* (−)-*trans*-carveol dehydrogenase

cisoid conformation), (4) a second ionization, (5) cyclization at C6–C1 to generate an α-terpinyl cation, and (6) deprotonation to yield primarily 4*S*-(−)-limonene (94 % of produced monoterpenes; [70]; Fig. 2, upper panel). Side reactions lead to the formation of myrcene (2 %), by premature deprotonation prior to step (5), and (−)-α-pinene/(−)-β-pinene (4 %) following a second ring closure between C2 and C8 of the geranyl diphosphate precursor and subsequent deprotonation. An analogous mechanism can be formulated for the (+)-limonene synthases of *Citrus* and caraway, but the reaction sequence in these enzymes proceeds via 3*R*-linalyl diphosphate (Fig. 2, lower panel, Fig. 3).

Limonene synthase sequences feature a conserved aspartate-rich motif (DDxxD) for binding the diphosphate moiety, which is released from geranyl diphosphate via divalent metal ions (Mg^{2+} or Mn^{2+}). A second metal cofactor binding region, termed the NSE/DTE motif [1], is also present in all limonene synthases. Despite a relatively low level of overall sequence conservation among monoterpene synthases, even when the nonconserved plastidial targeting sequence is excluded (45 % identity

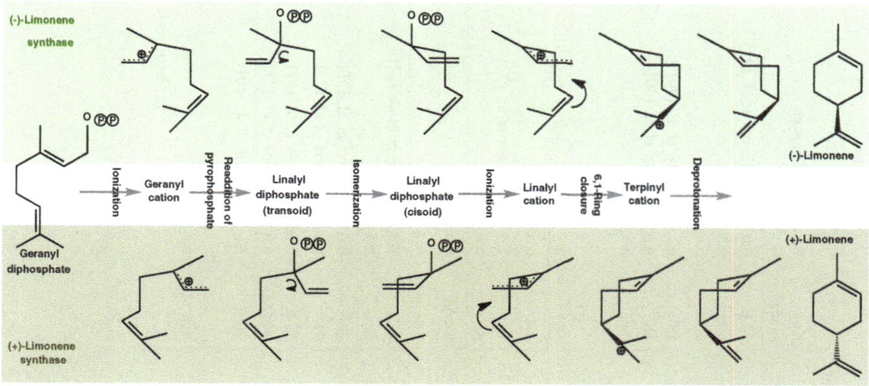

Fig. 2 Proposed reaction mechanism for (−)-limonene synthase in mint (*upper panel*) and (+)-limonene synthase in *Citrus* and caraway (*lower panel*). A helical folding of reaction intermediates is depicted. A mechanism involving extended conformations has also been proposed [36] but is not shown here

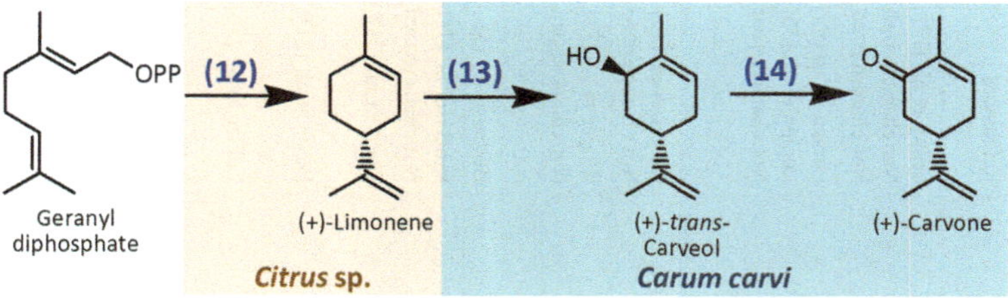

Fig. 3 *p*-Menthane monoterpene biosynthesis in *Citrus* and caraway (*Carum carvi* L.). The enzymes involved in this pathway are *12* (+)-limonene synthase; *13* (+)-limonene 6-hydroxylase; and *14* (+)-*trans*-carveol dehydrogenase

between (−)-limonene synthase (spearmint) and (+)-limonene synthase (orange); [52], the tertiary structures are remarkably similar. Spearmint (−)-limonene synthase shares a common fold, which consists mostly of α-helices with short connecting loops and turns, with all other members of the type I terpene synthases [36]. Limonene synthases have K_m values for geranyl diphosphate in the low micromolar range (12.6 μM for (−)-limonene synthase and 0.7 μM for (+)-limonene synthase) and a low catalytic rate constant ($K_{cat} = 0.037\ s^{-1}$ for (−)-limonene synthase; Table 2). The activity of (+)-limonene synthase in caraway has been detected [6] but the enzyme was not further characterized. All limonene synthases tested thus far generate only one enantiomer (either *d*- or *l*-limonene [22]).

Monoterpenoid Hydroxylases. *p*-Menthane monoterpenes are often accumulated as oxygenated metabolites. Hydroxylation reactions on monoterpenoid scaffolds are typically catalyzed by regiospecific cytochrome P450-dependent oxygenases localized to the endoplasmic reticulum (ER). In peppermint, the prevalent reaction is an oxygenation at C3 by (−)-limonene 3-hydroxylase, whereas in

Table 2 Genes and enzymes of *p*-menthane monoterpene biosynthesis

Annotation (Numbers used in figures given in parentheses)	Species	Gene information		Enzyme properties			
		ID	References	ID	Km (μM)	Kcat (s^{-1})	References
(−)-Limonene synthase (1)	*Mentha spicata* L. *Mentha* x *piperita* L. *Mentha arvensis* L.	L13459 EU108697 EF426463	Colby et al. [17] Not characterized Not characterized	EC 4.2.3.16 EC 4.2.3.16 EC 4.2.3.16	12.6 n.a. n.a.	0.037 n.a. n.a.	Williams et al. [91] n.a. n.a.
(−)-Limonene-3-hydroxylase (2)	*Mentha* x *piperita* L. *Mentha* x *piperita* L. *Mentha arvensis* L.	AF124817 (CYP71D13) AF124816 (CYP71D15) EF546776	Lupien et al. [55] Lupien et al. [55] Not characterized	EC 1.14.13.47 EC 1.14.13.47 EC 1.14.13.47	18 18 n.a.	n.a. n.a. n.a.	Karp et al. [40] Karp et al. [40] n.a.
(−)-*trans*-Isopiperitenol dehydrogenase (3)	*Mentha* x *piperita* L. *Mentha arvensis* L.	AY641428 AY641428	Ringer et al. [73] Not characterized	EC 1.1.1.223 EC 1.1.1.223	20 n.a.	n.a. n.a.	Ringer et al. [73] n.a.
(−)-*trans*-Isopiperitenone reductase (4)	*Mentha* x *piperita* L. *Mentha arvensis* L.	AY300162 AY300162	Ringer et al. [72] Not characterized	EC 1.3.1.82 EC 1.3.1.82	1.0 n.a.	1.3 n.a.	Ringer et al. [72] n.a.
(+)-*cis*-Isopulegone isomerase (5)	*Mentha* x *piperita* L.	n.a.	n.a.	n.a.	n.a.	n.a.	n.a.
(+)-Menthofuran synthase (6)	*Mentha* x *piperita* L.	AF346833	Bertea et al. [5]	EC 1.14.13.104	n.a.	n.a.	n.a.
(+)-Pulegone reductase (7)	*Mentha* x *piperita* L. *Mentha arvensis* L.	AY300163 EF426467	Ringer et al. [72] Not characterized	EC 1.3.1.81 EC 1.3.1.81	2.3 n.a.	1.8 n.a.	Ringer et al. [72] n.a.
(−)-Menthone:(−)-menthol reductase (8)	*Mentha* x *piperita* L.	AY288138	Davis et al. [21]	EC 1.1.1.207	3.0[a] 41[b]	0.6 n.a.	Davis et al. [21] Davis et al. [21]
(−)-Menthone:(+)-neomenthol reductase (9)	*Mentha* x *piperita* L.	AY288137	Davis et al. [21]	EC 1.1.1.208	674[a] >1,000[b]	0.06 n.a.	Davis et al. [21] Davis et al. [21]
(−)-Limonene-6-hydroxylase (10)	*Mentha spicata* L.	AF124815 (CYP71D18)	Lupien et al. [55]	EC 1.14.13.48	20.0	n.a.	Karp et al. [40]
(−)-*trans*-Carveol dehydrogenase (11)	*Mentha spicata* L.	n.a.	Ringer et al. [73]	EC 1.1.1.243	n.a.	n.a.	n.a.

(continued)

Table 2 (continued)

Annotation (Numbers used in figures given in parentheses)	Species	Gene information		Enzyme properties			
		ID	References	ID	Km (μM)	Kcat (s^{-1})	References
(+)-Limonene synthase (12)	*Citrus limon* L. Burm. F.	AF514287 (LS1)	Lücker et al. [52]	EC 4.2.3.20	0.7	n.a.	Lücker et al. [52]
	Citrus limon L. Burm. F.	AF514289 (LS2)	Lücker et al. [52]	EC 4.2.3.20	0.7	n.a.	Lücker et al. [52]
	Carum carvi L.	n.a.	n.a.	EC 4.2.3.20	n.a.	n.a.	n.a.
(+)-Limonene 6-hydroxylase (13)	*Carum carvi* L. *(annual)*	n.a.	n.a.	EC 1.14.13.48	11.4	n.a.	Bouwmeester et al. [7]
	Carum carvi L. *(biennial)*	n.a.	n.a.	EC 1.14.13.48	14.9	n.a.	Bouwmeester et al. [7]
(+)-*trans*-Carveol dehydrogenase (14)	*Carum carvi* L.	n.a.	n.a.	EC 1.1.1.275	n.a.	n.a.	n.a.
1,8-Cineole synthase (15)	*Eucalyptus sideroxylon*	n.a.	Keszei et al. [41]	EC 4.2.3.108	n.a.	n.a.	n.a.
Geraniol synthase (16)	*Eucalyptus sideroxylon*	n.a.	n.a.	EC 3.1.7.11	n.a.	n.a.	n.a.
Geraniol reductase (16)	*Eucalyptus sideroxylon*	n.a.	n.a.	n.a.	n.a.	n.a.	n.a.
Citronellol dehydrogenase (17)	*Eucalyptus sideroxylon*	n.a.	n.a.	n.a.	n.a.	n.a.	n.a.

Note that this table does not provide a comprehensive listing of all gene orthologues. *n.a.* not available
[a] Values given for (−)-(1*R*,4*S*)-menthone as substrate. [b] Values given for (+)-(1*R*,4*R*)-isomenthone as substrate

spearmint the predominant oxygenation occurs at C6 and is catalyzed by (−)-limonene 6-hydroxylase ([85]; Fig. 2). Two genes encoding (−)-limonene 3-hydroxylases, designated as CYP71D13 (1,497 bp) and CYP71D15 (1,503 bp), were cloned from peppermint and characterized following heterologous expression in both yeast [33] and the baculovirus-*Spodoptera* system ([55]; Table 2). The (−)-limonene 6-hydroxylase gene was cloned from spearmint (CYP71D18) and also functionally characterized by heterologous expression [55]. The enzyme has roughly the same size (~56 kDa) as the peppermint (−)-limonene 3-hydroxylase, but the sequences of these enzymes are only about 70 % identical. These hydroxylase sequences have the conserved features of typical cytochrome P450-dependent oxygenases, including an N-terminal membrane anchor for insertion into the ER, an oxygen-binding domain, a heme-binding motif, and a docking site for cytochrome P450 reductase (which is required for electron transfer from NADPH; [55]). The peppermint genome contains a C6 hydroxylase-like gene but the corresponding transcript is not expressed at appreciable levels. Spearmint expresses a (−)-limonene 3-hydroxylase-like gene but, compared to the 6-hydroxylase transcript, at fairly low levels (A. Ahkami, S. R. Johnson, N. Srividya, and B. M. Lange, unpublished results). The kinetic properties of the C3 and C6 hydroxylases (peppermint and spearmint, respectively) are very similar, with K_m values of 18–20 μM [40]. The gene encoding (+)-limonene 6-hydroxylase from caraway has not been cloned yet, but the kinetic properties (K_m of 11.4 μM for annual and 14.9 μM in biennial caraway) were very similar to those of (−)-limonene 6-hydroxylase from spearmint (Fig. 3; [7]). In contrast to limonene synthases characterized thus far, which generate either *d*- limonene or *l*-limonene stereospecifically [22], both the limonene C3 and C6 hydroxylases accept *d*-limonene as an unnatural substrate (thereby generating (+)-*cis*-carveol as a major product), albeit at lower affinity as natural *l*-limonene [40, 92].

The complex essential oil of peppermint contains, in addition to the major products *l*-menthone and *l*-menthol, (+)-(*R*)-menthofuran as a side product that is preferentially formed under stress conditions (Fig. 1; [10, 16, 74, 75, 89]). The enzyme responsible for the biosynthesis of this metabolite from the main pathway intermediate (+)-(1*R*)-pulegone is also a cytochrome P450-dependent oxygenase of the CYP71 family. The (+)-menthofuran synthase (MFS) gene (also known as (+)-pulegone 9-hydroxylase) was cloned from peppermint (1,479 nucleotides) and encodes a protein of 493 residues (~55 kDa). The function of this gene was assessed by functional expression/characterization in both *E. coli* and yeast [5]. Its closest homologues are as yet uncharacterized CYP71A family members of the Lamiaceae and Solanaceae (55–58 % identity). Both (+)-(1*R*)-pulegone and (+)-(*R*)-menthofuran are toxic to humans [98] and low levels in essential oils are generally preferable. However, because both metabolites impart a desirable and characteristic camphor-like note if present in small amounts, but carry an off-note in higher amounts, adjustments in the concentrations of these metabolites are critical for the flavor and fragrance industries.

Monoterpenoid Dehydrogenases/Reductases. C3 and C6 hydroxylated monoterpenes are further oxidized in both mint and caraway (Figs. 1 and 3). A cDNA clone with a 795 bp ORF (corresponding to a 265-residue protein) of the short-chain dehydrogenase/reductase (SDR) superfamily was cloned from peppermint and, following heterologous expression in *E. coli*, demonstrated to encode a 27 kDa protein under denaturing conditions (native protein likely a homodimer or homotetramer). The recombinant protein had relatively high affinity for (−)-(3*S*,4*R*)-*trans*-isopiperitenol and NAD^+ as cosubstrates (K_m values of 72 and 410 μM, respectively) but catalyzes a comparatively slow reaction (K_{cat} of 0.002 s^{-1}; Table 2; [73]). $NADP^+$ was a suitable alternative redox cofactor, but the activity of this dehydrogenase (for which the acronym ISPD was introduced) dropped to about $^1/_{10}$ of that with NAD^+. Interestingly, the affinity of peppermint ISPD for the spearmint pathway intermediate (−)-(4*R*,6*S*)-*trans*-carveol, when compared to the endogenous (−)-(3*S*,4*R*)-*trans*-isopiperitenol substrate, was significantly higher (K_m of 1.8). Furthermore, the reaction velocity with the unnatural substrate was also higher as with the endogenous pathway intermediate (K_{cat} of 0.02 s^{-1}). Various other alcohols occurring in members of the mint family were tested as substrates (NAD^+ as redox cofactor) but the ISPD activity was either very low or undetectable, indicating a notable specificity of this dehydrogenase. The reaction catalyzed by ISPD was irreversible (no reductase activity with either (−)-(4*R*)-*trans*-isopiperitenone or *l*-carvone as substrates).

The identity between putative ISPDs within the genus *Mentha* is >98 %, whereas the next closest SDR sequences in other plant families have approximately 60 % identity. Peppermint ISPD was localized to mitochondria of secretory cells within glandular trichomes using immunohistochemistry, with very little label found in other cell types [85]. An enzymatic activity for the NAD^+-dependent oxidation of various carveol isomers was detected in crude extracts of caraway fruit (highest specific activities for (+)-(4*R*,6*R*)-*trans*-carveol and (−)-(4*R*,6*R*)-*cis*-carveol), but the corresponding gene still awaits identification. The ISPD gene expression levels in secretory cells of both peppermint and spearmint glandular trichomes are relatively high, when compared to all other *p*-menthane pathway genes (A. Ahkami, S. R. Johnson, N. Srividya, and B. M. Lange, unpublished results), and the substrates of ISPD enzymes do not accumulate in the oil, indicating that this step is not rate-limiting for *p*-menthane monoterpene biosynthesis. The dehydrogenase reactions complete the pathway for the biosynthesis of the major spearmint and caraway monoterpenes, whereas further conversions, detailed in the following paragraphs, occur in peppermint.

The C3 oxidation of (−)-(3*S*,4*R*)-*trans*-isopiperitenol to (−)-(4*R*)-*trans*-isopiperitenone in peppermint is followed by a stereospecific reduction of the C_1–C_2 double bond to yield (+)-(1*R*,4*R*)-*cis*-isopulegone (Fig. 1). This reaction is catalyzed by (−)-*trans*-isopiperitenol reductase (ISPR), which is also a member of the SDR superfamily. The ISPR gene (942 bp ORF) codes for a cytosolic protein [86] with 314 residues and a size of approximately 34 kDa, which likely functions as a monomer [72]. ISPR shares fairly high sequence identity with (−)-menthone: (+)-neomenthol reductase (MNR; 64 % at the deduced amino acid level) and (−)-menthone:(−)-menthol reductase (MMR; 63 %) from peppermint (more details below), but is only distantly related to the ISPD gene (30 % identity). The native

ISPR protein, which was partially purified from peppermint glandular trichomes, is selective for NADPH as a redox cofactor (rather than NADH), and shows a high level of regioselectivity (reduction of C_1–C_2 double bond in *p*–menthadien-3-ones) as well as stereoselectivity (hydride attack from *si*-face of C_1–C_2 double bond to yield a product with 1*R* configuration; [18]). The reverse desaturation reaction (with $NADP^+$ as cofactor and (+)-(1*R*,4*R*)-*cis*-isopulegone as substrate) was not detectable. Recombinant ISPR (expressed in and purified from *E. coli*) has high affinity for (−)-(4*R*)-*trans*-isopiperitenone and NADPH (K_m values of 1.0 and 2.2 μM, respectively), with a fairly high turnover rate (*k*cat of 1.3 s^{-1}; [72]). (+)-(1*R*,4*R*)-*cis*-Isopulegone does not accumulate to measurable levels in peppermint essential oil, which is probably due to a combination of the relatively high abundance of ISPR transcript in secretory cells of peppermint glandular trichomes [45] and the comparatively high catalytic efficiency of the corresponding enzyme [72].

Pursuant to the ISPR reaction, an isomerization (not described in more detail here because the corresponding gene has not been identified yet) leads to the formation of (+)-(1*R*)-pulegone, which is then further reduced by (+)-pulegone reductase (PR; Fig. 1). The gene encoding PR consists of an ORF of 1,026 nucleotides, corresponding to a peptide sequence of 342 amino acid residues and a predicted monomeric protein size of about 38 kDa [72]. PR belongs to the medium-chain dehydrogenase/reductase (MDR) superfamily and is highly similar (60–68 % sequence identity) to double-bond reductases (some functionally characterized to act on structurally and biosynthetically unrelated substrates) across the angiosperm lineage. There is only very low sequence identity (below 20 %) between this MDR and the SDRs (ISPD, ISPR, MMR, and MNR) of peppermint *p*-menthane monoterpene biosynthesis. Based on immunohistochemical evidence, PR is localized to the cytosol of glandular secretory cells [85]. Partially purified, native PR (eluting as 45 kDa protein from size-exclusion chromatography) generates a mixture of *l*-menthone (70 %) and *d*-isomenthone (30 %) with (+)-(1*R*)-pulegone as a substrate and NADPH as redox cofactor (no activity with NADH; [72]. Other double-bond–containing *p*-menthane monoterpenes (e.g., (+)-(1*R*,4*R*)-*cis*-isopulegone or (+)-(4*S*)/(−)-(4*R*)-piperitone) were not suitable substrates and the reverse reaction (oxidation of *l*-menthone or *d*-isomenthone with $NADP^+$ as oxidant) was not observed. Interestingly, the partially purified, recombinant PR (assayed under the same conditions as the native enzyme with (+)-(1*R*)-pulegone and NADPH) yields *l*-menthone and *d*-isomenthone in a 55:45 ratio. As possible reasons for this discrepancy, a lack of posttranslational modifications of the recombinantly expressed protein or the involvement of a yet to be characterized epimerase converting *d*-isomenthone to *l*-menthone, have been discussed before [19], but the issue has remained unresolved. PR has a high affinity for the substrate and redox cofactor (K_m values of 2.3 and 6.9 μM for (+)-(1*R*)-pulegone and NADPH, respectively) and its activity is characterized by a relatively high reaction rate (k_{cat} of 1.8 s^{-1}). (+)-(1*R*)-Pulegone accumulates in the oil of peppermint (usually at 0.5–3.0 % in field-grown plants) which, considering the rather high enzymatic efficiency, likely reflects the relatively low amount of PR enzyme (when compared to the preceding enzymes ISPD and ISPR) present in glandular trichomes [59].

The last set of reactions of the *p*-menthane monoterpene biosynthetic pathway in peppermint is catalyzed by NADPH-dependent monoterpenoid keto reductases (Fig. 1; [42]). MMR and MNR are of similar size (1,096 and 1,131 nucleotide ORFs; 311 and 324 residue polypeptides; molecular weights of approximately 34 and 36 kDa, respectively). The sequences of MMR and MNR share 73 % identity and, as mentioned before, are almost equally similar to ISPR (63 and 64 %, respectively). Recombinant MMR converts *l*-menthone to *l*-menthol (95 %) and (+)-(1*R*,3*S*,4*S*)-neomenthol (syn. *d*-neomenthol; 5 %; K_m values of 3.0 and 0.12 μM for the substrate and redox cofactor, respectively; k_{cat} value of 0.6 s^{-1}), whereas (+)-(1*R*,3*S*,4*R*)-neoisomenthol (syn. *d*-neoisomenthol; 87 %) and *d*-isomenthol (13 %) are formed from *d*-isomenthone as a substrate (K_m value of 41 μM; [21]). The reactions catalyzed by MMR are irreversible (no oxidation of alcohols in the presence of $NADP^+$).

MNR catalysis leads to an essentially opposite product profile, when compared to that of MMR, but with dramatically reduced substrate affinity and reaction velocity. When *l*-menthone is used as a substrate in in vitro assays with the recombinant MNR enzyme (K_m value of 647 μM; k_{cat} value of 0.06 s^{-1}), *d*-neomenthol is the major product (94 %) and *l*-menthol a side product (6 %). With *d*-isomenthone as a substrate (K_m value of >1 mM), *d*-isomenthol (86 %) and *d*-neoisomenthol (14 %) are generated [21]. MMR is localized to the cytosol and nuclei of secretory cells in glandular trichomes at the later stages of peppermint leaf development [86]. This is consistent with the gene expression and enzyme activity profiles, which both indicate a late appearance of MMR during essential oil accumulation [21, 45, 59]. A combination of substrate availability (concentration of *l*-menthone is significantly higher than that of *d*-isomenthone in secretory cells) and the catalytic properties of the keto reductases (enzymatic efficiency of MMR is substantially higher than that of MNR) lead to the accumulation of predominantly *l*-menthol in the essential oil of peppermint, whereas the other alcohols are only minor constituents [74, 75].

***Eucalyptus p*-Menthane Monoterpene Pathway**. The principal monoterpene of *Eucalyptus* essential oils is 1,8-cineole (Fig. 4), which accumulates to >50 % (v/v)

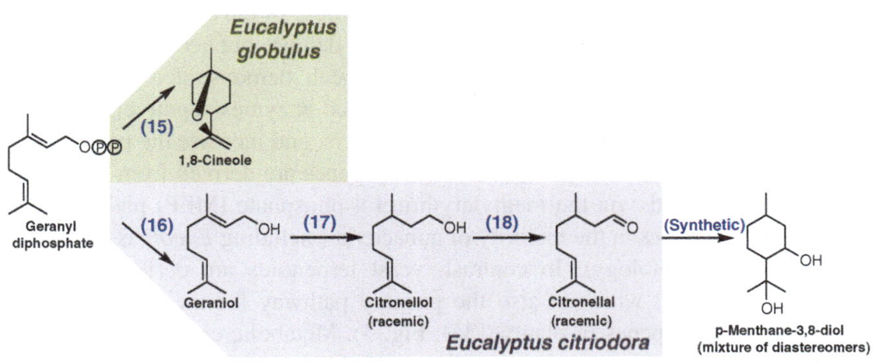

Fig. 4 *p*-Menthane monoterpene biosynthesis in *Eucalyptus* species. The enzymes involved in this pathway are *15* 1,8-cineole synthase; *16* geranyl diphosphate diphosphatase; *17* geraniol reductase; and *18* citronellol dehydrogenase

in commercial cultivars. A cDNA clone for a 1,8-cineole synthase was obtained recently and characterized by functional expression in *E. coli.* [41]. The recombinant protein released a mixture of monoterpenes with 1,8-cineole and linalool as major products, and α-terpineol, limonene, myrcene, sabinene, and ocimene (in decreasing amounts) as minor products. Interestingly, these metabolites also accumulate in the oil, although not in the same proportions as produced by recombinant 1,8-cineole synthase [41]. Lemon eucalyptus accumulates high levels of a mixture of (+)-(*R*)- and (−)-(*S*)-citronellal. Commercially, these metabolites are converted, by synthetic chemistry, into insecticidal *p*-menthane-3,8-diol (PMD; [97]. As a first step, the biosynthetic pathway involves the hydrolysis of the diphosphate group from geranyl diphosphate (Fig. 4). Geraniol synthases catalyzing this reaction have been cloned and characterized from various angiosperms [37, 38, 81, 94] but not within the Myrtaceae. The following double-bond reduction is catalyzed by geraniol reductase, the activity of which has been detected in the Rosids and Vitales [3, 26, 51], but the gene coding for this enzyme is still awaiting discovery. Finally, the (+)-(*R*)/(−)-(*S*)-citronellol mixture is oxidized to the corresponding aldehydes by citronellol dehydrogenase. Such an activity has been detected in several members of the genus *Pseudomonas*, which can utilize citronellal and other monoterpenes as carbon sources [30], but to date has not been described in plants.

4 Metabolic Engineering of the *p*-Menthane Monoterpene Biosynthetic Pathway

Fundamental Challenges with Monoterpene Storage. Larger amounts of volatile and semivolatile plant terpenoids (C_5 to C_{20} metabolites) are generally accumulated in intraorganellar structures that are surrounded by a membrane (e.g., oil bodies; [68]), larger extracellular storage structures (e.g., resin ducts, laticifers, secretory cavities, or glandular trichomes; [28]), or are emitted into the atmosphere [31]. These sequestration mechanisms likely evolved because of the cytotoxicity of many terpenoids, which can play important roles in plant defense but need to be separated from cellular metabolism. Monoterpenes have been demonstrated to affect the properties of both soluble and membrane-associated enzymes negatively, interfere with the action of intra- and intercellular transporters, and increase the permeability of membranes [80]. In most plant cells, monoterpenes are derived from precursors synthesized in plastids via the methylerythritol 4-phosphate (MEP) pathway. The same pathway operates in the majority of eubacteria (including *E. coli* as a common host for synthetic biology). In contrast, yeast terpenoids are derived from the mevalonate pathway, which is also the primary pathway for the biosynthesis of sterols and sesquiterpenes in plants ([34]; Fig. 5). Metabolic crosstalk, that is, the exchange of terpenoid intermediates between plastidial and cytosolic pathways, has been described in plants, but the extent, direction of carbon flux, and relevance of their intracellular transport for specific terpenoid end products is highly dependent on the plant species, the tissue of the cell type under investigation, and the environmental context [34].

The functionalization of monoterpene hydrocarbons usually commences with oxygenation reactions, often catalyzed by ER-localized cytochrome P450-dependent oxygenases [47]. Further modifications, including, among others, redox, desaturation, and conjugation reactions, occur mostly through catalysis by cytosolic or ER-associated enzymes (or less commonly mitochrondrial proteins). The final step, which involves either the volatilization or secretion/storage of monoterpenes, is only poorly understood [47]. In summary, there are three critical issues that need to be addressed when attempting to increase the amounts of specific monoterpenes in plants or microbes using biotechnological approaches: (1) volatility/storage, (2) cytotoxicity, and (3) precursor supply/compartmentation. In the following paragraphs, I highlight successes, but also mention limitations, of metabolic engineering efforts to increase the accumulation of monoterpenes.

Transformation of Commercial Cultivars to Accumulate High-Value *p*-Menthane Monoterpenes. Methods for the transformation of plant cultivars that naturally accumulate *p*-menthane monoterpenes were first developed in the 1990s. Most of these protocols (only early pioneering studies of gene introduction are listed here) involved the cocultivation of explants with *Agrobacterium* and the subsequent regeneration of transgenic plants, and were successful with *Mentha* [4, 11, 23, 64], *Carum* [44], *Citrus* [61, 88], and *Eucalyptus* [35, 62, 63]. Transient transformations using biolistic treatments were also reported [11, 76, 79, 96]. Although these methods are viable on an experimental scale, the transformation efficiency with these plant genera is universally low and the regeneration of transgenic plants with desirable traits fairly slow.

Despite these challenges, significant progress has been made with engineering mint for improved essential oil characteristics. Genes involved in precursor supply and at the interface toward monterpene end products (MEP and early *p*-menthane pathways) (Fig. 5) were overexpressed to increase flux and thereby increase oil yield. For several genes this approach was successful (1-deoxy-D-xylulose 5-phosphate reductoisomerase (DXR), isopentenyl diphosphate isomerase, geranyl diphosphate synthase), whereas increased transcript abundance in transgenic plants was not associated with yield gains for other genes (1-deoxy-D-xylulose 5-phosphate synthase and (−)-limonene 3-hydroxylase; Table 3; [24, 46, 56, 57]; all in peppermint). Results for (−)-limonene synthase overexpression were variable, with some authors reporting yield enhancements ([24]; in peppermint), whereas others did not detect improvements in yield (Table 3; [43], peppermint; [24], cornmint; [46], peppermint). The Croteau laboratory pursued compositional adjustments by decreasing the expression levels of genes with relevance for the formation of undesirable side products. The expression of the MFS gene (gene (6) in Fig. 1) in antisense orientation (MFS7A line) led to a decrease in the concentrations of, as expected, (+)-(*R*)-menthofuran but also, unexpectedly, (+)-(1*R*)-pulegone in transgenic peppermint plants (Table 3; [56]). It seemed counterintuitive that the concentration of the substrate for a downregulated enzyme would decrease and a

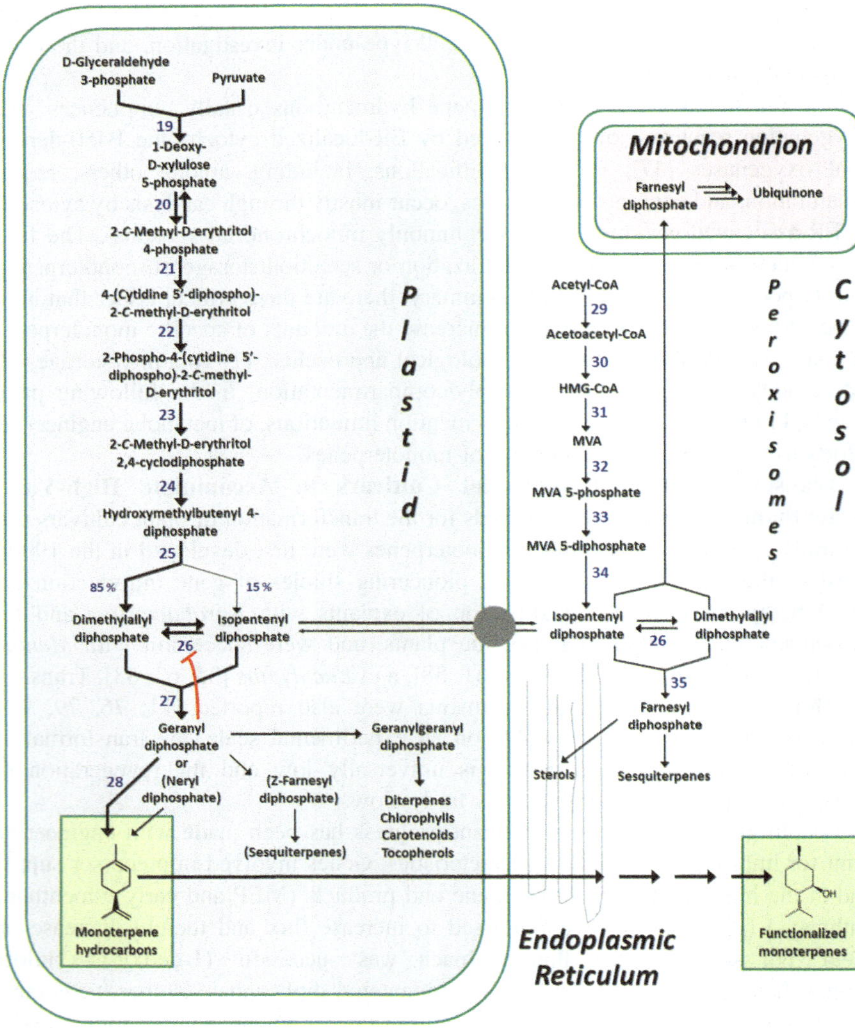

plausible explanation did not emerge until my laboratory developed a kinetic mathematical model of monoterpenoid essential oil biosynthesis [74]. Simulations indicated that the observed monoterpene profiles could potentially be explained if (+)-(*R*)-menthofuran acted as a competitive inhibitor of (+)-pulegone reductase, an hypothesis that was subsequently confirmed by experimental testing [74]. A second surprising finding was that the MFS7A line had an increased essential oil yield [56], although this gene was the only one found to be modulated in MFS7A and appeared to be unrelated to overall carbon flux into monoterpenes. Simulations with a second-generation mathematical model, which we had developed also to assess the determinants of yield, suggested that differences in the size distribution of glandular trichomes could explain the observed yield gain [75].

◄**Fig. 5** Outline of terpenoid biosynthesis in plants with an emphasis on the cellular context of *p*-menthane monoterpene formation. The generally higher flux observed for the plastidial methylerythritol phosphate (MEP) pathway, in comparison to the mevalonate (MVA) pathway, which is localized the cytosol, ER, and peroxisomes, is indicated by *thicker reaction arrows* (peroxisome not depicted for clarity). A transporter for the exchange of isoprenoid intermediates across the plastidial membrane is indicated by a *gray circle*. Abbreviations for metabolites: *CoA* coenzyme A; *HMG-CoA* 3-hydroxy-3-methylglutaryl-CoA; *MVA* mevalonate. Enzyme numbering: *19* 1-deoxy-D-xylulose 5-phosphate synthase; *20* 1-deoxy-D-xylulose 5-phosphate reductoisomerase; *21* 2-C-methyl-D-erythritol 4-phosphate cytidyltransferase; *22* 4-(cytidine 5′-diphospho)-2-C-methyl-D-erythritol kinase; *23* 2-C-methyl-D-erythritol 2,4-cyclodiphosphate synthase; *24* (*E*)-4-hydroxy-3-methyl-but-2-enyl diphosphate synthase; *25* (*E*)-4-hydroxy-3-methylbut-2-enyl diphosphate reductase; *26* isopentenyl diphosphate isomerase; *27* geranyl diphosphate synthase; *28* monoterpene synthase; *29* acetoacetyl-CoA thiolase; *30* 3-hydroxy-3-methyl-glutaryl-CoA synthase; *31* 3-hydroxy-3-methylglutaryl-CoA reductase; *32* mevalonate kinase; *33* phosphomevalonate kinase; *34* mevalonate diphosphate decarboxylase; *35* (*E,E*)-farnesyl diphosphate synthase. Plastidial pathways for the biosynthesis of monoterpenes via neryl diphosphate and certain sesquiterpenes via (Z,Z)-farnesyl diphosphate, as recently identified in the Solanaceae, are depicted as well. However, the relevance of these pathways for the biosynthesis of commercially harvested *p*-menthane monoterpenes has not yet been established [47]. Note that the specialized cell types involved in the biosynthesis of monoterpenes in glandular trichomes and secretory cavities generally have, with only a few exceptions, nongreen plastids that do not produce chlorophylls and carotenoids from plastidial geranylgeranyl diphosphate at the time of terpenoid accumulation

Follow-up experiments demonstrated that critical aspects of the pattern of glandular trichome development were altered in MFS7A, where these structures were more numerous and matured earlier than in wild-type controls, thus leading to higher essential oil amounts independent of gene expression patterns within secretory cells of glandular trichomes [75]. Essential oil yield gains did not correlate directly with the expression patterns of biosynthetic genes but were achieved through the initiation of more glandular trichomes, possibly by an as yet uncharacterized feedback mechanism resulting from gene overexpression. By transforming peppermint with two genes, one demonstrated before to affect yield (DXR) positively and one with beneficial effects on both yield and composition (MFS), we then generated transgenic plants with significant yield increases (up to 80 % higher than wild-type) and highly desirable oil profiles (high *l*-menthone and *l*-menthol; low (+)-(*R*)-menthofuran and (+)-(1*R*)-pulegone), as demonstrated in multiyear field trials (Table 3; [46]).

We have also employed peppermint as a host for biotechnological efforts to accumulate other *p*-menthane monoterpenes that occur only at very low levels in this plant. We were particularly interested to investigate if metabolic engineering was a viable approach to eliminate oxygenated constituents from the oil, thereby generating mixtures of monoterpene hydrocarbons with physicochemical properties desirable for the development of drop-in biofuels. Our strategy took advantage of a line (termed PM20) in which the limonene 3-hydroxylase gene was cosuppressed, thus resulting in an overaccumulation of *l*-limonene in the oil [57]. The PM20 line was transformed with additional constructs for the overexpression of various terpene synthases to generate transgenic lines with elevated levels of (*S*)-(+)-linalool, (+)-(1*R*,5*R*)-sabinene, and α/β-pinene [46]. Further advances can be expected from the use of glandular trichome-specific promoters (to increase the strength and specificity of transcript abundance in the cell types that are most relevant for essential oil biosynthesis) and global regulators of the *p*-menthane monoterpene pathway (to activate more than one or two genes at a time). Early work in this area has recently been summarized, and comments on opportunities and current constraints have been provided elsewhere [48].

The vast majority of studies with transgenic *Citrus* plants were performed to enhance the resistance to biotic and abiotic stresses [82], which is outside the scope of this review chapter. Very little effort has been spent on modulating the essential oils accumulated in fruit peel. One of the reasons is that, depending on the *Citrus* cultivar, the oil is already enriched in a desirable, high-value *p*-menthane monoterpene (*d*-limonene at up to 97 % of the oil). It is also important to note that monoterpenoid fruit volatiles play critical roles in the interaction between plants and beneficial as well as disease-causing organisms [78]. Interestingly, when the (+)-limonene synthase gene was downregulated in orange- using antisense technology, the resulting transgenic plants, which accumulated dramatically reduced peel oil, were significantly more resistant (when compared to appropriate controls) to a bacterial and a fungal pathogen, and males of the citrus pest medfly were less attracted to the fruit, thus decreasing feeding damage [77].

Ohara et al. [66] compared the effects of transforming *Eucalyptus camaldulensis* with two different constructs for the expression of a (−)-limonene synthase gene from *Perilla frutescens* (Table 3). One construct was designed to direct the gene product to the cytosol, and the other targeted the enzyme to plastids. As mentioned before, the precursors for monoterpenes, geranyl diphosphate (or neryl diphosphate in some plants), are generated in plastids [48]. The essential oil of *Eucalyptus* accumulates primarily 1,8-cineole, which is generated by a plastidial monoterpene synthase [41]. One would thus expect that a plastidial localization of an introduced monoterpene synthase should result in the generation of higher amounts of the corresponding monoterpene when compared to a cytosolic localization that would require translocation of geranyl diphosphate to the cytosol). Surprisingly, Ohara et al. [66] observed a higher (−)-limonene production with the construct for cytosolic localization of the gene product (Table 3). Even more surprisingly, it was also reported that enhanced (−)-limonene concentrations in transgenic plants strongly correlated with an increased accumulation of the plastidially generated *p*-menthane monoterpenes 1,8-cineole and α-pinene [66]. The causes for these highly unexpected findings were not further investigated.

Accumulation of High-Value *p*-Menthane Monoterpenes in Heterologous Plant Hosts. Although some plants that accumulate valuable essential oils might be amenable to metabolic engineering, the transformation efficiencies, as mentioned above, are generally fairly low and alternative plant hosts have been evaluated for biotechnological efforts to produce specific *p*-menthane monoterpenes. A (−)-limonene synthase gene from *Perilla frutescens* was transformed into tobacco using three different constructs designed to direct the gene product to either plastids, the

Table 3 Metabolic engineering of *p*-menthane monoterpene biosynthesis in commercial cultivars (representative examples)

Engineered species	Introduced gene	Promoter	Targeting of gene product	Primary *p*-menthane monoterpene targets	Outcomes	References
Mentha x *piperita* L.	1-Deoxy-D-xylulose 5-phosphate reductoisomerase	CaMV 35S (ubiquitous)	Plastid	(−)-(1*R*,4*S*)-Menthone (−)-(1*R*,3*R*,4*S*)-Menthol	44 % oil yield increase	Mahmoud et al. [56]
	Isopentenyl diphosphate isomerase	CaMV 35S (ubiquitous)	Plastid	(−)-(1*R*,4*S*)-Menthone (−)-(1*R*,3*R*,4*S*)-Menthol	26 % oil yield increase	Lange et al. [46]
	Geranyl diphosphate synthase	CaMV 35S (ubiquitous)	Plastid	(−)-(1*R*,4*S*)-Menthone (−)-(1*R*,3*R*,4*S*)-Menthol	18 % oil yield increase	Lange et al. [46]
	(+)-(*R*)-Menthofuran synthase (antisense orientation)	CaMV 35S (ubiquitous)	n.a.	(+)-(*R*)-Menthofuran (decrease)	35 % oil yield increase (+)-(*R*)-Menthofuran < 3 % of oil (+)-(1*R*)-Pulegone < 1 % of oil	Mahmoud et al. [56]
	1-deoxy-D-xylulose 5-phosphate reductoisomerase and (+)-(*R*)-Menthofuran synthase (antisense orientation)	CaMV 35S (ubiquitous) CaMV 35S (ubiquitous)	Plastid n.a.	(−)-(1*R*,4*S*)-Menthone (−)-(1*R*,3*R*,4*S*)-Menthol (+)-(*R*)-Menthofuran (decrease)	61 % oil yield increase (+)-(*R*)-Menthofuran < 2 % of oil (+)-(1*R*)-Pulegone < 1 % of oil	Lange et al. [46]
	(−)-Limonene 3-hydroxylase (cosuppression)	CaMV 35S (ubiquitous)	n.a.	(−)-(4*S*)-Limonene	(−)-(4S)-Limonene 79 % of oil (1–2 % in wild-type controls)	Mahmoud et al. [57]
Citrus sinensis L.	(−)-Limonene synthase (antisense orientation)	CaMV 35S (ubiquitous)	n.a.	(−)-(4*S*)-Limonene (decrease)	Enhanced resistance against *Penicillium digitatum*	Rodríguez et al. [77]

(continued)

Table 3 (continued)

Engineered species	Introduced gene	Promoter	Targeting of gene product	Primary *p*-menthane monoterpene targets	Outcomes	References
Eucalyptus camaldulensis Dehnh.	(−)-Limonene synthase	CaMV 35S (ubiquitous)	Cytosol	(−)-(4*S*)-Limonene	(−)-(4*S*)-Limonene 0.033 % of fresh weight (control: 0.001 %) 1,8-Cineole 0.16 % of fresh weight (control: 0.02 %) α-Pinene 0.016 % of fresh weight (control: 0.002 %)	Ohara et al. [66]
		CaMV 35S (ubiquitous)	Plastid	(−)-(4*S*)-Limonene	(−)-(4*S*)-Limonene 0.02 % of fresh weight (control: 0.001 %) 1,8-Cineole 0.08 % of fresh weight (control: 0.02 %) α-Pinene 0.016 % of fresh weight (control: 0.02 %)	

CaMV cauliflower mosaic virus; *n.a.* not applicable

cytosol, or ER membranes (Table 4; [65]). Transgenic plants carrying the construct for the plastidial localization of the enzyme had the highest (−)-limonene levels with up to 143 ng/g fresh weight (corresponding to roughly 0.0001 % of dry weight biomass). Lücker et al. [53] transformed tobacco with three terpene synthase genes from *Citrus* encoding (+)-limonene synthase, γ-terpinene synthase, and (−)-β-pinene synthase. The corresponding monoterpenes, (+)-limonene, γ-terpinene, and (−)-β-pinene, were emitted by transgenic lines (termed TERLIMPIN) at 70–1,200 ng/g fresh weight within a 24-h period (depending on developmental stage) (Table 4). Over longer periods of time, substantial amounts of monoterpenes were produced in these transgenic plants but, due to their volatilization, would not be harvestable in a commercially meaningful fashion. When TERLIMPIN plants were additionally transformed with a putative limonene 3-hydroxylase gene from curly mint, the 3-hydroxylated metabolite (+)-*trans*-isopiperitenol was emitted from flowers at 400 ng/g fresh weight within a 24-h period (Table 4) [54]. Notably, various other *p*-menthane monoterpenes, particularly *p*-cymene and isomers of menthatriene and isopiperitenone, were also formed, by as yet uncharacterized conversions, in these transgenic plants.

In lemon eucalyptus, geraniol is a biosynthetic precursor of (+)-(*R*)/(−)-(*S*)-citronellal, which itself is converted synthetically to PMD. Two independent attempts to produce geraniol in a heterologous plant host have been published. A geraniol synthase gene from sweet basil was transformed into tomato and expressed under control of the fruit-specific polygalacturonidase promoter [20]. A larger number of monoterpenes, including some that are derived from geraniol (e.g., geraniol, geranic acid, citronellol, neral) and others that are not (e.g., myrcene, limonene, and β-ocimene), were produced in fruit of transgenic plants at about 0.0004 % of fruit fresh weight (Table 4). In addition to releasing metabolic turnover products, the transgenic fruit were deficient in carotenoids, most likely due to a diversion of flux from a common precursor (geranyl diphosphate) to monoterpenes (Fig. 5; [20]).

A geraniol synthase from *Lippia dulcis* was expressed under the control of the ubiquitin promoter in transgenic maize plants [95]. Novel volatiles were not generated at detectable levels, but various glycosylated, acetylated, malonylated, and oxidized derivatives of geraniol accumulated (only geranoyl-6-*O*-malonyl-β-D-glucose was quantified at 0.0017 % of leaf fresh weight; Table 4). Fischer et al. [29] expressed a geraniol synthase from sweet basil in several different plant hosts. In *Arabidopsis thaliana*, geraniol was the only newly formed monoterpene in transgenic plants (0.0002 % of fresh weight). In tobacco, linalool, and nerol were found in addition to geraniol, whereas grapevine produced citronellol and nerol as side products (Table 4; [29]). The authors hypothesized that the terpene synthase activity can be modulated depending on the cellular context within the host organism. However, because side products accumulated only at very low levels, it is also conceivable that other endogenous activities were induced in transgenic plants (which was not investigated). In summary, these results indicate that in heterologous plant hosts expressing terpene synthases the initially formed catalytic products can be subjected to as yet unpredictable metabolic turnover.

Table 4 Metabolic engineering of *p*-menthane monoterpene biosynthesis in heterologous plant hosts

Engineered species	Introduced gene	Promoter	Targeting of gene product	Primary *p*-menthane monoterpene targets	Outcomes	References
Nicotiana tabacum L.	(−)-Limonene synthase	CaMV 35S (ubiquitous)	Cytosol	(−)-(4*S*)-Limonene	(−)-(4*S*)-Limonene 0.000004 % of fresh weight (control: n.d.)	Ohara et al. [65]
		CaMV 35S (ubiquitous)	Plastid	(−)-(4*S*)-Limonene	(−)-(4*S*)-Limonene 0.00001 % of fresh weight (control: n.d.)	
		CaMV 35S (ubiquitous)	ER	(−)-(4*S*)-Limonene	(−)-(4*S*)-Limonene n.d.	
	(+)-Limonene synthase	CaMV 35S (ubiquitous)	Plastid	(+)-Limonene	1,200 ng g^{-1} 24 h^{-1} (max. emission)	Lücker et al. [53]
	γ-Terpinene synthase	CaMV 35S (ubiquitous)	Plastid	γ-Terpinene	630 ng g^{-1} 24 h^{-1} (max. emission)	
	(−)-β-Pinene synthase	CaMV 35S (ubiquitous)	Plastid	(−)-β-Pinene	380 ng g^{-1} 24 h^{-1} (max. emission)	
	(+)-Limonene synthase	CaMV 35S (ubiquitous)	Plastid	(+)-Limonene	Roughly the same as in Lücker et al. [53]	Lücker et al. [54]
	γ-Terpinene synthase	CaMV 35S (ubiquitous)	Plastid	γ-Terpinene	Roughly 66 % lower as in Lücker et al. [53]	
	(−)-β-Pinene synthase	CaMV 35S (ubiquitous)	Plastid	(−)-β-Pinene	Roughly the same as in Lücker et al. [53]	
	limonene 3-hydroxylase	CaMV 35S (ubiquitous)	ER	(+)-*trans*-Isopiperitenol	400 ng g^{-1} 24 h^{-1} (max. emission) (other monoterpenes also formed)	
	Geraniol synthase	CaMV 35S (ubiquitous)		Geraniol	Geraniol at 0.001 % of fresh weight (linalool and nerol formed)	Fischer et al. [29]

(continued)

Table 4 (continued)

Engineered species	Introduced gene	Promoter	Targeting of gene product	Primary *p*-menthane monoterpene targets	Outcomes	References
Solanum lycopersicum L.	Geraniol synthase	Polygalacturonase (fruit-specific)	Plastid	Geraniol	Geraniol and other monoterpenes formed at 0.0004 % of fresh weight; carotenoid deficiency	Davidovich-Rikatani et al. [20]
Zea mays subsp. *mays* L.	Geraniol synthase	Ubiquitin (ubiquitous)	Plastid	Geraniol	Only geraniol derivatives (but not geraniol) detected (metabolites mostly not quantified);	Yang et al. [95]
Arabidopsis thaliana L. Heynh.	Geraniol synthase	CaMV 35S (ubiquitous)		Geraniol	Geraniol at 0.0002 % of fresh weight	Fischer et al. [29]
Vitis vinifera L.	Geraniol synthase	CaMV 35S (ubiquitous)		Geraniol	Geraniol at 0.0005 % of fresh weight	Fischer et al. [29]

CaMV cauliflower mosaic virus; *ER* endoplasmic reticulum

Accumulation of High-Value *p*-Menthane Monoterpenes in Microbial Hosts Tremendous progress has been made in recent years to develop microbial platform strains for the introduction of plant pathways, including those leading to terpenoids [93]. Considering the challenges with the high-level production of monoterpenes in engineered plants, microbial hosts are an obvious choice for monoterpene biotechnology. The genomes of typical engineering strains of *E. coli* and baker's yeast, which are the most common hosts due to the availability of a whole range of tools for synthetic pathway construction, are devoid of genes for the biosynthesis of monoterpenes. The first introduction of genes for an entire plant pathway into *E. coli* was performed by Carter et al. [12]. The authors generated cassettes for the single and combined expression (as polygenic operons) of genes encoding geranyl diphosphate synthase (N-terminally truncated version from *Abies grandis*), (−)-limonene synthase (N-terminally truncated version from *Mentha spicata*), (−)-limonene 6-hydroxylase (from *Mentha spicata*) fused to NADPH:cytochrome P450 reductase (from *Mentha* x *piperita*), and (−)-*trans*-carveol dehydrogenase (*Mentha spicata*), thus transferring the pathway used for the biosynthesis of *l*-carvone in spearmint. An engineered *E. coli* strain expressing only the geranyl diphosphate synthase and (−)-limonene synthase genes released *l*-limonene at 5 mg/l into the culture medium (Table 5; [12]). Interestingly, when all genes were coexpressed in engineered *E. coli*, the results remained the same (5 mg/l *l*-limonene). The authors hypothesized that two factors may have contributed to these undesirable results: (1) the *l*-limonene substrate might not be formed at sufficiently high levels (40 μM at the end of a 24-h growth period) and (2) *l*-limonene is secreted, which further decreases the amounts of available substrate for the hydroxylase. *l*-Limonene was then added to the culture medium (at 1 mM) and small amounts of *l*-carvone (2 μM) were produced. Although a proof-of-concept stage was achieved, the data indicated that secretion of *l*-limonene remained the most problematic challenge [12].

For the higher-level production of a functionalized monoterpene, Alonso-Gutierrez et al. [2] implemented several strain improvements. The supply of terpenoid precursors was enhanced through the introduction of a synthetic MVA pathway (with genes from various sources), thereby evading the tight regulation of the endogenous MEP pathway in the *E. coli* host (Fig. 6). In addition to a construct carrying codon-optimized variants of the genes encoding MVA pathway enzymes, geranyl diphosphate synthase (N-terminally truncated version from *Abies grandis*) and (−)-limonene synthase (N-terminally truncated version from *Mentha spicata*), the authors also introduced a second cassette carrying genes coding for a cytochrome P450-dependent oxygenase previously demonstrated to hydroxylate *l*-limonene at C7 [87], ferredoxin, and ferredoxin reductase (as electron carriers; all genes from *Mycobacterium* HXN 1500; Fig. 6).

In a rich medium containing 1 % glucose as the carbon source, the strain produced *l*-limonene at 400 mg/l and (−)-(4*S*)-perillyl alcohol at 100 mg/l (Table 5; [2]). These levels were significantly higher than those achieved before but the high concentrations of *l*-limonene indicate that secretion of this intermediate and/or inefficient turnover to the desired end product were still limiting factors. For the *E. coli* based production of simple monoterpene hydrocarbons as potential biofuels,

Table 5 Metabolic engineering of *p*-menthane monoterpene biosynthesis in microbial hosts

Engineered species	Introduced genes (Source)	Modifications	Primary *p*-menthane monoterpene targets	Outcomes	References
Escherichia coli	Geranyl diphosphate synthase (*Abies grandis*)	5′-Truncated (gene product lacks plastidial targeting sequence)	(−)-(4*R*)-*trans*-Carvone	(−)-(4*S*)-Limonene at 5 mg/l (no (−)-(4*R*)-*trans*-carvone)	Carter et al. [12]
	(−)-Limonene synthase (*Mentha spicata*)	5′-Truncated (gene product lacks plastidial targeting sequence)			
	(−)-Limonene 6-hydroxylase (*Mentha spicata*)–NADPH:cytochrome P450 reductase (*Mentha* x *piperita*)	Gene fusion			
	(−)-*trans*-carveol dehydrogenase (*Mentha spicata*)				
	Acetoacetyl-CoA thiolase (*Escherichia coli*)		(−)-(4*S*)-Perillyl alcohol	(−)-(4*S*)-Limonene at 400 mg/l; (−)-(4*S*)-Perillyl alcohol at 100 mg/l	van Beilen et al. [87]
	3-Hydroxy-3-methyl-glutaryl-CoA synthase (*S. aureus*)	Codon-optimized			
	3-Hydroxy-3-methylglutaryl-CoA reductase (*S. aureus*)	5′-Truncated; (gene product soluble rather than ER-localized) Codon-optimized			
	Mevalonate kinase (*Saccharomyces cerevisiae*)	Codon-optimized			
	Phosphomevalonate kinase (*S. cerevisiae*)	Codon-optimized			

(continued)

Table 5 (continued)

Engineered species	Introduced genes (Source)	Modifications	Primary *p*-menthane monoterpene targets	Outcomes	References
	Mevalonate diphosphate decarboxylase (*S. cerevisiae*)	Codon-optimized			
	Isopentenyl diphosphate synthase (*E. coli*)				
	Geranyl diphosphate synthase (*Abies grandis*)	5′-Truncated; Codon-optimized			
	(−)-Limonene synthase (*Mentha spicata*)	5′-truncated; Codon-optimized			
	CYP153 (Mycobacterium HXN 1500)				
	Ferredoxin (Mycobacterium HXN 1500)				
	Ferredoxin reductase (Mycobacterium HXN 1500)				
Saccharomyces cerevisiae	Farnesyl diphosphate synthase (*S. cerevisiae*) Geraniol synthase (*Ocimum basilicum*)	Mutated gene sequence	Geraniol	Geraniol at 5 mg/l	Oswald et al. [67] Fischer et al. [29]

Dunlop et al. [25] screened a library of efflux pumps for the secretion of various hydrophobic molecules. In this case, a secretion of nonfunctionalized terpenoids is actually desired. A construct very similar to that mentioned above (carrying genes coding for the MVA pathway enzymes, geranyl diphosphate synthase and *l*-limonene synthase) was combined with the expression of a suitable efflux pump (from *Alcanivorax borkumensis*). Controls (same constructs but lacking the pump) released *l*-limonene at 35 mg/l, whereas strains that also expressed the gene for the efflux pump produced the metabolite at 55 mg/l, indicating that product export is a viable strategy for the increased accumulation of monoterpene hydrocarbons (Table 5). The Karst laboratory explored yeast as an expression host for plant monoterpene biosynthetic genes. To accomplish this goal, a mutant farnesyl diphosphate synthase (FPPS) was developed that releases, in addition to the natural product farnesyl diphosphate, the shorter-chain geranyl diphosphate and derived hydrolysis products (geraniol and linalool; [13]. The release of geraniol from a yeast strain carrying the mutant FPPS was further increased by introducing a geraniol synthase gene from sweet basil [67]. Further optimization of the FPPS to produce high amounts of geranyl diphosphate, while maintaining sufficient farnesyl diphosphate levels for continued growth, resulted in strains that released geraniol at up to 5 mg/l (with linalool and citronellol as side products; Table 5) [29]. Brennan et al. [8] reported on two-phase extractive fermentation systems that trap monoterpenes in an organic phase, thus mitigating issues with the toxicity of these metabolites. It is now instructive to test if this approach is also viable when employed with engineered monoterpene-producing strains.

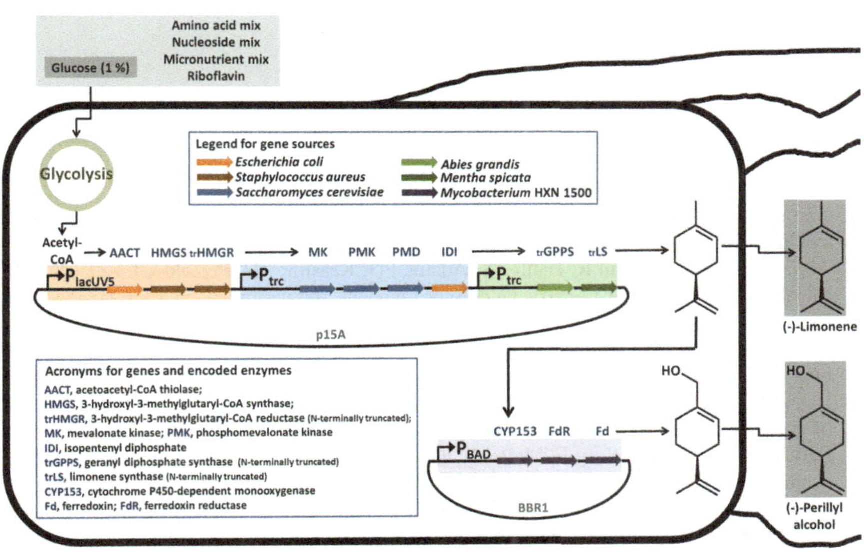

Fig. 6 Metabolic engineering of oxygenated *p*-menthane monoterpene biosynthesis in *E. coli* (modified after Alonso-Gutierrez et al. [2])

5 Conclusions

Numerous approaches have been tested to develop plant and microbial platform cultivars/strains for the high-level accumulation of terpenoids. After decades of research, several trends have emerged that are relevant for the production of *p*-menthane monoterpenes as targets:

- The yields of engineered terpenoids in the native producer (particularly if it is a plant with specialized anatomical structures for terpenoid storage) are generally higher than those in heterologous plant hosts (e.g., *Arabidopsis* or tobacco).
- Metabolic turnover (redox reactions and/or glycosylation) is a common challenge when novel terpenoids are accumulated in heterologous plant hosts. A second problem relates to the volatility of monoterpenes. Unless stored in a specialized anatomical structure (e.g., glandular trichomes or resin ducts) or biochemically conjugated (e.g., as glycosides), terpenoids are likely to be volatilized and are thus not harvestable on a commercial scale.
- Promoters that allow a transgene to be expressed specifically in specialized anatomical structures (e.g., glandular trichomes) have been employed but have not yet delivered a step change in the accumulation of terpenoids by metabolic engineering (for more details see [48]).
- Simple terpenoids can be accumulated to high levels in microbial hosts. However, the yields of highly functionalized plant terpenoids in engineered microbial hosts are typically fairly low. Nevertheless, because of the availability of versatile tools and relative speed of progress, microbial engineering appears to be the most promising near-term alternative for producing *p*-menthane monoterpenes.

References

1. Aaron JA, Christianson DW (2010) Trinuclear metal clusters in catalysis by terpenoid synthases. Pure Appl Chem 82:1585–1597
2. Alonso-Gutierrez J, Chan R, Batth TS, Adams PD, Keasling JD, Petzold CJ, Lee TS (2013) Metabolic engineering of Escherichia coli for limonene and perillyl alcohol production. Metab Eng 19:33–41
3. Banthorpe DV, LePatourel GN, Francis MJ (1983) Biosynthesis of geraniol and nerol and beta-D-glucosides in *Pelargonium graveolens* and *Rosa dilecta*. Biochem J 130:1045–1054
4. Berry C, van Eck JM, Kitto SL, Smigocki A (1996) *Agrobacterium*-mediated transformation of commercial mints. Plant Cell Tiss Org Cult 44:177–181
5. Bertea CM, Schalk M, Karp F, Maffei M, Croteau R (2001) Demonstration that menthofuran synthase of mint (*Mentha*) is a cytochrome P450 monooxygenase: cloning, functional expression, and characterization of the responsible gene. Arch Biochem Biophys 390:279–286
6. Bouwmeester HJ, Gershenzon J, Konings MC, Croteau R (1998) Biosynthesis of the monoterpenes limonene and carvone in the fruit of caraway. I. Demonstration of enzyme activities and their changes with development. Plant Physiol 117:901–912

7. Bouwmeester HJ, Konings MCJM, Gershenzon J, Karp F, Croteau R (1999) Cytochrome P-450 dependent (+)-limonene-6-hydroxylation in fruits of caraway (*Carum carvi*). Phytochemistry 50:243–248
8. Brennan TC, Turner CD, Krömer JO, Nielsen LK (2012) Alleviating monoterpene toxicity using a two-phase extractive fermentation for the bioproduction of jet fuel mixtures in *Saccharomyces cerevisiae*. Biotechnol Bioeng 109:2513–2522
9. Brophy JJ, Southwell IA (2003) Eucalyptus: the genus Eucalyptus. In: Coppen JJW (ed) Taylor and Francis, London, pp 102–160
10. Burbott AJ, Loomis WD (1967) Effects of light and temperature on the monoterpenes of peppermint. Plant Physiol 42:20–28
11. Caissard JC, Faure O, Jullien F, Colson M, Perrin A (1996) Direct regeneration *in vitro* and transient GUS expression in *Mentha* x *piperita*. Plant Cell Rep 16:67–70
12. Carter OA, Peters RJ, Croteau R (2003) Monoterpene biosynthesis pathway construction in *Escherichia coli*. Phytochemistry 64:425–433
13. Chambon C, Ladeveze V, Oulmouden A, Servouse M, Karst F (1990) Isolation and properties of yeast mutants affected in farnesyl diphosphate synthetase. Curr Genet 18:41–46
14. Clark GS (1994) Aroma chemical profile: menthone. Perf Flavor 19:41–45
15. Clark GS (2007) Aroma chemical profile: menthol. Perf Flavor 32:38–47
16. Clark RJ, Menary RC (1980) Environmental effects on peppermint (*Mentha piperita* L.). I. Effect of daylength, photon flux density, night temperature and day temperature on the yield and composition of peppermint oil. Funct Plant Biol 7:685–692
17. Colby SM, Alonso WR, Katahira EJ, McGarvey DJ, Croteau R (1993) 4S-limonene synthase from the oil glands of spearmint (*Mentha spicata*). cDNA isolation, characterization, and bacterial expression of the catalytically active monoterpene cyclase. J Biol Chem 268:23016–23024
18. Croteau R, Venkatachalam KV (1986) Metabolism of monoterpenes: demonstration that (+)-cis-isopulegone, not piperitenone, is the key intermediate in the conversion of (−)-isopiperitenone to (+)-pulegone in peppermint (*Mentha piperita*). Arch Biochem Biophys 249:306–315
19. Croteau RB, Davis EM, Ringer KL, Wildung MR (2005) (−)-Menthol biosynthesis and molecular genetics. Naturwissenschaften 92:562–577
20. Davidovich-Rikanati R, Sitrit Y, Tadmor Y, Iijima Y, Bilenko N, Bar E, Carmona B, Fallik E, Dudai N, Simon JE, Pichersky E, Lewinsohn E (2007) Enrichment of tomato flavor by diversion of the early plastidial terpenoid pathway. Nat. Biotechnol 25:899–901
21. Davis EM, Ringer KL, McConkey ME, Croteau R (2005) Monoterpene metabolism. Cloning, expression, and characterization of menthone reductases from peppermint. Plant Physiol 137:873–881
22. Degenhardt J, Köllner TG, Gershenzon J (2009) Monoterpene and sesquiterpene synthases and the origin of terpene skeletal diversity in plants. Phytochemistry 70:1621–1637
23. Diemer F, Jullien F, Faure O, Moja S, Colson M, Matthys-Rochon E, Caissard JC (1998) High efficiency transformation of peppermint (*Mentha* x *piperita* L.) with *Agrobacterium tumefaciens*. Plant Sci 136:101–108
24. Diemer F, Caissard JC, Moja S, Chalchat JC, Jullien F (2001) Altered monoterpene composition in transgenic mint following the introduction of 4S-limonene synthase. Plant Physiol Biochem 39:603–614
25. Dunlop MJ, Dossani ZY, Szmidt HL, Chu HC, Lee TS, Keasling JD, Hadi MZ, Mukhopadhyay A (2011) Engineering microbial biofuel tolerance and export using efflux pumps. Mol Syst Biol 7:487
26. Dunphy PJ, Allock C (1972) Isolation and properties of a monoterpene reductase from rose petals. Phytochemistry 11:1887–1891
27. Etzold B, Jess A, Nobis M (2009) Epimerisation of menthol stereoisomers: kinetic studies of the heterogeneously catalysed menthol production. Catal Today 140:30–36
28. Fahn A (1988) Secretory structures in plants. New Phytol 108:229–257

29. Fischer MJ, Meyer S, Claudel P, Bergdoll M, Karst F (2011) Metabolic engineering of monoterpene synthesis in yeast. Biotechnol Bioeng 108:1883–1892
30. Förster-Fromme K, Jendrossek D (2010) Catabolism of citronellol and related acyclic terpenoids in pseudomonads. Appl Microbiol Biotechnol 87:859–869
31. Gershenzon J, Dudareva N (2007) The function of terpene natural products in the natural world. Nat Chem Biol 3:408–414
32. Goodger JQ, Heskes AM, Mitchell MC, King DJ, Neilson EH, Woodrow IE (2010) Isolation of intact sub-dermal secretory cavities from *Eucalyptus*. Plant Methods 6:20
33. Haudenschild C, Schalk M, Karp F, Croteau R (2000) Functional expression of regiospecific cytochrome P450 limonene hydroxylases from mint (*Mentha* spp.) in *Escherichia coli* and *Saccharomyces cerevisiae*. Arch Biochem Biophys 379:127–136
34. Hemmerlin A, Harwood JL, Bach TJ (2012) A raison d'être for two distinct pathways in the early steps of plant isoprenoid biosynthesis? Prog Lipid Res 51:95–148
35. Ho CK, Chang SH, Tsay JY, Chiang VL, Chen ZZ (1998) *Agrobacterium tumefaciens*-mediated transformation of *Eucalyptus camaldulensis* and production of transgenic plants. Plant Cell Rep 17:675–680
36. Hyatt DC, Youn B, Zhao Y, Santhamma B, Coates RM, Croteau RB, Kang C (2007) Structure of limonene synthase, a simple model for terpenoid cyclase catalysis. Proc Natl Acad Sci USA 104:5360–5365
37. Iijima Y, Gang DR, Fridman E, Lewinsohn E, Pichersky E (2004) Characterization of geraniol synthase from the peltate glands of sweet basil. Plant Physiol 134:370–379
38. Ito M, Honda G (2007) Geraniol synthases from perilla and their taxonomical significance. Phytochemistry 68:446–453
39. Kamatou GP, Vermaak I, Viljoen AM, Lawrence BM (2013) Menthol: a simple monoterpene with remarkable biological properties. Phytochemistry 96:15–25
40. Karp F, Mihaliak CA, Harris JL, Croteau R (1990) Monoterpene biosynthesis: specificity of the hydroxylations of (−)-limonene by enzyme preparations from peppermint (*Mentha piperita*), spearmint (*Mentha spicata*), and perilla (*Perilla frutescens*) leaves. Arch Biochem Biophys 276:219–226
41. Keszei A, Brubaker CL, Carter R, Köllner T, Degenhardt J, Foley WJ (2010) Functional and evolutionary relationships between terpene synthases from Australian Myrtaceae. Phytochemistry 71:844–852
42. Kjonaas R, Martinkus-Taylor C, Croteau R (1982) Metabolism of monoterpenes: conversion of l-menthone to l-menthol and d-neomenthol by stereospecific dehydrogenases from peppermint (*Mentha piperita*) leaves. Plant Physiol 69:1013–1017
43. Krasnyanski S, Ball TM, Sink KC (1998) Somatic hybridization in mint: identification and characterization of *Mentha piperita* (+) *M. spicata* hybrid plants. Theor Appl Genet 96:683–687
44. Krens FA, Keizer LCP, Capel IEM (1997) Transgenic caraway, *Carum carvi* L.: a model species for metabolic engineering. Plant Cell Rep 17:39–43
45. Lange BM, Wildung MR, Stauber EJ, Sanchez C, Pouchnik D, Croteau R (2000) Probing essential oil biosynthesis and secretion by functional evaluation of expressed sequence tags from mint glandular trichomes. Proc Natl Acad Sci USA 97:2934–2939
46. Lange BM, Mahmoud SS, Wildung MR, Turner GW, Davis EM, Lange I, Baker RC, Boydston RA, Croteau RB (2011) Improving peppermint essential oil yield and composition by metabolic engineering. Proc Natl Acad Sci USA 108:16944–16949
47. Lange BM, Turner GW (2013) Terpenoid biosynthesis in trichomes—current status and future opportunities. Plant Biotechnol J 11:2–22
48. Lange BM, Ahkami A (2013) Metabolic engineering of plant monoterpenes, sesquiterpenes and diterpenes—current status and future opportunities. Plant Biotechnol J 11:169–196
49. Lawrence BM (2009) A preliminary report on the world production of some selected essential oils and countries. Perf Flavor 34:38–44
50. Lawrence BM (2013) The story of India's mint oils and menthol. Perf Flavor 38:26–35

51. Luan F, Hampel D, Mosandl A, Wüst M (2004) Enantioselective analysis of free and glycosidically bound monoterpene polyols in *Vitis vinifera* L. cvs. Morio Muscat and Muscat Ottonel: evidence for an oxidative monoterpene metabolism in grapes. J Agric Food Chem 52:2036–2041
52. Lücker J, El Tamer MK, Schwab W, Verstappen FW, van der Plas LH, Bouwmeester HJ, Verhoeven HA (2002) Monoterpene biosynthesis in lemon (*Citrus limon*). cDNA isolation and functional analysis of four monoterpene synthases. Eur J Biochem 269:3160–3171
53. Lücker J, Schwab W, van Hautum B, Blaas J, van der Plas LH, Bouwmeester HJ, Verhoeven HA (2004) Increased and altered fragrance of tobacco plants after metabolic engineering using three monoterpene synthases from lemon. Plant Physiol 134:510–519
54. Lücker J, Schwab W, Franssen MC, Van Der Plas LH, Bouwmeester HJ, Verhoeven HA (2004) Metabolic engineering of monoterpene biosynthesis: two-step production of (+)-trans-isopiperitenol by tobacco. Plant J 39:135–145
55. Lupien S, Karp F, Wildung M, Croteau R (1999) Regiospecific cytochrome P450 limonene hydroxylases from mint (*Mentha*) species: cDNA isolation, characterization, and functional expression of (−)-4S-limonene-3-hydroxylase and (−)-4S-limonene-6-hydroxylase. Arch Biochem Biophys 368:181–192
56. Mahmoud SS, Croteau RB (2001) Metabolic engineering of essential oil yield and composition in mint by altering expression of deoxyxylulose phosphate reductoisomerase and menthofuran synthase. Proc Natl Acad Sci USA 98:8915–8920
57. Mahmoud SS, Williams M, Croteau R (2004) Cosuppression of limonene-3-hydroxylase in peppermint promotes accumulation of limonene in the essential oil. Phytochemistry 65:547–554
58. Malhotra SK (2012) Caraway. In: Peter KV (ed) Handbook of herbs and spices. Woodhead Publishing, Philadelphia, pp 225–248
59. McConkey ME, Gershenzon J, Croteau RB (2000) Developmental regulation of monoterpene biosynthesis in the glandular trichomes of peppermint. Plant Physiol 122:215–224
60. McCoy M (2010) Hot market for a cool chemical. Chem Eng News 88:15–16
61. Moore GA, Jacono CC, Neidigh JL, Lawrence SD, Cline K (1992) Agrobacterium-mediated transformation of *Citrus* stem segments and regeneration of transgenic plants. Plant Cell Rep 11:238–242
62. Moralejo M, Rochange F, Boudet AM, Teulières C (1998) Generation of transgenic *Eucalyptus globulus* plantlets through *Agrobacterium tumefaciens* mediated transformation. Funct Plant Biol 25:207–212
63. Mullins KV, Llewellyn DJ, Hartney VJ, Strauss S, Dennis ES (1997) Regeneration and transformation of *Eucalyptus camaldulensis*. Plant Cell Rep 16:787–791
64. Niu X, Lin K, Hasegawa PM, Bressan RA, Weller SC (1998) Transgenic peppermint (*Mentha × piperita* L.) plants obtained by cocultivation with *Agrobacterium tumefaciens*. Plant Cell Rep 17:165–171
65. Ohara K, Ujihara T, Endo T, Sato F, Yazaki K (2003) Limonene production in tobacco with *Perilla* limonene synthase cDNA. J Exp Bot 54:2635–2642
66. Ohara K, Matsunaga E, Nanto K, Yamamoto K, Sasaki K, Ebinuma H, Yazaki K (2010) Monoterpene engineering in a woody plant *Eucalyptus camaldulensis* using a limonene synthase cDNA. Plant Biotechnol J 8:28–37
67. Oswald M, Fischer M, Dirninger N, Karst F (2007) Monoterpenoid biosynthesis in *Saccharomyces cerevisiae*. FEMS Yeast Res 7:413–421
68. Pateraki I, Andersen-Ranberg J, Hamberger B, Heskes AM, Martens HJ, Zerbe P, Bach SS, Møller BL, Bohlmann J, Hamberger B (2014) Manoyl oxide (13R), the biosynthetic precursor of forskolin, is synthesized in specialized root cork cells in Coleus forskohlii. Plant Physiol 164:1222–1236
69. Peier AM, Moqrich A, Hergarden AC, Reeve AJ, Andersson DA, Story GM, Earley TJ, Dragoni I, McIntyre P, Bevan S, Patapoutian A (2002) A TRP channel that senses cold stimuli and menthol. Cell 108:705–715

70. Rajaonarivony JI, Gershenzon J, Croteau R (1992) Characterization and mechanism of (4S)-limonene synthase, a monoterpene cyclase from the glandular trichomes of peppermint (*Mentha* x *piperita*). Arch Biochem Biophys 296:49–57
71. Reitsema RH (1958) Nitrosochloride syntheses and preparation of carvone. J Org Chem 23:2038–2039
72. Ringer KL, McConkey ME, Davis EM, Rushing GW, Croteau R (2003) Monoterpene double-bond reductases of the (−)-menthol biosynthetic pathway: isolation and characterization of cDNAs encoding (−)-isopiperitenone reductase and (+)-pulegone reductase of peppermint. Arch Biochem Biophys 418:80–92
73. Ringer KL, Davis EM, Croteau R (2005) Monoterpene metabolism. Cloning, expression, and characterization of (−)-isopiperitenol/(−)-carveol dehydrogenase of peppermint and spearmint. Plant Physiol 137:863–872
74. Rios-Estepa R, Turner GW, Lee JM, Croteau RB, Lange BM (2008) A systems biology approach identifies the biochemical mechanisms regulating monoterpenoid essential oil composition in peppermint. Proc Natl Acad Sci USA 105:2818–2823
75. Rios-Estepa R, Lange I, Lee JM, Lange BM (2010) Mathematical modeling-guided evaluation of biochemical, developmental, environmental, and genotypic determinants of essential oil composition and yield in peppermint leaves. Plant Physiol 152:2105–2119
76. Rochange F, Serrano L, Marque C, Teuliéres C, Boudet AM (1995) DNA delivery into *Eucalyptus globulus* zygotic embryos through biolistics: optimization of the biological and physical parameters of bombardment for two different particle guns. Plant Cell Rep 14:674–678
77. Rodríguez A, San Andrés V, Cervera M, Redondo A, Alquézar B, Shimada T, Gadea J, Rodrigo MJ, Zacarías L, Palou L, López MM, Castañera P, Peña L (2011) Terpene down-regulation in orange reveals the role of fruit aromas in mediating interactions with insect herbivores and pathogens. Plant Physiol 156:793–802
78. Rodríguez A, Alquézar B, Peña L (2013) Fruit aromas in mature fleshy fruits as signals of readiness for predation and seed dispersal. New Phytol 197:36–48
79. Serrano L, Rochange F, Semblat JP, Marque C, Teuliéres C, Boudet AM (1995) Genetic transformation of *Eucalyptus globulus* through biolistics: complementary development of procedures for organogenesis from zygotic embryos and stable transformation of corresponding proliferating tissue. J Exp Bot 47:285–290
80. Sikkema J, de Bont JA, Poolman B (1995) Mechanisms of membrane toxicity of hydrocarbons. Microbiol Rev 59:201–222
81. Simkin AJ, Miettinen K, Claudel P, Burlat V, Guirimand G, Courdavault V, Papon N, Meyer S, Godet S, St-Pierre B, Giglioli-Guivarc'h N, Fischer MJ, Memelink J, Clastre M (2013) Characterization of the plastidial geraniol synthase from Madagascar periwinkle which initiates the monoterpenoid branch of the alkaloid pathway in internal phloem associated parenchyma. Phytochemistry 85:36–43
82. Singh S, Rajam MV (2009) *Citrus* biotechnology: Achievements, limitations and future directions. Physiol Mol Biol Plants 15:3–22
83. Tranchida PQ, Bonaccorsi I, Dugo P, Mondello L, Dugo G (2011) Analysis of *Citrus* essential oils: state of the art and future perspectives. A review. Flav Fragr J 27:98–123
84. Turner GW, Gershenzon J, Nielson EE, Froehlich JE, Croteau R (1999) Limonene synthase, the enzyme responsible for monoterpene biosynthesis in peppermint, is localized to leucoplasts of oil gland secretory cells. Plant Physiol 120:879–886
85. Turner GW, Croteau R (2004) Organization of monoterpene biosynthesis in *Mentha*. Immunocytochemical localizations of geranyl diphosphate synthase, limonene-6-hydroxylase, isopiperitenol dehydrogenase, and pulegone reductase. Plant Physiol 136:4215–4227
86. Turner GW, Davis EM, Croteau RB (2012) Immunocytochemical localization of short-chain family reductases involved in menthol biosynthesis in peppermint. Planta 235:1185–1195
87. van Beilen JB, Holtackers R, Lüscher D, Bauer U, Witholt B, Duetz WA (2005) Biocatalytic production of perillyl alcohol from limonene by using a novel *Mycobacterium* sp. cytochrome

P450 alkane hydroxylase expressed in *Pseudomonas putida*. Appl Environ Microbiol 71:1737–1744
88. Vardi A, Bleichman S, Aviv D (1990) Genetic transformation of *Citrus* protoplasts and regeneration of transgenic plants. Plant Sci 69:199–206
89. Voirin B, Brun N, Bayet C (1990) Effects of daylength on the monoterpene composition of leaves of *Mentha* x *piperita*. Phytochemistry 29:749–755
90. Voo SS, Grimes HD, Lange BM (2012) Assessing the biosynthetic capabilities of secretory glands in *Citrus* peel. Plant Physiol 159:81–94
91. Williams DC, McGarvey DJ, Katahira EJ, Croteau R (1998) Truncation of limonene synthase preprotein provides a fully active 'pseudomature' form of this monoterpene cyclase and reveals the function of the amino-terminal arginine pair. Biochemistry 37:12213–12220
92. Wüst M, Croteau RB (2002) Hydroxylation of specifically deuterated limonene enantiomers by cytochrome p450 limonene-6-hydroxylase reveals the mechanism of multiple product formation. Biochemistry 41:1820–1827
93. Xu P, Bhan N, Koffas MA (2013) Engineering plant metabolism into microbes: from systems biology to synthetic biology. Curr Opin Biotechnol 24:291–299
94. Yang T, Li J, Wang HX, Zeng Y (2005) A geraniol-synthase gene from *Cinnamomum tenuipilum*. Phytochemistry 66:285–293
95. Yang T, Stoopen G, Yalpani N, Vervoort J, de Vos R, Voster A, Verstappen FW, Bouwmeester HJ, Jongsma MA (2011) Metabolic engineering of geranic acid in maize to achieve fungal resistance is compromised by novel glycosylation patterns. Metab Eng 13:414–425
96. Yao JL, Wu JH, Gleave AP, Morris BAM (1996) Transformation of citrus embryogenic cells using particle bombardment and production of transgenic embryos. Plant Sci 113:175–183
97. Yuasa Y, Tsuruta H, Yuasa Y (2000) A practical and efficient synthesis of p-menthane-3,8-diols. Org Proc Res Developm 4:159–161
98. Zhou S, Koh HL, Gao Y, Gong ZY, Lee EJD (2004) Herbal bioactivation: the good the bad and the ugly. Life Sci 74:935–968

Adv Biochem Eng Biotechnol (2015) 148: 355–390
DOI 10.1007/10_2014_288

Published Online: 11 January 2015

Isoprenoid Drugs, Biofuels, and Chemicals—Artemisinin, Farnesene, and Beyond

Kevin W. George, Jorge Alonso-Gutierrez, Jay D. Keasling and Taek Soon Lee

Abstract Isoprenoids have been identified and used as natural pharmaceuticals, fragrances, solvents, and, more recently, advanced biofuels. Although isoprenoids are most commonly found in plants, researchers have successfully engineered both the eukaryotic and prokaryotic isoprenoid biosynthetic pathways to produce these valuable chemicals in microorganisms at high yields. The microbial synthesis of the precursor to artemisinin—an important antimalarial drug produced from the sweet wormwood *Artemisia annua*—serves as perhaps the most successful example of this approach. Through advances in synthetic biology and metabolic engineering, microbial-derived semisynthetic artemisinin may soon replace plant-derived artemisinin as the primary source of this valuable pharmaceutical. The richness and diversity of isoprenoid structures also make them ideal candidates for advanced biofuels that may act as "drop-in" replacements for gasoline, diesel, and jet fuel. Indeed, the sesquiterpenes farnesene and bisabolene, monoterpenes pinene and limonene, and hemiterpenes isopentenol and isopentanol have been evaluated as fuels or fuel precursors. As in the artemisinin project, these isoprenoids have been produced microbially through synthetic biology and metabolic engineering efforts. Here, we provide a brief review of the numerous isoprenoid compounds that have found use as pharmaceuticals, flavors, commodity chemicals, and, most importantly, advanced biofuels. In each case, we highlight the metabolic engineering strategies that were used to produce these compounds successfully in microbial

K.W. George · J. Alonso-Gutierrez · J.D. Keasling · T.S. Lee (✉)
Joint BioEnergy Institute, 5885 Hollis St. 4th floor, Emeryville, CA 94608, USA
e-mail: tslee@lbl.gov

K.W. George · J. Alonso-Gutierrez · J.D. Keasling · T.S. Lee
Physical Biosciences Division, Lawrence Berkeley National Laboratory, Berkeley, CA, USA

J.D. Keasling
Department of Bioengineering, University of California, Berkeley, CA, USA

J.D. Keasling
Department of Chemical and Biomolecular Engineering, University of California, Berkeley, CA, USA

hosts. In addition, we present a current outlook on microbial isoprenoid production, with an eye towards the many challenges that must be addressed to achieve higher yields and industrial-scale production.

Keywords Isoprenoids · Antimalarial · Artemisinin · Biofuel · Sesquiterpene · Monoterpene · Isopentenol

Contents

1 Introduction

Isoprenoids are the largest and most diverse group of natural products, composed of over 50,000 compounds including primary metabolites such as sterols, carotenoids, and quinines, and secondary metabolites that are often used for medical purposes [1–3]. Chemists have long marveled at the structural diversity of terpenes in natural products and have engineered their biosynthetic pathways to develop numerous isoprenoid-derived commercial drugs. Recently, advanced biofuels have garnered attention as global climate change has driven the development of carbon-neutral energy sources. The chemical structure of isoprenoids confers several beneficial aspects as fuel compounds. For example, methyl branching is a common structural feature of isoprenoids that lowers the freezing point significantly. Also, cyclic structures, which are frequently observed in isoprenoids, increase energy density and are generally considered valuable features for jet fuels. In recent years, some isoprenoids have been tested and produced as potential gasoline, diesel, and jet fuels due to their favorable energy content, cold weather properties, and high octane/cetane numbers (Table 1) [4–11].

Isoprenoids are usually classified into groups according to the number of carbons: hemiterpenoids (C5), monoterpenoids (C10), sesquiterpenoids (C15), diterpenoids (C20), and triterpenoids (C30). Isopentenyl diphosphate (IPP) and its isomer dimethylallyl diphosphate (DMAPP) are the two universal C5 building blocks to synthesize isoprenoids. These starting precursors can be synthesized by two routes: the 1-deoxy-D-xylulose-5-phosphate (DXP) or 2-methyl-d-erythrito-4-phosphate (MEP) pathway and the mevalonate (MVA) pathway (Fig. 1).

Often found in bacteria and plant plastids, the MEP or non-mevalonate pathway forms IPP and DMAPP from 2-methyl-D-erythrito-4-phosphate (MEP) via the condensation of pyruvate and glyceraldehyde 3-phosphate (G3P). The mevalonate (MVA) pathway is present in archaea, fungi, plant cytoplasm, and other eukaryotes including mammalian cells where IPP and DMAPP are formed from the condensation of three molecules of acetyl-CoA to mevalonate. For decades, the mevalonate pathway has been studied to understand the production of isoprenoids inasmuch as it is responsible for cholesterol biosynthesis in humans and other mammals. In plants, these two isoprenoid biosynthesis pathways exist in parallel for primary and secondary metabolism, which could be useful for environmental adaptation and more efficient carbon utilization [12].

The first step in the MEP pathway is the formation of 1-deoxy-D-xylulose 5-phosphate (DXP) by the condensation of pyruvate and D-glyceraldehyde 3-phosphate, catalyzed by DXP synthase (DXS) encoded by the *dxs* gene (Fig. 1). This step is crucial and known as the rate-limiting step of the entire pathway. The second step is catalyzed by DXP reductoisomerase (encoded by the *dxr* gene) to convert DXP to MEP. MEP is then converted to 4-(cytidine-5′-diphospho)-2-methyl-D-erythritol (CDP-ME), 2-phospho-CDP-ME (CDP-ME2P), 2-methyl-D-erythritol 2,4-cyclodiphosphate (cMEPP), and IPP and DMAPP via the series of enzymatic reactions.

The mevalonate pathway initiates with the condensation of two acetyl-CoAs by thiolase to extend the carbon backbone to produce acetoacetyl-CoA. Subsequently, another acetyl-CoA is condensed with acetoacetyl-CoA to produce 3-hydroxy-3-methylglutaryl-CoA (HMG-CoA), which is physiologically almost irreversible. Then, mevalonic acid, a stable precursor of isoprenoid/sterol biosynthesis, is produced by the HMG-CoA reductase (HMGR) from HMG-CoA using primarily NADPH as a cofactor. Two kinases, mevalonate kinase (MK) and phosphomevalonate kinase (PMK), catalyze the phosphorylation of mevalonate to produce mevalonate 5-phosphate and mevalonate 5-diphosphate, respectively. The final step of the mevalonate pathway to produce IPP is the ATP-driven decarboxylative elimination reaction catalyzed by mevalonate diphosphate decarboxylase (PMD). Following the production of IPP, isomerization by Idi leads to DMAPP formation

Table 1 Properties of conventional fuels and isoprenoid-based biofuels

Fuel Name	Density (range) (kg/m^3)	Flash point (°C) (min)	Freezing point (°C) (max)	Net heat of combustion (MJ/kg) (min)	Cetane number (min 40)	References
Amorphane						
Jet-A (ASTM)	775–840	38	−40	42.8	N.D.	[9]
Jet-A	811	43	−47	43.42	N.D.	
20 % Amorphane[a]/JetA	818	49	−48	43.1	N.D.	
50 % Amorphane[a]/JetA	846	60	−53	42.97	N.D.	
100 % Amorphane[a]	880	113	<−52	42.79	N.D.	
Farnesene						
#2 Diesel	864.6	73	N.D.	42.4	41.6	[7]
5 % AMD-200/#2 Diesel	859.5	75	N.D.	N.D.	41.7	
20 % AMD-200/#2 Diesel	845.4	78	N.D.	42.8	45.2	
50 % AMD-200/#2 Diesel	820.4	86	N.D.	43.2	50.7	
100 % AMD-200 (Farnesane)	773.7	109	N.D.	44.2	58.6	
Monoterpene dimers						
Alpha-Pinene_Dimer	935	N.D.		42.047	N.D.	[136]
Beta-Pinene_Dimer	938	N.D.		42.118	N.D.	
Limonene_Dimer	914	N.D.		41.906	N.D.	
JP-5	820	N.D.	<−46.15	42.4	N.D.	
JP-10	940	N.D.	<−79.15	42.1	N.D.	
RJ-5	1,080	N.D.	>−18.15	41.6	N.D.	
Gasoline	740	N.D.	<−100.15	43.6	N.D.	
Diesel no. 2	850	N.D.	−12.15 to −6.15	42.4	N.D.	
Biodiesel	880	N.D.	>−10.15	37.4	N.D.	
β-Pinene	860	N.D.	−61.15	42.9	N.D.	

(continued)

Table 1 (continued)

Fuel Name	Density (range) (kg/m^3)	Flash point (°C) (min)	Freezing point (°C) (max)	Net heat of combustion (MJ/kg) (min)	Cetane number (min 40)	References
α-Pinene	860	N.D.	−64.15	42.9	N.D.	
Bicyclic monoterpenes						
Jet-A ASTM	775–840	38	−40	42.8	N.D.	[8]
Jet-A	811	43	−47	43.4192	N.D.	
20 % AMJ-400[b]/Jet-A	820.4	43	−51	43.0534	N.D.	
50 % AMJ-400[b]/Jet-A	834.9	44	−56.5	42.9881	N.D.	
100 % AMJ-400[b] (pinane)	860.3	43	<−70	42.8011	N.D.	
Hydrogenated limonene and myrcene						
Diesel		58.3	N.D.	N.D.	45.6	[10]
Hydrogenated limonene (limonane)						
10 % in diesel	N.D.	58.9	N.D.	N.D.	42.8	
Hydrogenated myrcene						
2 % in diesel	N.D.	63.9	N.D.	N.D.	44.9	
5 % in diesel	N.D.	61.7	N.D.	N.D.	44.3	
10 % in diesel	N.D.	60	N.D.	N.D.	44.7	
Bisabolane						
D2 diesel	850	60–80	N.D.	N.D.	40–55	[6]
Biodiesel	880	100–170	N.D.	N.D.	48–65	
Bisabolane	820	111	N.D.	N.D.	41.9	

[a] Partially mixed with nonhydrogenated amorphadiene
N.D. Not determined
[b] 98.7 % of pinane

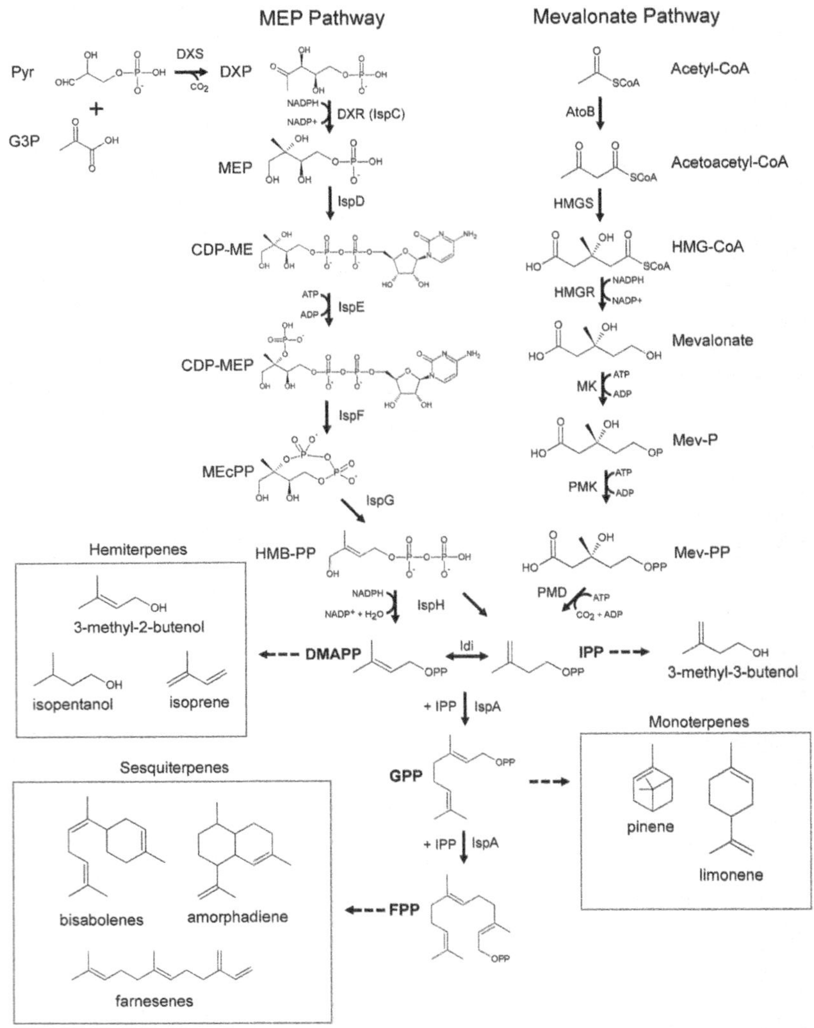

Fig. 1 Isoprenoid biosynthetic pathways. The MEP pathway (or DXP pathway) and mevalonate (*MVA*) pathway are two major isoprenoid biosynthetic pathways. Both pathways produce IPP and DMAPP as five-carbon building blocks for isoprenoid biosynthesis. The MEP pathway initiates with the condensation of pyruvate and glyceraldehyde 3-phopshate by DXS and an additional six steps transform DXP to IPP and DMAPP. In the mevalonate pathway, three molecules of acetyl-CoA condense to form HMG-CoA and an additional four steps transform HMG-CoA to IPP, which is isomerized to DMAPP by the isomerase (Idi). The condensation of DMAPP with one or two molecules of IPP leads to monoterpene or sesquiterpene production, respectively

and the initiation of isoprenoid precursor chain elongation (Fig. 1). The first successful engineering of a complete mevalonate (MVA) pathway in a heterologous host was accomplished about a decade ago for the biosynthesis of pharmaceutical isoprenoids [13]. The sesquiterpene precursor of the antimalarial drug artemisinin was produced at high titers using a heterologous mevalonate pathway in *E. coli*, and this work is extensively reviewed in Sect. 2.

In this review, we present current efforts and future outlooks towards the biosynthesis of various isoprenoid drugs, biofuels, and chemicals. In Sect. 2, we briefly review the general engineering strategy for the construction and optimization of a heterologous mevalonate pathway for isoprenoid biosynthesis. In particular, we focus on the biosynthesis of the antimalarial drug artemisinin, which is perhaps the most successful example of isoprenoid biosynthetic pathway engineering. We then further review the advances in this platform technology for biofuel and chemical production. Sesquiterpene biofuel precursors including farnesene and bisabolene are reviewed in Sect. 3, and monoterpene biofuel precursors such as limonene and pinene are discussed in Sect. 4. Other important sesquiterpenoids and monoterpenoids are also discussed. In Sect. 5, hemiterpenoid fuels and chemicals such as isopentenol are briefly reviewed along with the issue of IPP toxicity, which is a crucial engineering factor in high titer isoprenoid production.

2 Mevalonate Pathway Assembly and Artemisinin Production

2.1 Efforts Towards Microbial Production of Isoprenoids

As the most numerous and structurally diverse family of natural products, isoprenoids have a variety of commercial uses as flavors, fragrances, and pharmaceuticals [2]. Due to the small quantities of isoprenoids commonly produced by plants and naturally occurring microorganisms, there has been considerable interest in the development of engineered microbial platforms for isoprenoid production in large-scale fermentations [14]. Research in this area has focused primarily on the production of pharmaceutically important compounds including carotenoids, sterols, the anticancer drug Taxol, and the antimalarial artemisinin [15–18]. Rather than review the vast numbers of studies in each of these areas, we focus this section on efforts leading to the development of a microbial platform for semisynthetic artemisinin production (Fig. 2). This decade-long project covers the initial assembly of a high-flux isoprenoid pathway in both *E. coli* and *S. cerevisiae* and details numerous pathway optimizations. The advances in metabolic engineering and synthetic biology achieved during this project have ultimately influenced current efforts to produce diverse terpenoid compounds including sesquiterpene (Sect. 3), monoterpene (Sect. 4), and hemiterpene (Sect. 5) biofuels.

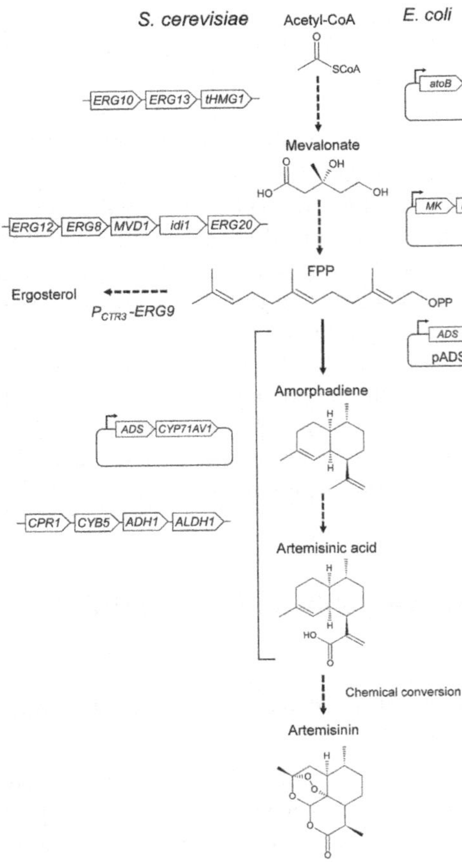

Fig. 2 Artemisinin strain engineering. An overview of important engineering breakthroughs in *S. cerevisiae* (*left*) and *E. coli* (*right*) for the production of amorphadiene, artemisinic acid, and semisynthetic artemisinin. Genes that were overexpressed on plasmids (*circular*) or on the chromosome (*straight line*) are highlighted. In the case of *S. cerevisiae*, ergosterol biosynthesis was also downregulated

2.2 Assembly and Optimization of Mevalonate Pathway for Large-Scale Production of Semisynthetic Artemisinin

Artemisinin is a potent antimalarial drug naturally produced by the sweet wormwood *Artemisia annua*, a plant long recognized for its medicinal properties [19]. Artemisinin and its derivatives have been designated as first-line antimalarial drugs [20] and are currently key components of antimalarial combination therapies (ACTs). Unfortunately, artemisinin availability and price has fluctuated over the past decade, largely due to the 18-month lag between planting, harvesting, and eventual supply. Although chemical synthesis could provide a stable supply of artemisinin, this is not a feasible or cost-effective option [21]. Chemically, artemisinin is an isoprenoid containing 15 carbon atoms (sesquiterpene) derived from farnesyl diphosphate (FPP). The first committed step in artemisinin biosynthesis is the conversion of FPP to amorphadiene, a step catalyzed by amorphadiene synthase (ADS). Because isoprenoids are readily produced in nature, the biosynthesis of amorphadiene and, eventually, artemisinic acid, served as an attractive alternative to chemical synthesis.

In nature, the two common building blocks of isoprenoids, isopentenyl diphosphate and dimethylallyl diphosphate, are produced either from the mevalonate pathway or the methylerythritol phosphate pathway [22]. The MVA pathway is generally present in eukaryotes and archaea, whereas the MEP is active in most bacteria including *E. coli*. Though the MEP pathway has recently been used to produce high levels of taxadiene, an isoprenoid precursor to the anticancer drug Taxol (paclitaxel; [15]), early efforts to engineer a high-flux MEP pathway in *E. coli* were met with limited success. The endogenous regulation of the MEP pathway was suspected to be a reason for this intractability and to bypass this suspected limitation, expression of the *S. cerevisiae* mevalonate pathway was engineered in *E. coli* [13]. This approach provided a high-flux route to produce IPP and DMAPP and thus the longer chain terpene FPP, the precursor to amorphadiene in an *E. coli* host.

The heterologous mevalonate pathway was initially divided into a three-enzyme "top" portion (MevT) responsible for the conversion of acetyl-CoA to mevalonate and a five-enzyme "bottom" portion (MBIS) that transformed mevalonate into FPP. The MevT operon was made up of acetoacetyl-CoA thiolase from *E. coli* (*atoB*), along with HMG-CoA synthase (*HMGS*) and reductase (*HMGR*) from *S. cerevisiae*. The MBIS operon consisted of *S. cerevisiae*-derived *MK*, *PMK*, and *PMK*, along *idi* and FPP synthase (*ispA*) from *E. coli*. Enzymes in these two operons were expressed under an IPTG-inducible *lac* promoter in two plasmids. To make amorphadiene from FPP, an *E. coli* codon-optimized amorphadiene synthase gene (*ADS*) from *A. annua* was synthesized and expressed under a *trc* promoter in a high copy plasmid (pTrc99A). Coexpression of MevT and MBIS in *E. coli* DH10B complemented an MEP pathway mutant even in the absence of mevalonate, confirming the functional expression of both operons. By feeding mevalonate to a strain expressing only MBIS and ADS, the authors showed that flux from the MBIS operon did not limit amorphadiene production at the highest mevalonate concentration used (40 mM). If MBIS was induced without ADS, growth inhibition increased with the amount of added mevalonate, suggesting that FPP is toxic. This effect was more extreme in a truncated bottom pathway expressing only MK, PMK, and PMK (pMevB), suggesting that IPP accumulation is even more deleterious [13]. With a complete mevalonate pathway (pMevT + pMBIS) and ADS, 3.1 μg caryophyllene equivalent/mL/OD_{600} of amorphadiene was produced in a shake-flask culture in LB medium after 9 h, a 36-fold improvement over the native MEP pathway. A glycerol-amended culture reached higher biomass yields and prolonged amorphadiene production into the stationary phase. When accounting for loss of amorphadiene to the headspace, a total production of 112 mg/L was calculated from the LB + 0.8 % glycerol culture. The heterologous mevalonate pathway assembled in this work proved far superior to earlier efforts to engineer the MEP pathway and laid the groundwork for assembly of a highly efficient platform for isoprenoid production [13].

A number of follow-up studies resulted in greatly improved amorphadiene yields in *E. coli.* Due to volatility, amorphadiene titers were severely underestimated in earlier work [13]. Use of a hydrophobic dodecane overlay relieved this issue and resulted in a significantly improved titer of 281.4 mg/L [23]. Other improvements included the use of TB medium and the "pulsing" of glycerol and other carbon sources during the stationary phase. With a dodecane overlay and increased carbon and complex nutrients, amorphadiene titers reached 480 mg/L, a 20-fold improvement over the original process.

The precursor pathway was improved through a variety of metabolic engineering techniques. The accumulation of HMG-CoA, later shown to inhibit fatty acid biosynthesis [24], was identified as a metabolic bottleneck in the MevT operon that limited mevalonate production [25]. Through growth analysis and LC-MS quantification of pathway intermediates, the authors demonstrated that improved expression of tHMGR, *N*-terminally truncated HMG-CoA reductase, relieved growth inhibition, reduced HMG-CoA accumulation, and improved mevalonate production threefold. This requirement for a balanced metabolic pathway was further emphasized in a related study [26]. In this work, the authors used a standardized vector system to alter gene dosage and identify MK and ADS as rate-limiting enzymes. Through modulation of additional parameters such as codon usage, promoter strength, and plasmid copy number, the authors assembled a pathway in *E. coli* DH1 that was sevenfold more efficient than the original strain [13].

Gene variants of *HMGS* and *HMGR* derived from *Staphylococcus aureus* (*mvaS* and *mvaA*, respectively) were used to more than double amorphadiene production [27]. This work also developed an effective process for high-density fermentation. A defined medium was used instead of TB, and parameters including carbon feed rate and nitrogen concentration were systematically changed to reduce acetate formation, curtail carbon flow towards protein synthesis, and improve amorphadiene yield. With a carefully balanced dual restriction of glucose and nitrogen, DH1 harboring the optimized pathway produced 27.4 g/L of amorphadiene, the highest-reported titer in *E. coli* to date.

In order to produce artemisinin from the sesquiterpene precursor, amorphadiene must first be biologically converted to artemisinic acid which can be easily processed to the final product via chemical synthesis. To accomplish this conversion in vivo, the *A. annua* native cytochrome P450 *CYP71AVI* was expressed in *E. coli* coexpressing the heterologous MVA pathway and ADS [28]. By changing expression vectors and the host to DH1 instead of DH10B, fully oxidized artemisinic acid was produced at a titer of 105 mg/L. This work served as the first demonstration of in vivo production of functionalized terpenoids with native plant P450s in *E. coli.*

Despite the high yields and titers of amorphadiene attained in *E. coli*, the artemisinin project was ultimately carried out using a refactored mevalonate pathway in

S. cerveisiae due primarily to difficulties in achieving efficient artemisinic acid production in the *E. coli* host [29]. Prior to the switch, Ro and coauthors initially devised a *S. cerevisiae* strain capable of high flux to artemisinic acid through a variety of host modifications [18]. To increase FPP production in *S. cerevisiae, tHMGR* was overexpressed along with the transcriptional regulator UPC2, which is involved in the biosynthesis of sterols. In addition, *ERG9*, which encodes squalene synthase, was downregulated to prevent carbon loss from FPP. By combining these modifications with the expression of ADS, 153 mg/L of amorphadiene was produced. Next, the authors sought to produce artemisinic acid from amorphadiene. To accomplish this goal, the genes responsible for this conversion in *A. annua* [cytochrome P450 *CYP71AV1* and NADPH:cytochrome P450 oxidoreductase (*CPR*)] were isolated and overexpressed in *S. cerevisiae* (Fig. 2). This work also developed an efficient purification technique for synthesized artemisinic acid. Artemisinic acid was efficiently transported out of the cell and remained bound to the cell surface while protonated. Thus, treatment with an alkaline buffer removed >96 % of artemisinic acid from the cell pellet. Using this feature, the authors developed a one-step purification method that routinely yielded >95 % pure artemisinic acid. Altogether, artemisinic acid was produced at a titer of 115 mg/L.

Westfall and coauthors focused on further improving amorphadiene and artemisinic acid productivity, achieving >40 g/L of amorphadiene in a fermentation process [30]. A variety of changes in the host, pathway, and fermentation process facilitated these improvements. The endogenous consumption of galactose, commonly used as an inducer, was eliminated through a *GAL1* deletion. This modification allowed for the production of amorphadiene and artemisinic acid using glucose as a sole carbon source and galactose as an inducer. In the pathway, heterologous genes derived from *A. annua* were codon-optimized, and the orientation of the *GAL1* promoters was altered to prevent recombination. These modifications were reconstructed in *S. cerevisiae* CEN.PK2, a better-characterized strain, rather than S288C. In the final production strain, every enzyme of the mevalonate pathway up to *ERG20* was transcribed from galactose-regulated, divergent *GAL1/GAL10* promoters, and three copies of *tHMG1* were integrated in the chromosome. *GAL80*, encoding the negative regulator of the galactose regulon, was also deleted to obviate the use of galactose entirely. With these optimizations, 41 g/L of amorphadiene were produced after 116 h at a yield of 16.98 Cmol %. Because yields of biosynthetic artemisinic acid were considerably less than amorphadiene, a three-step chemical conversion was also developed in this report that achieved a 48.4 % yield of artemisinic acid from amorphadiene.

Most recently, a highly efficient, complete biosynthetic process to artemisinic acid was constructed and allowed for the high-level, semisynthetic synthesis of artemisinin [31]. The combination of multiple components including an engineered host for an improved precursor pathway, artemisinic acid biosynthesis pathway, fermentation process engineering, and chemical conversion to artemisinin was required to accomplish this extraordinary feat. A copper-regulated *CTR3* promoter was used to downregulate *ERG9*, allowing for less-expensive $CuSO_4$ to be used as a repressor instead of methionine. To increase cell viability, expression of *CPR1*, the cognate reductase of *A. annua* amorphadiene oxidase *CYP71AV1,* was reduced. Cytochrome *b5* from *A. annua* (*CYB5*) was integrated into the yeast chromosome, yielding higher production of artemisinic acid and artemisinic aldehyde. *A. annua* artemisinic aldehyde dehydrogenase (*ALDH1*) was expressed, further increasing artemisinic acid production. To complete the artemisinic acid pathway and reduce artemisinic alcohol production, a putative *A. annua* alcohol dehydrogenase (*ADH1*) was expressed in conjunction with *CYP71AV1*, *CPR1*, *CYB5*, and *ALDH1*. With the complete biosynthetic pathway, artemisinic acid was produced as a crystalline extracellular precipitate that complicated sampling procedures. To overcome this complication, the authors used extractive fermentation with isopropyl myristate (IPM) oil and developed a method to extract artemisinic acid from IPM at high purity. With all of these optimizations, 25 g/L of artemisinic acid were produced, a >10-fold improvement over earlier efforts. Finally, an inexpensive chemical process based on singlet oxygen was used to convert artemisinic acid into semisynthetic artemisinin at a high yield. Remarkably, the strains and processes described in this work are currently being used in an industrial process for artemisinin production and distribution [29].

2.3 Tangential Development of Tools for Metabolic Engineering

A number of promising metabolic engineering tools were constructed in association with the artemisinin project. Many of these tools were developed in *E. coli*, where the heterologous mevalonate pathway served as a flexible model system. Some of the most promising tools served the purpose of balancing the expression of multiple genes, a reoccurring theme in the artemisinin project. A method for randomly tuning the expression of multiple genes within an operon by altering the sequences of intergenic regions was described and applied to the MevT operon (*atoB*, *HMGS*, *HMGR*; [32]). Tunable intergenic regions (TIGRs) containing posttranscriptional control elements including hairpins and RNase E sites were combinatorially generated and inserted in the MevT operon. Unexpectedly, reduced expression of HMGS and HMGR was shown to yield a sevenfold increase in mevalonate concentration. This observation highlights the strength of this methodology, as an intentional reduction in HMGS and HMGR expression would not have been a priority in rational pathway design.

MevT also served as a testbed for the development of synthetic protein scaffolds, another approach to tune expression and provide modular control over metabolic flux [33]. To test this system, AtoB, HMGS, and HMGR in the MevT operon were tagged with the metazoan protein–protein interaction ligands GBD, SH3, and PDZ. A synthetic scaffold containing the cognate binding domains for these tags was then coexpressed in this same system, allowing the tagged proteins to colocalize on the scaffold. By altering the number and/or the order of binding sites, enzyme stoichiometry could be predictably altered. Using this scaffold, a 77-fold increase in mevalonate yield was observed [33].

The application of targeted proteomics to metabolic pathway optimization provided a third method for tuning pathway expression [34]. A selected reaction monitoring (SRM) proteomics method was applied to the entire mevalonate pathway for amorphadiene production and used to measure concentrations of all nine pathway proteins. Protein levels of HMGR, MK, and PMK were shown to be low and were thus hypothesized to limit pathway flux. Codon-optimization of these genes and the addition of a supplemental promoter upstream of MK improved protein expression and facilitated a threefold increase in amorphadiene titer.

The toxicity of prenyl diphosphates such as IPP and FPP [13] provided a means to develop novel screening methods. Because IPP and FPP cause growth inhibition, overexpression of a terminal enzyme (such as a terpene synthase) that consumes these toxic precursors restores growth. For example, using this concept, a library of 19,000 clones harboring fragments of *B. subtilis* genomic DNA was screened for hemiterpenoid production in an IPP-accumulating *E. coli* strain. Two genes, identified as *yhfR* and *nudF*, overcame IPP toxicity and restored growth. Further analysis of *nudF* showed that it was capable of converting IPP into isopentenol, a potential biofuel (see Sect. 4.2). The toxicity of prenyl diphosphates was further exploited to program dynamic pathway expression using stress–response promoters [35]. In this work, a microarray was performed on *E. coli* harboring a truncated mevalonate pathway ending in the terminal production of FPP. A subset of promoters from genes that were differentially expressed under FPP accumulation was subsequently used to drive the mevalonate pathway. Specifically, the promoter for *gadE*, a gene downregulated during FPP accumulation, was used to drive expression of mevalonate pathway enzymes (MevT–MBIS operon) up to FPP synthase, whereas the promoter from *rstA*, a gene upregulated during FPP accumulation, was used to drive terpene synthase *ADS* that consumes FPP. In this manner, the cell would dynamically adjust expression to maintain FPP levels below a toxic threshold: high FPP concentrations would simultaneously slow its production (downregulate MevT–MBIS) and increase its consumption (upregulate ADS). Compared to the inducible pathway, this dynamic pathway produced twofold more amorphadiene (~ 1.5 g/L), accumulated less acetate, and reached higher ODs. Since this strategy does not require the development of biosensors or synthetic transcription factors, it should be broadly applicable to other metabolic pathways.

2.4 Future Work

The artemisinin project has served as a compelling example of the potential of metabolic engineering and synthetic biology to deliver real-world results. Semi-synthetic artemisinin is functionally identical to the plant-derived drug [31], has been approved by the World Health Organization (WHO), and is currently being produced on an industrial scale by the multinational pharmaceutical company

Sanofi [29]. Given the remarkable diversity of isoprenoid compounds, it is likely that a similar engineering approach can yield a variety of medicinally relevant molecules such as paclitaxel, prostratin, and lovastatins [29]. The remainder of this review focuses primarily on an additional application: isoprenoid-derived advanced biofuels.

3 Sesquiterpenoid Biofuels and Chemicals

3.1 Sesquiterpenoids Properties and Chemical Diversity

Sesquiterpenoids are one of the largest groups (>7,000 compounds) of isoprenoid natural products and perform important physiological functions in a wide range of organisms including plants, insects, and fungi [36]. This chemical diversity is derived from the flexibility of terpene synthases, enzymes that catalyze the conversion of FPP into a large variety of sesquiterpene skeletons that can be further modified into more complex molecules with increased functionality [36, 37]. Functionally, sesquiterpenes have a wide range of activities from antimicrobial agents (such as phytoalexins capsidiol [38]) to alarm pheromones (such as farnesene [39]). Structurally, sesquiterpenes can be acyclic, monocyclic, bi- or even tricyclic structures depending on the identity of the terpene synthase (Fig. 3).

Acyclic sesquiterpenoids can be found in essential oils and insect pheromones. They include farnesene and isomeric alcohols such as nerolidol and farnesol, and often act as important compounds for the fragrance, flavoring, and pharmaceutical industries [40–43]. In addition to these uses, acyclic sesquiterpenoids such as farnesane have been proposed as potential diesel and jet-fuel alternatives [7, 42, 44, 45].

Monocyclic sesquiterpenoids also include compounds for the pharmaceutical and perfumery industries. For example, humulene has antiallergic and anti-inflammatory properties [46, 47]. Elemol, zingiberene, and bisabolene occur in many essential oils and fragrances [40] and recently the hydrogenation product of bisabolene (i.e., bisabolane) has been proposed as a promising diesel replacement [6].

Bi- or tricyclic sesquiterpenoids are often valuable chemicals. Nootkatone, for example, found in low concentrations in diverse plant essential oils, may be used in foods, cosmetics, and pharmaceuticals due to its distinct grapefruit-like odor or as an insect repellent or insecticide [40, 48, 49]. Due to its low natural abundance, nootkatone has been produced by sequential oxidations of valencene, a cheaper sesquiterpene produced from citrus fruit, and also from engineered microbes [50, 51].

There are many sesquiterpenoid lactones (SLs) with chemical and structural properties favorable for selective activity towards tumor and cancer stem cells. These compounds function by targeting specific signaling pathways and characteristics present in cancerous cells and are thus lead compounds in cancer clinical trials.

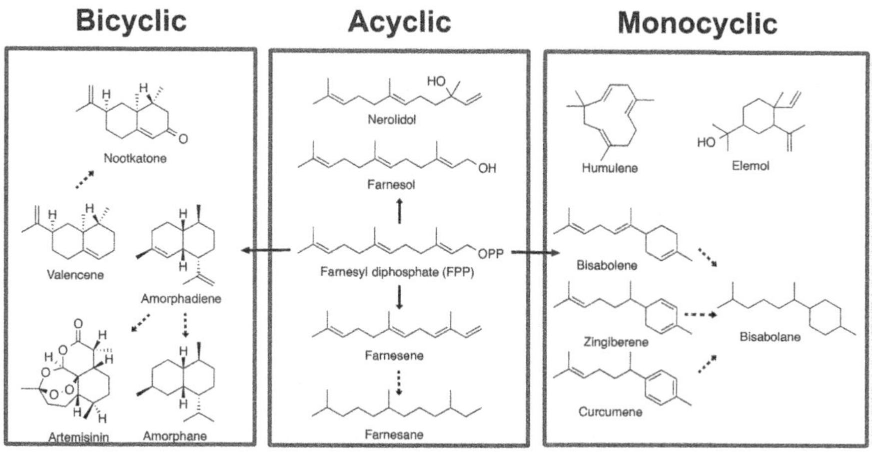

Fig. 3 Examples of sesquiterpenoids. All sesquiterpenoids are produced from the C15 precursor FPP (farnesyl diphosphate). According to the nature of terpene synthase, they can be either acyclic, monocyclic, or bi- or tricyclic compounds. The solid arrows show enzymatic processes, and the *dotted arrows* represent downstream conversion by either chemical or enzymatic processes

Artemisinin, thapsigargin, parthenolide, and many of their synthetic derivatives have been identified and tested for anticancer activities (Ghantous et al. 2010; [52–54]).

As mentioned above, farnesane and bisabolane have been identified and tested as diesel fuel alternatives. D2 Diesel, the fossil fuel for compression ignition engines, is a mixture of linear, branched, and cyclic alkanes with an average carbon length of 16. Sesquiterpenes are hydrocarbons of 15 carbons, close to the average length of diesel (C16), but with a branched—rather than a straight-chain structure. From a fuel performance point of view, the branching degree of the isoprenoid translates into greater molecular stability under high pressure, reduced premature ignition, increased octane number, and lowered freezing points through reduced molecule stacking [55]. However, having a slightly branched hydrocarbon rather than a straight-chain alkane also lowers the quality of combustion (i.e., reduced cetane number) in diesel engines. In the next section, we review these two biodiesel alternatives and the efforts to produce them in microbial hosts.

3.2 Farnesene and Bisabolene

The C15 isoprenoids farnesane and bisabolane have cetane numbers of 58 and 52, respectively, putting them within the expected range for diesel fuels (40–60; [55]). In addition, they display better cold properties, with cloud points of −78 and −25 °C compared with D2 diesel's cloud point of −3 °C. The ring portion of bisabolane increases the density of the fuel (0.88 g/mL), which will increase the energy density per volume of fuel [6]. Farnesane has a lower density (0.77 g/mL) than bisabolane, however, it has a better cetane number and is the closest to commercialization [7].

Although plants are the natural source of bisabolene and farnesene, which are sesquiterpene precursors of bisabolane and farnesane, respectively, engineered microbial platforms may be the most convenient and cost-effective means to produce these compounds [6, 56, 57]. Microbial production of bisabolene and farnesene has been explored, but an efficient biological route for the hydrogenation of these sesquiterpenes to produce the corresponding biofuels has not been established despite promising initial work [58]. As a result, isoprenoid biofuels can be produced through a hybrid process, using a microbial platform for sesquiterpene overproduction and then a chemical route to produce the fully reduced fuel.

Previously, the mevalonate pathway was engineered in both *E. coli* and *S. cerevisiae* to overproduce FPP and, potentially, any sesquiterpene for which the corresponding terpene synthase is known. The highest reported titers of any isoprenoid are those of the sesquiterpene amorphadiene: ~25 and ~40 g/L amorphadiene have been obtained by overexpression of the MVA pathway in *E. coli* and *S. cerevisiae,* respectively [27, 31]. The flexibility of the *E. coli* and *S. cerevisiae* FPP-overproducing platforms allowed scientists to switch rapidly from the production of amorphadiene to the production of bisabolene [6] and farnesene [7].

Farnesene is the generic name for a series of sesquiterpene isomers that in nature act as chemical signaling molecules with diverse functions in numerous organisms [16], playing roles as attractants in pollination [59] and predation response [60] in plants or as alarm pheromones in insects [61].

In heterologous hosts such as *E. coli* and *S. cerevisiae*, farnesene has been produced for use as a precursor for renewable fuels and chemicals from FPP via heterologous expression of farnesene synthase [7, 16, 62]. Farnesene synthases have been isolated from different sources including *Mentha piperita* [39], *A. annua* [63], *Picea abies* [64], *Zea mays* [60], or *Citrus junos* [65].

The molecular structure of *trans-β*-farnesene (commercialized under the name Biofene® by the biotech company, Amyris, based in Emeryville, California) makes it attractive as a scaffold for specialty chemical applications such as solvents, emollients, and vitamins (Amyris website at www.amyris.com). The fully reduced form of farnesene (farnesane) is being pursued as an alternative biosynthetic diesel and is the closest of the isoprenoid-based biofuels to commercialization [57].

Using the previously described gene expression systems of the MVA pathway [13, 66] with farnesene synthase from *A. annua* and *P. abies,* Amyris described the bioproduction of farnesene using both *E. coli* and yeast (up to 1.1 g/L farnesene in *E. coli* expressing the MVA pathway after 120 h, and 728 mg/ L in yeast after 72 h) and methods to hydrogenate the biologically produced sesquiterpene into farnesane and other derivatives in a two-step semisynthetic process [7]. Farnesene is currently produced by Amyris from sugarcane using laboratory-evolved strains of the industrial yeast *S. cerevisiae* PE-2. By iterations of random mutagenesis, analysis, and selection, evolved strains have produced farnesene at >50 % of theoretical mass yield [16]. According to the last public report of Amyris at the end of 2010, total titers reached 104.3 g/L of farnesene with a productivity of 16.9 g/L/d and a recovery of 95 % (source: http://www.biomassboard.gov/pdfs/biomass_tac_todd_pray_09_29_2010.pdf). Current titers are probably higher, but the company has not made the numbers public. Amyris began industrial manufacturing of farnesene in Brazil in December 2012. The plant is located at the Paraíso Bioenergia mill, where microbes convert sugarcane into Biofene that is then processed into specific renewable products. Amyris, currently selling renewable diesel in metropolitan areas in Brazil, has plans for additional production facilities to meet growing demand for its renewable products (www.amyris.com). Very recently, Amyris partnered with TOTAL (the French oil company) and the Brazilian airline GOL to fly the first commercial flight with farnesane and attained industry approval for this renewable jet fuel. Also, Amyris has patented amorphane (the hydrogenated form of the artemisinin precursor amorphadiene) as a component in a jet-fuel replacement [9, 27].

To produce the biofuel precursor bisabolene in *E. coli*, bisabolene synthases from various plant sources were coupled with the heterologous mevalonate pathway [6]. A bisabolene synthase from *Abies grandis* was the most promising, where an *E. coli* codon-optimized variant produced ~400 mg/L of bisabolene. Metabolic engineering of the precursor pathway to improve flux to FPP led to final bisabolene titers of ~900 mg/L in both *E. coli* and *S. cervisiae* [6]. Using carotenoid production as a visual phenotype, genes that affected isoprenoid synthesis in yeast were identified and knocked-out to increase terpene production. Combinations of these deletions and other pathway modifications improved titers of bisabolene more than 20-fold to 800 mg/L in flask and 5.2 g/L in a fermentation process [67]. The discovery of bisabolene synthase as the limiting factor of bisabolene synthesis in these studies prompted the identification of the crystal structure of the most efficient bisabolene synthase (from *A. grandis*) to aid in its engineering for increased microbial bisabolene production [68].

3.3 Farnesol

Farnesol ($C_{15}H_{26}O$) is an acyclic sesquiterpenoid alcohol derived from FPP. It is found in plant essential oils and is commercially important in the flavor and fragrance industries. In addition, farnesol is pharmaceutically relevant as an antimicrobial [69], anticancer drug precursor [43, 70], and is useful in agriculture as a biopesticide [45, 71]. Furthermore, this branched chain alcohol has also been considered as a diesel or jet-fuel substitute due to its low water solubility, high-energy content, and relatively low volatility [45, 57].

Farnesol has been microbially produced using some naturally occurring microbes such as *Candida albicans*, which uses farnesol as a quorum sensing molecule [72]. However, the best production of farnesol has been achieved by the dephosphorylation of FPP in engineered *S. cerevisiae* and *E. coli* overexpressing MVA pathway genes [16, 45]. Efforts have been directed towards increasing the FPP pool size through pathway engineering and redirecting FPP flux to farnesol formation by downregulating competing pathways [45]. A titer of 135 mg/L was reported in *E. coli* following heterologous expression of the MVA pathway and augmentation of IspA (FPP synthase; [42]). In developed strains of *S. cerevisiae*, titers up to ~5 g/L farnesol have been achieved after 215 h in 1-L fed-batch fermentations [73]. This farnesol production strain contained a mutation in the ERG9 gene that presumably led to an increased FPP pool and therefore increased farnesol production by nonspecific endogenous phosphatases, such as LPP1 and DPP1 [74, 75]. Further modifications included higher expression of HMG-CoA reductase (HMGR), found to be a key element in improved farnesol production [76].

FPP dephosphorylation is carried out by some endogenous promiscuous phosphatases, but can also be done by more specific sesquiterpene synthases [74, 75, 77]. However, the promiscuous phosphatases have high Km and low kcat values towards FPP due to their broad substrate specificity, and the farnesol synthases (usually of plant origin [77]) have poor expression in the microbial production host which limit the use of these enzymes. Protein engineering to improve the specificity of dephosphorylation and the soluble expression of farnesol synthases is likely needed to improve microbial farnesol production further [42].

3.4 Future Work

Microbial production of sesquiterpenes appears to be an attractive alternative to extraction from plants or chemical synthesis from petroleum-derived materials. However, the biological production of sesquiterpenes is still far from optimum because overexpression of the isoprenoid production pathway in the microbial host leads to toxicity and metabolic stress. Even though most sesquiterpenes are not toxic to the producing host, the toxicity of the intermediates leads to the reduced carbon flux to the final product and eventually leads to final yields quite lower than the theoretical maximum. Many studies have already identified many different pathway bottlenecks such as MK expression [34], HMG-CoA accumulation [25], and low activity of the terpene synthase [6, 67, 78] in microbial sesquiterpene production. Although partially addressed [25, 34, 79, 80], the unbalanced expression of the pathway is still a primary cause of low isoprenoid yields [81]. Therefore, further efforts to balance enzyme expression and enzyme screening/evolution seem a good strategy to follow in order to achieve higher yield of the final product.

4 Monoterpene Fuels and Chemicals

4.1 Monoterpene Properties and Chemical Diversity

Monoterpenes are C10 compounds built from two C5 isoprenoid units (one IPP and one DMAPP). Monoterpenes are generally synthesized in the glandular structures of plants and are thus common components of essential oils. As with sesquiterpenes, advances in metabolic engineering now allow for the microbial production of diverse, monoterpene-derived oils, flavors, and pharmaceuticals [57]. Moreover, recent success in the microbial production of limonene and pinene have expanded the microbial platform to include jet biofuels [10, 78, 82].

Monoterpenes can be divided into three major subgroups based on structural features (Fig. 4): acyclic, monocyclic, and bicyclic monoterpenes. Acyclic monoterpenes include compounds such as myrcene and ocimene, which are hydrocarbons very similar to classical petrochemicals. Both are valuable compounds for the perfume industry, and myrcene is an especially versatile starting material for flavors, fragrances, cosmetics, vitamins, and pharmaceuticals due to its reactive diene structure [83]. The hydrogenated forms of these acyclic monoterpenoids (i.e., 2,6-dimethyloctane) have been proposed as alternative biofuels [10]. Hydroxylated acyclic monoterpenoids such as linalool and geraniol are also potential biofuels in addition to being important components in fragrances and pesticides [37, 84–86].

Cyclic monoterpenoids include a huge variety of molecules, most of which are derived from the α-terpinyl cation. As a result, many important flavor and medicinal compounds have the same carbon skeleton as limonene. Most notable derivatives of monocyclic monoterpenes are oxygenated compounds, such as α-terpineol, perillyl alcohol, carveol, carvone, and menthol [87]. The fully hydrogenated form of limonene, limonane, is considered a promising jet-fuel replacement [10], and recently limonene itself has been evaluated as a jet-fuel additive [88]. Among bicyclic monoterpenoids, there are pinenes, which are the main components of turpentine produced from tapping trees [10, 78, 89]. Many valuable compounds that are used in the fragrance, flavor, and drug industries are derived from pinenes. Pinene-derived verbenone, for instance, is used in perfumery, as a cough suppressant, and in insect control [84, 90, 91], and camphor is used as a flavor compound, plasticizer, cosmetic, and pharmaceutical [92]. Other bicyclic monoterpenoids include sabinene and related compounds such as thujone, thujene, and umbellulone which are valuable as flavoring agents, perfume components, antimicrobials, and potential biofuels [41, 93, 94], however, toxicity of these compounds might limit their use [95–97].

Although isoprenoid alternatives to diesel fuel have been identified and microbially produced, metabolic engineering for the production of high-energy density tactical fuels (jet and missile fuels, e.g.) has lagged behind. Existing biosynthetic jet fuels derived from natural oils have been used to power aircraft in 50:50 blends with Jet-A fuel [98, 99], but lack the specifications required to replace jet fuels such as JP-10 [78].

4.2 Pinene and Limonene

α, β-Pinenes and limonene are widely used in fragrances, drugs, and as commodity chemicals. Recently, the demand for these compounds has risen due to their suitability as renewable, high-density jet fuels [78, 98]. Attaining the volumetric energy

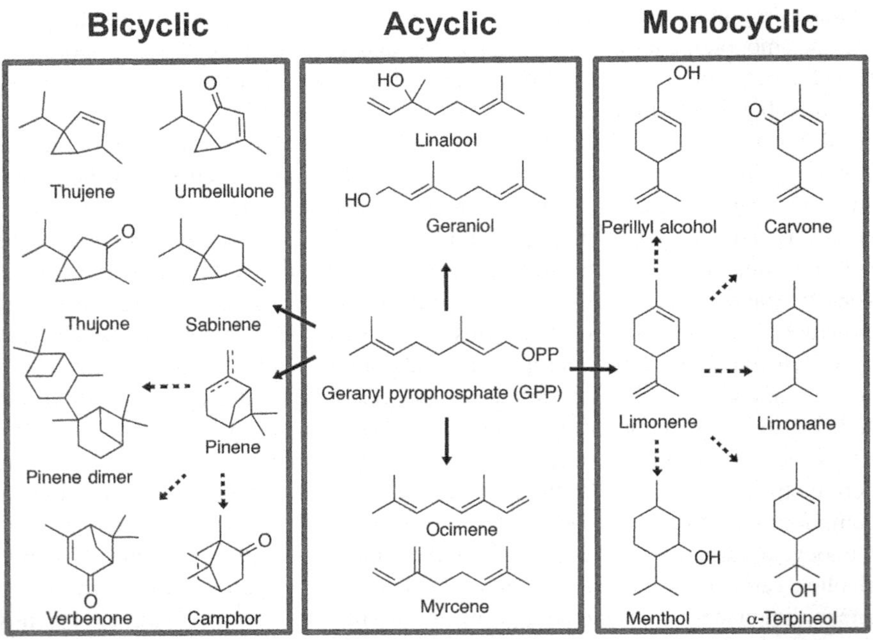

Fig. 4 Examples of monoterpenoids. All monoterpenoids are produced from the C10 precursor GPP (geranyl diphosphate). Depending on the terpene synthase, they can be acyclic, mono-, or bicyclic compounds. The *solid arrows* represent enzymatic processes, and the *dotted arrows* show downstream conversion by either chemical or enzymatic processes

content necessary for jet fuels requires mimicking the strained ring systems found in JP-10 [78]. Hydrogenated pinene dimers, synthesized via chemical dimerization of the bicyclic terpenes (pinenes), have been shown to contain high-volumetric energy similar to that found in JP-10 [4]. Similarly, the hydrogenated form of limonene has been reported to have favorable properties for next-generation jet biofuels and fuel additives that enhance cold-weather performance [10].

Currently, α-pinene and (+)-D-limonene are mainly obtained from the plant biomass of tapping trees (turpentine) or as a byproduct of orange juice production, respectively. However, fluctuations in their production from natural producers (i.e., plants) and subsequent cost limit their use as biofuels and chemical feedstock even though the demand for these monoterpenes is increasing [57, 78, 82, 87, 89]. Therefore, it is necessary to seek sustainable technologies for monoterpene production.

Although microbes can produce various isoprenoids through either MEP or MVA pathways to supply the essential metabolites DMAPP and IPP, they are usually unable to produce monoterpenes due to the lack of efficient geranyl diphosphate synthases (GPPS) and adequate monoterpene synthases. With the growing interest in these compounds, many metabolic engineers have explored biosynthetic methods for monoterpene production. A decade ago, the production of the monoterpenes 3-carene and limonene was demonstrated through heterologous expression of carene synthase and limonene synthase (LS) of plant origin in *E. coli* [100, 101]. However, yields were low: production levels of 3 μg/L/OD600 and 5 mg/L were achieved for 3-carene and limonene, respectively. In these early works, the authors hypothesized that low precursor availability (i.e., IPP and DMAPP) was the primary cause of low titers [13, 100, 101]. Although the endogenous MEP pathway in *E. coli* may not be sufficient for high flux supply of IPP and DMAPP, the heterologous expression of the MVA pathway in *E. coli* considerably improves flux to isoprenoid precursors [13].

Two recent publications managed to produce pinene in *E. coli* at reasonably high titers by using a heterologous MVA pathway. In the first study [89], α-pinene production was significantly enhanced in *E. coli* by assembling a heterologous MVA pathway, codon-optimized GPPS2 from *A. grandis*, and codon-optimized pinene synthase (PS) Pt30 from *Pinus taeda*. The final pinene-producing strain, YJM28, accumulated α-pinene up to 5.44 mg/L in a shake-flask and 0.97 g/L under the fed-batch fermentation conditions. In more recent work, the authors used a previously engineered *E. coli* strain for the overproduction of IPP and DMAPP and combinatorially screened PS and GPPS enzymes to improve flux through the last two steps of the pathway [78, 102]. By combining expression of three PS and three GPPS from conifers, they achieved about 28 mg/L of pinene using GPPS and PS from *A. grandis*. Furthermore, they designed GPPS-PS protein fusions to reduce GPP product inhibition and toxicity by substrate channeling, producing 32.4 mg/L of pinene [78].

A platform for high titer production of limonene was also recently described [82]. As explained above for pinene production, a heterologous MVA pathway was used to provide IPP and DMAPP precursors and combined with GPPS from *A. grandis* and LS from *Mentha spicata*. In this study, a series of engineering steps yielded much higher titers of limonene than previously reported: the limonene titer was over 450 mg/L, comparable to those achieved for sesquiterpenes (500–1,000 mg/L). Despite using a similar platform, this titer is also considerably higher than that of pinene ($\sim$32 mg/L), probably because the LS enzyme expresses better and/or has higher efficiency than any of the PS enzymes tested.

Compared to sesquiterpenes, one of the main differences and concerns in the microbial overproduction of monoterpenes is the toxicity of monoterpene products [103, 104]. In fact, efforts to detect and overexpress efflux pumps to counteract monoterpene toxicity has proven useful in the improvement of yields [105].

4.3 Acyclic Monoterpenes

Efforts to produce acyclic monoterpenes microbially have focused primarily on geraniol due to its value in the fragrance, agrochemical, and pharmaceutical industries [43, 86, 106]. As a biofuel compound, geraniol ($C_{10}H_{18}O$) has similar properties to farnesol, a potential sesquiterpenoid biofuel alcohol [57]. Geraniol is produced by dephosphorylation of GPP and has recently been produced in an engineered *E. coli* strain [86]. As practiced in other successful efforts, the authors engineered *E. coli* to have an exogenous MVA pathway for improved precursor production. They used an engineered GPPS mutated from native *E. coli* FPP synthase for the biosynthesis of GPP, and a truncated geraniol synthase from *O. basilicum* for efficient conversion of GPP to geraniol. After eliminating competing pathways, they achieved a final production titer of 182.5 mg/L of geraniol.

The hydrogenated products of acyclic monoterpenes such as myrcene and ocimene are also considered good biofuel replacements [10]. Microbial production of these acyclic monoterpenes, which originally are found in plants, has been reported in the product mixture of an engineered *E. coli* strain that produced various terpenes [101]. To our knowledge, however, specific microbial metabolic engineering efforts to overproduce these acyclic monoterpenes have not been reported. Because terpene synthases that synthesize these acyclic monoterpenes have been described [107], it is likely that microbial production with high product specificity will be achieved soon.

4.4 Future Work

In order to be considered to be economically competitive alternatives to petroleum-based fuels, near-theoretical yields would be required for biofuel production. To produce the monoterpene limonene, for example, it is known that the required microbial yields and titers significantly exceed current microbial limonene toxicity limits [105]. Clearly it will be necessary to remove the product continuously from the producing culture broth and/or engineer more tolerant strains.

There are various ways to overcome the product toxicity, but one of the most promising approaches for monoterpene production was the use of efflux pumps [105]. Efflux pumps are membrane transporters that recognize and export toxic compounds from the cell using the proton motive force. The best-studied solvent tolerance pumps are those from the resistance–nodulation–division (RND) family in gram-negative bacteria, but many other organisms also harbor efflux pumps [108]. Recent experiments have demonstrated that heterologously expressed RND efflux pumps can improve tolerance to biofuels [105]. Expression of efflux pumps is a promising engineering strategy to engineer tolerance for many isoprenoid biofuels, but there are several avenues for future research in this area. Directed evolution may help produce designer pumps that are specific to a particular biofuel, and fine-tuned efflux pump expression is also necessary inasmuch as it is well known that overexpression of membrane proteins can be detrimental [109, 110].

Though targeted efflux pump expression has been proven successful as a tolerance mechanism, traditional strain improvement methods such as chemostat-mediated adaptation, chemical mutagenesis, directed evolutionary engineering, genome shuffling [111], or targeted recombineering methodologies [112] constitute powerful tools for rapid evolution of tolerance mechanisms [105, 113]. It is also important to recognize that the tolerance engineering strategies are not universal and what works for one biofuel in one host strain may not work for others [104]. Another alternative to withstand toxicity would be to engineer the pathway in alternative hosts such as *Pseudomonas* that are known to have better mechanisms to respond to stress [104, 114, 115].

5 Hemiterpenoid Fuels and Chemicals

5.1 Hemiterpenoid Properties and Chemical Diversity

Hemiterpenes and hemiterpenoids are C5 compounds derived from a single isoprene subunit. Efforts to produce hemiterpenoids in microbial hosts have primarily targeted isoprene, a volatile monomer of significance to the rubber industry. Recently, there has been interest in the biosynthetic production of C3–C5 alcohols as advanced biofuels [57, 116]. One successful approach for microbial C3–C5 alcohol production utilizes refactored amino acid biosynthesis pathways [117]. Broadly speaking, this strategy relies on the overexpression of a promiscuous 2-keto-acid decarboxylase (KDCs) and alcohol dehydrogenase (ADHs) to produce a diverse range of alcohols from amino acid precursors. The leucine biosynthesis pathway, for example, was used to produce 3-methyl-1-butanol or isopentanol at high titers (Fig. 5a; [118]). Isoprenoid biosynthesis pathways provide additional routes to C5 alcohol production, namely isopentenols (3-methyl-3-butenol and 3-methyl-2-butenol) and isopentanol (3-methyl-1-butanol; Fig. 5b). It was recently determined that these alcohols have energy content and octane numbers that make them potential gasoline replacements [119]. In addition to favorable energy content, the alcohols 3-methyl-3- and 3-methyl-2-butenol were recently shown to function as ideal antiknock additives in spark ignition engines [5].

5.2 Isopentenols

Isopentenols are derived from the dephosphorylation of IPP and DMAPP, which form 3-methyl-3- and 3-methyl-2-butenol, respectively (Fig. 5b). Although there have been numerous studies concerning the engineering of both the MEP and MVA pathways for the production of C10, C15, and longer-chained terpenes, work on isoprenoid-derived short-chain alcohols was comparatively minimal until their characterization as biofuel candidates.

Isopentenol was first detected in *E. coli* cultures following the overexpression of the native MEP pathway and was suspected to result from IPP dephosphorylation [120]. More recent work has identified genes capable of catalyzing this dephosphorylation reaction [121]. Using a heterologous MVA pathway and a screening

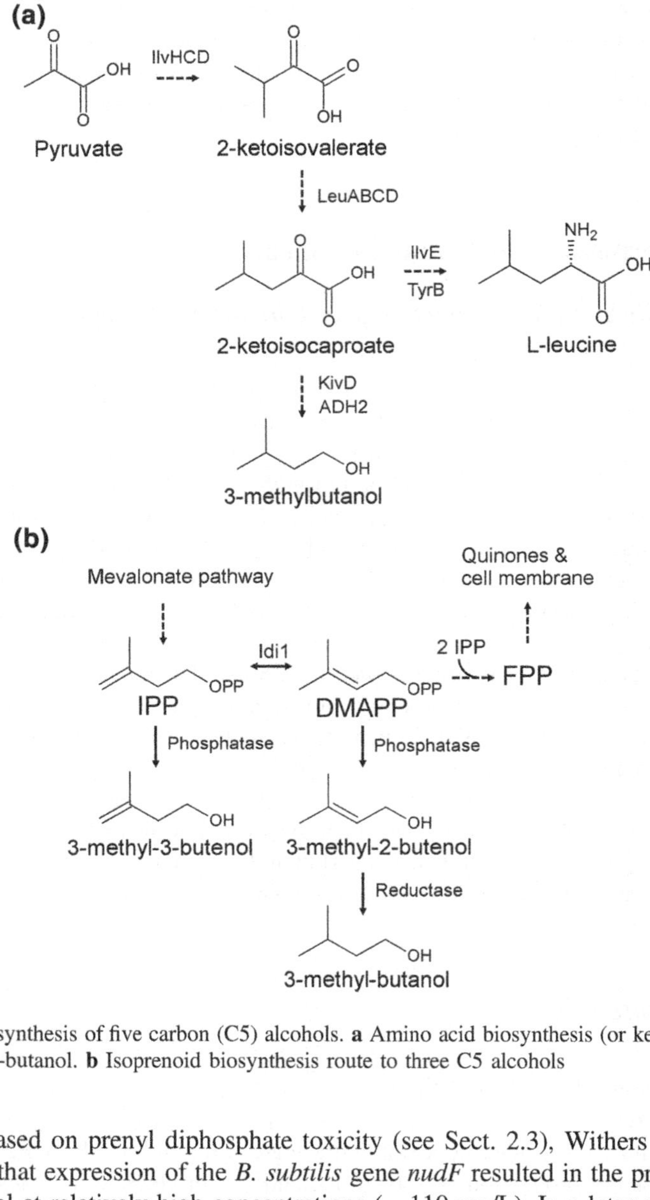

Fig. 5 Biosynthesis of five carbon (C5) alcohols. **a** Amino acid biosynthesis (or keto-acid) route to 3-methyl-butanol. **b** Isoprenoid biosynthesis route to three C5 alcohols

method based on prenyl diphosphate toxicity (see Sect. 2.3), Withers et al. demonstrated that expression of the *B. subtilis* gene *nudF* resulted in the production of isopentenol at relatively high concentrations (~110 mg/L). In a later study, NudF was compared with alternative phosphatases in *E. coli* in an effort to assemble a synthetic pathway for C5 alcohol production [122]. In this work, efficient IPP phosphatase activity was discovered by screening HAD [123] and Nudix [124] superfamilies of enzymes. Out of 23 HAD-like phosphatases and 13 Nudix hydrolases in *E. coli*, the Nudix hydrolase NudB was found to be the most efficient enzyme. By expressing NudB on a high-copy vector with pMevT and pMevB [13],

3-methyl-3-butenol was produced at 8.3 % of theoretical yield. Using a fusion protein of Idi1 and NudB, IPP-derived 3-methyl-3-butenol and DMAPP-derived 3-methyl-2-butenol were produced. The total C5 alcohol content was reduced from the strain expressing only NudB, however, and was hypothesized to result from competition with FPP biosynthesis. By expressing a promiscuous *E. coli* reductase (NemA) with the fusion protein, 3-methyl-2-butenol could be reduced to 3-methyl-1-butanol, resulting in the production of three C5 alcohols (Fig. 5b).

Optimizations to the heterologous mevalonate pathway have resulted in significantly higher yields of isopentenol in *E. coli.* Replacement of *HMGS* and *HMGR* from *S. cerevisiae* with versions of *Enterococcus faecalis* improved mevalonate production 45-fold [125]. Expression of this upper pathway with *MK*, *PMK*, *PMD*, and *nudF* from *B. subtilis* resulted in the production of 1.3 g/L of 3-methyl-3-butenol at a yield of 12 % [125]. When *idi* was also expressed, a mixture of 3-methyl-3- and 3-methyl-2-butenol was produced, albeit at a reduced total C5 alcohol yield as observed previously [122]. With NudF from *E. coli*, 3-methyl-2-butenol was produced with higher specificity, but at a significantly lower yield [125].

More recently, 3-methyl-3-butenol was produced at a yield of 46 % following extensive pathway engineering [81]. Using a systematic method, rationally constructed variants of the precursor pathway were assayed for pathway protein levels, glucose, acetate, and isopentenol. According to a correlation analysis, variations in the level of HMGS and MK explained the majority of pathway behavior. Using this information, a conceptual model of isopentenol pathway function was developed and used to guide engineering efforts. With the expression of *E. coli*-derived NudB, the most efficient pathway produced 1.5 g/L of 3-methyl-3-butenol at 46 % yield, consumed all available glucose, and produced low amounts of the waste-product acetate.

5.3 IPP Toxicity

Inasmuch as IPP is the immediate precursor of isopentenol, preventing or overcoming IPP toxicity is an important issue to consider when engineering isopentenol pathways. IPP toxicity was first identified a decade ago, when a high-flux mevalonate pathway was assembled in *E. coli* [13]. The kinetics of NudB, the protein used to produce the highest reported titer of isopentenol [81], show that this enzyme is quite slow, with a turnover rate significantly less than other mevalonate pathway enzymes [122]. When this enzyme was paired with a high-flux precursor pathway, IPP accumulation and toxicity were observed [81]. Specifically, cell growth and glucose consumption were attenuated when IPP concentrations were high. Expression of an additional copy of *nudB* alleviated this toxicity, albeit at a significant increase to metabolic burden. A similar reduction in OD was also observed using NudF from *B. subtilis*, though IPP measurements were not taken to confirm its accumulation [125]. Due to this toxicity, the development of more efficient hydrolases to catalyze isopentenol production is needed to maximize productivity. In addition, the mechanism of IPP toxicity should be studied to aid in tolerance engineering.

5.4 Future Work

The initial progress in the microbial production of hemiterpenes is encouraging: in shake-flask experiments, yields of hemiterpene alcohols approach or exceed the reported yields of monoterpene and sesquiterpene biofuels. In addition, the prospect of creating hemiterpene derivatives with attractive properties appears relatively straightforward and has yet to be explored. Through expression of alcohol acetyl transferases, for example, esters of C5 alcohols with promising fuel properties such as isoamyl acetate [126–128] can be readily produced.

Additional work is needed to increase the specificity of alcohol production. Fully reduced isopentanol, the most attractive target from a biofuel perspective, can currently only be produced as a component of a three-alcohol mixture at very low yields. Mitigation of carbon loss to FPP through a genetic knockdown of *ispA* may improve yields, and further engineering of the previously described Idi1 ~ NudB fusion [122] may improve flux to 3-methyl-2-butenol and subsequently isopentanol. Altering the substrate preference of NudB or NudF, both of which have native substrates that differ significantly from IPP and DMAPP [124], could also increase yields by providing improved product specificity. Finally, an improved reductase may be capable of increasing the conversion of 3-methyl-2-butenol to isopentanol.

6 Outlook

In this review chapter, we have summarized some of the engineering efforts towards the microbial synthesis of isoprenoid-based drugs, chemicals, and fuel compounds. The incredible diversity of isoprenoid compounds provides nearly countless target molecules for biological engineers. Though many challenges exist, even complex medicinal isoprenoids such as artemisinin and Taxol may eventually be produced entirely in vivo. The complete biosynthesis of these medicinal isoprenoids frequently requires several oxidations of the original terpene backbone, mostly by cytochrome P450 enzymes as shown for artemisinin and Taxol [15, 18]. Because it is important to carry out these P450-based oxidations in the engineered host, further study of P450 enzymes is critical to build a complete biological route for these compounds. P450s are frequently found as membrane-bound enzymes and require colocalized CPR (cytochrome P450 reductase) as a reductase partner. Although *E. coli* has been useful as a high-titer platform for terpene biosynthesis, achieving efficient P450 activity in this host has been challenging. The complex folding of P450 proteins, coupled with *E. coli*'s requirement for membrane-unbound, soluble

P450s, has severely limited progress. However, recent progress on the directed evolution of bacterial P450 enzymes has yielded P450 candidates that are functional in *E. coli* and have specific activity towards terpene targets.

Considerable progress has been made in the microbial production of isoprenoid-based advanced biofuels. With a few exceptions, however, most compounds are far from commercialization. There are several risks in commercialization such as capital risk, technology risk, market risk/volatility, and operational risk. To overcome these risks and prove economic feasibility are the most important considerations in commercialization. The technoeconomic analysis on feedstock to fermentable sugar and to biofuel suggests that there are many variables to consider such as feedstock prices, biomass depolymerization costs, the yield of the microbial process for biofuel production, and the scalability [129]. For example, when we assume a break-even price of sugar at the mill to be close to $0.10/lb, which is close to the long-term nominal price of the commodity, a rough calculation of the theoretical price of microbial sesquiterpene biofuel gives about $6/gal of biofuel based on data for ethanol production when we assume near 100 % theoretical yield for sesquiterpene production [6]. Due to these challenging economics, high product yields (generally at least 85 % of the theoretical maximum yield) are absolutely essential if biofuels are to be competitive with nonrenewable, petroleum-based fuels [57].

Optimization of the isoprenoid pathway itself is perhaps the most direct way to improve titer, yield, and productivity. A common theme that has emerged in isoprenoid pathway optimization is the requirement for balanced enzyme expression. In recent years, analytical tools such as targeted proteomics [34, 81] and metabolomics have paired with genetic standardization and modern cloning techniques to make this task far easier. In most cases, pathway balancing has led directly to improvements in titer [15, 25, 34, 81]. For C10 and C15 terpenoids, the slow kinetics of most terpene synthases poses another significant barrier to economical production. The construction of more efficient terpene synthases through protein engineering or directed evolution will undoubtedly provide further improvements in yield and productivity. Though pathway optimization is crucial, interactions with the microbial host must also be considered: competing pathways must be eliminated and perturbations to cellular central metabolism, redox balance, and energetics must be minimized. Systems biology and metabolic modeling hold particular promise for addressing these challenges.

Another important aspect to address is the scalability of microbial production inasmuch as it has a huge impact on the overall economics and feasibility of isoprenoid biofuel production. Production data from high-volume, fed-batch fermentations are necessary to assess the feasibility of large-scale production. At large scale, the addition of exogenous antibiotics or inducers becomes cost-prohibitive. Recently developed technologies that alleviate the requirement for both antibiotics [130] and inducers [35] should thus be used to improve pathway stability and lower production costs. Because successful scale-up is critical—and because the conditions for optimal production may change when fermentation takes place in a bioreactor—dynamic promoters and expression systems that respond to environmental queues are particularly desirable. Ideally, the cell should be able to adjust the pathway activity according to the metabolic status of the host or the concentrations of key pathway intermediates [131]. Although a successful example was recently reported for sesquiterpenoid production [35], further applications of dynamic regulation could potentially eliminate the accumulation of toxic intermediates, increase strain stability, and improve pathway efficiency.

The high cost of carbon sources such as purified glucose is detrimental to biofuel cost and must be reduced to achieve economical production. Cheaper, more abundant carbon sources are clearly required. Lignocellulosic biomass is perhaps the most promising due to its availability, and work is underway to construct microbes that efficiently metabolize plant-derived sugars. Sucrose serves as another promising alternative to purified glucose. In countries such as Brazil and Australia, sucrose is currently used as a major carbon source in the biobased chemical and fuel industry. Most native *E. coli* hosts, however, cannot metabolize sucrose, and engineering this capability is necessary. More recently, photosynthetic autotrophic microbes have been studied for isoprene production, and methanotrophic bacteria are being investigated as shale gas is increasingly available as a cheap and abundant carbon source.

Although we have focused primarily on simple isoprenoids in this review, there are more complex natural terpenoids that have important medicinal value and are potential targets for biological and metabolic engineers. Indeed, it may soon be possible to produce valuable new classes of complex isoprenoid derivatives biosynthetically due to recent advances in genomics and related disciplines. Meroterpenoids, oxygenated medicinal terpenoids biosynthesized from polyketide and terpenoid precursors [132], serve as attractive examples. As genomic and metagenomic sequencing data become more abundant and broadly applicable to natural products discovery, there have been efforts to understand and engineer meroterpenoid biosynthesis in bacteria and fungi [133, 134]. Though less genetically tractable, fungal systems are particularly promising for meroterpenoid production as they usually have a larger range of functional P450 enzymes and contain endogenous polyketide synthases and isoprenoid pathways [135]. With proper engineering, these fungal hosts will provide a promising route to improved terpene and medicinal meroterpenoid production in non-*E. coli*, nonyeast hosts.

References

1. Breitmaier E (2006) Terpenes flavors, fragrances, pharmaca, pheromones. Wiley, New York
2. Connolly J, Hill R (1991) Dictionary of terpenoids. Chapman and Hall, London
3. Davies FK, Jinkerson RE, Posewitz MC (2014) Toward a photosynthetic microbial platform for terpenoid engineering. Photosynth Res
4. Harvey BG, Wright ME, Quintana RL (2010) High-density renewable fuels based on the selective dimerization of pinenes. Energy Fuels 24:267–273
5. Mack JH, Rapp VH, Broeckelmann M, Lee TS, Dibble RW (2014) Investigation of biofuels from microorganism metabolism for use as anti-knock additives. Fuel 117:939–943

6. Peralta-Yahya P, Ouellet M, Chan R, Mukhopadhyay A, Keasling J, Lee T (2011) Identification and microbial production of a terpene-based advanced biofuel. Nat Commun 2:483
7. Renninger N, McPhee D (2008) Fuel compositions comprising farnesane and method of making and using same. US20080098645 A1
8. Renninger N, Ryder J, Fisher K (2011) Jet fuel compositions and methods of making and using same. US Patent 7,671,245. US 7942940 B2
9. Ryder JA (2012) Jet fuel compositions and methods of making and using same. US 8106247B2
10. Tracy NI, Chen DC, Crunkleton DW, Price GL (2009) Hydrogenated monoterpenes as diesel fuel additives. Fuel 88:2238–2240
11. Yang Y, Dec J, Dronniou N, Simmons B (2010) Characteristics of isopentanol as a fuel for HCCI engines. SAE Int J Fuels Lubr 3:725–741
12. Wagner A, Krab K (1995) The alternative respiration pathway in plants: role and regulation. Physiol Plant 95(2):318–325
13. Martin V, Pitera D, Withers S, Newman J, Keasling J (2003) Engineering a mevalonate pathway in *Escherichia coli* for production of terpenoids. Nat Biotechnol 21:796–802
14. McCaskill D, Croteau R (1997) Prospects for the bioengineering of isoprenoid biosynthesis. Adv Biochem Eng Biotechnol 55:107–146
15. Ajikumar PK, Xiao W-H, Tyo KEJ, Wang Y, Simeon F, Leonard E, Mucha O, Phon TH, Pfeifer B, Stephanopoulos G (2010) Isoprenoid pathway optimization for Taxol precursor overproduction in Escherichia coli. Science 330:70–74
16. Chandran SS, Kealey JT, Reeves CD (2011) Microbial production of isoprenoids. Process Biochem 46:1703–1710
17. Klein-Marcuschamer D, Ajikumar PK, Stephanopoulos G (2007) Engineering microbial cell factories for biosynthesis of isoprenoid molecules: beyond lycopene. Trends Biotechnol 25:417–424
18. Ro DK, Paradise EM, Ouellet M, Fisher KJ, Newman KL, Ndungu JM, Ho KA, Eachus RA, Ham TS, Kirby J, Chang MC, Withers ST, Shiba Y, Sarpong R, Keasling JD (2006) Production of the antimalarial drug precursor artemisinic acid in engineered yeast. Nature 440:940–943
19. Hsu E (2006) Reflections on the "discovery" of the antimalarial qinghao. Br J Clin Pharmacol 61:666–670
20. World Health Organisation (2010) Guidelines for the treatment of malaria. 2nd edn. World Health Organization, Geneva
21. White NJ (2008) Qinghaosu (artemisinin): the price of success. Science 320:330–334
22. Kuzuyama T (2002) Mevalonate and nonmevalonate pathways for the biosynthesis of isoprene units. Biosci Biotechnol Biochem 66:1619–1627
23. Newman JD, Marshall J, Chang M, Nowroozi F, Paradise E, Pitera D, Newman KL, Keasling JD (2006) High-level production of amorpha-4,11-diene in a two-phase partitioning bioreactor of metabolically engineered *Escherichia coli*. Biotechnol Bioeng 95:684–691
24. Kizer L, Pitera DJ, Pfleger BF, Keasling JD (2008) Application of functional genomics to pathway optimization for increased isoprenoid production. Appl Environ Microbiol 74:3229–3241
25. Pitera DJ, Paddon CJ, Newman JD, Keasling JD (2007) Balancing a heterologous mevalonate pathway for improved isoprenoid production in *Escherichia coli*. Metab Eng 9:193–207
26. Anthony JR, Anthony LC, Nowroozi F, Kwon G, Newman JD, Keasling JD (2009) Optimization of the mevalonate-based isoprenoid biosynthetic pathway in Escherichia coli for production of the anti-malarial drug precursor amorpha-4,11-diene. Metab Eng 11:13–19
27. Tsuruta H, Paddon CJ, Eng D, Lenihan JR, Horning T, Anthony LC, Regentin R, Keasling JD, Renninger NS, Newman JD (2009) High-level production of amorpha-4,11-diene, a precursor of the antimalarial agent artemisinin, in *Escherichia coli*. PLoS One 4:e4489

28. Chang MCY, Eachus RA, Trieu W, Ro D-K, Keasling JD (2007) Engineering Escherichia coli for production of functionalized terpenoids using plant P450s. Nat Chem Biol 3:274–277
29. Paddon CJ, Keasling JD (2014) Semi-synthetic artemisinin: a model for the use of synthetic biology in pharmaceutical development. Nat Rev Microbiol 1–13
30. Westfall PJ, Pitera DJ, Lenihan JR, Eng D, Woolard FX, Regentin R, Horning T, Tsuruta H, Melis DJ, Owens A, Fickes S, Diola D, Benjamin KR, Keasling JD, Leavell MD, McPhee DJ, Renninger NS, Newman JD, Paddon CJ (2012) Production of amorphadiene in yeast, and its conversion to dihydroartemisinic acid, precursor to the antimalarial agent artemisinin. Proc Natl Acad Sci USA 109:E111–E118
31. Paddon CJ, Westfall PJ, Pitera DJ, Benjamin K, Fisher K, McPhee D, Leavell MD, Tai A, Main A, Eng D, Polichuk DR, Teoh KH, Reed DW, Treynor T, Lenihan J, Fleck M, Bajad S, Dang G, Dengrove D, Diola D, Dorin G, Ellens KW, Fickes S, Galazzo J, Gaucher SP, Geistlinger T, Henry R, Hepp M, Horning T, Iqbal T, Jiang H, Kizer L, Lieu B, Melis D, Moss N, Regentin R, Secrest S, Tsuruta H, Vazquez R, Westblade LF, Xu L, Yu M, Zhang Y, Zhao L, Lievense J, Covello PS, Keasling JD, Reiling KK, Renninger NS, Newman JD (2013) High-level semi-synthetic production of the potent antimalarial artemisinin. Nature 496:528–532
32. Pfleger BF, Pitera DJ, Smolke CD, Keasling JD (2006) Combinatorial engineering of intergenic regions in operons tunes expression of multiple genes. Nat Biotechnol 24:1027–1032
33. Dueber JE, Wu GC, Malmirchegini GR, Moon TS, Petzold CJ, Ullal AV, Prather KLJ, Keasling JD (2009) Synthetic protein scaffolds provide modular control over metabolic flux. Nat Biotechnol 27:753–759
34. Redding-Johanson AM, Batth TS, Chan R, Krupa R, Szmidt HL, Adams PD, Keasling JD, Lee TS, Mukhopadhyay A, Petzold CJ (2011) Targeted proteomics for metabolic pathway optimization: application to terpene production. Metab Eng 13:194–203
35. Dahl RH, Zhang F, Alonso-Gutierrez J, Baidoo E, Batth TS, Redding-Johanson AM, Petzold CJ, Mukhopadhyay A, Lee TS, Adams PD, Keasling JD (2013) Engineering dynamic pathway regulation using stress-response promoters. Nat Biotechnol:1–10
36. Misawa N (2011) Pathway engineering for functional isoprenoids. Curr Opin Biotechnol 22:627–633
37. Kirby J, Keasling JD (2009) Biosynthesis of plant isoprenoids: perspectives for microbial engineering. Annu Rev Plant Biol 60:335–355
38. Maldonado-Bonilla LD, Betancourt-Jiménez M, Lozoya-Gloria E (2008) Local and systemic gene expression of sesquiterpene phytoalexin biosynthetic enzymes in plant leaves. Eur J Plant Pathol 121:439–449
39. Crock J, Wildung M, Croteau R (1997) Isolation and bacterial expression of a sesquiterpene synthase cDNA clone from peppermint (Mentha x piperita, L.) that produces the aphid alarm pheromone (E)-beta-farnesene. Proc Natl Acad Sci USA 94:12833–12838
40. Ansari HR, Curtis AJ (1974) Sesquiterpenes in the perfumery industry. J Soc Cosmet Chem 25:203–231
41. Arctander S (1969) Perfume and flavor chemicals: aroma chemicals, perfume and flavor chemicals: aroma chemicals. Allured Publishing Corporation, Vienna
42. Wang C, Yoon SH, Shah AA, Chung YR, Kim JY, Choi ES, Keasling JD, Kim SW (2010) Farnesol production from *Escherichia coli* by harnessing the exogenous mevalonate pathway. Biotechnol Bioeng 107:421–429
43. Wiseman D, Werner S, Crowell P (2007) Cell cycle arrest by the isoprenoids perillyl alcohol, geraniol, and farnesol is mediated by p21Cip1 and p27Kip1 in human pancreatic adenocarcinoma cells. J Pharmacol Exp Ther 320:1163–1170
44. Rude MA, Schirmer A (2009) New microbial fuels: a biotech perspective. Curr Opin Microbiol 12:274–281
45. Wang C, Kim J, Choi E, Kim S (2011) Microbial production of farnesol (FOH): current states and beyond. Process Biochem 46:1221–1229

46. Fernandes ES, Passos GF, Medeiros R, da Cunha FM, Ferreira J, Campos MM, Pianowski LF, Calixto JB (2007) Anti-inflammatory effects of compounds alpha-humulene and (-)-*trans*-caryophyllene isolated from the essential oil of *Cordia verbenacea*. Eur J Pharmacol 569:228–236
47. Passos GF, Fernandes ES, da Cunha FM, Ferreira J, Pianowski LF, Campos MM, Calixto JB (2007) Anti-inflammatory and anti-allergic properties of the essential oil and active compounds from *Cordia verbenacea*. J Ethnopharmacol 110:323–333
48. Dolan MC, Jordan RA, Schulze TL, Schulze CJ, Manning MC, Ruffolo D, Schmidt JP, Piesman J, Karchesy JJ (2009) Ability of two natural products, nootkatone and carvacrol, to suppress Ixodes scapularis and Amblyomma americanum (Acari: Ixodidae) in a Lyme disease endemic area of New Jersey. J Econ Entomol 102:2316–2324
49. Furusawa M, Hashimoto T, Noma Y, Asakawa Y (2005) Highly efficient production of nootkatone, the grapefruit aroma from valencene, by biotransformation. Chem Pharm Bull (Tokyo) 53:1513–1514
50. Girhard M, Machida K, Itoh M, Schmid RD, Arisawa A, Urlacher VB (2009) Regioselective biooxidation of (+)-valencene by recombinant E. coli expressing CYP109B1 from Bacillus subtilis in a two-liquid-phase system. Microb Cell Fact 8:36
51. Wriessnegger T, Augustin P, Engleder M, Leitner E, Müller M, Kaluzna I, Schürmann M, Mink D, Zellnig G, Schwab H, Pichler H (2014) Production of the sesquiterpenoid (+)-nootkatone by metabolic engineering of *Pichia pastoris*. Metab Eng 24C:18–29
52. Denmeade SR, Jakobsen CM, Janssen S, Khan SR, Garrett ES, Lilja H, Christensen SB, Isaacs JT (2003) Prodrug as targeted therapy for prostate cancer 95
53. Drew DP, Krichau N, Reichwald K, Simonsen HT (2009) Guaianolides in apiaceae: perspectives on pharmacology and biosynthesis. Phytochem Rev 8:581–599
54. Shanmugam R, Kusumanchi P, Appaiah H, Cheng L, Crooks P, Neelakantan S, Peat T, Klaunig J, Matthews W, Nakshatri H, Sweeney CJ (2011) A water soluble parthenolide analogue suppresses in vivo tumor growth of two tobacco associated cancers, lung and bladder cancer, by targeting NF-κB and generating reactive oxygen species. Int J Cancer 128 (10):2481–2494
55. Peralta-Yahya PP, Keasling JD (2010) Advanced biofuel production in microbes. Biotechnol J 5:147–162
56. Keasling JD (2010) Manufacturing molecules through metabolic engineering. Science 80, 330:1355–1358
57. Peralta-Yahya PP, Zhang F, del Cardayre SB, Keasling JD (2012) Microbial engineering for the production of advanced biofuels. Nature 488:320–328
58. Kung Y, McAndrew R, Xie X, Liu C, Pereira J, Adams P, Keasling JD (2014) Constructing tailored isoprenoid products by structure-guided modification of geranylgeranyl reductase. Structure 22:1028–1036
59. Nieuwenhuizen NJ, Green S, Atkinson RG (2010) Floral sesquiterpenes and their synthesis in dioecious kiwifruit 61–63
60. Köllner TG, Gershenzon J, Degenhardt J (2009) Molecular and biochemical evolution of maize terpene synthase 10, an enzyme of indirect defense. Phytochemistry 70:1139–1145
61. Bowers W, Nault L, Webb R (1972) Aphid alarm pheromone : isolation, identification, synthesis. Science 80:1–2
62. Wang C, Yoon SH, Jang HJ, Chung YR, Kim JY, Choi ES, Kim SW (2011) Metabolic engineering of Escherichia coli for alpha-farnesene production. Metab Eng 13:648–655
63. Picaud S, Brodelius M, Brodelius PE (2005) Expression, purification and characterization of recombinant (E)-beta-farnesene synthase from Artemisia annua. Phytochemistry 66:961–967
64. Martin D, Fäldt J, Bohlmann J (2004) Functional characterization of nine Norway spruce TPS genes and evolution of gymnosperm terpene synthases of the TPS-d subfamily. Plant Physiol 135:1908–1927
65. Maruyama T, Ito M, Honda G (2001) Molecular cloning, functional expression and characterization of (E)-beta farnesene synthase from *Citrus junos*. Biol Pharm Bull 24: 1171–1175

66. Ro D-K, Newman KL, Paradise E, Keasling JD, Ouellet M, Eachus RA, Ho K, Ham T (2007) Polynucleotides encoding isoprenoid modifying enzymes and methods to use thereoff. US20100218283 A1
67. Ozaydin B, Burd H, Lee TS, Keasling JD (2013) Carotenoid-based phenotypic screen of the yeast deletion collection reveals new genes with roles in isoprenoid production. Metab Eng 15:174–183
68. McAndrew RP, Peralta-Yahya PP, DeGiovanni A, Pereira JH, Hadi MZ, Keasling JD, Adams PD (2011) Structure of a three-domain sesquiterpene synthase: a prospective target for advanced biofuels production. Structure 19:1876–1884
69. Inoue Y, Shiraishi A, Hada T, Hirose K, Hamashima H, Shimada J (2004) The antibacterial effects of terpene alcohols on Staphylococcus aureus and their mode of action. FEMS Microbiol Lett 237:325–331
70. Goldberg L, Haklai R, Bauer V, Heiss A, Kloog Y (2009) New derivatives of farnesylthiosalicylic acid (salirasib) for cancer treatment: farnesylthiosalicylamide inhibits tumor growth in nude mice models. J Med Chem 52:197–205
71. Grace MH (2002) Chemical composition and biological activity of the volatiles of Anthemis melampodina and Pluchea dioscoridis. Phytother Res 16:183–185
72. Hornby JM, Nickerson KW (2004) Enhanced production of Farnesol by Candida albicans treated with four azoles. Antimicrob Agents Chemother 48(6):2305–2307
73. Millis J, Maurina-brunker J, McMullin TW (2003) Production of farnesol and geranylgeraniol. US2003/0092144A1
74. Faulkner A, Chen X, Horazdovsky B, Charles J, Carman GM, Paul C, Chem JB, Rush J, Waechter CJ, Sternweis PC (1999). Carbohydrates, lipids, and other natural products: the LPP1 and DPP1 gene products account for most of the isoprenoid phosphate phosphatase activities in saccharomyces cerevisiae. J Biol Chem 274:14831–14837
75. Song L (2006) A soluble form of phosphatase in *Saccharomyces cerevisiae* capable of converting farnesyl diphosphate into E. E-farnesol Appl Biochem Biotechnol 128:149–158
76. Ohto C, Muramatsu M, Obata S, Sakuradani E, Shimizu S (2009) Overexpression of the gene encoding HMG-CoA reductase in Saccharomyces cerevisiae for production of prenyl alcohols. Appl Microbiol Biotechnol 82:837–845
77. Cheng A-X, Xiang C-Y, Li J-X, Yang C-Q, Hu W-L, Wang L-J, Lou Y-G, Chen X-Y (2007) The rice (E)-beta-caryophyllene synthase (OsTPS3) accounts for the major inducible volatile sesquiterpenes. Phytochemistry 68:1632–1641
78. Sarria S, Wong B, Martín H, Keasling J, Peralta-Yahya P (2014) Microbial synthesis of pinene. ACS Synth Biol 3(7):466–475
79. Dahl RH (2011) Engineering a responsive, heterologous mevalonate pathway in *Escherichia coli* using microarrays. University of California at Berkeley, Berkeley
80. Ma SM, Garcia DE, Redding-Johanson AM, Friedland GD, Chan R, Batth TS, Haliburton JR, Chivian D, Keasling JD, Petzold CJ, Lee TS, Chhabra SR (2011) Optimization of a heterologous mevalonate pathway through the use of variant HMG-CoA reductases. Metab Eng 13:588–597
81. George KW, Chen A, Jain A, Batth TS, Baidoo EEK, Wang G, Adams PD, Petzold CJ, Keasling JD, Lee TS (2014) Correlation analysis of targeted proteins and metabolites to assess and engineer microbial isopentenol production. Biotechnol Bioeng 111:1648–1658
82. Alonso-Gutierrez J, Chan R, Batth TS, Adams PD, Keasling JD, Petzold CJ, Lee TS (2013) Metabolic engineering of *Escherichia coli* for limonene and perillyl alcohol production. Metab Eng 19:33–41
83. Behr A, Johnen L (2009) Myrcene as a natural base chemical in sustainable chemistry: a critical review. Chem Sus Chem 2(12):1072–1095
84. Papachristos DP, Karamanoli KI, Stamopoulos DC, Menkissoglu-Spiroudi U (2004) The relationship between the chemical composition of three essential oils and their insecticidal activity against Acanthoscelides obtectus (Say). Pest Manag. Sci. 60:514–520
85. Rastogi SC, Heydorn S, Johansen JD, Basketter DA (2001) Fragrance chemicals in domestic and occupational products. Contact Dermatitis 45:221–225

86. Zhou J, Wang C, Yoon S-H, Jang H-J, Choi E-S, Kim S-W (2014) Engineering *Escherichia coli* for selective geraniol production with minimized endogenous dehydrogenation. J Biotechnol 169:42–50
87. Duetz WA, Bouwmeester H, van Beilen JB, Witholt B (2003) Biotransformation of limonene by bacteria, fungi, yeasts, and plants. Appl Microbiol Biotechnol 61:269–277
88. Chuck CH, Donnelly J (2014) The compatibility of potential bioderived fuels with Jet A-1 aviation kerosene. Appl Energy 118:83–91
89. Yang J, Nie Q, Ren M, Feng H, Jiang X, Zheng Y, Liu M, Zhang H, Xian M (2013) Metabolic engineering of *Escherichia coli* for the biosynthesis of alpha-pinene. Biotechnol Biofuels 6:60
90. Breu F, Guggenbichler S, Wollmann J (2008) Verbenone (4,6,6-trimethyl-bicyclo (3.1.1) hept-3-en-2-one) (128986) fact sheet. Vasa 2–3
91. Mafra-Neto A, Lame F, de Fettig C, Munson A, Perring T, Stelinski L, Stoltman L, Mafra L, Borges R, Vargas R (2014) Manipulation of insect behavior with specialized pheromone and lure application technology (SPLAT®). In: Pest management with natural products. American Chemical Society
92. Chen W, Vermaak I, Viljoen A (2013) Camphor—a fumigant during the black death and a coveted fragrant wood in ancient Egypt and Babylon—a review. Molecules 18:5434–5454
93. Jirovetz L, Buchbauer G, Denkova Z, Slavchev A, Stoyanova A, Schmidt E (2006) Chemical composition, antimicrobial activities and odor descriptions of various Salvia sp. and Thuja sp. essential oils. Ernährung/Nutrition 30(4):152–159
94. Zhang H, Liu Q, Cao Y, Feng X, Zheng Y, Zou H, Liu H, Yang J, Xian M (2014) Microbial production of sabinene—a new terpene-based precursor of advanced biofuel. Microb Cell Fact 13:20
95. Höld KM, Sirisoma NS, Ikeda T, Narahashi T, Casida JE (2000) Alpha-thujone (the active component of absinthe): gamma-aminobutyric acid type A receptor modulation and metabolic detoxification. Proc Natl Acad Sci USA 97:3826–3831
96. MacGregor JT, Layton LL, Buttery RG (1974) California bay oil. II. Biological effects of constituents. J Agric Food Chem 22:777–780
97. Nassini R, Materazzi S, Vriens J, Prenen J, Benemei S, De Siena G, la Marca G, Andrè E, Preti D, Avonto C, Sadofsky L, Di Marzo V, De Petrocellis L, Dussor G, Porreca F, Taglialatela-Scafati O, Appendino G, Nilius B, Geppetti P (2012) The "headache tree" via umbellulone and TRPA1 activates the trigeminovascular system. Brain 135:376–390
98. Braukus M (2013) NASA begins flight research campaign using alternate jet fuel [WWW Document]. URL http://www.nasa.gov/home/hqnews/2013/mar/. Accessed 20 Sep 2013
99. Trimbur D, Im C-S, Dillon H, Day A, Franklin S, Coragliotta A (2011) Renewable chemicals and fuels from oleginous yeast. US20110190522 A1
100. Carter OA, Peters RJ, Croteau R (2003) Monoterpene biosynthesis pathway construction in *Escherichia coli*. Phytochemistry 64:425–433
101. Reiling K, Yoshikuni Y, Martin V, Newman J, Bohlmann J, Keasling J (2004) Mono and diterpene production in *Escherichia coli*. Biotechnol Bioeng 87:200–212
102. Bokinsky G, Peralta-Yahya PP, George A, Holmes BM, Steen EJ, Dietrich J, Soon Lee T, Tullman-Ercek D, Voigt CA, Simmons BA, Keasling JD (2011) Synthesis of three advanced biofuels from ionic liquid-pretreated switchgrass using engineered *Escherichia coli*. Proc Natl Acad Sci USA 108:19949–19954
103. Brennan TCR, Krömer JO, Nielsen LK (2013) Physiological and transcriptional responses of Saccharomyces cerevisiae to d-limonene show changes to the cell wall but not to the plasma membrane. Appl Environ Microbiol 79:3590–3600
104. Dunlop MJ (2011) Engineering microbes for tolerance to next-generation biofuels. Biotechnol Biofuels 4:32
105. Dunlop MJ, Dossani ZY, Szmidt HL, Chu HC, Lee TS, Keasling JD, Hadi MZ, Mukhopadhyay A (2011) Engineering microbial biofuel tolerance and export using efflux pumps. Mol Syst Biol 7:487

106. Carnesecchi S, Bras-Gonçalves R, Bradaia A, Zeisel M, Gossé F, Poupon M-F, Raul F (2004) Geraniol, a component of plant essential oils, modulates DNA synthesis and potentiates 5-fluorouracil efficacy on human colon tumor xenografts. Cancer Lett 215:53–59
107. Dudareva N, Martin D, Kish CM, Kolosova N, Gorenstein N, Fäldt J, Miller B, Bohlmann J (2003) (E)-β-Ocimene and myrcene synthase genes of floral scent biosynthesis in snapdragon: function and expression of three terpene synthase genes of a new terpene synthase subfamily 15:1227–1241
108. Putman M, Veen HW, Van Konings WN, Hendrik W (2000) Molecular properties of bacterial multidrug transporters. Mol Prop Bact Multidrug Transporters 64(4):672–693
109. Dunlop MJ, Keasling JD, Mukhopadhyay A (2010) A model for improving microbial biofuel production using a synthetic feedback loop. Syst Synth Biol 4:95–104
110. Wagner S, Baars L, Ytterberg AJ, Klussmeier A, Wagner CS, Nord O, Nygren PA, van Wijk KJ, de Gier JW (2007) Consequences of membrane protein overexpression in Escherichia coli. Mol. Cell. Proteomics 6:1527–1550
111. Patnaik R, Louie S, Gavrilovic V, Perry K, Stemmer WPC, Ryan CM, del Cardayré S (2002) Genome shuffling of Lactobacillus for improved acid tolerance. Nat Biotechnol 20:707–712
112. Wang HH, Isaacs FJ, Carr PA, Sun ZZ, Xu G, Forest CR, Church GM (2009) Programming cells by multiplex genome engineering and accelerated evolution. Nature 460:894–898
113. Patnaik R (2008) Engineering complex phenotypes in industrial strains. Biotechnol Prog 24:38–47
114. Mirata MA, Heerd D, Schrader J (2009) Integrated bioprocess for the oxidation of limonene to perillic acid with Pseudomonas putida DSM 12264. Process Biochem 44:8
115. Rühl J, Schmid A, Blank LM (2009) Selected Pseudomonas putida strains able to grow in the presence of high butanol concentrations. Appl Environ Microbiol 75:4653–4656
116. Atsumi S, Liao JC (2008) Metabolic engineering for advanced biofuels production from Escherichia coli. Curr Opin Biotechnol 19:414–419
117. Atsumi S, Hanai T, Liao JC (2008) Non-fermentative pathways for synthesis of branched-chain higher alcohols as biofuels. Nature 451:86–89
118. Cann AF, Liao JC (2008) Production of 2-methyl-1-butanol in engineered *Escherichia coli*. Appl Microbiol Biotechnol 81:89–98
119. Hull A, Golubkov I, Kronberg B, Marandzheva T, van Stam J (2006) An alternative fuel for spark ignition engines. Int J Engine Res 7:203–214
120. Rohdich F, Hecht S, Gärtner K, Adam P, Krieger C, Amslinger S, Arigoni D, Bacher A, Eisenreich W (2002) Studies on the nonmevalonate terpene biosynthetic pathway: metabolic role of IspH (LytB) protein. Proc Natl Acad Sci USA 99:1158–1163
121. Withers ST, Gottlieb SS, Lieu B, Newman JD, Keasling JD (2007) Identification of isopentenol biosynthetic genes from *Bacillus subtilis* by a screening method based on isoprenoid precursor toxicity. Appl Environ Microbiol 73:6277–6283
122. Chou HH, Keasling JD (2012) Synthetic pathway for production of five-carbon alcohols from isopentenyl diphosphate. Appl Env Microbiol 78:7849–7855
123. Kuznetsova E, Proudfoot M, Gonzalez CF, Brown G, Omelchenko MV, Borozan I, Carmel L, Wolf YI, Mori H, Savchenko AV, Arrowsmith CH, Koonin EV, Edwards AM, Yakunin AF (2006) Genome-wide analysis of substrate specificities of the *Escherichia coli* haloacid dehalogenase-like phosphatase family. J Biol Chem 281:36149–36161
124. McLennan AG (2006) The Nudix hydrolase superfamily. Cell Mol Life Sci 63:123–143
125. Zheng Y, Liu Q, Li L, Qin W, Yang J, Zhang H, Jiang X, Cheng T, Liu W, Xu X, Xian M (2013) Metabolic engineering of Escherichia coli for high-specificity production of isoprenol and prenol as next generation of biofuels. Biotechnol Biofuels 6:57
126. Fortman JL, Chhabra S, Mukhopadhyay A, Chou H, Lee TS, Steen E, Keasling JD (2008) Biofuel alternatives to ethanol: pumping the microbial well. Trends Biotechnol 26:375–381
127. Horton CE, Huang K-X, Bennett GN, Rudolph FB (2003) Heterologous expression of the Saccharomyces cerevisiae alcohol acetyltransferase genes in Clostridium acetobutylicum and *Escherichia coli* for the production of isoamyl acetate. J Ind Microbiol Biotechnol 30:427–432

128. Singh R, Vadlani PV, Harrison ML, Bennett GN, San K-Y (2008) Aerobic production of isoamyl acetate by overexpression of the yeast alcohol acetyl-transferases AFT1 and AFT2 in *Escherichia coli* and using low-cost fermentation ingredients. Bioprocess Biosyst Eng 31:299–306
129. Brown TR, Brown RC (2013) Techno-economics of advanced biofuel pathways. RSC Adv 3:5758–5764
130. Tyo KEJ, Ajikumar PK, Stephanopoulos G (2009) Stabilized gene duplication enables long-term selection-free heterologous pathway expression. Nat Biotechnol 27:760–765
131. Holtz WJ, Keasling JD (2010) Engineering static and dynamic control of synthetic pathways. Cell 140:19–23
132. Geris R, Simpson TJ (2009) Meroterpenoids produced by Fungi. Nat Prod Rep Aug 26(8):1063–1094
133. Cane D, Ikeda H (2012) Exploration and mining of the bacterial terpenome. Acc Chem Res 45(3):463–472
134. Itho T, Tokunaga K, Matsuda Y, Fujii I, Abe I, Ebizuka Y, Kushiro T (2010) Reconstitution of fungal meroterpenoid biosynthesis reveals the involvement of a novel family of terpene cyclases. Nat Chem 2(10):858–864
135. Chooi YH, Hong YJ, Cacho RA, Tantillo DJ, Tang Y (2013) A cytochrome P450 serves as an unexpected terpene cyclase during fungal meroterpenoid biosynthesis. J Am Chem Soc 135(45):16805–16808
136. Meylemans HA, Quintana RL, Harvey BG (2012) Efficient conversion of pure and mixed terpene feedstocks to high density fuels. Fuel 97:560–568

Adv Biochem Eng Biotechnol (2015) 148: 391–404
DOI 10.1007/10_2014_279

Published Online: 19 October 2014

Nootkatone

Robin-Hagen Leonhardt and Ralf G. Berger

Abstract The continuing interest in the sesquiterpene ketone (+)-nootkatone is stimulated by its strong grapefruit-like odor and numerous further bioactivities. Also numerous were the attempts to chemosynthesize or biotechnologically produce the compound. Cytochrome P_{450} enzymes from bacteria and fungi were intensively studied and expressed in *Escherichia coli* and in more food compatible hosts, such as *Saccharomyces cerevisiae*. The lipoxygenase-catalyzed generation was demonstrated using an enzyme from several *Pleurotus* species. Laccases required artificial mediators for an efficient catalysis. More recently, plant valencene synthases were expressed in microbial hosts. Combined with an endogenous farnesyl diphosphate delivery pathway and a valencene oxidase, this approach opened access to high yields of nootkatone possessing the appreciated attribute of "natural" according to present food legislation. Little biochemical engineering was carried out on the novel recombinant strains, leaving many options for future improved bioprocesses.

Keywords Natural flavor · Insecticide · CYP450 · Lipoxygenase · Laccase · Valencene synthase · Valencene oxidase

Contents

R.-H. Leonhardt (✉) · R.G. Berger
Institut für Lebensmittelchemi, Gottfried Wilhelm Leibniz Universität Hannover, Callinstraße 5, 30167 Hannover, Germany
e-mail: rg.berger@lci.uni-hannover.de

1 Introduction

The compound (+)-nootkatone [2(3H)-Naphthalenone, 4,4a,5,6,7,8-Hexahydro-4,4a-dimethyl-6-(1-methylethenyl)-, (4R,4aS,6R)–eremophila-1(10),11-dien-2-one; CAS # 4674-50-4; Fig. 1] is a flavor compound with a grapefruit-like odor, an odor perception threshold in the range of 1 μg/L water, and a slightly bitter taste. First isolated from the heartwood of Alaska cedar [*Cupressus nootkatensis* (D. Don 1824); syn. *Chamaecyparis nootkatensis* (D. Don) Spach 1841, *Callitropsis nootkatensis* (D. Don) Oersted 1864, *Xanthocyparis nootkatensis* (D. Don) Farjon & Hiep 2002], nootkatone was found in essential oils of the genera *Citrus* (Rutaceae), *Alpinia* (*Zingiberaceae*) and other *Pinales*; in Java (*Cyperus rotundus*) and Vetiver grass (*Vetiveria* sp.); and in many other plant essential oils.The biogenetic precursor of nootkatone, (+)-Valencene, is just two steps away from the sesquiterpene starter, the farnesyl cation, and is quickly produced through a germacryl intermediate (Fig. 1). This explains the widespread occurrence of the (+)-valencene bicyclus and its accompanying allylic oxidation products, the stereoisomer *α*- and *β*-nootkatols, and (+)-nootkatone. Valencene shows a distinct and pleasant orange odor, whereas

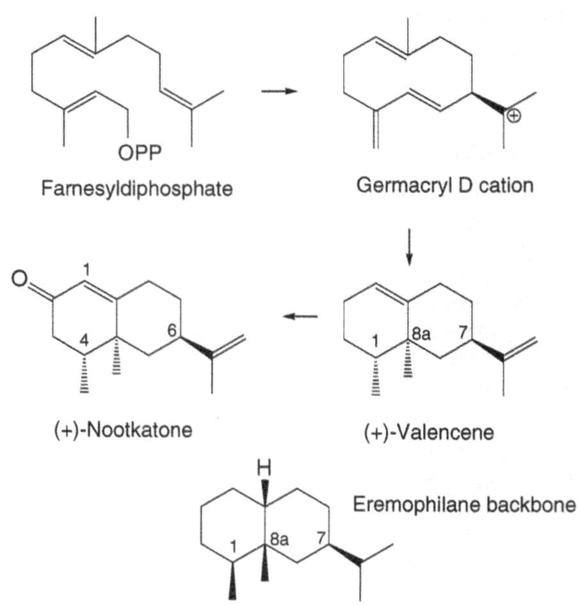

Fig. 1 The proposed biogenetic path from farnesyldiphosphate to the eremophilanes (+)-valencene and (+)-nootkatone

α- and β-nootkatol (= 2*R* and 2*S*-hydroxyvalencene), with lower odor intensity, are reminiscent of grapefruit. Single compounds or mixtures of this group can be used to flavor food, fragrances, cosmetics, and detergents. The worldwide consumption of flavors and fragrances is increasing, as indicated by almost doubled total sales in the past decade. The estimated turnover of the flavor and fragrance industries in 2013 amounted to around $24 billion (www.leffingwell.com/top_10.htm). Patents also claim the use of (+)-nootkatone as an insecticide and repellent.

Because (−)-nootkatone shows an odor threshold at least 1,000 times higher than the (+)-isomer and has less overall bioactivity, it is not of commercial interest. In this chapter, the term *nootkatone* refers only to the more bioactive (+)-form of the molecule.

2 Chemical Analyses

Nootkatone is quite amenable to both gas chromatography (GC) and liquid chromatography (LC). A thorough gas chromatographic reanalysis of an Alaska cedar steam distillate identified additional nootkatone-related compounds, such as the cycloheptatrienone nootkatin and the triene nootkatene [1]. Apart from its better resolution, GC offers the advantage of delivering sensory data on each separated compound, if coupled with a sniff-port for online olfactometric assessment. Ultrafast LC coupled to a time-of-flight mass spectrometer (MS) identified nootkatone in *Citrus grandis* (i.e., *Citrus maxima*, pomelo) [2], whereas classic reverse-phase LC was applied to analyze fruits of *Alpinia oxyphylla* (a relative of blue ginger) [3]. A particularly detailed picture of complex essential oils was obtained by the combination of high-resolution LC for prefractionation and two-dimensional GC coupled to MS [4]. This kind of approach produces a number of less crowded gas chromatograms and creates better spectra for a safer identification of compounds, which is a generally difficult task for samples rich in oligoisoprenoids with often very similar fragmentation patterns. Quantitative analyses have confirmed that the concentration of nootkatone in essential oils is low. The total yield of hydro-distilled oil obtained from various *Citrus* fruits ranged from 0.13 to 0.53 %, and nootkatone had a share of 1.6–2.5 % thereof [5]. A concentration of roughly 100 mg of nootkatone per kilogram of fresh fruit at best is calculated from these data. Because of the high sensory potency, even ultralow concentrations impart the typical grapefruit impression. For commercial production, however, natural plant sources of nootkatone are not rich enough.

3 Biological Activities

Apart from the sensory properties, *Citrus* essential oils were associated with numerous biological activities, such as antioxidant, antimicrobial, anti-inflammatory, and more [6]. Frequently, whole distillates or extracts were applied, making it impossible to confidently ascribe the activity observed to a single constituent of the sample. The case is different for nootkatone and some of its derivatives and analogues. A recent patent claimed various nootkatone carriers and their application as an insecticide [7]. The efficiency of nootkatone against termites [8] and ticks (arachnid parasites) [9] was experimentally proven, explaining the superior tolerance of cedar wood against insect infestation. Because nootkatone is regularly consumed by mammals, it was concluded that its application would pose no threat to livestock or human consumers. It was suggested to inhibit an insect infestation by onsite inoculation of a nootkatone-transforming microorganism, such as *Paenibacillus alginolyticus,* exploiting this "fermentation" to produce advanced oxidation products of nootkatone with anti-insecticidal activities [8].

The primary mode of action of nootkatone remains uncertain. A recent study [10] disproved earlier assumptions that nootkatone could inhibit acetylcholinesterase (ACE). When compared to the known ACE inhibitor carbaryl, neither nootkatone nor carvacrol, a monoterpenoid also abundant in cedar wood, showed more than a weak activity in the ACE assay.

Human metabolic pathways appear to be affected by nootkatone because of its cardiovascular [11], antiproliferative [12], and antiplatelet effects [13]. The positive effect on survival of septic mice was attributed to the induction of a heme oxygenase [14]. The expression of this enzyme inhibits, in turn, nitric oxide (NO) synthase and NO production, eventually resulting in the observed anti-inflammatory effect. The stimulation of an Adenosin monophosphate-activated protein kinase (AMPK) was recently substantiated [15]. The resulting acceleration of the energy metabolism was supposed to aid in the treatment of human obesity. Upstream of AMPK lays liver kinase B1 (LKB1) or renal carcinoma antigen NY-REN-19, another protein kinase involved in regulating cell polarity and development. LKB1 exerts growth suppressing effects by activating an entire group of kinases, and it is thus a key player in preventing carcinogenesis. Nootkatone also activated this kinase [16].

4 Chemosynthesis

Chemosyntheses of nootkatone were developed to meet industrial demand. Because the immediate precursor valencene is abundant in orange peel oil, its allylic oxidation using chromate (VI) or manganate (VII) ions provided easy access to the target compound. Advanced versions are still under investigation using phase transfer catalysis [17] or a metal mixed oxide catalyst and *tert*-butyl hydroperoxide as a less problematic oxidant [18]. Introducing two of the three isoprene units with the substrate molecule, routes from the easily available substrate β-pinene were also explored [19]. Among the hazardous chemicals required for this synthesis were alkenyl chlorides, benzene, heavy metal salts, and again manganate (VII).

Increasing environmental concerns and costs for the handling and disposal of toxic chemicals are considered in novel approaches that rely on less critical oxidants, with molecular oxygen being one of those. Orange juice develops a grapefruit note after opening and prolonged exposure to air, indicating that autoxidation of valencene occurred in the acidic medium. Accelerated autoxidation of valencene was achieved in a continuous-flow microreactor [20]. The reactor consisted of heated Pyrex spiral glass tubing, and the reaction proceeded rapidly with high yields and in the absence of a catalyst, radical initiator, or solvent.

5 Biosynthesis

Nootkatone produced by a concerted and deliberate chemical reaction must not be labelled "natural", but this would be the highly preferred quality for food flavors. Effective European law (EG 1334/2008) defines a "natural flavouring substance" as a compound "obtained by appropriate physical, enzymatic or microbiological processes from material of vegetable, animal or microbiological origin." The U.S. Food and Drug Administration Code of Federal Regulation likewise emphasizes the need for the presence of a biocatalyst, using the terms "enzymolysis" and "fermentation."

The distinction between a naturally generated chemical with oxidizing properties and an enzyme with a metal ion coordinated in a high positive redox state is not sharp. However, processes based on an enzymatically generated oxidant are more likely to become approved. Molecular oxygen is a natural co-substrate of dioxygenases and laccases; hydrogen peroxide, the natural co-substrate of peroxidases and peroxygenases, is easily generated, such as by using glucose oxidase in glucose-containing foods. Is a linoleic acid hydroperoxide formed by a soybean lipoxygenase a natural oxidant? This reagent was generated in the presence of valencene, and a subsequent co-oxidation of the hydrocarbon in excellent yields was reported [21]. While the first step is clearly enzyme catalyzed, the second is an inevitable, but deliberate, chemical reaction. Both the forced autoxidation and the lipoxygenase-mediated process benefit from the preferred reactivity of the allylic C3 of valencene (Fig. 1), which helps to prevent the formation of larger amounts of unwanted side products.

5.1 Biosynthesis Based on CYP450

Cytochrome P450 enzymes (CYP450) are a large group of hemoproteins occurring in animals, plants, fungi, bacteria, archaea, and even in viruses. The name-giving property is an absorption maximum at $\lambda = 450$ nm when in the reduced state and complexed with CO. Lipophilic compounds, such as lipids and steroids, and also terpenes are preferably accepted. The typical reaction catalyzed is mono-oxygenation—that is, the introduction of an oxygen atom into the hydrocarbon part of the substrate (HC), while a second oxygen atom is reduced to water:

$$\mathrm{HC} + \mathrm{O_2} + \mathrm{NADPH} + \mathrm{H^+} \xrightarrow{\text{Monooxygenase}} \mathrm{HC-OH} + \mathrm{H_2O} + \mathrm{NADP^+}$$

High catalytic efficiency and pronounced substrate promiscuity of CYP450 suggested these enzymes for the conversion of valencene to hydroxyvalencene. A subsequent oxidation yielded the desired carbonyl nootkatone. In fact, many of the successful transformations reported may be attributed to CYP450 activity [22], although in most cases its actual involvement was merely concluded from the structure of the products. Cellular systems, rather than purified enzymes, were used to meet the demand for protein co-factors. The (often) soluble bacterial enzymes, for example, depend on a ferredoxin reductase and a ferredoxin to deliver the electrons for the final reduction of molecular oxygen.

Among the enzymes used for the hydroxylation of valencene is CYP450 $_{BM-3}$ from *Bacillus megaterium*. Originally found involved in terminal or subterminal hydroxylations of long-chain fatty acids, it presents the particular feature of being a fusion protein between a CYP domain and an electron donor. Mutants of this enzyme were created and compared with the wild-type enzyme [23]. Activity as measured by the consumption of NADPH and the generation of nootkatone were evaluated. Another versatile CYP450 was found in *Bacillus subtilis* [24]. Here, the CYP450 activity was obtained by co-expression with genes for a putidaredoxin reductase (PdR) and putidaredoxin (Pdx) from *Pseudomonas putida* in *E. coli* to obtain both nootkatols and nootkatone. Typical of many CYP450 catalysed reactions, a large number of side products were observed (Fig. 2). A better product yield of 120 mg/L of odor-active compounds was finally achieved in a two-phase system that dissolved the primary products into the organic phase, thus protecting them from further oxidation. A CYP450 from *Nicotiana* and a P450-reductase from *Arabidopsis* were cloned and expressed in *Saccharomyces cerevisiae* [25]. Product yields were low, but the chances to transfer such a process based on a food-grade recombinant host into industrial application are higher than with *E. coli* or other potentially toxin-forming bacteria. Consumers know and appreciate bakers′ yeast; although nootkatone produced by a recombinant cell will quite obviously not contain remains of inactivated or living genetically modified organisms, the food industry has set out to deal respectfully with public concerns.

While products of valencene overoxidation are too polar to be of interest as flavors, they may be useful for therapeutic applications. A CYP450 from *Sorangium cellulosum*, a myxococcus, accepted nootkatone as a substrate, and also the norisoprenoids α- and β-ionone [26]. Antiproliferative properties of the nootkatone derivatives were examined, and the structural background of the regioselective hydroxylation was elucidated using a homology model. A similar study supplemented cell cultures of Ascomycetes, such as *Aspergillus*, *Chaetomium* or *Fusarium*, with nootkatone. The product profile showed mono- and diols, triols, and epoxides indicating active CYP450s, but the identity of the fungal enzymes was not elucidated [12].

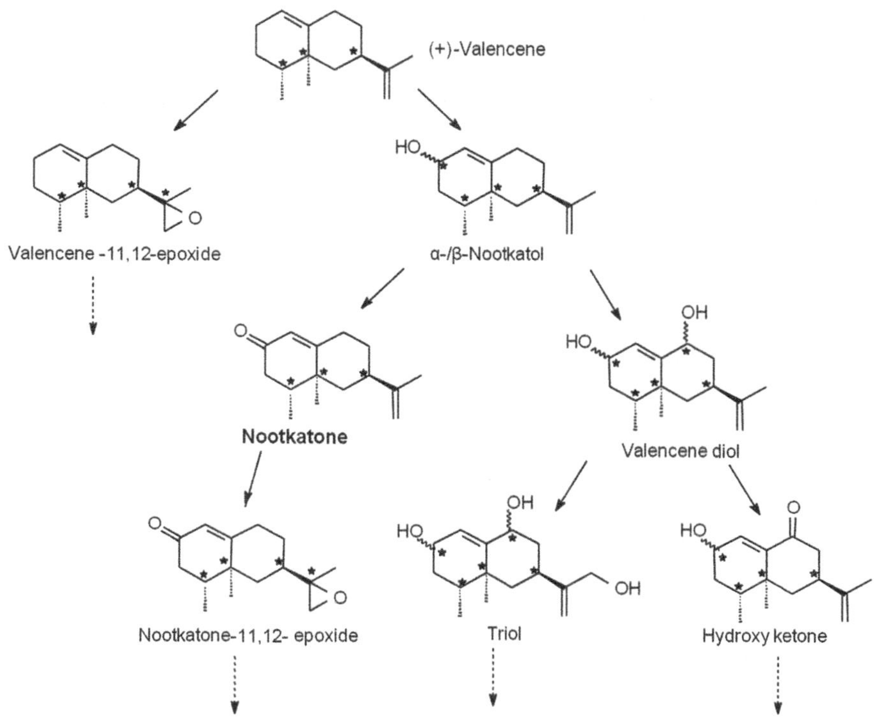

Fig. 2 Overoxidation of nootkatone yields multiple nonflavor compounds with antiproliferative and possibly other bioactivities

It was estimated that product peak yields of 1 g/L or productivities of 0.1 g/L per hour will be required to cross the profitability threshold for a food flavor. Bioprocesses based on CYP450 have not yet reached this stage. The reasons are numerous:

- The terpene hydrocarbon substrate is often insufficiently soluble in water.
- Emulsions complicate the system due to the presence of the cytotoxic emulsifier.
- Both the terpene hydrocarbon substrate and terpenoid product may be toxic for the native or recombinant producer cell or inhibit the enzymes involved.
- Reactive primary metabolites, such as epoxides, give reason to side-product formation.
- Substrate and product accumulate in lipophilic cell compartments and are not secreted.
- Substrate and product will be stripped out of aerated bioreactors.
- The second oxidation step to the often more odor-active carbonyl is not catalyzed by CPY450s.

5.2 Biosynthesis Based on Lipoxygenases

During extended screenings for microbial producers of nootkatone, the basidiomycete *Pleurotus sapidus*, a member of the order *Agaricomycetes* and a close relative of the edible oyster mushroom *Pleurotus ostreatus*, stood out for a selective and highly efficient oxidation of valencene to nootkatone [27]. Although the occurrence of both nootkatols suggested another CYP450-driven reaction, the responsible enzyme, after laborious purification, showed homologies of around 50 % to putative lipoxygenases from *Aspergillus fumigatus* and *Laccaria bicolor*, and 26 % homology to the well-known commercial lipoxygenase-1 of soy bean (*Glycine max*). Because no lipoxygenase from a basidiomycetous source was known on the molecular level, detailed metabolite analyses were carried out using cold on-column GC-MS and APCI-LC-MS methods. The data suggested a lipoxygenase-type oxidation of valencene via secondary and tertiary hydroperoxides [28]:

$$\text{Valencene} + O_2 \xrightarrow{\text{Dioxygenase}} \text{4a-}(R,S)\text{-Val-OO-H} \xrightarrow{\text{Schenkrearrangement (Peroxo - shift)}} 3-(R,S)\text{-Val-OO-H} \xrightarrow{\text{Disproportionation}} \text{Val-OH} + \text{Nootkatone}$$

The gene was cloned and heterologously expressed in *E. coli* [29]. Soluble recombinant protein was gained by cold shock expression, chaperone co-expression, and employment of mutant *E. coli* strains. The expression in *P. pastoris* SMD1168 was carried out using a pPIC9K vector construct [30]. In this construct, the α-factor signal sequence provided by the original vector was replaced by insertion of a second Kozak sequence between the signal sequence and the lipoxygenase gene [30]. The recombinant enzymes, both purified using the N-terminal His tag, showed the catalytic properties of the wild-type enzyme, as was confirmed by the LC-MS analysis of hydroperoxide intermediates and GC-MS analysis of the volatile products. As with the wild-type lipoxygenase, optimal activity was found at pH 7 at 30–35 °C. Conversion of linoleic acid gave high yields of 13*S*-hydroperoxy-9*Z*,11*E*-octadecadienoic acid (94 % ee), as was confirmed by chiral high-performance LC analysis of the hydroperoxides [30, 31].

A follow-up study on a number of *Pleurotus* species resulted in lipoxygenase genes from five strains. The genes were amplified, functionally expressed in the proven *E. coli* system, and characterized [32]. All sequences coded for proteins of 643 amino acids, sharing similarities greater than 95 % among each other. The ligands that determined iron coordination and stereochemistry were fully conserved. Similar lipoxygenase activities were found when linoleic acid served as the substrate. The conversion of valencene, however, differed between two clusters of highly homologous sequences. The future engineering of these dioxygenases could create catalysts accepting other terpenes and alkenes as precursors for novel flavor compounds and possibly bioactives. In clear contrast to CYP450 enzymes, oxygen is the only co-substrate required. Moreover, the situation regarding substrate and product inhibition as well as side product formation appears to be more favorable with these fungal dioxygenases.

5.3 Biosynthesis Based on Laccases

Laccases comprise a large group of well-described oxidases that are frequently used in different applications and processes throughout the food industry [33]. These enzymes are easily available, either by purification from culture supernatants, or in larger amounts by heterologous expression of the respective genes in host strains, such as *P. pastoris*. A patent from 2001 [34] described the use of laccases for the industrial-scale production of nootkatone by oxidation of valencene. The work focused on a mediator-based oxidation, with most of the mentioned mediators being nonnatural laccase substrates (i.e. 1-hydroxybenzotriazole or 2,2′-azino-bis(3-ethylbenzothiazoline-6-sulphonic acid). This conflicts with the legal definition of natural flavors; however, optimal reaction conditions gave an impressive molar conversion rate of around 40 %. Because laccases preferably oxidize diphenolic compounds [35], the exact reaction mechanism responsible for the oxidation of the sesquiterpene valencene is disputable. Due to the radicals formed during the enzymatic reaction and the various possible oxidations sites of the precursor molecule, side reactions to other products could be envisaged, as they were found for chemosynthetic or CYP450 catalyzed reactions (Fig. 2). Similar processes based on natural precursors and mediators were successfully optimized [36], but no further data on the mediated oxidation of valencene have been published since.

5.4 Biogenesis Based on Plant Enzymes Involved in Terpene Metabolism

There is no generally accepted definition of the term *synthetic biology*, but a short definition could be "to make use of the advantages of biocatalysis using engineering concepts." Because the easily cultivated pro- and eukaryotic cells of classical bioprocesses do not possess a pronounced terpene metabolism, plant genes have to be implanted into recombinant "engineered" hosts. The concept has been patented several times by flavor, food, and even rubber companies to produce terpene hydrocarbons; sometimes terpenoids were also claimed. A recent example was a patchoulol synthase from *Pogostemon cablin* (Indian patchouli) cloned in *E. coli* [37]. To prevent formation of inclusion bodies, the protein of interest was fused to a thioredoxin-tag. Codon optimization improved the yield of soluble enzyme significantly. The product spectrum as analyzed by GC-MS showed farnesol isomers from the hydrolysis of the substrate and germacra-3,9,11-triene A as the major components and the target compound (−)-patchoulol.

The building block molecule of Amyris Inc. for flavors is *β*-(*E*)-farnesene, called "biofene." One of the company's fundamental patents described the engineering of isoprenoid pathways in host cells [38]. By incorporation of an acetoacetyl-CoA synthase gene and other enzymes of the mevalonate pathway, the host cell became independent from the activated diphosphate precursors used, such as in the patchoulol study [37]. Branching pathways leading to unwanted intermediates were blocked by disrupting the responsible genes. Also considered was the issue of redox balance in the heterologous host [38].

Another option is to equip the heterologous host with an additional synthase gene, which is a strategy followed by Allylix (www.allylix.com) and Isobionics (http://www.isobionics.com). The cyclisation of farnesyl diphosphate (FPP) to valencene was achieved using enzymes from *C. nootkatensis* [39–41], from *Citrus* species [42] or from *Vitis vinifera* (grape). According to the patent [42], the genes may have been engineered by changing, adding, or deleting codons to obtain variants that are more suitable than the wild-type enzyme. Once the genes were optimized, they were introduced into a tailored yeast strain. This strain was engineered to produce endogenous FPP. The enzyme from Alaska cedar was robust and efficient in vitro, but not well expressed in the usual microbial hosts. When incorporated in *Saccharomyces cerevisiae*, a valencene concentration of around 1.4 mg/L was obtained; a heterologous *Rhodococcus sphaeroides*, under an n-dodecane accumulation layer and upon co-expression of a mevalonate operon, eventually reached remarkable 352 mg valencene per liter [40].

To arrive at nootkatone, a valencene converting CPY450 enzyme from *Cichorium intybus* (chicory) was co-expressed with a synthase gene in a nonspecified yeast [43]. Because chicory contains neither valencene nor nootkatone, this activity can again be related to the frequently observed substrate promiscuity of this type of enzyme. However, the nootkatone levels were low, and it was not clear if this second oxidation step was the result of the activity of other enzymes [Alcohol dehydrogenases (ADHs)] present in the host cell. *Pichia pastoris* was complemented using the premnaspirodiene oxygenase of *Hyoscyamus muticus* (a member of the *Solanaceae*) and the *Arabidopsis thaliana* reductase to hydroxylate valencene [44]. Intracellular production of valencene was achieved by co-expression of valencene synthase from *C. nootkatensis*. Biphasic cultivations of *P. pastoris* resulted in the accumulation of trans-nootkatol, which was oxidized to nootkatone by a genuine activity of *P. pastoris*. Further genetic fine-tuning enhanced the nootkatone yield to 208 mg/L in bioreactor cultivations.

The two approaches are different, but both suffer from inherent problems: To establish an entirely new pathway in a heterologous host is feasible, but the additional load of new genes and enzymes may overburden a simple prokaryotic host. Eukaryotes, on the other hand, are more complicated to transform, grow more slowly, and might more certainly recognize the new genes as "foreign," resulting in the rapid degradation of transcribed or translated metabolites. To implement one or two key enzymes (valencene synthase and oxidase) appears to be more straightforward, but it must cope with the high hydrolytic sensitivity of the obligatory precursor FPP. This activated compound cannot be isolated in sufficient yields from natural sources; thus, the conversion of the chemosynthetic precursor by cell cultures or enzymes cannot yield legally "natural" valencene and nootkatone. As with all enzyme-based attempts, cytotoxic or inhibitory effects of metabolites accumulating in abundance must be expected. Moreover, the poor qualitative and quantitative predictability of heterologous attempts presents a severe technical drawback.

Table 1 Chemical and biochemical generation of nootkatone

Process/organism	Nootkatone (maximum yield)	References
Cr^{VI} oxidation of valencene	47 % from valencene	[48]
Oxidation using lipoxygenases (patented)	60 g L^{-1} (24 h)	[49]
Laccase-mediated oxidation (patented)	50 % from valencene	[34]
Transformation using *Chlorella spec.*	~330 mg L^{-1} (7 d)	[50]
Cultured plant cells (*Gynostemma pentaphyllum*)	~650 mg L^{-1} (20 d)	[51]
CYP450 expression in *E. coli*/two-phase system	120 mg L^{-1} (8 h)	[24]
Oxidation using a lipoxygenase from *Pleurotus sapidus*	~300 mg L^{-1} (16 h)	[28]
	~250 mg L^{-1} (4 h)	[R-HL unpublished]
CYP450 expression in *P. pastoris*	7 mg L^{-1} (24 h)	[25]
Autoxidation in a microreactor	316 g L^{-1} h^{-1}	[20]
Cultivation of recombinant *P. pastoris*	208 mg L^{-1} (15 h)	[44]

In this situation, the idea to engineer a transgenic plant is self-suggesting [45]. Essential oil plants are distinguished by special morphological features, such as storage hairs or cavities, which provide protection to terpenoid compounds from evaporation and overoxidation. Moreover, any inhibitory effects are minimized due to the spatial compartmentalization of producing enzyme and enzyme product. A severe argument speaks against this solution: the continuing aversion of the public toward transgenic plants.

6 Outlook

Significant progress was made in recent years to characterize and apply biocatalysts for the production of the sought-after flavor compound and insecticide nootkatone [46]. However, the former molecule of the month [47] still remains a kind of "holy grail" of flavor biotechnology because the maximum yields reported for processes resulting in a product categorized as "natural" are not yet sufficient to set up an industrial process [48–51] (Table 1). Novel and more efficient enzymes and improved systems to heterologously express them are the focus of current research. The development of alternative bioengineering solutions is a neglected field. The tremendous progress achieved with simple two-phase systems [24, 43] shows that the hidden potential is far from being exhausted. New impetus for the field could come from medical research. If one of the claimed bioactivities on human metabolism, such as antiproliferative, could be verified, the intensified interest in a high-quality nootkatone would induce increased efforts for its cost-effective production through bioprocesses.

Acknowledgments RGB thanks his group members for their contributions to the work included in this chapter. Financial support came from the Deutsche Bundesstiftung Umwelt (ChemBioTec AZ13187), Forschungskreis der Ernährungsindustrie via AiF and BMWi (299 ZN) and the BMBF cluster Biokatalyse 2021 (FKZ0315172B).

References

1. Khasawneh MA, Xiong Y, Peralta-Cruz J, Karchesy JJ (2011) Biologically important eremophilane sesquiterpenes from Alaska cedar heartwood essential oil and their semi-synthetic derivatives. Molecules 16:4775–4785
2. Li P, Liu M, Hu J, Su W (2014) Systematic chemical profiling of *Citrus grandis* 'Tomentosa' by ultra-fast liquid chromatography/diode-array detector/quadrupole time-of-flight tandem mass spectrometry. J Pharm Biomed Anal 90:167–179
3. Li Y, Lai W, Huang S, Wang J, Wang Y, Chen F, Song W, Zhang J (2013) RP-HPLC determination of nootkatone and yakuchinone A in fruits of *Alpinia oxyphylla* at different growth phases. Yaowu Fenxi Zazhi 33:808–811
4. Tranchida PQ, Zoccali M, Bonaccorsi I, Dugo P, Mondello L, Dugo G (2013) The off-line combination of high performance liquid chromatography and comprehensive two-dimensional gas chromatography-mass spectrometry: a powerful approach for highly detailed essential oil analysis. J Chromatogr A 1305:276–284
5. Rana V, Blazquez MA (2012) Compositions of the volatile oils of *Citrus macroptera* and *C. maxima*. Nat Prod Commun 7:1371–1372
6. Hamdan D, Ashour ML, Mulyaningsih S, El-Shazly A, Wink M (2013) Chemical composition of the essential oils of variegated pink-fleshed lemon (*Citrus* × *limon* L. Burm. f.) and their anti-inflammatory and antimicrobial activities. Z Naturforsch C: J Biosci 68:275–284
7. Goldblum S, Warren CB (2014) Nootkatone in various carriers and formulations as an insecticide and insect repellent. PCT Int Appl, WO 2014031790 A1 20140227. Assignee: Allylix, Inc., USA
8. Zhao K (2013) A nootkatone-producing *Bacillus alginolyticus*. Faming Zhuanli Shenqing CN 103289923 A 20130911
9. Flor-Weiler LB, Behle RW, Stafford KC (2011) Susceptibility of four tick species, *Amblyomma americanum, Dermacentor variabilis, Ixodes scapularis*, and *Rhipicephalus sanguineus* (Acari: Ixodidae), to nootkatone from essential oil of grapefruit. J Med Entomol 48:322–326
10. Anderson JA, Coats JR (2012) Acetylcholinesterase inhibition by nootkatone and carvacrol in arthropods. Pestic Biochem Physiol 102:124–128
11. Jung IS, Seo EJ, Lee DU (2012) Nootkatone for preventing and treating cardiovascular disease. Korean Kongkae Taeho Kongbo KR 2012021545 A 20120309
12. Gliszczynska A, Lysek A, Janeczko T, Switalska M, Wietrzyk J, Wawrzenczyk C (2011) Microbial transformation of (+)-nootkatone and the antiproliferative activity of its metabolites. Bioorg Med Chem 19:2464–2469
13. Seo EJ, Lee DU, Kwak JH, Lee SM, Kim YS, Jung YS (2011) Antiplatelet effects of *Cyperus rotundus* and its component (+)-nootkatone. J Ethnopharmacol 135:48–54
14. Tsoyi K, Jang HJ, Lee YS, Kim YM, Kim HJ, Seo HG, Lee JH, Kwak JH, Lee DU, Chang KC (2011) (+)-Nootkatone and (+)-valencene from rhizomes of *Cyperus rotundus* increase survival rates in septic mice due to heme oxygenase-1 induction. J Ethnopharmacol 137:1311–1317
15. Murase T, Misawa K, Haramizu S, Minegishi Y, Hase T (2010) Nootkatone, a characteristic constituent of grapefruit, stimulates energy metabolism and prevents diet-induced obesity by activating AMPK. Amer J Physiol 299:E266–E275
16. Misawa K, Murase T (2010) Nootkatone as LkB1 activator. Jpn. Kokai Tokkyo Koho JP 2010280598 A 20101216

17. Mucha JK, Munck M, Grant KE, Karchesy J (2013) Oxidation of valencene to nootkatone with potassium permanganate (KMnO4) facilitated by crown ether phase transfer catalysis. In: Abstracts, 68th northwest regional meeting of the American Chemical Society, Corvallis, Or, United States, 21–24 July, NORM-186
18. Garcia-Cabeza AL, Marin-Barrios R, Moreno-Dorado FJ, Ortega MJ, Massanet GM, Guerra FM (2014) Allylic oxidation of alkenes catalyzed by a copper-aluminum mixed oxide. Org Lett 16:1598–1601
19. Sauer AM, Crowe WE, Henderson G, Laine RA (2009) An Efficient and Economic Asymmetric Synthesis of (+)-Nootkatone, Tetrahydronootkatone, and Derivatives. Org Lett 11:3530–3533
20. Neuenschwander U, Jensen KF (2014) Olefin autoxidation in flow. Ind Eng Chem Res 53:601–608
21. Djuris M, Stefuca V, Eghbaldar M, van den Schaft P. Edited by Blank I, Wuest M, Yeretzian C (2010) Biocatalysis based transformation of valencene into nootkatone. In: Expression of multidisciplinary flavour science, proceedings of the weurman symposium, 12th, Interlaken, Switzerland, 1–4 July 2008, pp 332–335
22. Asakawa Y, Hashimoto T, Noma Y, Furusawa M (2013) Modification of valencene by bio- and chemical transformation. Nat Prod Commun 8:859–862
23. Meng F, Yu C, Li H, Xie T (2012) Production of (+)-nootkatone from (+)-valencene by biocatalysis. Shengwu Jishu Tongbao 8:194–198
24. Girhard M, Machida K, Itoh M, Schmid RD, Arisawa A, Urlacher VB (2009) Regioselective biooxidation of (+)-valencene by recombinant E. coli expressing CYP109B1 from *Bacillus subtilis* in a two-liquid-phase system. Microb Cell Fact 8:0. doi:10.1186/1475-2859-8-36
25. Gavira C, Hofer R, Lesot A, Lambert F, Zucca J, Werck-Reichhart D (2013) Challenges and pitfalls of P450-dependent (+)-valencene bioconversion by *Saccharomyces cerevisiae*. Metab Eng 18:25–35
26. Ly TTB, Khatri Y, Zapp J, Hutter MC, Bernhardt R (2012) CYP264B1 from *Sorangium cellulosum* So ce56: a fascinating norisoprenoid and sesquiterpene hydroxylase. Appl Microbiol Biotechnol 95:123–133
27. Fraatz MA, Riemer SJL, Stoeber R, Kaspera R, Nimtz M, Berger RG, Zorn H (2009) A novel oxygenase from *Pleurotus sapidus* transforms valencene to nootkatone. J Mol Cata B: Enzym 61:202–207
28. Kruegener S, Krings U, Zorn H (2010) Berger RG (2009) A dioxygenase of *Pleurotus sapidus* transforms (+)-valencene regio-specifically to (+)-nootkatone via a stereo-specific allylic hydroperoxidation. Bioresour Technol Volume Date 101:457–462
29. Zelena K, Krings U, Berger RG (2012) Functional expression of a valencene dioxygenase from *Pleurotus sapidus* in *E. coli*. Bioresour Technol 108:231–239
30. Kelle S, Zelena K, Krings U, Linke D, Berger RG (2014) Expression of soluble recombinant lipoxygenase from *Pleurotus sapidus* in *Pichia pastoris*. Protein Expr Purif 95:233–239
31. Plagemann I, Zelena K, Arendt P, Ringel PD, Krings U, Berger RG (2013) LOXPsa1, the first recombinant lipoxygenase from a basidiomycete fungus. J Mol Cata B: Enzym 87:99–104
32. Leonhardt RH, Plagemann I, Linke D, Zelena K, Berger RG (2013) Orthologous lipoxygenases of *Pleurotus* spp.—A comparison of substrate specificity and sequence homology. J Mol Cata B: Enzym 97:189–195
33. Minussi RC, Pastore GM, Duran N (2002) Potential applications of laccase in the food industry. Trends Food Sci Technol 13:205–216
34. Huang R, Christenson PA, Labuda IM (2001) Process for the preparation of nootkatone by laccase catalysis. Google Patents, US5847226 A
35. Calcaterra A, Galli C, Gentili P (2008) Phenolic compounds as likely natural mediators of laccase: a mechanistic assessment. J Mol Cata B: Enzym 51:118–120
36. Canas AI, Camarero S (2010) Laccases and their natural mediators: Biotechnological tools for sustainable eco-friendly processes. Biotechnol Adv 28:694–705

37. Hartwig S, Frister T, Alemdar S, Li Z, Krings U, Berger RG, Scheper T, Beutel S (2014) Expression, purification and activity assay of a patchoulol synthase cDNA variant fused to thioredoxin in *Escherichia coli*. Protein Expr Purif 97:61–71
38. Gardner TS, Hawkins KM, Meadows AL, Tsong AE, Tsegaye Y (2013) Fermentation of acetyl-coenzyme A-derived isoprenoids using transgenic microorganisms. US 8415136 B1 20130409
39. Sonke T, De Jong RM (2012) Valencene synthase from *Chamaecyparis nootkatensis* with improved activity and specificity for use in converting farnesyl diphosphate to valencene. PCT Int. Appl., WO 2012177129 A2 20121227
40. Beekwilder J, van Houwelingen A, Cankar K, van Dijk ADJ, de Jong RM, Stoopen G, Bouwmeester H, Achkar J, Sonke T, Bosch D (2014) Valencene synthase from the heartwood of Nootka cypress (*Callitropsis nootkatensis*) for biotechnological production of valencene. Plant Biotechnol J 12:174–182
41. Achkar J, Sonke T. (2011) Valencene synthase from Chamaecyparis nootkatensis with improved activity and specificity for use in converting farnesyl diphosphate to valencene. PCT Int Appl WO 2011074954 A2 20110623
42. Amick JD, Park E, Julien Bryan N, Burlingame RP (2012). Modified valencene synthase polypeptides, encoding nucleic acid molecules and uses for terpene biosynthesis. PCT Int Appl WO 2012058636 A1 20120503
43. Cankar K, van Houwelingen A, Bosch D, Sonke T, Bouwmeester H, Beekwilder J (2011) A chicory cytochrome P450 mono-oxygenase CYP71AV8 for the oxidation of (+)-valencene. FEBS Lett 585:178–182
44. Wriessnegger T, Augustin P, Engleder M, Leitner E, Muller M, Kaluzna I, Schurmann M, Mink D, Zellnig G, Schwab H, Pichler H (2014) Production of the sesquiterpenoid (+)-nootkatone by metabolic engineering of *Pichia pastoris*. Metab Eng 24:18–29
45. Lange BM, Rios-Estepa R, Turner GW (2010) Production of terpenes and terpenoids in transgenic glandular trichome-bearing plants. PCT Int. Appl., WO 2010111571 A1 20100930
46. Fraatz MA, Berger RG, Zorn H (2009) Nootkatone—a biotechnological challenge. Appl Microbiol Biotechnol 83:35–41
47. Zviely M (2009) Molecule of the month: nootkatone. Perfumer Flavorist 34(20):22
48. Wilson CW, Shaw PE (1978) Synthesis of nootkatone from valencene. J Agric Food Chem 26:1430–1432
49. Muller B, Dean C, Schmidt C, Kuhn JC (1998) Process for the preparation of nootkatone. Google Patents, US5847226 A
50. Furusawa M, Hashimoto T, Noma Y, Asakawa Y (2005) Highly efficient production of nootkatone, the grapefruit aroma from valencene, by biotransformation. Chem Pharm Bull 53:1513–1514
51. Sakamaki H, Itoh K, Taniai T, Kitanaka S, Takagi Y, Chai W, Horiuchi CA (2005) Biotransformation of valencene by cultured cells of *Gynostemma pentaphyllum*. J Mol Catal B: Enzym 32(3):103–106

Adv Biochem Eng Biotechnol (2015) 148: 405–426
DOI 10.1007/10_2014_292

Published Online: 21 December 2014

Current and Emerging Options for Taxol Production

Yi Li, Guojian Zhang and Blaine A. Pfeifer

Abstract Paclitaxel (trademark "Taxol") is a plant-derived isoprenoid natural product that exhibits potent anticancer activity. Taxol was originally isolated from the Pacific yew tree in 1967 and triggered an intense scientific and engineering venture to provide the compound reliably to cancer patients. The choices available for production include synthetic and biosynthetic routes (and combinations thereof). This chapter focuses on the currently utilized and emerging biosynthetic options for Taxol production. A particular emphasis is placed on the biosynthetic production hosts including macroscopic and unicellular plant species and more recent attempts to elucidate, transfer, and reconstitute the Taxol pathway within technically advanced microbial hosts. In so doing, we provide the reader with relevant background related to Taxol and more general information related to producing valuable, but structurally complex, natural products through biosynthetic strategies.

Keywords Paclitaxel · *Taxus* · Metabolic engineering · *E. coli* · Yeast · Fungi

Abbreviations

BMS	Bristol-Myers Squibb
DOE	Design of experiments
DMAPP	dimethylallyl diphosphate
DXS	1-deoxy-D-xylulose 5-phosphate synthase
FDA	Food and Drug Administration
FPP	farnesyl diphosphate
FPPS	farnesyl diphosphate synthase
GGPPS	geranylgeranyl diphosphate synthase
HMG-CoA	3-hydroxy-3-methylglutaryl-CoA
IPP	isopentenyl diphosphate

Y. Li · G. Zhang · B.A. Pfeifer (✉)
Department of Chemical and Biological Engineering,
The State University of New York at Buffalo, Buffalo,
NY 14260, USA
e-mail: blainepf@buffalo.edu

ispD	4-diphosphocytidyl-2-*C*-methyl-D-erythritol synthase
ispF	2-*C*-methyl-D-erythritol 2,4-cyclodiphosphate synthase
IDI	isopentenyl diphosphate isomerase
IND	Investigational new drug
MEP	2*C*-methyl-D-erythritol-4-phosphate
MVA	Mevalonate
NCI	National Cancer Institute

Contents

1 Introduction

The isolation and elucidation of the Taxol compound began two separate journeys based upon the compound's medicinal value and impact. First, there was a clinical goal of formulating and administering the compound for anticancer application. In this capacity, the compound has seen tremendous success [1, 2]. The international market for Taxol production reached $1.6 billion in 2005 [3]. Such financial outcomes established Taxol as the worldwide best-selling anticancer drug, regardless of direct source or semisynthetic production routes [4]. Furthermore, as a cytotoxic chemotherapy drug with high response rates, Taxol is widely used for not only various types of solid tumors associated with breast, ovarian, lung, and bladder cancer but also in treatment therapies for Kaposi's sarcoma [5–8]. In parallel, a second effort also began towards process development for improved accessibility to the compound.

Originally, Taxol was isolated from the Pacific yew tree *Taxus brevifolia* through an ambitious program initiated by the NCI in 1958 to screen 35,000 plants for anticancer activity [9]. Once enough Taxol had been harvested to confirm medicinal potency fully and allow full chemical structural characterization [10], efforts began towards scalable production of the Taxol molecule. There are several options towards this end including a purely synthetic route [11]. However, the structural complexity of Taxol, like many other complex natural products, places limits on a synthetic production option that must also be affordable and scalable.

In other words, synthetic routes have been established, but the multiple steps needed and the subsequent loss in yield at each step complicate economic production.

Alternatively, the native biological host could be considered a vessel for production purposes. From this perspective, additional consideration would have to be given to the biological constraints of the native producer. Insurmountable or very challenging obstacles associated with native host production has spurred further and more recent production alternatives that include an approach termed heterologous biosynthesis [12, 13]. In this capacity, the biosynthetic steps required for Taxol formation are transferred from the native and typically more fastidious host and implemented in more technically friendly microbial hosts possessing both innate biological properties and engineering tools to facilitate pathway reconstitution and compound overproduction. However, the process of establishing heterologous biosynthesis also presents a set of technical challenges that must be overcome. As such, the following chapter is dedicated to summarizing Taxol production with an emphasis on options that are currently used and those that may one day supplant them.

2 Taxol Origins and Production Development

2.1 Natural Role and Therapeutic Utilization

Although clearly established as a cancer chemotherapeutic, the native role of Taxol is much less clear. The mechanism of anticancer activity is tubule stabilization and, hence, cell-cycle arrest [14, 15]. Hypothetically, this same activity could serve in a defensive role during events that threaten plant health. For example, Taxol showed particular activity towards invasive oomycetes across various taxonomic groups [16]. In studies designed to assess the impact of altering compound side-chain functionality, activity was also demonstrated against *Phytophthora capsici* and *Aphanomyces cochlioides*, two plant pathogens [17].

These same properties, however, endowed the compound as a strong neoplastic agent. *Taxus brevifolia* crude plant extract was found to be cytotoxic in cellular assays conducted in the early to mid-1960s [4]. Monroe Wall then isolated and named the active Taxol component from fresh *Taxus* samples in 1967 [9]. In early 1979, Susan Horwitz demonstrated that Taxol had a previously unknown mechanism of action involving the stabilization of microtubules [18]. Upon collection of 20,000 lbs of *Taxus* bark for compound isolation, animal toxicology studies were complete by June 1982, and in November, the NCI submitted an IND application to begin clinical trials [19]. Early clinical trials began in April 1984, and in March 1988 the FDA released the final results of Phase II trials against the most aggressive forms of ovarian cancer with a response rate that averaged 30 % [20]. The FDA first approved Taxol for the treatment for ovarian cancer in 1992 [21]. FDA approval for breast cancer and Kaposi's sarcoma was accomplished in 1994 and 1997, respectively [22, 23].

2.2 Production Based upon Plant Organisms

The identification of Taxol as a potent anticancer agent provided the impetus for early production efforts. Assays and tests designed to characterize and assess the Taxol compound further could rely on the simplest isolation routes that involved direct extraction of the compound from the bark of the yew tree. Once these studies confirmed the potential of the molecule, more refined approaches were required for the processes needed to enable widespread distribution.

It readily became evident that bark extraction could not economically support large-scale production attempts. Furthermore, the environmental impact for such a strategy was unacceptable. To treat one patient effectively, the bark of two to three yew trees was required with each fully grown tree only able to provide 0.5 g of Taxol [24]. Considering that each extraction process was destructive to the tree source and mature yews required 200 years to develop fully, this form of direct plant isolation was clearly inefficient and nonsustainable.

However, a compromise was reached regarding direct yew tree Taxol isolation. Namely, a key precursor from the Taxol pathway, Baccatin III (Fig. 1), was readily extracted from yew tree needles in a manner that was nondamaging to the overall health of the tree [25, 26]. Although this intermediate did not provide the same therapeutic value of Taxol, the late-stage precursor could be readily converted to the full Taxol molecule through separate synthetic chemistry transformations [27, 28].

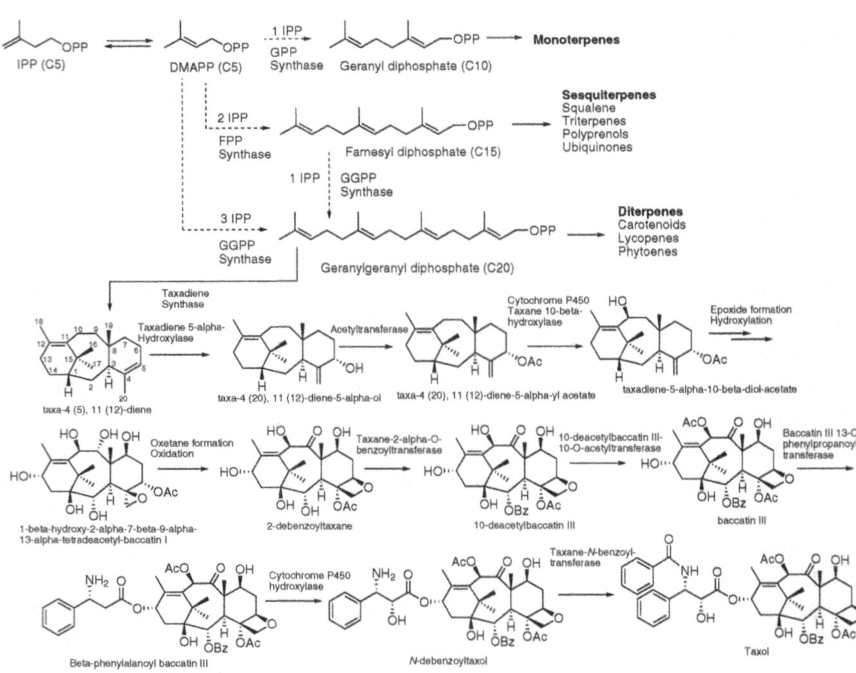

Fig. 1 Biosynthetic path to Taxol

Hence, a semisynthetic route to Taxol was achieved. Indeed, the strategy proved effective in large-scale Taxol production. As such, BMS became the industrial partner involved with mass production and led an effort towards commercialization by signing a cooperative research and development agreement with Robert Holton, who developed the semisynthetic route to Taxol, at Florida State University [28].

Over the course of the research to establish the semisynthetic route to Taxol, there remained interests in alternative production options. In particular, although the semisynthetic strategy proved viable, there was still the need to rely on macroscopic tree components and direct extraction from such sources. One option more compliant with developing bioreactor strategies for biological products was plant cell culture. In this context, early attempts at cell suspensions were initiated by conditioning macroscopic fragments of *Taxus calli* for liquid-phase growth, which led to culture-based production and optimization strategies [29, 30]. The first reported *T. baccata* callus induction and culture was reported in 1973 [31].

Cell culture-based process development strategies allowed the transition to a unit operation-based form of production and the associated application of process engineering strategies as outlined in Table 1. Cell culture variability and low product yields are two limitations for plant cell culture technology [32]. In addition, secondary metabolite accumulation has been correlated with negative cellular effects resulting in culture growth inhibition and overall production instability. At the cellular population level, such phenotypic pressure results in groups of aggregates and heterogeneity, contributing to production variability [33, 34]. To help address these issues and effectively increase paclitaxel production, strategies such as precursor and substrate feeding, media composition modification [35], two-phase partitioned organic solvent extraction, elicitor addition (including agents such as chitosan, silver ion, methyl jasmonate), and cellular dispersion (ultrasound) have been attempted. Other strategies utilizing solid adsorbants (e.g., polymeric resins of the nonionic Amberlite XAD series [36]) as well as self-immobilized aggregates of *Taxus cuspidata* [37] have also been tested as ways to improve overall production.

For larger-scale production of paclitaxel, optimized bioreactor designs and configurations are favored. Elicitation and immobilization strategies had been applied to bioreactors operated in stirred, airlift, pneumatic, and wave formats [46]. Currently, all paclitaxel production for BMS uses plant-cell fermentation technology developed by the Ithaca, New York biotechnology company Phyton Biotech, Inc. and carried out at their production plant in Germany [42]. Phyton Biotech currently employs up to 75,000 L bioreactors for commercial production of paclitaxel from cell culture [47].

2.3 Taxol Pathway Elucidation

Plant-derived Taxol biosynthesis remains a viable option for production purposes. However, there was both a basic interest and applied rationale for identifying the dedicated pathway required for biosynthesis. Scientifically, the enzymology behind

Table 1 Plant-cell culture-based production strategies

Cellular source	Engineering strategy	Vessel format	Compound titer (mg/L)	Reference
Taxus baccata	Immobilization within Ca^{2+}—alginate beads	Stirred bioreactor	43.43 at day 16	[38]
		Wave bioreactor	20.79 at day 8	
	Methyl jasmonate-induced and silver thiosulfate (antiethylene) addition	Large-scale two-stage process	295	[39]
	Jasmonate-induced	Shake flask	48.3 at day 14	[40]
Taxus chinensis and endophytic fungi	Reparable separation membrane	Cobioreactor	25.63 at day 15	[41]
Taxus chinensis	Combination of enhancement agents and medium exchange	Fed batch	900	[42]
	Chitosan, methyl jasmonate-induced, and Ag^+ addition	Shake flask	25	[43]
	Mechanical stimulus, ultrasound, methyl jasmonate-induced, in situ solvent extraction	Shake flask	33–35	[44]
Taxus cuspidata	Self-immobilization aggregate	Shake flask	4.9 at day 40	[37]
	Medium composition modification	Shake flask	431 at day 55	[45]

biosynthesis would contribute to a growing number of understood complex natural product pathways and the mechanisms behind individual intermediate transformations. Such insight then lends itself to applied efforts to modify or utilize resulting information for unique production purposes. For example, an understanding of the dedicated biosynthetic steps would allow intervention for the purpose of altering pathway steps and eventual final products, which may, in turn, possess new biological activity.

Alternatively, complete knowledge of the Taxol biosynthetic pathway would allow the potential for heterologous biosynthesis through an alternative production host. Notwithstanding the production options offered by plant systems, the constraints of original production routes suggest an opportunity to harness production through technically amenable microbial hosts. This basic theme is the genesis for most interests and efforts in heterologous biosynthesis.

Towards this end, Rodney Croteau's group at Washington State University took the lead in elucidating the steps required for Taxol biosynthesis [48–53]. The important outcome of this research was an outline of a hypothetical path to Taxol biosynthesis that could be utilized in the context of basis studies to understand enzymatic conversion steps and applied attempts to heterologously reconstitute the biosynthetic pathway. However, towards the latter goal, a significant challenge beyond the obvious technical steps in designing reconstitution is the fact that several steps remain unknown in the biosynthetic path to Taxol (Fig. 1). In particular, genes encoding taxoid 1β-hydroxylase, taxoid 9-keto-oxidase, β-phenylalanoyl-CoA ligase, taxoid 2′-hydroxylase, epoxidation reactions at C4 and C5, and those involved in oxetane formation are still to be identified. Thus, more research is required to identify these steps before reconstitution efforts can be planned with complete confidence.

2.4 Microbial Production

With the partial elucidation of the Taxol biosynthetic pathway came early attempts towards heterologous biosynthesis. As a basis for such efforts, the new hosts must provide the prerequisite universal isoprenoid substrates needed to support downstream compound biosynthesis. The two routes available for this purpose include the MVA and nonmevalonate (or MEP) pathways (Fig. 2) [54]. The MVA pathway is generally found in eukaryotic organisms; whereas, the MEP pathway is commonly associated with prokaryotic metabolism [55, 56]. Before biosynthesis can commence, a substrate support pathway must be available and, in most cases, engineered to support isoprenoid overproduction.

However, the engineering dedicated to pathway support must be completed with care. Otherwise, the buildup of intermediates will lead to global cellular toxicity. For example, prenyl pyrophosphate intermediates such as IPP, DMAPP, and FPP involved in the biosynthetic pathway towards Taxol have demonstrated intracellular toxicity within microbial systems [57]. In addition, HMG-CoA from the alternative MVA pathway has resulted in negative cellular effects upon fatty acid biosynthesis [58].

Upon completing the cellular engineering to support broad isoprenoid biosynthesis, the dedicated enzymes required for product formation will subsequently be introduced. From this perspective, the downstream steps to isoprenoid formation can be subdivided into those reactions required for terpene synthesis (the hydrocarbon portion of the final isoprenoid compound) and those steps required for terpene tailoring. The first series of reactions can include both isoprene elongation steps, which dictate the classification of the final isoprenoid compound (Fig. 1), and a terpene synthase step that cyclizes the elongated terpene chain into a configurational core structure [59]. The isoprenoid tailoring reactions then proceed to decorate the resulting terpene compound with functional units that confer key bioactivity. These reactions, particularly featured in Taxol formation, include

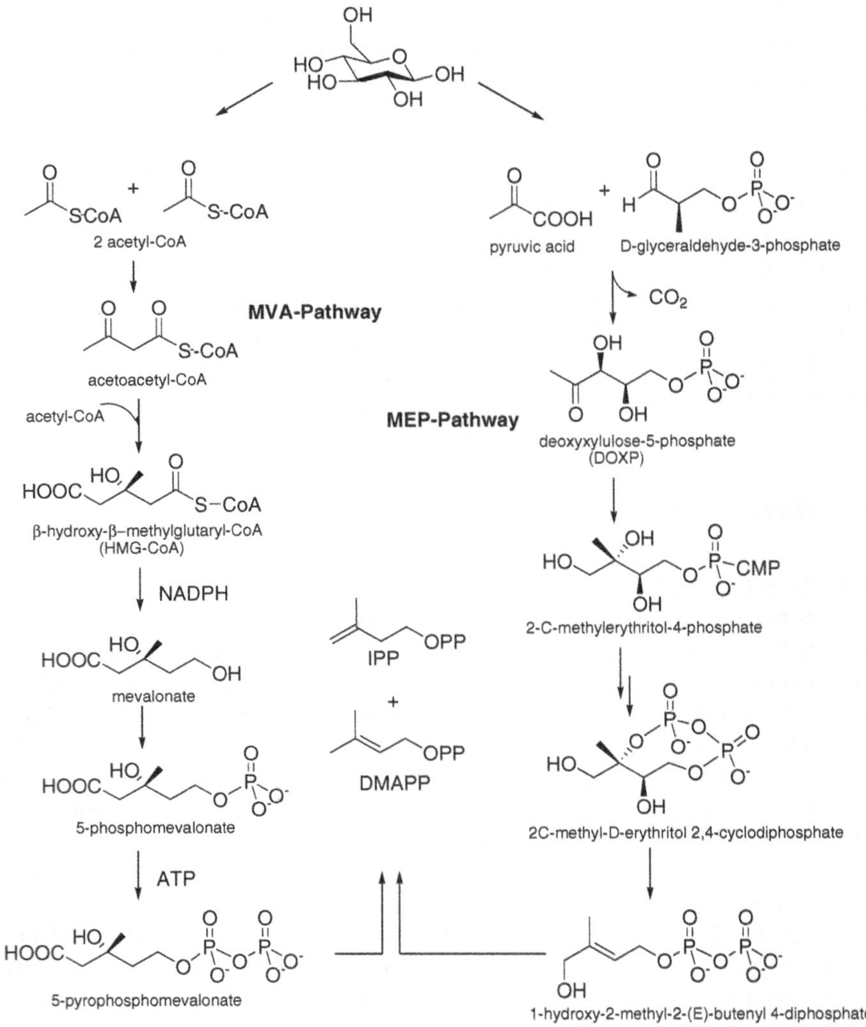

Fig. 2 MVA and nonmevalonate (or MEP) pathways

hydroxylation, transacetylation, stereospecific oxidation, benzoylations, epoxidation, esterification, and oxetane ring formation. In particular, hydroxylation reactions are common and are catalyzed by dedicated pathway P450 enzymes [60].

In step with other early efforts in the heterologous production of isoprenoid compounds, *E. coli* was a first option in efforts to generate Taxol intermediates [61, 62]. The choice to use *E. coli* coincided with the technical convenience associated with the host, namely, a rapid growth rate and well-established molecular biology protocols. The host natively utilizes the MEP pathway for cellular isoprenoid production purposes that include tRNA prenylation and steps in cell-wall and quinone formation [63]. As such, this intrinsic support pathway was utilized to provide the universal IPP and DMAPP substrates for heterologous biosynthesis. Pathway genes associated with bottleneck steps in precursor support were identified as *dxs*, *idi*, *ispD*, and *ispF* and have since been the targets of many engineering efforts (including those of Taxol heterologous biosynthesis attempts) to improve intracellular precursor availability [64, 65].

Building upon contemporary work to generate isoprenoid compounds through *E. coli* in the 1990s [66–68], Croteau's group attempted to generate the first dedicated intermediate of the Taxol pathway, taxadiene [48, 51, 62, 69]. Given the Croteau group's pioneering work in Taxol pathway elucidation, the putative genetic material to support heterologous biosynthesis was available to test in the context of initial gene expression design. Interestingly, as advanced as these attempts were, success was only realized upon manipulation of the *E. coli* MEP pathway to improve precursor support by targeting bottleneck steps (introduced above) [70]. Even so, taxadiene titer levels achieved were modest (1.3 mg/L) [62]. Furthermore, little progress was made in generating sufficient quantities of later pathway intermediates when using *E. coli* prior to renewed efforts that began in the middle 2000s. By this time, numerous and high-titer examples of isoprenoid heterologous biosynthesis had been established through *E. coli* [71, 72]. Based upon this renewed level of success and emerging synthetic biology and metabolic engineering experimental tools, a more concerted effort was made in establishing early Taxol pathway biosynthesis through *E. coli* [61, 73, 74].

First, optimized strains designed for ample precursor supply were used as a foundation for introducing the Taxol biosynthetic pathway. Next, knowledge accumulated in gene expression design allowed successful activity from the plant-derived downstream enzymes needed to generate taxadiene and the subsequent hydroxylated intermediate. When combined, improved or new production titers of each intermediate were recorded. From this point, production was next optimized through a careful balancing of the upstream precursor support and downstream biosynthetic components of the cellular design. The final optimized product titers were further boosted through the application of small-scale bioreactor strategies [61].

Production of Taxol intermediates (and isoprenoids more generally) was also extended to yeast systems that natively possess the MVA pathway. Regarding Taxol, a strong motivation for pursuing an alternative heterologous host was the multiple steps requiring P450 hydroxylation [59, 75]. Although a bacterial host provides convenience in genetic manipulation steps, the eukaryotic nature of the yeast host provides a better cellular match to the native plant cellular environment for the Taxol biosynthetic pathway [76, 77]. As such, support mechanisms, especially in the context of the P450 enzymes, were available in this surrogate host. For example, the yeast cell could natively provide a required P450 reductase to assist in the hydroxylation reactions associated with the Taxol P450 biosynthetic steps [78]. The cellular structure of the yeast system also provides physical localization of P450 enzymes (to the endoplasmic reticulum) [79]. Hence, an emerging trend in the heterologous production of isoprenoid compounds requiring P450 activity is the use of yeast systems [77, 80, 81]. In the context of Taxol efforts, both the Croteau and

Jennewein groups tested production using yeast with successful biosynthesis of the second dedicated Taxol pathway intermediate, which required the activity of a P450 enzyme [75, 78, 82].

Finally, another potential option for microbial-based Taxol production is the use of fungal organisms. This approach is supported by interesting observations of Taxol production by endophytic fungi associated with Taxol-producing plants [83]. Several follow-up studies examine the capabilities of these organisms to produce Taxol; however, production levels were not high enough, relative to competing technology, to support production processes [84]. A recent study has also questioned endophytic fungal biosynthesis altogether, based upon the lack of compelling biosynthetic sequence identification [85]. Still, the results and observations associated with Taxol-producing fungi support the prospect of production through industrially relevant and continually developing fungal host systems such as *Aspergillus niger*, *A. nidulans*, *A. oryzae*, and *A. terreus* [86, 87].

2.5 Engineering Across Host Systems

Engineering strategies have been developed continually for Taxol biosynthetic production. Strategies vary depending on the host system and the objective. In some cases, improvements associated with final compound production have been initiated towards immediate benefits in the form of more economical and timely production routes. Longer-term research efforts, heavily featuring microbial hosts, have been dedicated to heterologous biosynthesis and post-reconstitution optimization.

In plant systems, several process engineering strategies have been summarized in Table 1. This has been the primary avenue to improving and optimizing production levels to this stage. However, more recent developments have opened the possibility of cellular-based engineering. For example, *Taxus* culture transformation methods have recently been established [88, 89]. Such advances are necessary for themes in genetic and metabolic engineering to commence. In this regard, direct pathway engineering has been initiated via overexpression of the *txs* and *dbat* biosynthetic genes coding for taxadiene synthase and 10-deacetylbaccatin III-10 β-O-acetyltransferase, respectively, in transgenic *T. chinesis*, resulting in increased production titers [90].

Engineering within microbial cells can be subdivided across different scales: genetic, metabolic, and process. The genetic level requires close attention to the details used in establishing active gene expression and the pathways responsible for biosynthesis. Viewing the wild-type cell as the basis, any engineering completed regarding gene expression must account for intrinsic challenges and resulting problems that may exist as a result of the manipulations made. Gene expression must be designed to perturb native metabolism minimally. At the same time, tuning gene expression to maximize an alternative objective relative to the cellular status quo is clearly important. Thus, the gene expression process, and the growing number of tools available to program expression, must be designed to find an optimum of the overall objective. Key components of the process include the following.

2.5.1 Gene Source

Genetic content information (i.e., the gene sequence) must be available to initiate the design process. The native source genomic material is the obvious location for obtaining such information. Especially relevant to genes that must be introduced heterologously, the content of the gene reading frame may experience significant expression problems within a new host. This issue can be addressed by codon optimizing the gene of interest or introducing rare tRNA molecules to the new host. These considerations and others related to proper gene expression design are built into the plan of microbial biosynthesis using established and emerging tools of genetic engineering and synthetic biology [91]. The reconfigured gene or genes required for biosynthesis then provide the template for the remaining steps in pathway reconstitution.

2.5.2 Gene Expression

The gene expression process can be influenced at multiple levels. First, there is now a wealth of promoters, repressors, terminators, and other expression parameters available to design and tune gene expression (together with emerging analytical technologies to measure the resulting gene expression finely) [92]. These parameters can be combined with gene copy number and process conditions (such as temperature modulation) to optimize expression of a particular gene, which will limit detrimental metabolic imbalance. Similarly, microbial systems offer the opportunity for gene expression format variation in the way of operon designs, which offer an alternative means of modulating the potential for expression burden.

2.5.3 Expression Duration

Expression duration will also influence overall production potential. Linked to mRNA levels, extended gene expression will be closely correlated with production and will be influenced by some of the same parameters introduced above. For example, promoter type and strength will be a key determinant of mRNA content. As one example, the powerful T7 promoter has been used in several heterologous production efforts because of the strong and processive nature of the associated T7 RNA polymerase that has been built into certain microbial strains [93, 94]. Alternatively, weaker promoters stand as alternatives, as do copy number formats, to modify transcript levels [61].

Another option in transcriptional influence is the type of expression regulation. Numerous microbial systems are inducible with the *lac*, *tet*, and *ara* operators [95, 96]. Constitutive or continuous expression is the counter option to inducible systems and offers a steadier level of mRNA over time [97]. The choice of using one system versus the other is driven by experience and the specific goals of the production process. It may be desirable to attempt to decouple biosynthesis from cellular growth. In this case, an inducible expression design will allow the potential to delay the biosynthetic process until a suitable cell density has been achieved. However, in this format, inducible expression may then completely imbalance metabolism and shorten the overall production timeframe; whereas, a more measured level of constitutive expression may allow continuous and extended production while minimizing metabolic burden, potentially leading to improved production metrics. It may be difficult to choose correctly between these two expression formats in the initial design phase, but heuristics such as the likelihood of heterologous pathway products contributing to detrimental cellular effects would suggest an option of inducible expression. Doing so would allow a balance of robust biomass accumulation and detrimental side effects of biosynthesis.

The transcriptional approaches indicated above relate to mRNA generation, but influencing degradation patterns also provides a means of stabilizing transcript levels and expression over time. Particularly in bacteria, mRNA decay rates can differ by two orders of magnitude, thus, posttranscriptional processing plays a key role in controlling gene expression in cases not limited by ribosome availability [98]. For example, by introducing DNA cassettes encoding mRNA secondary structures, carotenoid biosynthesis could be modulated, resulting in target compound yield increases [99]. In addition, varying the distance between promoter and gene initiation was implicated as another influence on transcript levels and overall isoprenoid biosynthesis [100].

Finally, it should be mentioned that the majority of genetic systems indicated above were not designed with optimal pathway reconstitution in mind. Instead, most derive from attempts at single gene overexpression for the purpose of purification and characterization of the final protein product (usually in the context of biochemical characterization or protein structural determination). Thus, such systems will most likely not provide sustained and measured gene expression and subsequent product formation. More recent efforts have begun, however, to reconfigure traditional expression formats for multistep biosynthetic pathways. Commercial examples include the pET–Duet vectors from Novagen and additional plasmids designed for pathway reconstruction [101]. In other approaches, pathways are constructed using recombination-based assembly methods [102–105].

As indicated above, gene expression design is intrinsically linked to metabolic analysis and engineering. The development and continued maturity in characterization strategies, especially with regard to comprehensive and omic-type analyses, is capable of providing a complete readout of cellular states [106, 107]. With such information comes a clearer understanding of potential imbalances associated with production efforts. As indicated, such problems can be addressed at a pathway level through genetic design. However, pathway efforts must be viewed within the context of complete metabolism. Towards this end, computational metabolic engineering has proven beneficial in predicting and identifying interactions between native and engineered metabolisms that provide the basis for additional genetic engineering for improving overall production [108, 109]. For example, stoichiometric constraint-based modeling has been utilized in the context of predicting those native reactions contributing to or detracting from the flow of carbon and other metabolites required for optimal pathway operation [110, 111]. These predictions can then be readily incorporated through the utilization of the molecular biology protocols associated with model organisms [74, 112]. It must be emphasized that the hosts must be well-characterized with regard to genome sequencing and annotation and possess tractable genetic manipulation methods. If these conditions are met, the utilization of metabolic modeling can be readily implemented in cellular-based production efforts. As an alternative to metabolic modeling, the same characterization strategies that may indicate metabolic imbalance could be used in a comprehensive fashion to allow genetic alterations in a purely experimental format. However, such efforts may be prohibitively resource intensive, which is a significant reason why cost-effective modeling approaches have found widespread utilization.

Apart from the directed metabolic engineering strategies indicated above, there also exist nondirected efforts that take advantage of classical biological principles of perturbation and selection. For example, native production of various natural products have effectively utilized mutagenesis and screening protocols [113]. In particular, such approaches are viable options when production hosts are not well-characterized or genetically tractable. In such cases, genetic manipulations towards improved phenotypes are possible, but the genotypic changes made are not done so in a directed fashion and, therefore, there is less of an engineering aspect to these approaches. Nonetheless, they have been utilized with great success with several natural products [114–119].

Finally, at the process level, there exist additional engineering options to improve final production metrics further. First, there is the opportunity to scale production conditions from culture tubes and shake flasks to well-controlled bioreactors that allow precise maintenance of parameters such as pH, temperature, agitation, and dissolved oxygen content in addition to larger reaction volumes [120]. This improvement in process control and overall scale typically translates to improved final production titers [121, 122]. However, the advantage to utilizing nonaggregating, unicellular microbial hosts possessing simple media requirements and rapid growth kinetics has the potential to modify the typical batch bioreactor configuration to include a nutrient feed capable of greatly extending volumetric cell density and productivity. In this format, a highly concentrated nutrient stream is slowly fed to the primary culture, allowing for a finely controlled and extended growth profile and eliminating the carbon nutrient limitations that would normally accompany a batch reactor [123–125]. The fed-batch design therefore allows for effective cell density scaling with a modest increase in total volume, effectively allowing continued scaling of production without the need physically to increase the size of the bioreactor.

Although a fed-batch bioreactor design will eliminate carbon source limitation, there exists the danger of oxygen limitation. This concern can be addressed through the reconfiguration of bioreactor air ports or the introduction of pure oxygen [126, 127]. Related to nutrient input and other process parameters, the process production formats can also be optimized utilizing design of experiments (DOE) techniques [128]. In certain situations, such as media optimization, the approach may be best accomplished utilizing small-scale (i.e., 96-well) culture formats with subsequent scaled experiments used to confirm results [129, 130]. Alternatively, the same sort of optimization experiments could be applied to the controllable process parameters at the bioreactor scale [131].

3 Summary, Conclusions, and Outlook

Current Taxol production processes utilize plant sources in either semisynthetic or culture-based formats. Such routes have allowed for the distribution and medicinal use of this potent anticancer compound. However, relative advantages of microbial production hosts, together with emerging technology and examples of heterologously produced isoprenoid compounds, have initiated studies of reconstituted production efforts through host systems such as *E. coli* and *S. cerevisiae*. Regardless of the biological production host, engineering strategies have been implemented across several scales.

Plant-based production has seen a transition from direct (and nonsustainable) extraction from the native Pacific yew tree to industrial production dependent upon plant cell culture. Bridging these two production strategies was a semisynthetic route that first allowed larger-scale access to the Taxol compound. As more information and technology are gathered using plant-cell culture, the application of metabolic and process engineering is expected to continue to result in improved titers. However, the challenge of engineering plant cells has spurred parallel research in heterologous biosynthesis.

Engineering of microbial production hosts has been focused on two goals. First, there is the need to implement fully and reconstitute successfully the complete Taxol pathway (or enough of the pathway to allow semisynthesis). A number of genetic engineering strategies have been utilized to enable early-stage pathway intermediate production. These same tools stand as a way to enable the second objective which is to design production carefully for the purpose of streamlining later metabolic and process engineering. Combined, the available of emerging engineering options associated with the genetic, metabolic, and process scales will each need to be utilized and optimized to make compounds such as Taxol feasible for microbial production with the promise of even more widespread accessibility to this and similar compounds.

This vision is therefore a key future objective. Furthermore, successful Taxol heterologous biosynthesis would allow an additional frontier: the potential to engineer rationally the biosynthetic steps for molecular diversification. The goals associated with molecular diversification could range from improving the formulation and solubility of the Taxol compound to designing altered anticancer (and other forms of) bioactivity [132, 133].

The same themes summarized above for Taxol provide the basis for a broader future direction. Namely, a highly desired vision is to achieve the successful production, heterologous reconstitution, and manipulation of numerous other isoprenoid natural products, including those yet to be discovered. The success recorded thus far regarding the medicinal utility of isoprenoids represents only a small fraction of the possibilities available through nature. This chapter and ongoing work are dedicated to summarizing and projecting opportunities to identifying, producing, and benefiting from similar production efforts with increasingly successful frequency.

References

1. Cragg GM (1998) Paclitaxel (Taxol®): a success story with valuable lessons for natural product drug discovery and development. Med Res Rev 18:315–331
2. Skeel RT, Khleif SN (2011) Handbook of cancer chemotherapy, vol IV. Lippincott Williams & Wilkins, Philadelphia, p 803
3. Thayer A (2010) More than a supplier. Chem Eng News 88:25–27
4. Goodman J, Walsh V (2001) The story of taxol: nature and politics in the pursuit of an anticancer drug, vol 323(7304). Cambridge University Press, Cambridge
5. Voigt W, Kegel T, Weiss M, Mueller T, Simon H, Schmoll HJ (2005) Potential activity of paclitaxel, vinorelbine and gemcitabine in anaplastic thyroid carcinoma. J cancer Res Clin Oncol 131:585–590
6. Galsky MD (2005) The role of taxanes in the management of bladder cancer. Oncologist 10:792–798
7. Chougule PB, Akhtar MS, Rathore R, Koness J, McRae R, Nigri P, Radie-Keane K, Kennedy T, Wanebo HJ, Ready N (2008) Concurrent chemoradiotherapy with weekly paclitaxel and carboplatin for locally advanced head and neck cancer: Long-term follow-up of a Brown University Oncology Group Phase II Study (HN-53). Head Neck 30:289–296
8. Crown J, O'Leary M, Ooi WS (2004) Docetaxel and paclitaxel in the treatment of breast cancer: a review of clinical experience. Oncologist 2:24–32
9. Wani MC, Taylor HL, Wall ME, Coggon P, McPhail AT (1971) Plant antitumor agents. VI. The isolation and structure of taxol, a novel antileukemic and antitumor agent from *Taxus brevifolia*. J Am Chem Soc 93:2325–2327
10. Holton RA, Biediger RJ, Boatman PD (1995) Semisynthesis of taxol and taxotere. Taxol Science Appl:97–121
11. Danishefsky SJ, Masters JJ, Young WB, Link JT, Snyder LB, Magee TV, Jung DK, Isaacs RCA, Bornmann WG, Alaimo CA et al (1996) Total synthesis of baccatin III and Taxol. J Am Chem Soc 118:2843–2859
12. Zhang H, Boghigian BA, Armando J, Pfeifer BA (2011) Methods and options for the heterologous production of complex natural products. Nat Prod Rep 28:125–151
13. Ongley SE, Bian X, Neilan BA, Muller R (2013) Recent advances in the heterologous expression of microbial natural product biosynthetic pathways. Nat Prod Rep 30:1121–1138
14. De Brabander M, Geuens G, Nuydens R, Willebrords R, De Mey J (1981) Taxol induces the assembly of free microtubules in living cells and blocks the organizing capacity of the centrosomes and kinetochores. Proc Natl Acad Sci USA 78:5608–5612

15. Manfredi JJ, Parness J, Horwitz SB (1982) Taxol binds to cellular microtubules. J Cell Biol 94:688–696
16. Dostál V, Libusová L (2014) Microtubule drugs: action, selectivity, and resistance across the kingdoms of life. Protoplasma 1:1–15
17. Young DH, Michelotti EL, Swindell CS, Krauss NE (1992) Antifungal properties of taxol and various analogues. Experientia 48:882–885
18. Schiff PB, Fant J, Horwitz SB (1979) Promotion of microtubule assembly in vitro by taxol. Nature 277:665–667
19. Cragg GM, Schepartz SA, Suffness M, Grever MR (1993) The taxol supply crisis. New NCI policies for handling the large-scale production of novel natural product anticancer and anti-HIV agents. J Nat Prod 56:1657–1668
20. McGuire WP, Rowinsky EK, Rosenshein NB, Grumbine FC, Ettinger DS, Armstrong DK, Donehower RC (1989) Taxol: a unique antineoplastic agent with significant activity in advanced ovarian epithelial neoplasms. Ann Intern Med 111:273–279
21. Trimble E, Adams J, Vena D, Hawkins M, Friedman M, Fisherman J, Christian M, Canetta R, Onetto N, Hayn R (1993) Paclitaxel for platinum-refractory ovarian cancer: results from the first 1,000 patients registered to National Cancer Institute Treatment Referral Center 9103. J Clin Oncol 11:2405–2410
22. Suffness M, Wall M, Suffness M (2002) Taxol: Science and Applications, 1995. Mekhail TM Markman M Expert Opin Pharmacother 3:755
23. Wilson WH, Wittes R (1996) Taxol treatment of breast cancer. US5496846 A
24. Horwitz SB (1994) How to make taxol from scratch. Nature 367:593–594
25. Chun M-K, Shin H-W, Lee H (1996) Supercritical fluid extraction of paclitaxel and baccatin III from needles of *Taxus cuspidata*. J Supercrit Fluids 9:192–198
26. van Rozendaal EL, Lelyveld GP, van Beek TA (2000) Screening of the needles of different yew species and cultivars for paclitaxel and related taxoids. Phytochemistry 53:383–389
27. Kingston DG, Chaudhary AG, Gunatilaka A, Middleton ML (1994) Synthesis of taxol from baccatin III via an oxazoline intermediate. Tetrahedron Lett 35:4483–4484
28. Holton RA (1993) Semi-synthesis of taxane derivatives using metal alkoxides and oxazinones. EP0568203 A1:C07D265/06
29. Jaziri M, Zhiri A, Guo Y-W, Dupont J-P, Shimomura K, Hamada H, Vanhaelen M, Homès J (1996) *Taxus sp.* cell, tissue and organ cultures as alternative sources for taxoids production: a literature survey. Plant Cell Tissue Organ Cult 46:59–75
30. Gibson D, Ketchum R, Vance N, Christen A (1993) Initiation and growth of cell lines of *Taxus brevifolia* (Pacific yew). Plant Cell Rep 12:479–482
31. Rohr R (1973) Production of callus by the male gametophytes of *Taxus baccata L.* cultured on artificial medium. Study with light and electron microscopes. Caryologia 25:177–189
32. Roberts SC (2007) Production and engineering of terpenoids in plant cell culture. Nat Chem Biol 3:387–395
33. Kim BJ, Gibson DM, Shuler ML (2004) Effect of subculture and elicitation on instability of taxol production in *Taxus sp.* suspension cultures. Biotechnol Prog 20:1666–1673
34. Ketchum REB, Gibson DM (1996) Paclitaxel production in suspension cell cultures of Taxus. Plant Cell Tissue Organ Cult 46:9–16
35. Fett-Neto AG, Melanson SJ, Nicholson SA, Pennington JJ, DiCosmo F (1994) Improved taxol yield by aromatic carboxylic acid and amino acid feeding to cell cultures of *Taxus cuspidata*. Biotechnol Bioeng 44:967–971
36. Kwon IC, Yoo YJ, Lee JH, Hyun JO (1998) Enhancement of taxol production by in situ recovery of product. Process Biochem 33:701–707
37. Xu JF, Yin PQ, Wei XG, Su ZG (1998) Self-immobilized aggregate culture of *Taxus cuspidata* for improved taxol production. Biotechnol Tech 12:241–244
38. Bentebibel S, Moyano E, Palazon J, Cusido RM, Bonfill M, Eibl R, Pinol MT (2005) Effects of immobilization by entrapment in alginate and scale-up on paclitaxel and baccatin III production in cell suspension cultures of *Taxus baccata*. Biotechnol Bioeng 89:647–655

39. Tabata H (2004) Paclitaxel production by plant-cell-culture technology. Biomanufacturing 87:1–23
40. Yukimune Y, Tabata H, Higashi Y, Hara Y (1996) Methyl jasmonate-induced overproduction of paclitaxel and baccatin III in Taxus cell suspension cultures. Nat Biotechnol 14:1129–1132
41. Li YC, Tao WY, Cheng L (2009) Paclitaxel production using co-culture of Taxus suspension cells and paclitaxel-producing endophytic fungi in a co-bioreactor. Appl Microbiol Biotechnol 83:233–239
42. Bringi V, Kadkade PG, Prince CL, Roach BL (2007) Enhanced production of taxol and taxanes by cell cultures of Taxus species. C12N5/02
43. Zhang CH, Mei XG, Liu L, Yu LJ (2000) Enhanced paclitaxel production induced by the combination of elicitors in cell suspension cultures of *Taxus chinensis*. Biotechnol Lett 22:1561–1564
44. Wu J, Lin L (2003) Enhancement of taxol production and release in *Taxus chinensis* cell cultures by ultrasound, methyl jasmonate and in situ solvent extraction. Appl Microbiol Biotechnol 62:151–155
45. Fett-Neto AG, Melanson SJ, Sakata K, DiCosmo F (1993) Improved growth and taxol yield in developing calli of *Taxus cuspidata* by medium composition modification. Nat Biotechnol 11:731–734
46. Huang T-K, McDonald KA (2009) Bioreactor engineering for recombinant protein production in plant cell suspension cultures. Biochem Eng J 45:168–184
47. Kolewe ME, Gaurav V, Roberts SC (2008) Pharmaceutically active natural product synthesis and supply via plant cell culture technology. Mol Pharm 5:243–256
48. Hefner J, Rubenstein SM, Ketchum RE, Gibson DM, Williams RM, Croteau R (1996) Cytochrome P450-catalyzed hydroxylation of taxa-4(5),11(12)-diene to taxa-4(20),11(12)-dien-5alpha-ol: the first oxygenation step in taxol biosynthesis. Chem Biol 3:479–489
49. Hefner J, Ketchum RE, Croteau R (1998) Cloning and functional expression of a cDNA encoding geranylgeranyl diphosphate synthase from *Taxus canadensis* and assessment of the role of this prenyltransferase in cells induced for taxol production. Arch Biochem Biophys 360:62–74
50. Koepp AE, Hezari M, Zajicek J, Vogel BS, LaFever RE, Lewis NG, Croteau R (1995) Cyclization of geranylgeranyl diphosphate to taxa-4(5),11(12)-diene is the committed step of taxol biosynthesis in Pacific yew. J Biol Chem 270:8686–8690
51. Hezari M, Lewis NG, Croteau R (1995) Purification and characterization of taxa-4(5),11(12)-diene synthase from Pacific yew (*Taxus brevifolia*) that catalyzes the first committed step of taxol biosynthesis. Arch Biochem Biophys 322:437–444
52. Guerra-Bubb J, Croteau R, Williams RM (2012) The early stages of taxol biosynthesis: an interim report on the synthesis and identification of early pathway metabolites. Nat Prod Rep 29:683–696
53. Koksal M, Jin Y, Coates RM, Croteau R, Christianson DW (2011) Taxadiene synthase structure and evolution of modular architecture in terpene biosynthesis. Nature 469:116–120
54. Kuzuyama T (2002) Mevalonate and nonmevalonate pathways for the biosynthesis of isoprene units. Biosci Biotechnol Biochem 66:1619–1627
55. Kuzuyama T, Seto H (2003) Diversity of the biosynthesis of the isoprene units. Nat Prod Rep 20:171–183
56. Miziorko HM (2011) Enzymes of the mevalonate pathway of isoprenoid biosynthesis. Arch Biochem Biophys 505:131–143
57. Martin VJ, Pitera DJ, Withers ST, Newman JD, Keasling JD (2003) Engineering a mevalonate pathway in *Escherichia coli* for production of terpenoids. Nat Biotechnol 21:796–802
58. Dahl RH, Zhang F, Alonso-Gutierrez J, Baidoo E, Batth TS, Redding-Johanson AM, Petzold CJ, Mukhopadhyay A, Lee TS, Adams PD et al (2013) Engineering dynamic pathway regulation using stress-response promoters. Nat Biotechnol 31:1039–1046

59. Jennewein S, Wildung MR, Chau M, Walker K, Croteau R (2004) Random sequencing of an induced Taxus cell cDNA library for identification of clones involved in Taxol biosynthesis. Proc Nat Acad Sci USA 101:9149–9154
60. Jung ST, Lauchli R, Arnold FH (2011) Cytochrome P450: taming a wild type enzyme. Curr Opin Biotechnol 22:809–817
61. Ajikumar PK, Xiao WH, Tyo KEJ, Wang Y, Simeon F, Leonard E, Mucha O, Phon TH, Pfeifer B, Stephanopoulos G (2010) Isoprenoid Pathway Optimization for Taxol Precursor Overproduction in *Escherichia coli*. Science 330:70–74
62. Huang Q, Roessner CA, Croteau R, Scott AI (2001) Engineering *Escherichia coli* for the synthesis of taxadiene, a key intermediate in the biosynthesis of taxol. Bioorg Med Chem 9:2237–2242
63. Ma SM, Garcia DE, Redding-Johanson AM, Friedland GD, Chan R, Batth TS, Haliburton JR, Chivian D, Keasling JD, Petzold CJ et al (2011) Optimization of a heterologous mevalonate pathway through the use of variant HMG-CoA reductases. Metab Eng 13:588–597
64. Zhang C, Chen X, Zou R, Zhou K, Stephanopoulos G, Too HP (2013) Combining genotype improvement and statistical media optimization for isoprenoid production in *E. coli*. Plos One 8:10
65. Yuan LZ, Rouviere PE, Larossa RA, Suh W (2006) Chromosomal promoter replacement of the isoprenoid pathway for enhancing carotenoid production in *E. coli*. Metab Eng 8:79–90
66. Misawa N, Nakagawa M, Kobayashi K, Yamano S, Izawa Y, Nakamura K, Harashima K (1990) Elucidation of the Erwinia uredovora carotenoid biosynthetic pathway by functional analysis of gene products expressed in *Escherichia coli*. J Bacteriol 172:6704–6712
67. Harker M, Bramley PM (1999) Expression of prokaryotic 1-deoxy-D-xylulose-5-phosphatases in *Escherichia coli* increases carotenoid and ubiquinone biosynthesis. Febs Lett 448:115–119
68. Wang C-W, Oh M-K, Liao JC (1999) Engineered isoprenoid pathway enhances astaxanthin production. Biotechnol Bioeng 62:235–241
69. Huang KX, Huang QL, Wildung MR, Croteau R, Scott AI (1998) Overproduction, in *Escherichia coli*, of soluble taxadiene synthase, a key enzyme in the Taxol biosynthetic pathway. Protein Expr Purif 13:90–96
70. Jiang M, Stephanopoulos G, Pfeifer BA (2012) Toward biosynthetic design and implementation of *Escherichia coli*-derived paclitaxel and other heterologous polyisoprene compounds. Appl Environ Microbiol 78:2497–2504
71. Alper H, Miyaoku K, Stephanopoulos G (2005) Construction of lycopene-overproducing *E. coli* strains by combining systematic and combinatorial gene knockout targets. Nat Biotech 23:612–616
72. Newman JD, Marshall J, Chang M, Nowroozi F, Paradise E, Pitera D, Newman KL, Keasling JD (2006) High-level production of amorpha-4,11-diene in a two-phase partitioning bioreactor of metabolically engineered *Escherichia coli*. Biotechnol Bioeng 95:684–691
73. Boghigian BA, Salas D, Ajikumar PK, Stephanopoulos G, Pfeifer BA (2012) Analysis of heterologous taxadiene production in K- and B-derived *Escherichia coli*. Appl Microbiol Biotechnol 93:1651–1661
74. Boghigian BA, Armando J, Salas D, Pfeifer BA (2012) Computational identification of gene over-expression targets for metabolic engineering of taxadiene production. Appl Microbiol Biotechnol 93:2063–2073
75. Jennewein S, Long RM, Williams RM, Croteau R (2004) Cytochrome P450 taxadiene 5 alpha-hydroxylase, a mechanistically unusual monooxygenase catalyzing the first oxygenation step of taxol biosynthesis. Chem Biol 11:379–387
76. Jardine O, Gough J, Chothia C, Teichmann SA (2002) Comparison of the small molecule metabolic enzymes of *Escherichia coli* and *Saccharomyces cerevisiae*. Genome Res 12:916–929
77. Gruchattka E, Hadicke O, Klamt S, Schutz V, Kayser O (2013) In silico profiling of *Escherichia coli* and *Saccharomyces cerevisiae* as terpenoid factories. Microb Cell Fact 12:1475–2859

78. Engels B, Dahm P, Jennewein S (2008) Metabolic engineering of taxadiene biosynthesis in yeast as a first step towards Taxol (Paclitaxel) production. Metab Eng 10:201–206
79. Preuss D, Mulholland J, Kaiser CA, Orlean P, Albright C, Rose MD, Robbins PW, Botstein D (1991) Structure of the yeast endoplasmic reticulum: localization of ER proteins using immunofluorescence and immunoelectron microscopy. Yeast 7:891–911
80. Paddon CJ, Keasling JD (2014) Semi-synthetic artemisinin: a model for the use of synthetic biology in pharmaceutical development. Nat Rev Microbiol 12:355–367
81. Carlsen S, Ajikumar PK, Formenti LR, Zhou K, Phon TH, Nielsen ML, Lantz AE, Kielland-Brandt MC, Stephanopoulos G (2013) Heterologous expression and characterization of bacterial 2-C-methyl-D-erythritol-4-phosphate pathway in *Saccharomyces cerevisiae*. Appl Microbiol Biotechnol 97:5753–5769
82. Jennewein S, Park H, DeJong JM, Long RM, Bollon AP, Croteau RB (2005) Coexpression in yeast of Taxus cytochrome P450 reductase with cytochrome P450 oxygenases involved in Taxol biosynthesis. Biotechnol Bioeng 89:588–598
83. Stierle A, Strobel G, Stierle D (1993) Taxol and taxane production by *Taxomyces andreanae*, an endophytic fungus of Pacific yew. Science 260:214–216
84. Liu K, Ding X, Deng B, Chen W (2009) Isolation and characterization of endophytic taxol-producing fungi from *Taxus chinensis*. J Ind Microbiol Biotechnol 36:1171–1177
85. Heinig U, Scholz S, Jennewein S (2013) Getting to the bottom of Taxol biosynthesis by fungi. Fungal Diversity 60:161–170
86. Wang C, Wu J, Mei X (2001) Enhancement of taxol production and excretion in *Taxus chinensis* cell culture by fungal elicitation and medium renewal. Appl Microbiol Biotechnol 55:404–410
87. Zhao K, Ping W, Li Q, Hao S, Zhao L, Gao T, Zhou D (2009) Aspergillus niger var. taxi, a new species variant of taxol-producing fungus isolated from Taxus cuspidata in China. J Appl Microbiol 107:1202–1207
88. Ketchum RE, Wherland L, Croteau RB (2007) Stable transformation and long-term maintenance of transgenic Taxus cell suspension cultures. Plant Cell Rep 26:1025–1033
89. Vongpaseuth K, Nims E, St Amand M, Walker EL, Roberts SC (2007) Development of a particle bombardment-mediated transient transformation system for *Taxus spp.* cells in culture. Biotechnol Prog 23:1180–1185
90. Zhang P, Li S-T, Liu T-T, Fu C-H, Zhou P-P, Zhao C-F, Yu L-J (2011) Overexpression of a 10-deacetylbaccatin III-10 β-O-acetyltransferase gene leads to increased taxol yield in cells of *Taxus chinensis*. Plant Cell. Tissue Organ Cult (PCTOC) 106:63–70
91. Li Y, Pfeifer BA (2014) Heterologous production of plant-derived isoprenoid products in microbes and the application of metabolic engineering and synthetic biology. Curr Opin Plant Biol 19C:8–13
92. Vogel C, Marcotte EM (2012) Insights into the regulation of protein abundance from proteomic and transcriptomic analyses. Nat Rev Genet 13:227–232
93. Golomb M, Chamberlin M (1974) Characterization of T7-specific ribonucleic acid polymerase. IV. Resolution of the major in vitro transcripts by gel electrophoresis. J Biol Chem 249:2858–2863
94. McAllister WT, Morris C, Rosenberg AH, Studier FW (1981) Utilization of bacteriophage T7 late promoters in recombinant plasmids during infection. J Mol Biol 153:527–544
95. Khalil AS, Collins JJ (2010) Synthetic biology: applications come of age. Nat Rev Genet 11:367–379
96. Terpe K (2006) Overview of bacterial expression systems for heterologous protein production: from molecular and biochemical fundamentals to commercial systems. Appl Microbiol Biotechnol 72:211–222
97. Schlegel S, Rujas E, Ytterberg AJ, Zubarev R, Luirink J, de Gier J-W (2013) Optimizing heterologous protein production in the periplasm of *E. coli* by regulating gene expression levels. Microbial Cell Fact 12:24
98. Filipowicz W, Bhattacharyya SN, Sonenberg N (2008) Mechanisms of post-transcriptional regulation by microRNAs: are the answers in sight? Nat Rev Genet 9:102–114

99. Smolke CD, Martin VJ, Keasling JD (2001) Controlling the metabolic flux through the carotenoid pathway using directed mRNA processing and stabilization. Metab Eng 3:313–321
100. Nishizaki T, Tsuge K, Itaya M, Doi N, Yanagawa H (2007) Metabolic engineering of carotenoid biosynthesis in Escherichia coli by ordered gene assembly in Bacillus subtilis. Appl Environ Microbiol 73:1355–1361
101. Xu P, Vansiri A, Bhan N, Koffas MA (2012) ePathBrick: a synthetic biology platform for engineering metabolic pathways in *E. coli*. ACS Synth Biol 1:256–266
102. Yu D, Ellis HM, Lee E-C, Jenkins NA, Copeland NG (2000) An efficient recombination system for chromosome engineering in *Escherichia coli*. Proc Nat Acad Sci 97:5978–5983
103. Tyson GW, Chapman J, Hugenholtz P, Allen EE, Ram RJ, Richardson PM, Solovyev VV, Rubin EM, Rokhsar DS, Banfield JF (2004) Community structure and metabolism through reconstruction of microbial genomes from the environment. Nature 428:37–43
104. Wingler LM, Cornish VW (2011) Reiterative Recombination for the in vivo assembly of libraries of multigene pathways. Proce Nat Acad Sci 108:15135–15140
105. Gibson DG, Young L, Chuang RY, Venter JC, Hutchison CA 3rd, Smith HO (2009) Enzymatic assembly of DNA molecules up to several hundred kilobases. Nat Methods 6:343–345
106. Lee JW, Na D, Park JM, Lee J, Choi S, Lee SY (2012) Systems metabolic engineering of microorganisms for natural and non-natural chemicals. Nat Chem Biol 8:536–546
107. Schofield MM, Sherman DH (2013) Meta-omic characterization of prokaryotic gene clusters for natural product biosynthesis. Curr Opin Biotechnol 24:1151–1158
108. Alper H, Jin Y-S, Moxley J, Stephanopoulos G (2005) Identifying gene targets for the metabolic engineering of lycopene biosynthesis in *Escherichia coli*. Metab Eng 7:155–164
109. Feist AM, Palsson BØ (2008) The growing scope of applications of genome-scale metabolic reconstructions using *Escherichia coli*. Nat Biotechnol 26:659–667
110. Becker SA, Feist AM, Mo ML, Hannum G, Palsson BØ, Herrgard MJ (2007) Quantitative prediction of cellular metabolism with constraint-based models: the COBRA Toolbox. Nat Protoc 2:727–738
111. Boghigian BA, Shi H, Lee K, Pfeifer BA (2010) Utilizing elementary mode analysis, pathway thermodynamics, and a genetic algorithm for metabolic flux determination and optimal metabolic network design. BMC Syst Biol 4:0509–1752
112. Bordbar A, Monk JM, King ZA, Palsson BO (2014) Constraint-based models predict metabolic and associated cellular functions. Nat Rev Genet 15:107–120
113. Zhou X, Wang Z, Jiang K, Wei Y, Lin J, Sun X, Tang K (2007) Screening of taxol-producing endophytic fungi from *Taxus chinensis* var. *mairei*. Appl Biochem Microbiol 43:439–443
114. Santos CNS, Stephanopoulos G (2008) Combinatorial engineering of microbes for optimizing cellular phenotype. Curr Opin Chem Biol 12:168–176
115. Adrio JL, Demain AL (2006) Genetic improvement of processes yielding microbial products. FEMS Microbiol Rev 30:187–214
116. Kardos N, Demain AL (2011) Penicillin: the medicine with the greatest impact on therapeutic outcomes. Appl Microbiol Biotechnol 92:677–687
117. Demain AL, Sanchez S (2009) Microbial drug discovery: 80 years of progress. J Antibiot 62:5–16
118. Demain AL, Elander RP (1999) The beta-lactam antibiotics: past, present, and future. Antonie Van Leeuwenhoek 75:5–19
119. Demain AL (2006) From natural products discovery to commercialization: a success story. J Ind Microbiol Biotechnol 33:486–495
120. Srinivasan V, Pestchanker L, Moser S, Hirasuna TJ, Taticek RA, Shuler ML (1995) Taxol production in bioreactors: kinetics of biomass accumulation, nutrient uptake, and taxol production by cell suspensions of *Taxus baccata*. Biotechnol Bioeng 47:666–676
121. Navia-Osorio A, Garden H, Cusidó RM, Palazón J, Alfermann AW, Piñol MT (2002) Taxol® and baccatin III production in suspension cultures of *Taxus baccata* and *Taxus wallichiana* in an airlift bioreactor. J Plant Physiol 159:97–102

122. Alper H, Miyaoku K, Stephanopoulos G (2006) Characterization of lycopene-overproducing *E. coli* strains in high cell density fermentations. Appl Microbiol Biotechnol 72:968–974
123. Chattopadhyay S, Farkya S, Srivastava AK, Bisaria VS (2002) Bioprocess considerations for production of secondary metabolites by plant cell suspension cultures. Biotechnol Bioprocess Eng 7:138–149
124. Sandén AM, Prytz I, Tubulekas I, Förberg C, Le H, Hektor A, Neubauer P, Pragai Z, Harwood C, Ward A (2003) Limiting factors in Escherichia coli fed-batch production of recombinant proteins. Biotechnol Bioeng 81:158–166
125. Shiloach J, Fass R (2005) Growing *E. coli* to high cell density–a historical perspective on method development. Biotechnol Adv 23:345–357
126. Birol G, Ündey C, Çinar A (2002) A modular simulation package for fed-batch fermentation: penicillin production. Comput Chem Eng 26:1553–1565
127. Yamane Y, Higashida K, Nakashimada Y, Kakizono T, Nishio N (1997) Influence of Oxygen and Glucose on Primary Metabolism and Astaxanthin Production by *Phaffia rhodozyma* in Batch and Fed-Batch Cultures: Kinetic and Stoichiometric Analysis. Appl Environ Microbiol 63:4471–4478
128. Plackett RL, Burman JP (1946) The design of optimum multifactorial experiments. Biometrika:305–325
129. Betts JI, Baganz F (2006) Miniature bioreactors: current practices and future opportunities. Microb Cell Fact 5:21
130. Lye GJ, Ayazi-Shamlou P, Baganz F, Dalby PA, Woodley JM (2003) Accelerated design of bioconversion processes using automated microscale processing techniques. Trends Biotechnol 21:29–37
131. Sajc L, Grubisic D, Vunjak-Novakovic G (2000) Bioreactors for plant engineering: an outlook for further research. Biochem Eng J 4:89–99
132. Chaudhary AG, Rimoldi JM, Kingston DG (1993) Modified taxols. 10. Preparation of 7-deoxytaxol, a highly bioactive taxol derivative, and interconversion of taxol and 7-epi-taxol. J Organ Chem 58:3798–3799
133. Verweij J, Clavel M, Chevalier B (1994) Paclitaxel (TaxolTM) and docetaxel (TaxotereTM): not simply two of a kind. Ann Oncol 5:495–505

Adv Biochem Eng Biotechnol (2015) 148: 427–448
DOI 10.1007/10_2015_308

Published Online: 7 April 2015

Enzymes for Synthetic Biology of Ambroxide-Related Diterpenoid Fragrance Compounds

Philipp Zerbe and Jörg Bohlmann

Abstract Ambrox and related ambroxides are highly priced in the fragrance industry, and valued for their delicate odor and fixative properties. Historically, ambrox was obtained from ambergris, a waxy excretion produced by sperm whales, now an endangered species. Synthetic ambroxides have replaced ambergris in perfume manufacture. Plant labdane diterpenoids can serve as starting material for ambroxide synthesis. Among these, the diterpene alcohol sclareol is the major industrial precursor obtained from cultivated clary sage (*Salvia sclarea*). In plants, a large family of diterpene synthase (diTPS) enzymes controls key reactions in diterpenoid biosynthesis. Advanced metabolite profiling and high-throughput sequencing of fragrant and medicinal plants have accelerated discovery of novel diTPS functions, providing a resource for combinatorial synthetic biology and metabolic engineering approaches. This chapter highlights recent progress on the discovery, characterization, and engineering of plant diTPSs with potential uses in ambroxide production. It features biosynthesis of sclareol, *cis*-abienol, and diterpene resin acids, as sources of genes and enzymes for diterpenoid bioproducts.

Keywords Ambrox · Cytochrome P450 · Diterpenoid · Fragrance · Metabolic engineering · Terpene synthase

P. Zerbe (✉) · J. Bohlmann
Michael Smith Laboratories, University of British Columbia, 301-2185 East Mall, Vancouver, BC V6T 1Z4, Canada
e-mail: pzerbe@ucdavis.edu

J. Bohlmann
e-mail: bohlmann@msl.ubc.ca

P. Zerbe
Department of Plant Biology, University of California, Davis, One Shields Avenue, Davis, CA 95616, USA

Contents

1 General Introduction

Terpenoids are one of the largest and most diverse classes of natural products, comprising tens of thousands of different metabolites that share a common origin from five-carbon (C_5) building blocks [1, 2]. In plants, the chemical diversity of terpenoids is a reflection of their manifold biological functions [3–5]. Although some terpenoids, such as gibberellins or carotenoids, play important roles in general (primary) plant metabolism, the majority of compounds are specialized (secondary) metabolites mediating ecological interactions of plants with the environment. These specialized terpenoids are often confined to individual plant species, genera, or families, and may be formed or accumulated in specialized anatomical structures. For example, trees of the pine family (Pinaceae) produce large quantities of monoterpenoids and diterpene resin acids (DRAs) as a viscous oleoresin, which accumulates in resin blisters and ducts, and forms part of a chemical defense system against herbivores and pathogens [4, 6]. Similarly, many angiosperms (flowering plants) accumulate specialized terpenoids in glandular trichomes, epidermal protuberances on the surface of leaves and other organs [7]. In addition to their tissue-specific occurrence, these biologically active terpenoids may only be formed upon induction by biotic or abiotic stressors [2, 4, 5].

Owing to their different chemophysical properties that range from highly volatile to semi- and nonvolatile compounds, terpenoids and terpenoid-producing plants have a long history of practical uses as flavors (e.g., menthol), fragrances (e.g., linalool), pharmaceuticals (e.g., Taxol, artemisinin), industrial resins and coatings (e.g., DRAs), and more recently as biofuels (e.g., farnesene, bisabolene) [8–12]. The ancient and modern use of volatile terpenoids as fragrances draws on their wide range of scent qualities. Volatile monoterpenoids with applications in the fragrance industry include pinene, myrcene, and limonene, which are significant by-products of the forest and fruit industries and used as large-volume feedstocks for the production of numerous household products, such as soaps, air fresheners, cosmetics, and perfumes [8, 10]. Semi- or nonvolatile diterpenoids, such as the nor-labdane diterpenoid (-)-ambrox (trademarked by Firmenich) and related ambroxides (Fig. 1)

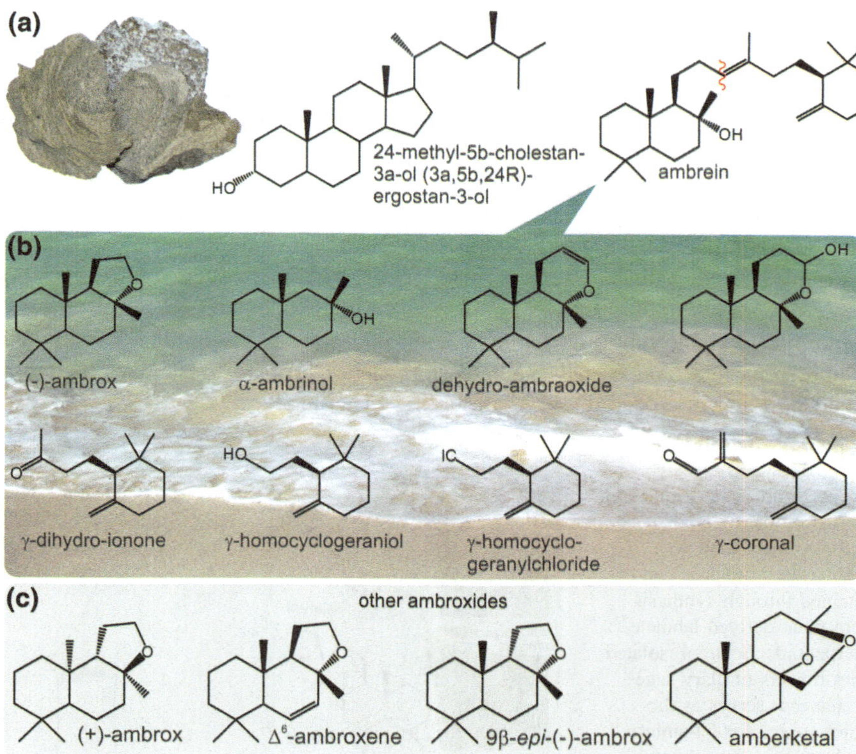

Fig. 1 Ambergris and ambroxide fragrances. Ambergris is an intestinal excretion produced by sperm whales, primarily comprised of cholestanol-type sterols and ambrein (**a**). After regurgitation by the animal, exposure to water and sunlight leads to the oxidation and degradation of ambrein to form (-)-ambrox and related compounds that define the odor characteristics of ambergris (**b**). Today, synthetic ambrox-related compounds produced primarily from plant diterpenoids are used in perfume fragrance production (**c**)

are valued for fragrance manufacture, due to their delicate amber, sweet, and woody odor characteristics and excellent fixative properties to preserve the evanescent top and heart notes of a perfume [13].

Similar to other perfume fixatives, such as musk and civet, (-)-ambrox was historically obtained from an animal product, ambergris or grey amber (Fig. 1). Ambergris is a waxy material formed as a biliary secretion of sperm whales (*Physeter macrocephalus*). Its major constituents are cholestanol-type sterols and the odorless triterpene ambrein. After the material has been regurgitated by the animals, ambrein undergoes oxidative degradation to form (-)-ambrox and numerous other compounds that contribute to the typical ambergris odor [14]. Knowledge of ambergris dates back to its use as ceremonial incense by the ancient Egyptians. Since the tenth century AD, ambergris was harvested on shores or from whale carcasses and traded across the Mediterranean and Europe for its use in perfumery. (-)-Ambrox remains a rare natural product, found only as a component of ambergris, in the essential oil of gum rockrose (*Cistus ladanifer*, Cistaceae), cypress (*Cupressus sempervirens*, Cupressaceae), and clary sage (*S. sclarea*, Lamiaceae), and the absolute of tobacco (*Nicotiana tabacum*, Solanaceae). It was not before the early 1950s that semi-synthetic ambroxide compounds replaced the scarce natural (-)-ambrox [15]. Today, synthetic (-)-ambrox is available under several trade names, including Ambrox® (Firmenich), Cetalox® (racemic mixture, Firmenich), Amberlyn® (Quest), Ambroxan® (Henkel), and Ambrofix® (Givaudan Roure). It is used as the base note in many popular perfumes, such as *Davidoff Cool Water®*, *Chanel No. 5®*, or *Georgi Armani Si®*. The demand for a reliable and economically viable supply of synthetic ambroxides has motivated the development of numerous total and partial synthetic approaches [16]. In most cases, naturally occurring plant diterpenoids serve as starting material (Fig. 2).

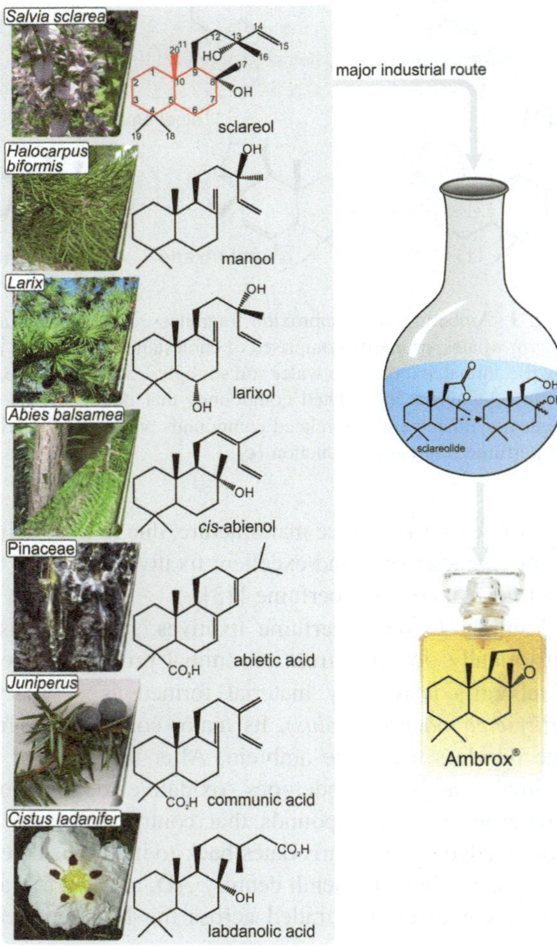

Fig. 2 Ambroxide synthesis from plant diterpenoids. Ambrox® and related ambroxide fragrances can be obtained through synthesis from plant-derived labdane diterpenoids. Sclareol isolated from flowers of clary sage (*S. sclarea*) serves as the major source for (-)-ambrox production in the perfume industry. In addition, a variety of alternative semi-synthetic routes from other labdane diterpene alcohols and acids have been established. For example, *cis*-abienol, labdanolic acid, and communic acid are starting materials for producing (-)-ambrox. Larixol can be converted to Δ^6-ambroxene, and manool is a precursor for the synthesis of amberketal. The *trans*-decalin backbone of labdane diterpenes is highlighted in *red*

Some of these diterpenoids such as DRAs are abundant in nature, however, others can only be obtained in low amounts from wild plant species. Lack of cultivation of many diterpenoid-producing plants and the requirement for protection of some wild species may constrain their use in the fragrance and other industries. Recent advances in transcriptome sequencing and functional genomics, metabolomics, and biochemistry of nonmodel plants have accelerated the discovery of natural product biosynthetic pathways in many previously unexplored species [17–19]. The portfolio of diverse biosynthetic genes and enzymes made accessible through these approaches has, in turn, inspired metabolic engineering and synthetic biology approaches for producing natural plant products, such as terpenoid fragrances, flavors, and pharmaceuticals [11, 17, 18, 20–26]. In this chapter, we highlight recent advances in the discovery, characterization, and metabolic engineering of diterpene biosynthetic pathway genes and enzymes with applications in ambroxide production. On the basis of three examples, sclareol, *cis*-abienol, and DRAs, we discuss new opportunities for combining chemical synthesis with metabolic engineering as alternatives to conventional methods for producing ambroxide fragrances.

2 Chemical Syntheses of Ambroxides

Since the first semi-synthesis of (-)-ambrox from the diterpenoid sclareol in 1950 by Stoll and Hinder [15], several total and partial syntheses toward ambroxide odorants have been established [13, 16]. Kawanobe and coworkers described the total synthesis of (-)-ambrox from β-ionone [27], and several methods for (-)-ambrox production via biomimetic polyene cyclization, for example, from homofarnesol, were described [28]. However, most synthetic routes represent partial syntheses using plant-derived labdane diterpenoids as starting material, because these natural metabolites readily provide the *trans*-decalin scaffold also found in (-)-ambrox [29–34] (Fig. 2). Labdane diterpenoids are widely distributed across the plant kingdom, comprising a large group of several thousand distinct bi-, tri-, or tetracyclic structures, most of which carry extensive functional modifications, including oxygenation, carboxylation, acetylation, methylation, and glycosylation [1, 2, 35].

As the many different synthetic approaches for converting diterpenoids into (-)-ambrox would exceed the scope of this review, only select plant labdane-related diterpenoids with use in ambroxide synthesis are highlighted here. For a detailed review on the underlying chemical procedures, we refer the reader to comprehensive reviews by Frater et al. [13] and Frija et al. [16]. The diterpene-diol sclareol, the primary constituent of clary sage, is currently the major industrial resource for (-)-ambrox production [23] (Fig. 2). Sclareol is isolated from inflorescences either by steam distillation or more efficiently by organic solvent extraction of dried plant material [21, 36]. Several approaches for the conversion of sclareol into (-)-ambrox have been demonstrated [15, 29, 37]. A critical step is the initial oxidative degradation of the carbon-9 side chain of sclareol to form a mixture of sclareolide diterpene lactones (Fig. 2). Further reduction to the respective diols followed by acid-driven cyclization affords (-)-ambrox. Although so far not industrially competitive, a range of other oxygenated labdane diterpenoids have been employed for the synthesis of (-)-ambrox and its equivalents (Fig. 2). Although the individual synthetic approaches differ, most proceed via reductive degradation of the C-9 side chain and subsequent cyclization as key steps. In particular, conifer trees including species of pine (*Pinus*), spruce (*Picea*), and fir (*Abies*) are a rich source of labdane-related diterpene alcohols and acids, which accumulate in copious amounts in the form of oleoresin [4]. These diterpenoids are obtained as large feedstocks by tapping resin from tree stems or as a by-product of the timber and pulp industry [8, 10]. For example, the tertiary diterpene alcohol, *cis*-abienol, is a feasible starting material obtained from the oleoresin of balsam fir (*Abies balsamea*) and a few other coniferous trees [38]. Barrero et al. [29] demonstrated a three-step conversion of *cis*- abienol into (-)-ambrox in high yield. In addition, synthesis of (-)-ambrox and amberketal (Fig. 1), a structural analogue with similar odor characteristics, from manool has been established [13]. Manool occurs naturally in several plant species and is commercially harvested from yellow pine (*Halocarpus biformis*, Podocarpaceae), a conifer tree endemic to New Zealand. Another diterpene alcohol, larixol, abundant in the hardwood of Dahurian larch (*Larix gmelini*) and other *Larix* species, has also been employed for producing ambroxides. Bolster et al. [32] reported the conversion of larixol into Δ^6-ambroxene, an unsaturated (-)-ambrox analogue, called superambrox for its superb odor qualities. Other diterpene acids with successful use in the synthesis of (-)-ambrox include abietic acid and levopimaric acid as major components of many conifer oleoresins, labdanolic acid from leaf exudates of *C. ladanifer*, and communic acid abundant in the wood and berries of *Juniperus sabina* (Cupressaceae) [30, 31, 33, 34].

In addition to chemical synthesis, bioconversion of natural products using bacteria, fungi, or plant cell cultures have been explored for the manufacture of natural or nature-like diterpenoids [16, 39]. Their advantage lies in the possibility to catalyze regio- and stereospecific reactions under milder conditions, and the modification of functional groups that are difficult to access with conventional chemical methods. For example, fungal cell cultures of *Hyphozyma roseoniger* (Ascomycota) and *Cryptococcus albidus* (Basidiomycota) have been used to convert sclareol into sclareolide and the respective diols as an alternative to the chemical degradation of the side chain of sclareol [40]. Moreover, fungal cultures of *Macrophomina phaseolina* (Ascomycota), as well as plant cell cultures of kiwi fruit (*Actinidia deliciosa*, Actinidiaceae) have been applied for the biotransformation of (-)-ambrox into a range of structural analogues with potentially novel or improved odor qualities [41, 42].

3 Diterpene Biosynthesis and Bioengineering of Ambroxide Precursors

3.1 Introduction

Despite the common utility of plant-derived labdane diterpenoids as starting materials for the semi-synthesis of (-)-ambrox, several factors limit their industrial use. Many natural diterpenoids occur in only minor quantities and often as part of complex mixtures *in planta*, requiring laborious and cost-intensive purification procedures for their application in chemical synthesis. Furthermore, access to wild diterpenoid-producing plant species can be limited, and plant cultivation can be affected by unfavorable environmental conditions. Discovery and engineering of diterpenoid-producing genes and enzymes offers an opportunity to improve and expand the availability of starting materials for ambroxide manufacture. Advanced high-throughput metabolite analysis and transcriptome sequencing are enabling the rapid and cost-efficient investigation of many diterpenoid-forming plant species not previously accessible to comprehensive gene discovery [17, 19]. Recent years have witnessed the identification and definite functional annotation of many diterpene synthases (diTPSs) and cytochrome P450 monooxygenases (P450s), two key enzyme classes in diterpenoid biosynthesis [17, 20, 21, 25, 43–49].

Generally, terpenoids share a common biosynthetic origin from two C_5 intermediates, isopentenyl diphosphate (IPP) and dimethylallyl diphosphate (DMAPP). Condensation of these building blocks through the activity of prenyl transferases affords a few prenyl diphosphate intermediates of distinct chain length that serve as central precursors for all terpenoids [50]. Diterpenoids derive from the C_{20} intermediate geranylgeranyl diphosphate (GGPP), which is converted by diTPSs through multistep carbocation-driven (cyclo)isomerization reactions to form various linear or cyclic scaffolds [35]. Functional modification of the diTPS products through the activity of P450s and a few other enzyme classes then affords the tremendous chemical diversity of plant diterpenoids [51].

The diversity of diTPS functions is based, in part, on their modular structure that is comprised of variations of three α-helical domains α, β, and γ [52]. The three major diTPS classes are monofunctional class I diTPS, monofunctional class II diTPSs, and bifunctional class I/II diTPSs [2, 35]. These three classes have different domain architectures and differ in the number of active sites and characteristic functional motifs. Class II diTPSs harbor an N-terminal active site in the βγ-domain accompanied by a common DxDD motif that facilitates the protonation-initiated cycloisomerization of GGPP into a variety of bicyclic prenyl diphosphate intermediates of distinct stereochemistry and regiospecific oxygenation. In contrast, class I diTPSs contain a C-terminal active site in the α-domain along with catalytic DDxxD and NSE/DTE motifs for binding the substrate's diphosphate moiety. Class I diTPSs facilitate the transformation of either GGPP or the transformation of products of class II diTPS. Class I diTPSs catalyze ionization-promoted cleavage of the substrate's diphosphate group, and subsequent rearrangement of resulting carbocations to yield a multitude of diterpene structures. Bifunctional class I/II diTPSs contain both class II and class I active sites and functionalities in one protein. To current knowledge, only mosses, the lycophyte *Selaginella moellendorffii*, and gymnosperms contain bifunctional class I/II diTPSs [53–55]. In angiosperms, the majority of diterpenoids, including all labdane-related compounds, are formed through the sequential activity of pairs of monofunctional class II and class I diTPSs [35]. Repeated gene duplications, followed by sub- and neofunctionalization led to the expansion and functional divergence of the diTPS gene families in the course of evolution. Recent studies in several species suggest that diterpenoid biosynthesis may be organized in the form of modular systems, in which different combinations of class I and class II enzymes can yield different diterpene scaffolds to further increase the chemical space of diterpenoid intermediates and products [17, 25, 44, 45, 56–59].

Information gleaned from the characterization of naturally occurring modular pathways of pairwise acting functionally distinct diTPSs *in planta* can be applied to develop metabolic engineering and synthetic biology strategies based on combinatorial expression of available diTPSs and possibly P450 enzymes [17, 23, 60–62]. Building on the experience gained with established metabolic engineering platforms for high-value terpenoids, such as the anticancer diterpenoid drug Taxol and the sesquiterpenoid artemisinin used in the treatment of malaria [24, 26, 63], new microbial and plant-based platforms for the production of natural and nonnatural diterpene bioproducts are currently being developed in several laboratories [17, 26, 64–66]. In the following sections, we describe three examples of diterpene biosynthesis, sclareol production in clary sage, *cis*-abienol formation in balsam fir, and DRA biosynthesis in the Pinaceae, to highlight recent research on the genetic and enzymatic diversity of diterpenoid metabolism and its application for the fragrance industry.

3.2 Sclareol

Sclareol is presently the most important biologically sourced feedstock for the commercial semi-synthetic production of (-)-ambrox and related ambroxide fragrances for high-end perfume manufacture. In nature, sclareol is found in only a few plant species, including *Cistus creticus* (Cistaceae), *Cleome spinosa* (Brassicaceae), *Nicotiana glutinosa* (Solanaceae), and clary sage (Lamiaceae) [21, 23]. In particular, clary sage accumulates large enough quantities of sclareol that it forms epi-cuticular crystals primarily on the surfaces of flower calyces and bracts [36]. Although sclareol was shown to have antimicrobial properties, its biological function *in planta* is unknown [67]. Owing to its high content of sclareol, clary sage is commercially cultivated in Europe (France, Hungary, Bulgaria), the northern United States, and China. However, annual production levels of sclareol from clary sage can fluctuate as a result of environmental variables, affecting availability and price. Consequently, research on the biosynthesis of sclareol in clary sage and the development of enzymatic production systems as an alternative to harvesting from plant material have attracted much interest in recent years.

In agreement with the modular architecture of diterpenoid biosynthesis in angiosperms, a pair of monofunctional class I and class II diTPSs is involved in the formation of sclareol [21]. Transcriptome analysis of clary sage calyx tissue led to the identification and functional characterization of the two relevant diTPSs, the class II diTPS labda-13-en-8-ol diphosphate synthase (*Ss*LPS) and the class I enzyme sclareol synthase (*Ss*SCS), which together form sclareol from GGPP [21] (Fig. 3). First, *Ss*LPS forms labda-13-en-8-ol diphosphate (LPP) through protonation-initiated cyclization of GGPP and subsequent regiospecific water capture of the intermediate carbocation at position C-8. *Ss*SCS then converts LPP through ionization of the diphosphate ester, followed by rearrangement of the resulting carbocation and secondary cyclization at C-13 to afford sclareol. In addition to sclareol, the coupled reaction of *Ss*LPS and *Ss*SCS yields small amounts of manool (Fig. 2), which may result from conversion of *ent*-copalyl diphosphate by *Ss*SCS. *In planta*, *ent*-copalyl diphosphate may either be recruited from the gibberellin plant hormone biosynthetic pathway or formed as a minor by-product of *Ss*LPS as observed in vitro. Manool formation through a side reaction of *Ss*LPS and *Ss*SCS is consistent with the low metabolite abundance in clary sage, whereas other *Salvia* species, such as *S. oligophylla* and *S. argentea* accumulate manool rather than sclareol [68].

The complete biosynthesis of sclareol and manool from GGPP is achieved by diTPSs and, unlike many other diterpenoid pathways, it does not require a cytochrome P450-dependent activity to introduce the alcohol functionality. Recent studies suggest that the capacity of diTPSs to facilitate oxygenation of the diterpene backbone is more common than was previously appreciated. Functionally orthologous LPS enzymes include the copal-8-ol diphosphate synthase from *C. creticus* and the recently identified LPSs from tobacco (Solanaceae), *Grindelia robusta* (Asteraceae), and *Coleus forskohlii* (Lamiaceae) [17, 43, 45, 69]. This broad taxonomic distribution indicates that the function of LPS enzymes evolved early in the speciation of angiosperms, but possibly after the split of monocotyledonous and dicotyledonous lineages, inasmuch as no LPSs were found among the large diTPS families in members of the grasses (Poaceae) [57, 70]. The *Ss*SCS-catalyzed transformation of LPP into sclareol via a regiospecific hydroxylation at C-13 is, to current knowledge, unique to clary sage. Other known diTPSs that facilitate oxygenation of the hydrocarbon backbone represent bifunctional class I/II enzymes from the nonseed plants *Physcomitrella patens* and *S. moellendorfii*, and the levopimaradiene/abietadiene synthases (LASs) from gymnosperm species of spruce, pine, and fir [20, 53, 54, 56, 71]. LAS enzymes were previously thought to produce nonoxygenated diterpenes from GGPP but were recently shown to form 13-hydroxy-(8,14)-abietene as an initial unstable product. In contrast to these class I/II bifunctional diTPSs, *Ss*SCS belongs to a group of monofunctional class I diTPSs that adopt an atypical βα-domain architecture through loss of the γ-domain, which is more commonly known from mono- and sesqui-TPSs, as well as casbene synthase-like macrocyclases [52]. Known members of this group are restricted to the

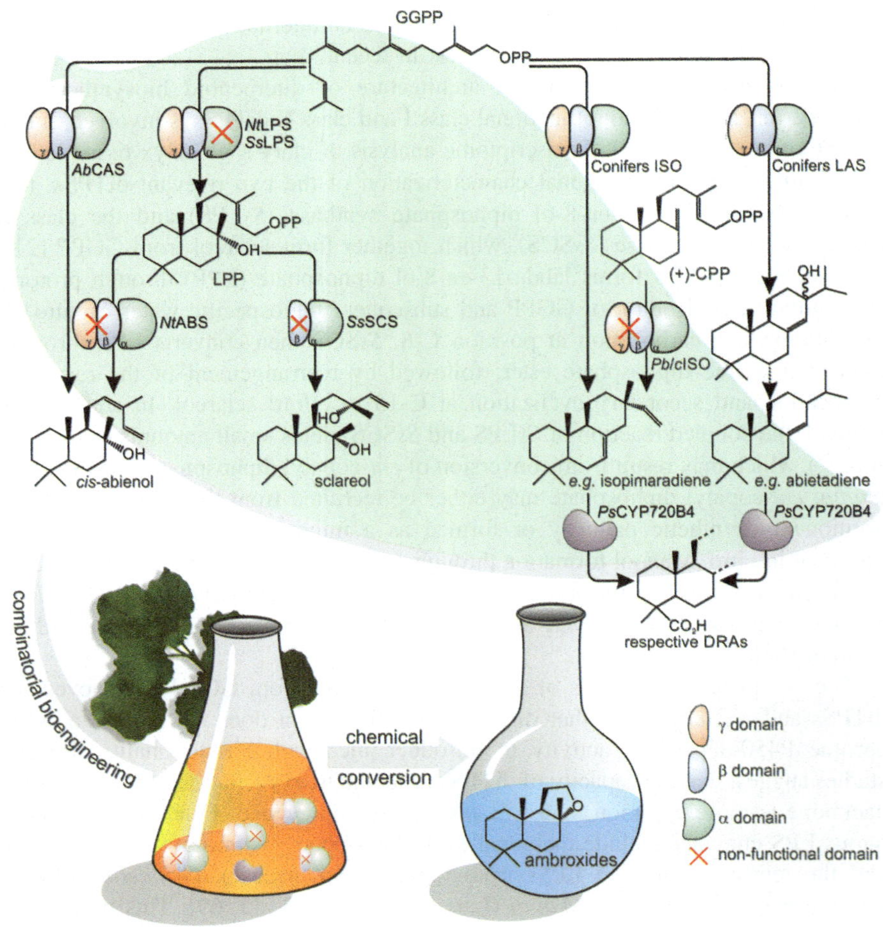

Fig. 3 Modularity of plant diterpenoid biosynthetic pathways and their utility for metabolic engineering of biological systems for ambroxide precursor production. A diverse portfolio of diterpene synthases and cytochrome P450 monooxygenases involved in the formation of plant-derived ambroxide precursors, such as sclareol, *cis*-abienol, and diterpene resin acids, has been discovered in recent years based on transcriptome sequencing and enzyme characterization in nonmodel fragrant plants. Single or combinatorial expression of diterpenoid-biosynthetic enzymes in microbial and plant host systems offers opportunity for metabolic engineering of ambroxide precursors and other diterpenoid bioproducts. These engineered systems enable the production of natural or novel diterpenoids in the form of single target compounds or mixtures of low complexity. Abbreviations: GGPP, geranylgeranyl diphosphate; LPP, labda-13-en-8-ol diphosphate; (+)-CPP, (+)-copalyl diphosphate; *Ab*CAS, *Abies balsamea cis*-abienol synthase; *Nt*LPS, *N. tabacum* LPP synthase; *Nt*ABS, *N. tabacum cis*-abienol synthase; *Ss*LPS, *S. sclarea* LPP synthase; *Ss*SCS, *S. sclarea* sclareol synthase; ISO, isopimaradiene synthase from different conifers; LAS, levopimaradiene/abietadiene synthase from different conifers; *Pb*/*c*ISO, *P. banksiana*/*P. contorta* isopimaradiene; *Ps*CYP720B4, *P. sitchensis* cytochrome P450 720B4; DRA, diterpene resin acid. A *red* X in the schematic depiction of diTPS indicates a nonfunctional active site in monofunctional enzymes

Lamiaceae and wheat (*Triticum aestivum*, Poaceae), with functions in specialized diterpenoid metabolism [21, 25, 45, 72]. Notably, the majority of these bidomain class I diTPSs show substrate promiscuity similar to *Ss*SCS, including manoyl oxide/miltiradiene synthases from *C. forskohlii* and *Rosmarinus officinalis*, and a multifunctional diTPS from *Marrubium vulgare* [21, 25, 44, 45].

As illustrated here for sclareol, the biosynthesis of oxygenated diterpenoids through a modular biosynthetic system of class II and class I diTPSs without requirement for additional P450-catalyzed oxidation steps may have advantages for metabolic engineering of microbial production systems. Recently, Ignea and colleagues demonstrated the successful production of sclareol by combinatorial overexpression of the *C. creticus* LPS with *Ss*SCS in engineered yeast achieving titers of ~400 mg/L [66]. In addition, researchers at Firmenich developed an engineered *E. coli* platform for commercial sclareol production to yield product amounts of up to 1.5 g/L [23]. Here, *Ss*LPS and *Ss*SCS were coexpressed with the bacterial GGPP synthase CrtE and two synthetic operons that encode key enzymes of the mevalonate pathway to improve precursor supply and conversion.

3.3 Cis-*Abienol*

Similar to sclareol, the tertiary diterpene alcohol *cis*-abienol occurs naturally only in a few plant species. For example, *cis*-abienol contributes to the flavor and aroma of cultivars of tobacco, where it accumulates in leaf glandular trichomes [43]. In addition, balsam fir (*A. balsamea*), a coniferous tree native to central and eastern Canada, is a rich source of *cis*-abienol, which accounts for up to 40 % of the aromatic oleoresin accumulating in resin ducts and blisters [20]. Traditionally, balsam fir oleoresin has been traded as "Canada Balsam" for uses as an antifouling agent in naval stores or as glue in the optical industry [38]. As described above, *cis*-abienol is also a suitable starting material for the semi-synthesis of (-)-ambrox [29].

Due to its potential application in the fragrance and flavor industry, details of the biosynthesis of *cis*-abienol *in planta* have been investigated in the last few years. Because biosynthesis of conifer diterpene resin components, such as *cis*-abienol, typically involves bifunctional class I/II diTPSs [55], we had hypothesized that *cis*-abienol formation in balsam fir would likewise be catalyzed by a bifunctional enzyme. To identify the proposed *cis*-abienol synthase (CAS), we generated transcriptome inventories for gene discovery by combining high-throughput transcriptome sequencing with metabolite profiling of balsam fir bark tissue [17, 20]. Querying the transcriptome assemblies against a database of TPS protein sequences resulted in the discovery of three bifunctional class I/II balsam fir diTPSs. Biochemical characterization showed that two candidates, *Ab*LAS and *Ab*ISO, were paralogous to LASs and isopimaradiene synthases (ISO) from other conifers, and the third more divergent enzyme, *Ab*CAS, showed a novel function, directly converting GGPP into *cis*-abienol [20] (Fig. 3). Analogous to angiosperm LPSs, *Ab*CAS catalyzes the cyclohydration of GGPP to form LPP in the enzyme's class II active site. LPP can then freely diffuse to the class I active site, where *cis*-abienol is formed by ionization of the diphosphate ester without further cyclization (Fig. 3). Formation of a bicyclic diterpenol was a newly discovered feature of gymnosperm bifunctional class I/II diTPSs, whereas all previously known conifer class I/II diTPS enzymes are involved in the formation of tricyclic scaffolds in DRA biosynthesis [17, 56, 73–76]. Notably, a recent structural-functional analysis of grand fir (*Abies grandis*) abietadiene synthase (*Ag*AS) demonstrated that substitution of a single aspartate residue in the enzyme's class II active site altered activity from producing (+)-CPP en route to DRAs to formation of LPP [77]. This aspartate is indeed conserved in *Ab*CAS and may have been critical in the functional evolution of *cis*-abienol biosynthesis, presumably by neofunctionalization of a duplicated diTPS of DRA biosynthesis in a balsam fir ancestor [20].

In parallel with the discovery of *Ab*CAS for *cis*-abienol production in balsam fir, Sallaud and coworkers elucidated an alternative *cis*-abienol biosynthetic pathway in tobacco [43]. In contrast to the bifunctional *Ab*CAS that facilitates class II and class I catalysis, tobacco contains a mechanistically analogous pair of two monofunctional diTPSs that act sequentially to form *cis*-abienol (Fig. 3). Much like sclareol biosynthesis, a class II diTPS (*Nt*LPS) catalyzes the formation of LPP, which is then transformed by a class I diTPS (*Nt*ABS) to afford *cis*-abienol. These two different systems of *cis*-abienol formation suggest that diTPSs involved in *cis*-abienol biosynthesis evolved independently in gymnosperm and angiosperm lineages, highlighting an intriguing example of parallel functional specialization of diTPS enzymes. A similar case of parallel evolution of specialized diterpenoid biosynthesis appears to be the biosynthesis of miltiradiene, which is catalyzed by a class I/II bifunctional diTPS in the primitive land plant *S. moellendorfii* and by pairs of monofunctional diTPSs in several Lamiaceae species [25, 44, 45, 78, 79].

A prototype yeast strain for *cis*-abienol production via coexpression of NtABS with a fusion protein of *C. creticus* LPS and a GGPP-producing variant of the yeast FPP synthase Erg20 has recently been described [66]. It remains to be examined if *Ab*CAS can offer improved *cis*-abienol production, as this bifunctional diTPS requires expression and optimization of only a single gene that provides the naturally evolved scaffold of both active sites, allowing for efficient channeling of reaction intermediates.

3.4 Diterpene Resin Acids

The biosynthesis of conifer DRAs has attracted much attention, due to their important roles in the defense of conifer trees against insects and pathogens, and as a renewable resource for industrial bioproducts [8, 10]. Building on pioneering work by Professors Robert Coates, Rodney Croteau, and their colleagues [80–82], much has been learned over the last two decades about the genomic, molecular, and enzymatic features that define DRA formation in conifers. Numerous bifunctional conifer class I/II diTPSs involved in DRA metabolism have been characterized, all of which share common catalytic principles in forming pimarane- or abietane-type compounds through conversion of GGPP into (+)-CPP and subsequent rearrangement of distinct secondary and tertiary carbocations (Fig. 3). During the last few years new insight has been gained on the mechanistic, evolutionary, and physiological underpinnings of conifer DRA biosynthesis, which suggest a more complex biosynthetic system than previously known. For example, detailed functional analysis of Norway spruce (*Picea abies*) LAS illustrated that this class I/II diTPS forms the epimeric diterpenol 13-hydroxy-(8,14)-abietene as the initial product, and not a mixture of diterpene olefins as described earlier [71] (Fig. 3). In vitro, and possibly in vivo, these tricyclic diterpene alcohol epimers dehydrate to form the described LAS products, abietadiene, palustradiene, levopimaradiene, and neoabietadiene. In subsequent studies, this reaction was also confirmed for LAS enzymes of balsam fir, jack pine (*Pinus banksiana*), lodgepole pine (*Pinus contorta*), and golden larch (*Pseudolarix amabilis*) and seems to be a specific feature of conifer LASs, whereas the closely related conifer ISO class I/II diTPS enzymes directly form the diterpene olefin product [17, 20, 56]. Recent work also showed that jack pine and lodgepole pine contain both bifunctional class I/II diTPS as well as, surprisingly, a group of monofunctional class I diTPSs for DRA metabolism [56]. The monofunctional diTPSs in pine lack a functional class II active site and convert (+)-CPP, but not GGPP, into pimaradiene and isopimaradiene, respectively. No monofunctional (+)-CPP synthase has been discovered in a gymnosperm species, however, in vitro assays showed that these diTPSs can "hijack" the freely diffusing (+)-CPP intermediate produced by bifunctional conifer diTPSs.

The abietane and pimarane diTPS products serve as substrates for P450s of the gymnosperm-specific CYP720B family of conifer DRA biosynthesis. Members of this family comprise approximately a dozen different genes in a given conifer species suggesting an ancient evolutionary diversification of the gene family [49, 83]. The two functionally characterized members of the CYP720B family, loblolly pine (*Pinus taeda*) CYP720B1 and Sitka spruce (*Picea sitchensis*) CYP720B4, are multifunctional and multisubstrate enzymes, catalyzing the three-step oxidation of all major diterpene olefins via the corresponding C18-alcohols and aldehydes to afford the respective DRAs (Fig. 3). CYP720B4 catalyzes an array of three different oxidation reactions on each of eight different diterpenes accounting for most major and minor DRA components in spruce. A role of CYP720B4 in conifer defense is underpinned by findings of its inducibility upon treatment with methyl jasmonate and high transcript abundance in epithelial cells of cortical resin ducts, the primary site of terpenoid biosynthesis in conifers [6, 49]. The recent publication of draft assemblies of the very large genomes of three conifer species, white spruce (*Picea glauca*), Norway spruce, and loblolly pine, provides a means to better understand the genomic background of DRA diversity in conifers [84–88].

With the large set of characterized diTPSs of DRA biosynthesis and the emerging characterization of diterpene-converting P450s, bioengineering platforms for the stereospecific production of individual DRAs may advance as an alternative to feedstocks from conifer oleoresin, which are complex mixtures of many different terpenoids and, when sourced as by-products of the forest industry, often contain other lipophilic compounds. Prototype yeast strains have been established that allow the formation of pimarane- and abietane-type DRAs through combinatorial expression of LAS and ISO enzymes with CYP720B4 and the yeast endogenous GGPP synthase [49, 83]. For example, microbially produced abietic acid and levopimaric acid may serve as feasible starting materials for the synthesis of (-)-ambrox [34] (Fig. 2). Other DRA-based biomaterials range from industrial inks and resins to possible biodegradable plastics and rubbers [8, 14, 89].

4 Future Challenges and Opportunities

The above examples illustrate some of the potential of combining genomics-enabled pathway discovery in plants with metabolic engineering and synthetic biology of diterpenoids, specifically for those compounds that may serve as precursors for commercial (-)-ambrox production. It is important to note that both naturally evolved production systems (i.e., plants), and metabolically engineered microbial systems, each have unique advantages and disadvantages. The most obvious advantages of using plants to harvest diterpenoids is the fact that biosynthesis is fueled by sunlight through photosynthesis and operates via optimized metabolic pathways and cell structures, which have evolved over millions of years under natural selection to produce diterpenoids optimized for functions in plant biology.

Some plants that are used for bulk extraction of diterpenoids, such as long-lived conifer trees, require no or only minimal human input for planting or cultivation. A large feedstock of diterpenoid-rich oleoresin is extracted as a by-product from conifers used primarily in the wood or pulp and paper industries. Other diterpenoid-producing plants, such as clary sage, the major source of sclareol, require more intensive crop management, and variation of environmental conditions may substantially affect yield and price of the desired diterpenoid. Finally, some plant systems that produce diterpenoids of interest may not be suitable at the present time for agricultural production, or would be at risk of over-harvesting from their natural environments. Considering all these scenarios, plants appear to remain, for now, the sole major source for all diterpenoids used in commercial ambroxide production.

Alternative production systems, such as engineered microorganisms, in which plant genes for diterpene biosynthesis are heterologously expressed, will have to be competitive in product yield, purity, and production cost. One common limitation inherent to plant production systems is the complexity of their natural product mixtures, especially with regard to mixtures of terpenoids. Plants typically do not produce a single target compound, such as, for example, *cis*-abienol or sclareol, but diverse arrays of metabolites of often similar structures and physicochemical properties, making it difficult to obtain a single compound of interest in pure form. Such arrays of metabolites are likely to have advantages for plants in their natural environments. For example, complex terpenoid defense systems may be harder to breach by pests and pathogens than a defense that is based on a single compound.

For diterpenoids, it is also now known that the suite of metabolites produced in an individual plant, such as the many diterpenoids of conifer oleoresin, are the result of biosynthetic grids with many possible products. Current knowledge suggests that plant diterpenoids are generally not derived from simple linear pathways, but there is growing evidence for dynamic, modular pathway systems, leading to dozens of similar products in a single plant. Considering the genomic (e.g., multigene diTPS and P450 gene families) and biochemical (multiproduct and modular diTPS enzymes, multisubstrate P450 enzymes) multiplicity and promiscuity of plant terpenoid metabolism, it would be difficult to apply traditional selection and plant breeding for the purpose of selectively producing a single target compound. Indeed, this goal may be impossible to achieve unless through use of genetically engineered plants in which existing terpenoid metabolism is redirected or focused to a single major product.

It is here where microbial production systems engineered with individual plant diTPSs or combinations of diTPSs and/or P450s may have the largest advantage. Expression of single enzymes or simple pathways can be optimized for producing one or a few target compounds of interest, thus avoiding or reducing the need for downstream product separation. In addition, engineered microbial systems may offer greater reproducibility, as they are independent of seasonal growth and variable environmental conditions that can affect agricultural or silvicultural production. Engineered microbial systems also have the advantage of (co)expressing optimized diTPS or P450 enzymes, and to explore the vast conceivable space of possible new combinations of diTPS and P450. Following the modular architecture of diterpenoid biosynthesis in nature, combinatorial engineering of diTPSs and P450s of different origin can be utilized for the enantiospecific production of a wide range of natural and nonnatural diterpenoid scaffolds [17, 18, 62].

Issues such as precursor supply, metabolic flux, potential toxicity, accumulation, and transport of diterpenoids in engineered microbial systems can pose limitations in productivity, and will have to be addressed case by case. Host toxicity and often low expression and catalytic efficiency of diTPSs and P450s are some of the major factors constraining diterpenoid production in metabolically engineered microorganisms [26, 90, 91]. However, the succesfull production of sclareol, as well as the sesquiterpenoids farnesene and artemisinin, in *E. coli* and yeast have shown that these types of limitations can be overcome [23, 24, 66, 92]. The scalability of engineered microbial and plant systems continues to improve through optimization of host systems and fermentation processes, as well as enhanced pathway efficiency by establishing a balanced metabolite flow between individual components [23, 26, 47, 61, 65, 90].

As engineered microbial production platforms, and potentially also engineered plant systems, are being advanced for diterpenoids of interest for the fragrance and other industries, plant genomics and plant biochemistry will continue to make important contributions to harness nature's rich resource of genes and enzymes of specialized diterpenoid metabolism [17, 19]. The opportunities and challenges here are inherent to the often high sequence identity and functional divergence of the large diTPS and P450 gene families, which make a priori and ad hoc functional annotations nearly impossible. In vitro and in vivo biochemical characterization of gene candidates remains an absolute requirement for the discovery of new diTPS and P450 enzymes [2, 17, 51]. Recent systems biology approaches focusing on plant specialized metabolism have already expanded the known catalytic landscape of diTPSs and terpenoid-modifying enzymes, such as P450s [17–19, 93], and the vast majority of plant species still remains to be explored.

Acknowledgments We acknowledge the funding support for our research on diterpenoids provided by the University of British Columbia (JB), the Natural Sciences and Engineering Research Council of Canada (JB), Genome Canada (JB), Genome British Columbia (JB), and the University of California at Davis (PZ). Some of the work discussed in this chapter resulted from the Tria Project (www.thetriaproject.ca), the Treenomix Project (www.treenomix.ca), the SMarTForests Project (www.smartforests.ca), and the PhytoMetaSyn Project (www.phytometasyn.ca). JB is a UBC Distinguished University Scholar.

References

1. Davis EM, Croteau R (2000) Cyclization enzymes in the biosynthesis of monoterpenes, sesquiterpenes, and diterpenes. Biosynthesis: Aromatic Polyketides. Isoprenoids, Alkaloids, Springer, pp 53–95
2. Chen F, Tholl D, Bohlmann J, Pichersky E (2011) The family of terpene synthases in plants: a mid-size family of genes for specialized metabolism that is highly diversified throughout the kingdom. Plant J 66:212–229
3. Gershenzon J, Dudareva N (2007) The function of terpene natural products in the natural world. Nat Chem Biol 3:408–414
4. Keeling CI, Bohlmann J (2006) Diterpene resin acids in conifers. Phytochemistry 67:2415–2423
5. Schmelz EA, Huffaker A, Sims JW, Christensen SA, Lu X, Okada K, Peters RJ (2014) Biosynthesis, elicitation and roles of monocot terpenoid phytoalexins. Plant J 79:659–678
6. Zulak KG, Bohlmann J (2010) Terpenoid biosynthesis and specialized vascular cells of conifer defense. J Integr Plant Biol 52:86–97
7. Schilmiller AL, Pichersky E, Last RL (2012) Taming the hydra of specialized metabolism: how systems biology and comparative approaches are revolutionizing plant biochemistry. Curr Opin Plant Biol 15:338–344
8. Bohlmann J, Keeling CI (2008) Terpenoid biomaterials. Plant J 54:656–669
9. Peralta-Yahya PP, Zhang F, del Cardayre SB, Keasling JD (2012) Microbial engineering for the production of advanced biofuels. Nature 488:320–328
10. Zerbe P, Bohlmann J (2014) Metabolic engineering of terpene synthases of conifer defense: bioproducts, biofuels and perfumes. Rec Adv Phytochem 44:85–107
11. Philippe RN, De Mey M, Anderson J, Ajikumar PK (2014) Biotechnological production of natural zero-calorie sweeteners. Curr Opin Biotechnol 26:155–161
12. Lange BM, Mahmoud SS, Wildung MR, Turner GW, Davis EM, Lange I, Baker RC, Boydston RA, Croteau RB (2011) Improving peppermint essential oil yield and composition by metabolic engineering. Proc Natl Acad Sci USA 108:16944–16949
13. Frater G, Bajgrowicz JA, Kraft P (1998) Fragrance chemistry. Tetrahedron 54:7633–7703
14. Sell C (1990) The chemistry of ambergris. Chem Ind 20:516–520
15. Stoll M, Hinder M (1950) Odeur et constitution III les substances bicyclohomofarnésiques. Helv Chim Acta 33:1251–1260

16. Frija LMT, Frade RFM, Afonso CAM (2011) Isolation, chemical, and biotransformation routes of labdane-type diterpenes. Chem Rev 111:4418–4452
17. Zerbe P, Hamberger B, Yuen MM, Chiang A, Sandhu HK, Madilao LL, Nguyen A, Hamberger B, Bach SS, Bohlmann J (2013) Gene discovery of modular diterpene metabolism in non-model systems. Plant Physiol 162:1073–1091
18. Facchini PJ, Bohlmann J, Covello PS, De Luca V, Mahadevan R, Page JE, Ro DK, Sensen CW, Storms R, Martin VJ (2012) Synthetic biosystems for the production of high-value plant metabolites. Trends Biotechnol 30:127–131
19. Higashi Y, Saito K (2013) Network analysis for gene discovery in plant-specialized metabolism. Plant Cell Environ 36:1597–1606
20. Zerbe P, Chiang A, Yuen M, Hamberger B, Hamberger B, Draper JA, Britton R, Bohlmann J (2012) Bifunctional *cis*-abienol synthase from *Abies balsamea* discovered by transcriptome sequencing and its implications for diterpenoid fragrance production. J Biol Chem 287:12121–12131
21. Caniard A, Zerbe P, Legrand S, Cohade A, Valot N, Magnard JL, Bohlmann J, Legendre L (2012) Discovery and functional characterization of two diterpene synthases for sclareol biosynthesis in *Salvia sclarea* (L.) and their relevance for perfume manufacture. BMC Plant Biol 12:119
22. Diaz-Chavez ML, Moniodis J, Madilao LL, Jancsik S, Keeling CI, Barbour EL, Ghisalberti EL, Plummer JA, Jones CG, Bohlmann J (2013) Biosynthesis of sandalwood oil: *Santalum album* CYP76F cytochromes P450 produce santalols and bergamotol. PLoS ONE 8:e75053
23. Schalk M, Pastore L, Mirata MA, Khim S, Schouwey M, Deguerry F, Pineda V, Rocci L, Daviet L (2012) Towards a biosynthetic route to sclareol and amber odorants. J Am Chem Soc 134:18900–18903
24. Paddon CJ, Westfall PJ, Pitera DJ, Benjamin K, Fisher K, McPhee D, Leavell MD, Tai A, Main A, Eng D, Polichuk DR, Teoh KH, Reed DW, Treynor T, Lenihan J, Fleck M, Bajad S, Dang G, Dengrove D, Diola D, Dorin G, Ellens KW, Fickes S, Galazzo J, Gaucher SP, Geistlinger T, Henry R, Hepp M, Horning T, Iqbal T, Jiang H, Kizer L, Lieu B, Melis D, Moss N, Regentin R, Secrest S, Tsuruta H, Vazquez R, Westblade LF, Xu L, Yu M, Zhang Y, Zhao L, Lievense J, Covello PS, Keasling JD, Reiling KK, Renninger NS, Newman JD (2013) High-level semi-synthetic production of the potent antimalarial artemisinin. Nature 496:528–532
25. Zerbe P, Chiang A, Dullat H, O'Neil-Johnson M, Starks C, Hamberger B, Bohlmann J (2014) Diterpene synthases of the biosynthetic system of medicinally active diterpenoids in *Marrubium vulgare*. Plant J 79:914–927
26. Ajikumar PK, Xiao WH, Tyo KE, Wang Y, Simeon F, Leonard E, Mucha O, Phon TH, Pfeifer B, Stephanopoulos G (2010) Isoprenoid pathway optimization for taxol precursor overproduction in *Escherichia coli*. Science 330:70–74
27. Kawanobe T, Kogami K, Matsui M (1986) New syntheses of (±)-ambrox, (±)-ambra oxide and their stereoisomers. Agric Biol Chem 50:1475–1480
28. Snowden RL (2008) Cetalox® and analogues: synthesis via acid-mediated polyene cyclizations. Chem Biodivers 5:958–969
29. Barrero AF, Alvarez-Manzaneda EJ, Altarejos J, Salido S, Ramos JM (1994) Synthesis of Ambrox® from (-)-sclareol and (+)-*cis*-abienol. Tetrahedron 49:10405–10412
30. Barrero AF, Mar Herrador M, Arteaga P, Arteaga JF, Arteaga AF (2012) Communic acids: occurrence, properties and use as chirons for the synthesis of bioactive compounds. Molecules 17:1448–1467
31. Bolster MG, Jansen BJM, de Groot A (2001) The synthesis of (-)-Ambrox® starting from labdanolic acid. Tetrahedron 57:5657–5662
32. Bolster MG, Jansen BJM, de Groot A (2001) The synthesis of Ambrox®-like compounds starting from (+)-larixol. Tetrahedron 57:5663–5679
33. Castro JM, Salido S, Altarejos J, Nogueras M, Sanchez A (2002) Synthesis of Ambrox® from labdanolic acid. Tetrahedron 58:5941–5949
34. Koyama H, Kaku Y, Ohno M (1987) Synthesis of (-)-Ambrox® from L-abietic acid. Tetrahedron Lett 28:2863–2866

35. Peters RJ (2010) Two rings in them all: the labdane-related diterpenoids. Nat Prod Rep 27:1521–1530
36. Caissard J-C, Olivier T, Delbecque C, Palle S, Garry PP, Audran A, Valot N, Moja S, Nicolé F, Magnard JL, Legrand S, Baudino S, Jullien F (2012) Extracellular localization of the diterpene sclareol in clary sage (*Salvia sclarea* L, Lamiaceae). PLoS ONE 7:e48253
37. Barton D, Parekh S, Taylor D, Tse C (1994) An efficient synthesis of (-)-dodecahydro-3a, 6, 6, 9a-tetramethylnaphthol [2,1-B] furan from (-)-sclareol. Tetrahedron Lett 35:5801–5804
38. Gray PS, Mills JS (1964) The isolation of abienol from Canada Balsam, the oleoresin of *Abies balsamea* (L) Mill. J Chem Soc 1:5822–5825
39. De Carvalho CCCR, da Fonseca MMR (2006) Biotransformation of terpenes. Biotechnol Adv 24:134–142
40. Cheetham PSJ (1993) The use of biotransformations for the production of flavours and fragrances. Trends Biotechnol 11:478–488
41. Musharraf SG, Naz S, Najeeb A, Khan S, Choudhary MI (2012) Biotransformation of perfumery terpenoids, (-)-Ambrox® by a fungal culture *Macrophomina phaseolina* and a plant cell suspension culture of *Peganum harmala*. Chem Cent J 6:82
42. Nasib A, Musharraf SG, Hussain S, Khan S, Anjum S, Ali S, Atta-Ur-Rahman AU, Choudhary MI (2006) Biotransformation of (-)-Ambrox® by cell suspension cultures of *Actinidia deliciosa*. J Nat Prod 69:957–959
43. Sallaud C, Giacalone C, Töpfer R, Goepfert S, Bakaher N, Rösti S, Tissier A (2012) Characterization of two genes for the biosynthesis of the labdane diterpene Z-abienol in tobacco (*Nicotiana tabacum*) glandular trichomes. Plant J 72:1–17
44. Brückner K, Božić D, Manzano D, Papaefthimiou D, Pateraki I, Scheler U, Ferrer A, de Vos RC, Kanellis AK, Tissier A (2014) Characterization of two genes for the biosynthesis of abietane-type diterpenes in rosemary (*Rosmarinus officinalis*) glandular trichomes. Phytochemistry 101:52–64
45. Pateraki I, Andersen-Ranberg J, Hamberger B, Heskes AM, Martens HJ, Zerbe P, Bach SS, Møller BL, Bohlmann J, Hamberger B (2014) Manoyl oxide (13*R*), the biosynthetic precursor of forskolin, is synthesized in specialized root cork cells in *Coleus forskohlii*. Plant Physiol 164:1222–1236
46. Gao W, Sun HX, Xiao H, Cui G, Hillwig ML, Jackson A, Wang X, Shen Y, Zhao N, Zhang L, Wang XJ, Peters RJ, Huang L (2014) Combining metabolomics and transcriptomics to characterize tanshinone biosynthesis in *Salvia miltiorrhiza*. BMC Genom 15:73
47. Guo J, Zhou YJ, Hillwig ML, Shen Y, Yang L, Wang Y, Zhang X, Liu W, Peters RJ, Chen X, Zhao ZK, Huang L (2013) CYP76AH1 catalyzes turnover of miltiradiene in tanshinones biosynthesis and enables heterologous production of ferruginol in yeasts. Proc Natl Acad Sci USA 110:12108–12113
48. Zi J, Peters RJ (2013) Characterization of CYP76AH4 clarifies phenolic diterpenoid biosynthesis in the Lamiaceae. Org Biomol Chem 11:7650–7652
49. Hamberger B, Ohnishi T, Hamberger B, Séguin A, Bohlmann J (2011) Evolution of diterpene metabolism: Sitka spruce CYP720B4 catalyzes multiple oxidations in resin acid biosynthesis of conifer defense against insects. Plant Physiol 157:1677–1695
50. Beale MH (1990) The biosynthesis of C5-C20 terpenoid compounds. Nat Prod Rep 7:25–39
51. Nelson D, Werck-Reichhart DA (2011) P450-centric view of plant evolution. Plant J 66:194–211
52. Gao Y, Honzatko RB, Peters RJ (2012) Terpenoid synthase structures: a so far incomplete view of complex catalysis. Nat Prod Rep 29:1153–1175
53. Hayashi K-I, Kawaide H, Notomi M, Sakigi Y, Matsuo A, Nozaki H (2006) Identification and functional analysis of bifunctional *ent*-kaurene synthase from the moss *Physcomitrella patens*. FEBS Lett 580:6175–6181
54. Mafu S, Hillwig ML, Peters RJ (2011) A novel labda-7,13*E*-dien-15-ol-producing bifunctional diterpene synthase from *Selaginella moellendorffii*. Chem Bio Chem 12:1984–1987

55. Keeling CI, Bohlmann J (2006) Genes, enzymes and chemicals of terpenoid diversity in the constitutive and induced defence of conifers against insects and pathogens. New Phytol 170:657–675
56. Hall DE, Zerbe P, Jancsik S, Quesada AL, Dullat H, Madilao LL, Yuen M, Bohlmann J (2012) Evolution of conifer diterpene synthases: Diterpene resin acid biosynthesis in lodgepole pine and jack pine involves monofunctional and bifunctional diterpene synthases. Plant Physiol 161:600–616
57. Xu M, Wilderman PR, Morrone D, Xu J, Roy A, Margis-Pinheiro M, Upadhyaya NM, Coates RM, Peters RJ (2007) Functional characterization of the rice kaurene synthase-like gene family. Phytochemistry 68:312–326
58. Morrone D, Hillwig ML, Mead ME, Lowry L, Fulton DB, Peters RJ (2011) Evident and latent plasticity across the rice diterpene synthase family with potential implications for the evolution of diterpenoid metabolism in the cereals. Biochem J 435:589–595
59. Zhou K, Xu M, Tiernan M, Xie Q, Toyomasu T, Sugawara C, Oku M, Usui M, Mitsuhashi W, Chono M, Chandler PM, Peters RJ (2012) Functional characterization of wheat ent-kaurene(-like) synthases indicates continuing evolution of labdane-related diterpenoid metabolism in the cereals. Phytochemistry 84:47–55
60. Yadav VG, De Mey M, Lim CG, Ajikumar PK, Stephanopoulos G (2012) The future of metabolic engineering and synthetic biology: towards a systematic practice. Metab Eng 14:233–241
61. Zhou YJ, Gao W, Rong Q, Jin G, Chu H, Liu W, Yang W, Zhu Z, Li G, Zhu G, Huang L, Zhao ZK (2012) Modular pathway engineering of diterpenoid synthases and the mevalonic acid pathway for miltiradiene production. J Am Chem Soc 134:3234–3241
62. Ignea C, Ioannou E, Georgantea P, Loupassaki S, Trikka FA, Kanellis AK, Makris AM, Roussis V, Kampranis SC (2015) Reconstructing the chemical diversity of labdane-type diterpene biosynthesis in yeast. Metab Eng 28:91–103
63. Ro D-K, Paradise EM, Ouellet M, Fisher KJ, Newman KL, Ndungu JM, Ho KA, Eachus RA, Ham TS, Kirby J, Chang MC, Withers ST, Shiba Y, Sarpong R, Keasling JD (2006) Production of the antimalarial drug precursor artemisinic acid in engineered yeast. Nature 440:940–943
64. Bach SS, King BC, Zhan X, Simonsen HT, Hamberger B (2014) Heterologous stable expression of terpenoid biosynthetic genes using the moss *Physcomitrella patens*. Methods Mol Biol 1153:257–271
65. Brückner K, Tissier A (2013) High-level diterpene production by transient expression in Nicotiana benthamiana. Plant Methods 9:46
66. Ignea C, Trikka FA, Nikolaidis AK, Georgantea P, Ioannou E, Loupassaki S, Kefalas P, Kanellis AK, Roussis V, Makris AM, Kampranis SC (2015) Efficient diterpene pro duction in yeast by engineering Erg20p into a geranylgeranyl diphosphate synthase. Metab Eng 27:65–75
67. Bailey JA, Carter GA, Burden RS, Wain RL (1975) Control of rust diseases by diterpenes from *Nicotiana glutinosa*. Nature 255:328–329
68. Couladis M, Tzakou O, Stojanovic D, Mimica-Dukic N, Jancic R (2001) The essential oil composition of *Salvia argentea* L. Flavour Fragr J 16:227–229
69. Falara V, Pichersky E, Kanellis AK (2010) A copal-8-ol diphosphate synthase from the angiosperm *Cistus creticus* subsp. *creticus* is a putative key enzyme for the formation of pharmacologically active, oxygen-containing labdane-type diterpenes. Plant Physiol 154:301–310
70. Wu Y, Zhou K, Toyomasu T, Sugawara C, Oku M, Abe S, Usui M, Mitsuhashi W, Chono M, Chandler PM, Peters RJ (2012) Functional characterization of wheat copalyl diphosphate synthases sheds light on the early evolution of labdane-related diterpenoid metabolism in the cereals. Phytochemistry 84:40–46
71. Keeling CI, Madilao LL, Zerbe P, Dullat HK, Bohlmann J (2011) The primary diterpene synthase products of *Picea abies* levopimaradiene/abietadiene synthase (PaLAS) are epimers of a thermally unstable diterpenol. J Biol Chem 286:21145–21153

72. Hillwig ML, Xu M, Toyomasu T, Tiernan MS, Wei G, Cui G, Huang L, Peters RJ (2011) Domain loss has independently occurred multiple times in plant terpene synthase evolution. Plant J 68:1051–1060
73. Peters RJ, Flory JE, Jetter R, Ravn MM, Lee HJ, Coates RM, Croteau RB (2000) Abietadiene synthase from grand fir (*Abies grandis*): characterization and mechanism of action of the 'pseudomature' recombinant enzyme. Biochemistry 39:15592–15602
74. Martin DM, Fäldt J, Bohlmann J (2004) Functional characterization of nine Norway Spruce TPS genes and evolution of gymnosperm terpene synthases of the TPS-d subfamily. Plant Physiol 135:1908–1927
75. Keeling CI, Weisshaar S, Ralph SG, Jancsik S, Hamberger B, Dullat HK, Bohlmann J (2011) Transcriptome mining, functional characterization, and phylogeny of a large terpene synthase gene family in spruce (Picea spp.). BMC Plant Biol 11:43
76. Schepmann HG, Pang J, Matsuda SP (2001) Cloning and characterization of *Ginkgo biloba* levopimaradiene synthase which catalyzes the first committed step in ginkgolide biosynthesis. Arch Biochem Biophys 392:263–269
77. Criswell J, Potter K, Shephard F, Beale MH, Peters RJ (2012) A single residue change leads to a hydroxylated product from the class II diterpene cyclization catalyzed by abietadiene synthase. Org Lett 14:5828–5831
78. Gao W, Hillwig ML, Huang L, Cui G, Wang X, Kong J, Yang B, Peters RJ (2009) A functional genomics approach to tanshinone biosynthesis provides stereochemical insights. Org Lett 11:5170–5173
79. Sugai Y, Ueno Y, Hayashi K, Oogami S, Toyomasu T, Matsumoto S, Natsume M, Nozaki H, Kawaide H (2011) Enzymatic (13)C labeling and multidimensional NMR analysis of miltiradiene synthesized by bifunctional diterpene cyclase in Selaginella moellendorffii. J Biol Chem 286:42840–42847
80. Coates RM (1976) Biogenetic-type rearrangements of terpenes. Fortschr Chem Org Naturst 33:73–230
81. Trapp SC, Croteau RB (2001) Genomic organization of plant terpene synthases and molecular evolutionary implications. Genetics 158:811–832
82. Bohlmann J, Croteau R (1999) Diversity and variability of terpenoid defences in conifers: molecular genetics, biochemistry and evolution of the terpene synthase gene family in grand fir (*Abies grandis*). Novartis Found Symp 223:132–145
83. Ro D-K, Bohlmann J (2006) Diterpene resin acid biosynthesis in loblolly pine (*Pinus taeda*): functional characterization of abietadiene/levopimaradiene synthase (PtTPS-LAS) cDNA and subcellular targeting of PtTPS-LAS and abietadienol/abietadienal oxidase (PtAO, CYP720B1). Phytochemistry 67:1572–1578
84. Birol I, Raymond A, Jackman SD, Pleasance S, Coope R, Taylor GA, Yuen MM, Keeling CI, Brand D, Vandervalk BP, Kirk H, Pandoh P, Moore RA, Zhao Y, Mungall AJ, Jaquish B, Yanchuk A, Ritland C, Boyle B, Bousquet J, Ritland K, Mackay J, Bohlmann J, Jones SJ (2013) Assembling the 20 Gb white spruce (*Picea glauca*) genome from whole-genome shotgun sequencing data. Bioinformatics 29:1492–1497
85. Nystedt B, Street NR, Wetterbom A, Zuccolo A, Lin YC, Scofield DG, Vezzi F, Delhomme N, Giacomello S, Alexeyenko A, Vicedomini R, Sahlin K, Sherwood E, Elfstrand M, Gramzow L, Holmberg K, Hällman J, Keech O, Klasson L, Koriabine M, Kucukoglu M, Käller M, Luthman J, Lysholm F, Niittylä T, Olson A, Rilakovic N, Ritland C, Rosselló JA, Sena J, Svensson T, Talavera-López C, Theißen G, Tuominen H, Vanneste K, Wu ZQ, Zhang B, Zerbe P, Arvestad L, Bhalerao R, Bohlmann J, Bousquet J, Garcia Gil R, Hvidsten TR, de Jong P, MacKay J, Morgante M, Ritland K, Sundberg B, Thompson SL, Van de Peer Y, Andersson B, Nilsson O, Ingvarsson PK, Lundeberg J, Jansson S (2013) The Norway spruce genome sequence and conifer genome evolution. Nature 497:579–584
86. Neale DB, Wegrzyn JL, Stevens KA, Zimin AV, Puiu D, Crepeau MW, Cardeno C, Koriabine M, Holtz-Morris AE, Liechty JD, Martínez-García PJ, Vasquez-Gross HA, Lin BY, Zieve JJ, Dougherty WM, Fuentes-Soriano S, Wu LS, Gilbert D, Marçais G, Roberts M, Holt C, Yandell M, Davis JM, Smith KE, Dean JF, Lorenz WW, Whetten RW, Sederoff R, Wheeler

N, McGuire PE, Main D, Loopstra CA, Mockaitis K, Dejong PJ, Yorke JA, Salzberg SL, Langley CH (2014) Decoding the massive genome of loblolly pine using haploid DNA and novel assembly strategies. Genome Biol 15:R59

87. De La Torre AR, Birol I, Bousquet J, Ingvarsson PK, Jansson S, Jones SJ, Keeling CI, MacKay J, Nilsson O, Ritland K, Street N, Yanchuk A, Zerbe P, Bohlmann J (2014) Insights into conifer giga-genomes. Plant Physiol 166:1724–1732
88. Zimin A, Stevens KA, Crepeau MW, Holtz-Morris A, Koriabine M, Marçais G, Puiu D, Roberts M, Wegrzyn JL, de Jong PJ, Neale DB, Salzberg SL, Yorke JA, Langley CH (2014) Sequencing and assembly of the 22-gb loblolly pine genome. Genetics 196:875–890
89. Hillwig ML, Mann FM, Peters RJ (2011) Diterpenoid biopolymers: new directions for renewable materials engineering. Biopolymers 95:71–76
90. Leonard E, Ajikumar PK, Thayer K, Xiao WH, Mo JD, Tidor B, Stephanopoulos G, Prather KL (2010) Combining metabolic and protein engineering of a terpenoid biosynthetic pathway for overproduction and selectivity control. Proc Natl Acad Sci USA 107:13654–13659
91. Keasling JD (2012) Synthetic biology and the development of tools for metabolic engineering. Metab Eng 14:189–195
92. Buijs NA, Siewers V, Nielsen J (2013) Advanced biofuel production by the yeast *Saccharomyces cerevisiae*. Curr Opin Chem Biol 17:480–488
93. Kitaoka N, Lu X, Yang B, Peters RJ (2014) The application of synthetic biology to elucidation of plant mono-, sesqui- and diterpenoid metabolism. Mol Plant. doi:10.1093/mp/ssu104

Adv Biochem Eng Biotechnol (2015) 148: 449–468
DOI 10.1007/10_2014_277

Published Online: 18 October 2014

Carotenoids of Biotechnological Importance

Gerhard Sandmann

Abstract Carotenoids are natural pigments with antioxidative functions that protect against oxidative stress. They are essential for humans and must be supplied through the diet. Carotenoids are the precursors for the visual pigment rhodopsin, and lutein and zeaxanthin must be accumulated in the yellow eye spot to protect the retina from excess light and ultraviolet damage. There is a global market for carotenoids as food colorants, animal feed, and nutraceuticals. Some carotenoids are chemically synthesized, whereas others are from natural sources. Microbial mass production systems of industrial interest for carotenoids are in use, and new ones are being developed by metabolic pathway engineering of bacteria, fungi, and plants. Several examples will be highlighted in this chapter.

Keywords Carotenoid biosynthesis · Carotenoid market · Carotenoid production · Carotenoid sources · Metabolic pathway engineering

Contents

G. Sandmann (✉)
Biosynthesis Group, Molecular Biosciences, Goethe University Frankfurt,
Max-von-Laue Str. 9, 60438 Frankfurt, Germany
e-mail: sandmann@bio.uni-frankfurt.de

1 Properties and Biosynthesis of Carotenoids

1.1 Carotenoid Structures and Chemical Features

Carotenoids comprise a large group of terpenoid pigments found in pro- and eukaryotic organisms. They are initially synthesised with a chain of 30 or 40 carbon atoms from prenyl pyrophosphates, and they may be extended at a later stage up to 50 carbon atoms [28]. Carotenoids can also be cleaved to shorter metabolites, resulting in apo carotenoids such as crocetin (the red pigment of saffron) and bixin (from annatto seeds) [61]. They are lipophilic compounds and have to be stored in a lipophilic environment. In addition to a small number of acyclic structures, most carotenoids carry an ε- or a β-ionone terminal ring. The latter end group may carry oxo substituents, such as 2-HO, 3-HO, 4-keto, or 5,6-epoxy groups [11]. The most important structural feature of the carotenoid molecule is the conjugated polyene system. These Π-electrons determine the spectral properties with light absorbance in a range from approximately 400–500 nm which is responsible for their yellow via orange to red color and influences the chemical reactivity.

Cis/trans isomers occur at the double bonds of the polyene chain. In general, they are in the trans configuration, but during biosynthesis some can be inserted in a cis configuration. In natural β-carotene apart from all-E, prominent cis isomers are 9-, 13-, and 15-Z (Fig. 1). During handling and storage, trans double bonds can isomerize, especially under illumination. Some carotenoids modified at the ionone rings exist in different optical isomeric forms. For example, astaxanthin with 4-HO- and 4-keto groups at both ionone rings can exist with 3S,3′S, 3R,3′S, and 3R,3′R chirality (Fig. 2). The first enantiomer is present in most organisms that synthesize this carotenoid. The 3R,3′R isomer is the form found in astaxanthin from *Phaffia rhodozyma* (in the sexual state, named *Xanthophyllomyces dendrorhous*) [2]. Astaxanthin exists not only in its free form but also as fatty acid ester, such as in *Haematococcus pluvialis* [40].

1.2 Carotenoid Functions

A universal function of carotenoids is protection at the cellular level against reactive oxygen species. This is highly important in organisms with photosynthesis, where carotenoids are essential. In the photosynthesis apparatus, specific carotenoids are integral components of pigment–protein complexes. In addition to their function in light harvesting, they quench the photosensitizer chlorophyll, preventing the formation of singlet oxygen and inactivate oxygen-derived radicals [36]. This antioxidative function is also exerted in nonphotosynthetic organisms, such as bacteria and fungi with photosensitizers other than chlorophyll. Carotenoids that are especially important in plants include 5,6-epoxy derivatives, which are the precursors for the synthesis of the phytohormone abscisic acid, which is an essential plant hormone regulating leaf

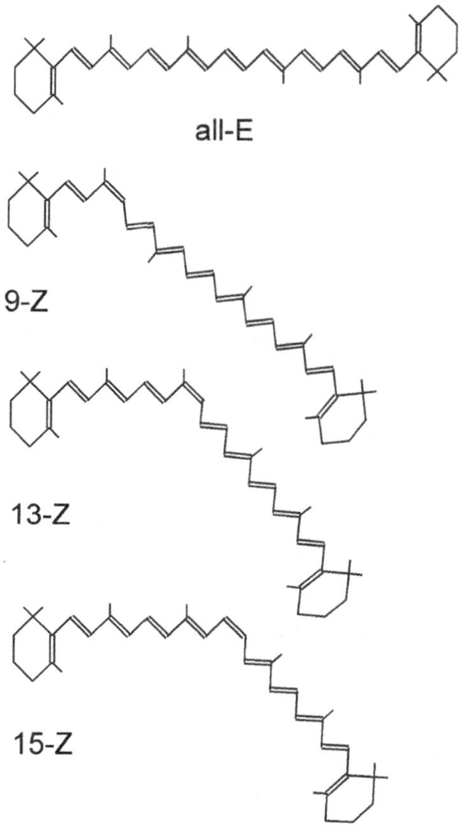

Fig. 1 Prominent geometrical isomers of β-carotene

abscission, seed dormancy, and adaptation to drought stress [71]. In plants, flowers and fruit are pigmented by carotenoids to attract animals for pollination and seed dispersal.

1.3 Biosynthesis

Carotenoids are a branch of terpenoid biosynthesis from precursors. In fungi, these originate from the mevalonate pathway; in bacteria, algae, and plants, they are from the 1-deoxy-D-xylulose 5-phosphate pathway (alternatively named the 2-*C*-methyl-D-erythritol 4-phosphate pathway). The specific pathway exclusive to C40 carotenoids starts from two molecules of geranylgeranyl pyrophosphate, which are condensed to phytoene, the first carotene in the pathway with three conjugated double bonds [58]. In the following desaturation steps, double bonds are inserted adjacent to the existing conjugated ones, alternatively at each side. This integrates an isolated double bond into the conjugated system. In contrast to fungi and non-photosynthetic bacteria, desaturations in algae and plants proceed via poly cis

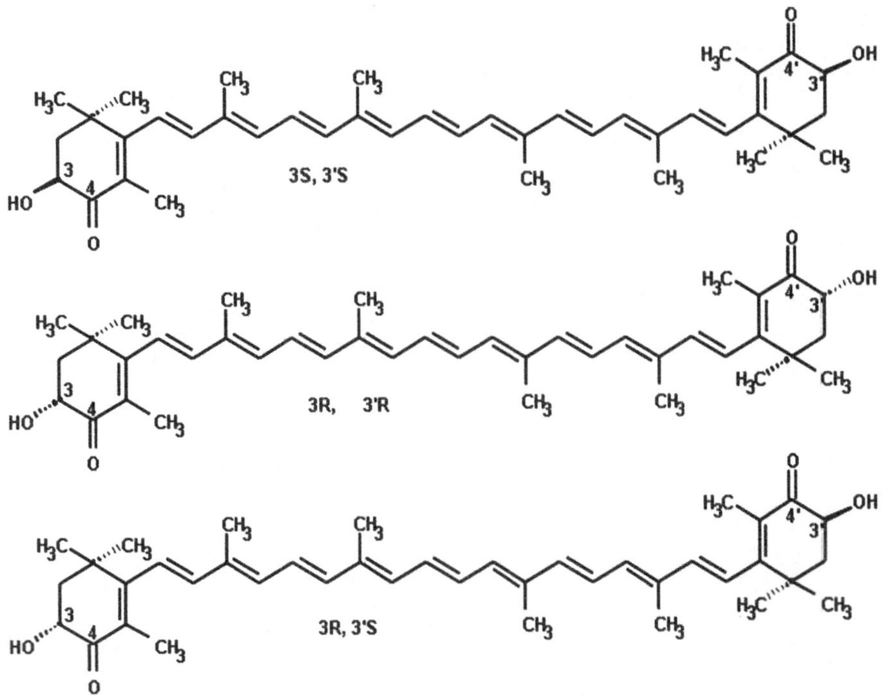

Fig. 2 Stereoisomers of astaxanthin

carotenes, with a subsequent isomerization reaction yielding all-trans lycopene [10]. Most carotenoids are derived from lycopene by cyclization of both ends, forming either two β-ionone rings or one β- and one ε-ionone ring. These end groups are the primary targets for modifications. The most important oxo groups found in carotenoids are 3-hydroxy, 4-keto, and 5,6-epoxy. The pathway of commercially important carotenoids is shown in Fig. 3.

Carotenoids with less than 40 carbon atoms mostly originated from C40 carotenoids by cleavage. However, in a few bacteria, C30 carotenoids are synthesized directly from two molecules of C15 farnesyl pyrophosphate instead of C20 geranylgeranyl pyrophosphate. The subsequent metabolic reactions are similar to those of the C40 pathway [52]. In addition to C40 carotenoids, a few bacteria synthesize C45 and C50 carotenoids from lycopene by adding one or two C5 units to each end of the acyclic molecule [37].

1.4 Importance for Human Health

Humans need carotenoids for their vision. The essential component of rhodopsin is the visual chromophore retinal. Precursors for retinal are carotenoids with an

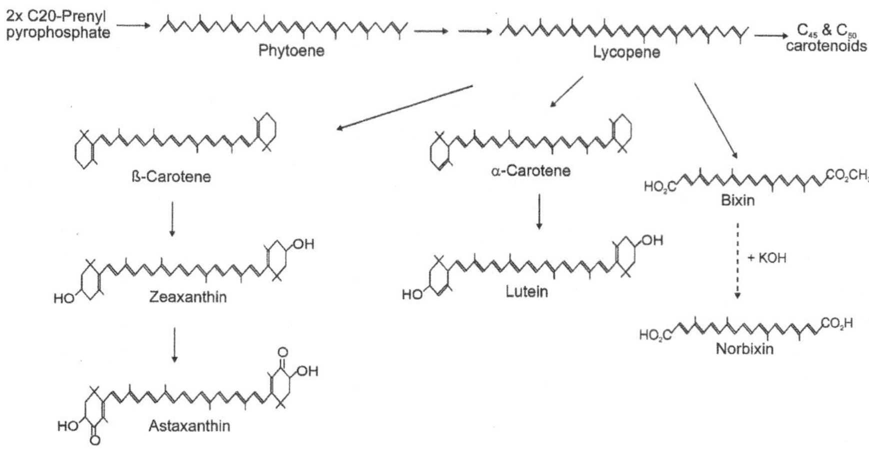

Fig. 3 Structures and biosynthesis of commercially interesting carotenoids. *Dotted arrow* indicates chemical conversion

unsubstituted β-ionone ring, which are all called provitamin A. Because humans are devoid of synthesizing carotenoids, a deficiency of provitamin A carotenoids leads to blindness. Vitamin A deficiency is the leading cause of blindness in children, which is a severe problem in developing countries [7]. It is a big challenge to counteract this problem by supplementation and food fortification.

In addition to carotenoid-related rhodopsin acting as a visual pigment, two carotenoids—lutein and zeaxanthin—are located in the eye in the macula lutea (the yellow eye spot). There, they protect the retina from excess light and ultraviolet (UV) damage, thus preventing cataracts and age-related macular degeneration, which leads to blindness [38, 57]. Like the provitamin A carotenoids, zeaxanthin and lutein have to be provided in the human diet because low levels of intake increase the risk of age-related macular degeneration.

Other health-beneficial effects of carotenoids are attributed to carotenoids. They include protection against chronic diseases, anti-inflammatory actions, and enhancement of the immune response. Carotenoids are also regarded as potential treatment for some types of cancer. Some of these functions may be due to the antioxidative properties of the carotenoids.

2 Sources, Uses, and Production

Carotenoids are essential pigments for all photosynthetic organisms. In addition, they are also found in nonphotosynthetic prokaryotes and eukaryotes or nonphotosynthetic plant tissues. Carotenogenic plant tissues include leaves, roots, flowers, and fruit. In addition to natural carotenoids, several carotenoids are chemically synthesized in bulk quantities.

2.1 The Market for Carotenoids

There is a global market for carotenoids, which was valued at approximately US $1 billion in 2010, with good growth potential [4]. The dominating carotenoid was β-carotene, with a 25 % market share, followed by astaxanthin and lutein with 23 % each. The latter carotenoid is regarded to have the highest growth potential. The carotenoid with the highest market price was astaxanthin, at around $2,000 per kg for the synthetic compound and about $7,000 per kg for natural astaxanthin [64]. However, due to new suppliers for chemically synthesized astaxanthin, its price has recently dropped considerably. The structures of commercially important carotenoids are shown in Fig. 3.

Several carotenoids are approved and widely used as food colorants [44]. As a lipophilic compound, β-carotene provides a yellow pigmentation for food (e.g. butter) or liquids (e.g. soft drinks) after appropriate formulation. For the yellow pigmentation of dairy products, water-soluble norbixin (annatto) is used, which is derived from bixin by alkaline hydrolysis of the methyl ester group [33]. For red pigmentation in food and beverages, canthaxanthin or lycopene are appropriate.

There are two major applications for carotenoids as feed additives. In salmon and trout farming, the addition of either astaxanthin or canthaxanthin is essential to obtain the pink coloration of the flesh. In poultry, carotenoids are responsible for the color of the skin and egg yolk [67]. According to the different market demands for pigmentation, a broad range of synthetic or natural carotenoids are applied to the feed. Natural carotenoids in use are plant extracts or powder, such as those from the marigold as a source for yellow lutein or from red pepper as a source of reddish capsanthin.

The carotenoids astaxanthin and colorless phytoene are now used as ingredients in cosmetic skin products marketed by AstaReal and IBR, respectively. Phytoene, a colorless and relatively light-stable carotenoid, is used in dietary nutricosmetics accumulating in the skin and for topical skin application to protect against oxidative skin damage (http://www.cosmeticsandtoiletries.com/research/techtransfer/38701807.html).

2.2 Production of Carotenoids

For a few carotenoids, including β-carotene, astaxanthin, canthaxanthin, lycopene, and (to a certain extent) zeaxanthin, chemical synthesis is established on an industrial scale (see [18] for description of the synthesis processes). The main producers are BASF and DSM. These chemically synthesized carotenoids dominate the market, although they are also available as natural products to a small extent. For other commercially important carotenoids, such as lutein, only natural carotenoids are available. Considerable attempts have been made to replace chemically synthesized carotenoids at least partially by natural ones from renewable resources.

To date, most of the carotenoids produced by plants and microorganisms cannot compete because they are less cost-efficient to produce. One obstacle is the low yield of biological systems. Nevertheless, some carotenoid products from plant raw materials are on the market, including lutein from the flowers of *Tagetes erecta*, lycopene from tomato fruit, β-carotene from carrot roots and the fruit of the oil palm tree, capsanthin from red pepper, and bixin from *Bixa orellana* seeds.

2.2.1 Plants

The best plant source for β-carotene is red palm oil from the African palm (*Elaeis guineensi*). The oil is produced from the outer mesocarp of the fruit covering the seeds. It contains about 40 mg/g β-carotene and about half this amount of α-carotene together with other minor carotenes [48]. World production of palm oil in 2006 was 14 million tons. The largest producers are Malaysia and Indonesia (http://www.soyatech.com/Palm_Oil_Facts.htm).

The tomato fruit has one of the richest lycopene tissues. In commercial varieties, one can find more than 100 μg/g fresh weight [34]. Natural lycopene from tomato competes with synthetic lycopene. Tomato lycopene is used as a nutraceutical and as red food colorant. The product Tomat-O-Red from LycoRed contains tomato lycopene extracted from oleoresin in a crystalline formulation to deliver a fine dispersion in water.

Lutein is the most abundant plant carotenoid. Marigold (*Tagetes erecta*) flowers are a commercially important lutein source. In the petals, lutein fatty acid esters accumulate to concentrations as high as 3–6 mg/g [51]. Marigold is grown on commercial plantations; oleoresin powder or extracts from the flowers are used as chicken feed to improve the color intensity of the egg yolk [67]. Lutein esters purified from the oleoresin or free lutein are also available as nutraceuticals.

2.2.2 Microbial Mass Production Systems for β-carotene and Astaxanthin of Industrial Interest

Two organisms have been developed for the industrial production of β-carotene: the mucor fungus *Blakesleea trispora* [13] and the alga *Dunaliella salina* (or *D. bardawil*) [47]. *D. salina* is a halophilic unicellular alga that accumulates high levels of β-carotene as a stress response. β-Carotene production depends on high salinity, high temperatures, and high light conditions [53]. Typically, production sites for β-carotene from *D. salina* are located in hot areas with high solar radiation and a cheap salt source nearby. However, optimum salinity for carotenoid production is beyond 27 % NaCl, whereas maximum cell growth is in the range of 18–22 % NaCl [8]. The extreme environment in which *D. salina* grows allows for economically favorable outdoor large-scale cultivation with low risk of contamination. Compared to open ponds, paddlewheel-driven raceway ponds yield substantially higher biomass. The global production of *D. salina* is estimated at about 1,200 tons per year (http://www.oilgae.com/non_fuel_products/betacarotene.html). The expected cellular β-carotene concentrations are in the range of 100 mg/g dry matter, or even higher [5]. Products on the market are *D. salina* powder as a β-carotene-rich supplement for human health and animal feed or extracted β-carotene in vegetable oil in concentrations of 1–20 %.

Alternatively, production processes have been developed with the fungus *B. trispora* by co-cultivation of (+) and (–) sexual mating types simultaneously in large-scale fermenters with plant materials as substrates, such as molasses [13]. Higher yields were obtained with improved mutant strains and with intersexual heterokaryons, which contain nuclei of opposite sex [43]. The typical concentration of β-carotene in fermenter-grown *B. trispora* mycelium is around 5 % [45]. From the cultured material, β-carotene is extracted and processed into a pure beta-carotene crystal. β-Carotene products in the market include CaroPure (DSM), with a purity of >97 %, and Lyc-O-Beta (LycoRed Company). This type of β-carotene material is approved and safe for use in dietary supplements and as food colorant. Because lycopene is the acyclic precursor of β-carotene, the industrial process of β-carotene formation with *B. trispora* can be turned into a lycopene-producing process by the application of lycopene cyclase inhibitors, such as nicotine [41].

Another unicellular algal commercially relevant carotenoid producer is *Haematococcus pluvialis,* which as a chlorophyte is related to *D. salina* [40]. Under light and nutritional stress, it produces astaxanthin during encystment into a dormant state. Most of the astaxanthin is esterified with fatty acids via its 3- and 3'-hydroxy groups. The majority (~70 %) exists as a monoester with 16:0, 18:1, and 18:2 fatty acids, and another 25 % represents diesters (Dore and Cysewski, Cyanotech Corporation at http://www.ruscom.com/cyan/web02/pdfs/naturose/nrtl09.pdf).

The production of astaxanthin containing *H. pluvialis* is a two-step process. In the first phase, cells are grown under conditions for optimal biomass production. In the second phase, *H. pluvialis* is stressed by deprivation of nutritional minerals and/or increased illumination [9]. Concentrations of astaxanthin and esters can reach up to 3 % of cell dry mass. Either bioreactors or outdoor systems are in use. Before the alga can be applied directly as feed, the dried encysted cells have to be cracked by milling to ensure the astaxanthin bioavailability. Some companies extract the astaxanthin derivatives and convert them to free astaxanthin by enzymatic hydrolysis. Products in the market include AstaREAL (Fuji Chemical Industry), AstaPure (Algatechnologies), and BioAstin (Cyanotech) [17].

2.3 Concepts for Improvement of Biological Carotenogenic Systems

Crop plants have been improved by conventional breeding and selection of useful traits. One of the targets is increased carotenoid content in carotenogenic tissue. For crossing and selection for higher pigmentation, the gene pools from elite lines, wild species, and interspecific hybrids can be used [34]. Identification of quantitative trait loci related to carotenogenesis facilitated the selection of the progenies. Successful attempts to breed crops with increased carotenoid content have been reported for tomato [35], potato [12], and cassava [14].

Screening of different varieties of a crop for a certain carotenoid may also be successful. This is the case for orange pepper fruit, which is enriched with zeaxanthin to concentrations of up to 17 μg/g fw [50]. Breeding marigold plants to shift the flower carotenoids from lutein to zeaxanthin has also been achieved.

In microorganisms, the conventional approach to strain improvement is random mutagenesis (chemical or UV-initiated) and selection for the desired phenotype. Although this is easier for heterotrophically growing organisms, the mutagenesis approach was successful for *H. pluvialis*, generating mutants with a 30 % higher astaxanthin production [29]. Great efforts have been made to increase the astaxanthin content in *X. dendrorhous* by classical strain development. Chemical mutagenesis resulted in high-yield astaxanthin mutants of up to 5 mg/g dw from wild-type strains with less than one tenth of this amount [55, 64]). An alternative approach to mutagenesis is the construction of carotenoid-hyperproducing hybrids by protoplast fusion. With this procedure, *X. dendrorhous* strains with 1.6 mg/g dw of total carotenoid were obtained [16]. High-carotene mutants were also obtained with *B. trispora* after chemical mutagenesis [43]. In these mutants, β-carotene production was about 100-fold higher than in the wild-type strain.

Considerable progress has been made in targeted genetic modification of carotenoid biosynthesis by engineering this pathway in plants and microorganisms. This was facilitated by a deeper understanding of the carotenoid biosynthesis pathway and its interaction with other terpenoid pathways and the development of appropriate molecular genetic tools, including the cloning of useful genes from this pathway [61]. Several strategies can be followed by engineering a carotenoid pathway [59]. One is the enforcement of the formation of an already existing carotenoid. For this approach, the limiting step(s) of the biosynthesis pathway have to be recognized and the gene for the corresponding enzyme has to be overexpressed. In carotenogenesis, these limiting enzymes are phytoene synthase [24], catalyzing the step leading to the first carotenoid and 1-deoxy-d-xylulose-5-phosphate synthase, the initial enzyme for the synthesis of prenyl pyrophosphates in bacteria and plant plastidic terpenoid biosynthesis [1, 19]. Other goals are accumulation of an intermediate that is present at low level or the extension of an existing pathway to a novel product. Especially in microorganisms, it is possible to construct entire carotenoid pathways in noncarotenogenic hosts.

2.3.1 Engineering of Carotenoid Composition in Edible Crop Plants

When carotenoids are desired for nutritional supplementation, they can be produced in a suitable organism and added during food processing. However, it is more straightforward to engineer the synthesis in a staple crop for direct consumption. Pioneering work has been carried out with the tomato for higher lycopene content

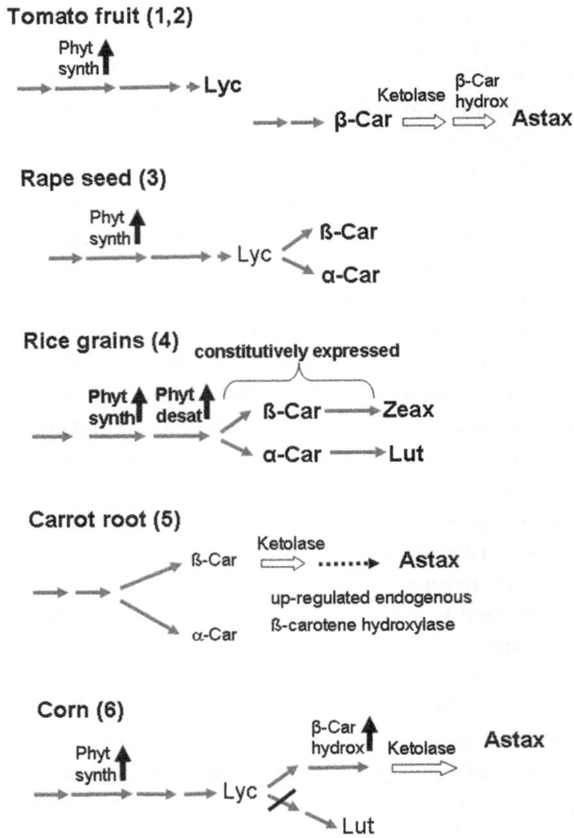

Fig. 4 Genetic engineering of carotenoid biosynthesis in plants. Vertical arrows indicate over-expression of endogenous genes, open arrows indicate pathway extension by expression of foreign genes. References: *1* [24], *2* [31], *3* Shewmaker et al. 1999, *4* [73], *5* [32], *6* [74] and [20]. *Lyc* lycopene; *β-Car* β-carotene; *α-Car* α-carotene; *Lut* lutein; *Zeax* zeaxanthin; *Astax* astaxanthin

(see [22] for a review) and with rape (*Brassica napus*) to increase carotenoid synthesis in seeds [65]. The cloning strategies followed with different plants for targeting the production of different carotenoids are outlined in Fig. 4.

During tomato fruit ripening, lycopene is accumulated due to an upregulation of phytoene synthase, making this enzyme a prominent target for genetic manipulation [24]. The first attempts to overexpress the phytoene synthase gene worked, but they severely affected the synthesis of terpenoid-related phytohormons in the chloroplast, which reduced the transgenic plants in stature. This problem was later avoided by the use of a fruit-specific promoter for the overexpression of the phytoene synthase gene [23]. Normal growing transgenic tomato lines were obtained with doubled fruit lycopene content of up to 5 mg/g dw. In an attempt to convert tomato lycopene into β-carotene, plastome transformation with a plant lycopene β-cyclase resulted in a β-carotene yield of 1 mg/g dw [3].

For the enrichment of β-carotene in rape seed oil, a similar strategy as for tomato was followed. A phytoene synthase gene was overexpressed in rape under a seed-specific promoter (Shewmaker et al. 1999). This led to a 50-fold increase in α-carotene and β-carotene, both with provitamin A activity. The rape seed oil obtained from these transformants showed a β-carotene concentration of 0.1–0.15 %, together with up to 0.1 % α-carotene. Further engineering attempts with tomato have been reviewed [22].

Golden rice has gained considerable attention. Rice grains are colorless and devoid of carotenoids. Nevertheless, a part of the pathway already exists, but the initial steps in the carotenogenesis are absent [63]. This limitation could be overcome by expressing the genes of a phytoene synthase and a phytoene desaturase in rice endosperm [73]. This genetic modification is sufficient to complete the biosynthesis pathway for the synthesis of α- and β-carotene and their hydroxylation products lutein and zeaxanthin (Fig. 4). The next generation of transgenic rice improved by gene optimization reached a maximum 37 mg/g of carotenoids with a high proportion of β-carotene [49].

Only a few plants accumulate zeaxanthin. Typically, it is an intermediate in the synthesis pathway to the epoxy carotenoid violaxanthin. Tuber-specific antisense inactivation or co-suppression of the zeaxanthin epoxidase in potato inhibited violaxanthin formation and resulted in many lines with higher levels of zeaxanthin [56]. The best tubers accumulated zeaxanthin up to 40 μg/g dw, which corresponds to a 130-fold increase of this carotenoid.

Carotenoid pathway modification and extension for improved animal feed has been carried out with maize. A combinatorial transformation technique has been developed for genome insertion of a combination of genes simultaneously. It was successfully employed for engineering of carotenoid biosynthesis. This led to a variety of corn transformants with high content of β-carotene and other yellow carotenoids or pathway extensions ([74]). One goal was the extension of the enhanced endogenous synthesis capacity to the formation of astaxanthin by overexpression of limiting phytoene synthase and β-carotene hydroxylase, the inactivation of the lycopene ε-cylase, and the expression of foreign β-carotene ketolase (Fig. 4). This finally led to an astaxanthin yield of up to 20 μg/g dw in the seeds [21]. In addition, a maize transformant was generated with improved nutritional value, with not only a multifold higher β-carotene content but also with an increased level of the other vitamins ascorbate and folate [46].

In carrot root, the carotenoid pathway ends by accumulation of α- and β-carotene. Engineering of a ketolase gene was sufficient to extend the pathway to astaxanthin, although an additional hydroxylase is necessary [32]. As a response to the expressed ketolase, an endogenous β-carotene hydroxylase was upregulated to complete the biosynthesis pathway from β-carotene to astaxanthin (Fig. 4). In the best resulting plants, the roots contained 90 μg/g fw astaxanthin. Also, tomato has been used to engineer astaxanthin synthesis [31]. Genetic modification was carried out with a yellow variety, which accumulates β-carotene instead of lycopene. Transformation with a ketolase and a hydroxylase gene resulted in the accumulation of astaxanthin (Fig. 4). The astaxanthin content in the transgenic tomato fruits was up to 16 mg/g dw, which is more than 80 % of total carotenoids. This concentration is about 1.8-fold higher than in the engineered carrot and is suitable for the direct use of freeze-dried tomato powder as feed additive for trout and salmon. Table 1 lists the highest yields of major carotenoids obtained by plant transformation.

Table 1 Engineered carotenoid synthesis in plants with the highest carotenoid yields

Plant	Carotenoid	Content (mg/g dw)	Engineering	Reference
Tomato	Lycopene	5	Phytoene synthase	[23]
Tomato	β-Carotene	1	Lycopene β-cyclase	[3]
Tomato	Astaxanthin	16	β-Carotene hydroxylase, β-Carotene ketolase	[31]
Potato	Zeaxanthin	0.04	Inactivation of zeaxanthin epoxidase	[56]

Details on carotenoid genetic pathway engineering of plants and resulting carotenoid yields have been reviewed elsewhere [27, 61].

2.3.2 Genetic Pathway Engineering with Microorganisms

Successful attempts have been made to engineer carotenogenesis in microorganisms, either by modification and enhancement of an existing pathway or by implementation of a carotenoid pathway into a noncarotenogenic host. The latter approach is much more problematic because of metabolic competition for precursors and interference with other well-balanced pathways. In addition, lipophilic sequestering systems must be present for substantial storage of carotenoids.

Bacteria

Escherichia coli was one of the first hosts for heterologous carotenoid synthesis. Initially, this bacterium was used for combinatorial synthesis of various and novel carotenoid structures by combining genes from different organisms with different carotenoid biosynthesis branches [60] to be tested as efficient lipophilic antioxidants. Recently, tremendous progress in the synthesis of carotenoids has been made by integration of carotenogenic genes into the *E. coli* chromosome. Concentrations of 6.2 mg/g dw of β-carotene, 1.4 mg/g dw astaxanthin [39], and approximately 30 mg/g dw of lycopene [15] have been obtained.

Some species of cyanobacteria synthesize zeaxanthin as the end product of their carotenoid biosynthesis. By overexpression of either the phytoene synthase gene or the β-carotene hydroxylase gene, zeaxanthin concentrations of 1.9 mg/g dw were reached [30].

Fungi

The biotechnological potential of fungi for the heterologous production of important carotenoids has already been reviewed [62]. The target carotenoids are mainly

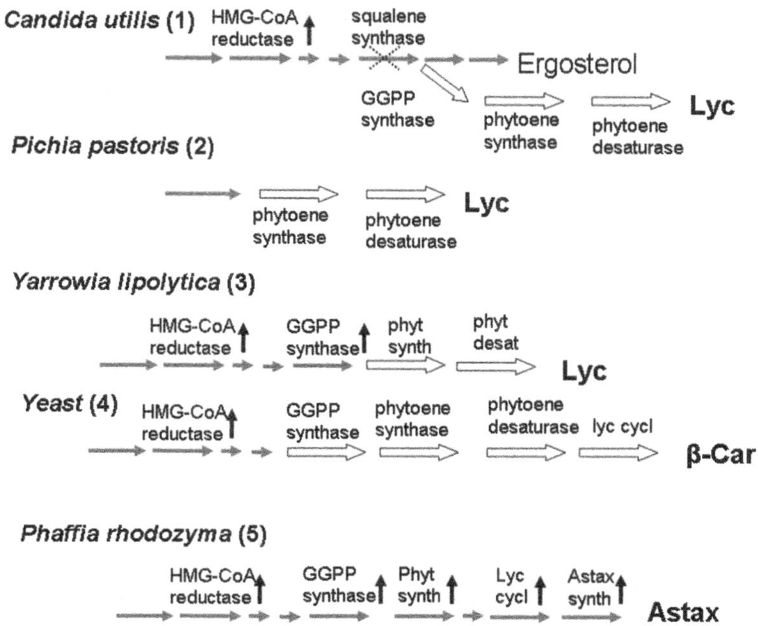

Fig. 5 Genetic engineering of carotenoid biosynthesis in fungi. Vertical arrows indicate over-expression of endogenous genes, open arrows indicate pathway extension by expression of foreign genes. References: *1* [66], *2* [6], *3* [42], *4* [69], *5* [25]. *Lyc* lycopene; *β-Car*, β-carotene; *Astax* astaxanthin

β-carotene and astaxanthin. The first attempts to engineer a carotenoid pathway into a non-carotenogenic fungus were made with yeast for the production of lycopene [72]. However, *Candida utilis* was the first fungus in which the carotenoid pathway was systematically engineered by genetic modification for the high-yield production of the carotenoids lycopene, β-carotene, and astaxanthin [66]. In addition to the genes for the establishment of carotenogenesis, the gene encoding 3-hydroxy-3-methylglutaryl–coenzyme A reductase from *C. utilis* was overexpressed and the gene for squalene synthase was partially inactivated, redirecting parts of ergosterol biosynthesis into carotenoid biosynthesis (Fig. 5). The resulting lycopene concentration of 7.8 mg/g dw was set as a benchmark for all future engineering attempts. Furthermore, some of the engineering strategies developed for *C. utilis* were used in the following years for other fungi. When later the engineering carotenoid pathway in yeast was extended for the synthesis of β-carotene, optimization of precursor supply by simultaneous overexpression of the genes encoding 3-hydroxy-3-methylglutaryl–coenzyme A reductase and geranylgeranyl pyrophosphate synthase (GGPP) (Fig. 5) resulted in a β-carotene yield of 5.9 mg/g dw [69]. This value for β-carotene synthesis in yeast could be further increased to 7.4 mg/g dw by the insertion of controllable promoters for the carotenogenic genes inserted as duplicates with galactose induction and glucose repression of β-carotene synthesis [70].

Table 2 Engineered carotenoid synthesis in microorganisms with the highest carotenoid yields

Strain	Carotenoid	Content (mg/g dw)*	Engineering	Reference
Escherichia coli	β-Carotene	6.2	Whole pathway	[39]
	Astaxanthin	1.4	Whole pathway	[39]
	Lycopene	30	Whole pathway	[15]
Synechococcus	Zeaxanthin	1.9	Hydroxylation	[30]
Candida utilis	Lycopene	7.8	Whole pathway + inactivation of parallel pathway	[66]
Pichia pastoris	Lycopene	4.6	Whole pathway	[6]
Yarrowia lipolytica	Lycopene	1.5	Whole pathway	[42]
Yeast	β-Carotene	7.4	Whole pathway	[70]
Phaffia rhodozyma	Astaxanthin	9	Mutant + pathway optimization	[25]

*In shake-flask cultures

A successful approach to increase β-carotene production in a genetically engineered yeast was by laboratory evolution [54]. Making use of the carotenoid antioxidative potential, the application of hydrogen peroxide as stressor resulted in β-carotene concentrations of 18 mg/g dw.

In *Pichia pastoris* engineered only for the synthesis of lycopene from GGPP without any further optimization of precursor supply, lycopene accumulated to 4.6 mg/g dw [6]. Another noncarotenogenic fungus used recently for carotenoid production is *Yarrowia lipolytica,* which is an oleaginous yeast that has a high storage capacity for lipophilic compounds, such as carotenoids. A strain already modified for high accumulation of lipid bodies was used for transformation with the heterologous genes. In addition to the genes for lycopene synthesis from GGPP, again the genes for 3-hydroxy-3-methylglutaryl–coenzyme A reductase and GGPP synthase were also overexpressed (Fig. 5). In laboratory cultures, synthesis of 1.5 mg/g dw of lycopene was achieved [42]. This accumulation could be increased to 16 mg/g dw under controlled fed-batch fermentation.

The only carotenogenic fungus optimized for carotenoid biosynthesis is the basidiomycetous yeast *P. rhodozyma,* which is one of the rare producers of astaxanthin; it can be genetically improved for carotenogenesis [68]. High-yield astaxanthin strains have been obtained by a combination of classical mutagenesis and genetic pathway engineering [26]. This approach started from chemically-induced mutants with a 15-fold higher astaxanthin content. The genes of limiting carotenogenic reactions were then expressed stepwise, resulting in an additional 6-fold increase and a final astaxanthin concentration of 6 mg/g dw [25].

3 Conclusion

Astaxanthin and β-carotene dominate the global market as feed additives and colorants. More than 90 % of these carotenoids are chemically synthesized. However, lutein is another feed additive that is progressing strongly in the market; it is mainly of plant origin. Other carotenoids, such as lycopene, zeaxanthin, and phytoene, are of minor economic importance.

It is a great challenge to develop biological systems for the production of astaxanthin and β-carotene, in particular, as well as other carotenoids. Although suitable producers exist, they currently serve only a niche market due to their inferior economic competitiveness compared to the synthetic products. Therefore, considerable attempts are being made to increase the carotenoid yields of various organisms. One challenge is the genetic pathway engineering of suitable hosts. This includes optimization of the entire pathway, from precursor supply to the enhancement of limiting carotenogenic reactions. Either noncarotenogenic organisms have been used for the establishment of the entire pathway, or carotenogenic organisms have been used to enforce their pathways or to extend them for novel carotenoids. To date, these engineering approaches have been especially successful with some bacteria and several fungi. Table 2 lists the best carotenoid yields that have been obtained due to genetic modifications in different organisms. These values were obtained with laboratory shake-flask cultures. With controlled fermenter cultivation, a 3- to 10-fold increase has been demonstrated to be possible compared to the shake-flask cultures [26, 42]. Therefore, optimized fermenter cultures of some of the genetically modified strains listed in Table 1 should be considered as the next step in developing cost-competitive bioprocesses.

References

1. Albrecht M, Misawa N, Sandmann G (1999) Metabolic engineering of the terpenoid biosynthetic pathway of *Escherichia coli* for production of the carotenoids ß-carotene and zeaxanthin. Biotechn Lett 21:791–795
2. Andrewes AG, Starr MP (1976) (3R,3'R)-Astaxanthin from the yeast *Phaffia rhodozyma*. Phytochem 15:1009–1011
3. Apel W, Bock R (2009) Enhancement of carotenoid biosynthesis in transplastomic tomatoes by induced lycopene-to-provitamin A conversion. Plant Physiol 151:59–66
4. BCC Research (2011). http://www.bccresearch.com/report/carotenoids-global-market-fod025d.html
5. Ben-Amotz A (1991) The biotechnology of cultivating *Dunaliella* for production of β-carotene rich algae. Bioresource Technology 38:233–235
6. Bhataya A, Schmidt-Dannert C, Lee PC (2009) Metabolic engineering of *Pichia pastoris* X-33 for lycopene production. Process Biochem 44:1095–1102
7. Black RE, Allen LH, Bhutta ZA, Caulfield LE, De Onis M, Ezzati M, Mathers C, Rivera J et al (2008) Maternal and child undernutrition: global and regional exposures and health consequences. Lancet 371:243–260

8. Borowitzka MA, Borowitzka LJ, Moulton TP (1984) The mass culture of *Dunaliella salina* for fine chemicals: from laboratory to pilot plant. Hydrobiologia 116(117):115–134
9. Boussiba S, Vonshak A (1991) Astaxanthin accumulation in the green alga *Haematococcus pluvialis*. Plant Cell Physiol 32:1077–1082
10. Breitenbach J, Sandmann G (2005) ζ-Carotene *cis* isomers as products and substrates in the plant poly-*cis* carotenoid biosynthetic pathway to lycopene. Planta 220:785–793
11. Britton G (1995) Structure and properties of carotenoids in relation to function. FASEB J. 9:1551–1558
12. Brown CR, Kim TS, Ganga Z, Haynes K, DeJong D, Jahn M, Paran I, DeJong W (2006) Segregation of total carotenoid in high level potato germplasm and its relationship to beta-carotene hydroxylase polymorphism. Americ J Potato Res 83:365–372
13. Böhme K, Richter C, Pätz R (2006) New insights into mechanisms of growth and β-carotene production in *Blakeslea trispora*. Biotechnol J 1:1080–1084
14. Ceballos H, Morante N, Sánchez T, Ortiz D, Aragón I, Chávez AL, Pizarro M, Calle F, Dufour D (2013) Rapid cycling recurrent selection for increased carotenoids content in cassava roots. Crop Sci 53:1–10
15. Chen YY, Shen HJ, Cui YY, Chen SG, Weng ZM, Zhao M, Liu JZ (2013) Chromosomal evolution of *Escherichia coli* for the efficient production of lycopene. BMC Biotechnol 13:6
16. Chun SB, Chin JE, Bai S, An GH (1992) Strain improvement of *Phaffia rhodozyma* by protoplast fusion. FEMS Microbiol Lett 93:221–226
17. Dufosse L, Galaup P, Yaron A, Arad SM, Blanc P, Chidambara Murthy KN, Ravishankar GA (2005) Microorganisms and microalgae as sources of pigments for food use: a scientific oddity or an industrial reality? Trends Food Sci Technol 16:389–406
18. Ernst H (2002) Recent advances in industrial carotenoid synthesis. Pure Appl Chem 74:1369–1382
19. Estévez JM, Cantero A, Reindl A, Reichler S, León P (2001) Deoxy-D-xylulose-5-phosphate synthase, a limiting enzyme for plastidic isoprenoid biosynthesis in plants. J Biol Chem 276:22901–22909
20. Farré G, Maiam Rivera S, Alves R, Vilaprinyo E, Sorribas A, Canela R, Naqvi S, Sandmann G, Capell T, Zhu C, Christou P (2013) Targeted transcriptomic and metabolic profiling reveals temporal bottlenecks in the maize carotenoid pathway that can be addressed by multigene engineering. Plant J 75:441–455
21. Farré G, Zhu C, Capell T, Sandmann G, Christou P (2014) Reconstruction and extension of the carotenoid pathway in corn a hybrid with high levels of ketocarotenoids, specially astaxanthin (manuscript in preparation)
22. Fraser PD, Enfissi EM, Bramley PM (2009) Genetic engineering of carotenoid formation in tomato fruit and the potential application of systems and synthetic biology approaches. Arch Biochem Biophys 483:196–204
23. Fraser PD, Römer S, Shipton CA, Mills PB, Kiano JW, Misawa N, Drake RG, Schuch W, Bramley PM (2002) Evaluation of transgenic tomato plants expressing an additional phytoene synthase in a fruit-specific manner. Proc Natl Acad Sci USA 99:1092–1097
24. Fray RG, Wallace A, Fraser PD, Valero D, Hedden P, Bramley PM, Grierson D (1995) Constitutive expression of a fruit phytoene synthase gene in transgenic tomatoes causes dwarfism by redirecting metabolites from the gibberellin pathway. Plant J 8:693–701
25. Gassel S, Breitenbach J, Sandmann G (2014) Genetic engineering of the complete carotenoid pathway towards enhanced astaxanthin formation in *Xanthophyllomyces dendrorhous* starting from a high-yield mutant. Appl Microbiol Biotechnol 98:345–350
26. Gassel S, Schewe H, Schmidt I, Schrader J, Sandmann G (2013) Multiple improvement of astaxanthin biosynthesis in *Xanthophyllomyces dendrorhous* by a combination of conventional mutagenesis and metabolic pathway engineering. Biotechnol Lett 35:565–569
27. Giuliano G, Tavazza R, Diretto G, Beyer P, Taylor MA (2008) Metabolic engineering of carotenoid biosynthesis in plants. Trends Biotechnol 26:139–145
28. Goodwin TW (1980) The biochemistry of the carotenoids, vol I. Plants, Chapman and Hall, London, New York

29. Gómez PI, Inostroza I, Pizarro M, Pérez J (2013) From genetic improvement to commercial-scale mass culture of a Chilean strain of the green microalga *Haematococcus pluvialis* with enhanced productivity of the red ketocarotenoid astaxanthin. AoB Plants. doi:10.1093/aobpla/plt026
30. Götz T, Windhövel U, Böger P, Sandmann G (1999) Protection of photosynthesis against UV-B radiation by carotenoids in transformants of the cyanobacterium *Synechococcus* PCC7942. Plant Physiol 120:599–604
31. Huang JC, Zhong YJ, Liu J, Sandmann G, Chen F (2013) Metabolic engineering of tomato for high-yield production of astaxanthin. Metab Eng 17:59–67
32. Jayaraj J, Devlin R, Punja Z (2008) Metabolic engineering of novel ketocarotenoid production in carrot plants. Transgenic Res 17:489–501
33. Kang EK, Campbell RE, Bastian E, Drake MA (2010) Annatto usage and bleaching in dairy foods. J Dairy Sci 93:3891–3901
34. Kavitha P, Shivashankara KS, Rao VK, Sadashiva AT, Ravishankar KV, Sathish GJ (2013) Genotypic variability for antioxidant and quality parameters among tomato cultivars, hybrids, cherry tomatoes and wild species. J Sci Food Agric 94:993–999
35. Kinkade MP, Foolad MR (2013) Validation and fine mapping of lyc12.1, a QTL for increased tomato fruit lycopene content. Theor Appl Genet 126:2163–2175
36. Krinsky NI (1989) Antioxidant functions of carotenoids. Free Radic Biol Med 7:617–635
37. Krubasik P, Kobayashi M, Sandmann G (2001) Expression and functional analysis of a gene cluster involved in the synthesis of decaprenoxanthin reveals the mechanisms for C_{50} carotenoid formation. Eur J Biochem 268:3702–3708
38. Landrum JT, Bone RA (2001) Lutein, zeaxanthin and the macular pigment. Arch Biochem Biophys 385:28–40
39. Lemuth K, Steuer K, Albermann C (2011) Engineering of a plasmid-free *Escherichia coli* strain for improved in vivo biosynthesis of astaxanthin. Microbial Cell Factories 10:29–40
40. Lorenz RT, Cysewski GR (2000) Commercial potential for *Haematococcus* microalgae as a natural source of astaxanthin. Trends Biotechnol 18:160–167
41. López-Nieto MJ, Costa J, Peiro E, Méndez E, Rodríguez-Sáiz M, de la Fuente JL, Cabri W, Barredo JL (2004) Biotechnological lycopene production by mated fermentation of *Blakeslea trispora*. Appl Microbiol Biotechnol 66:153–159
42. Matthäus F, Ketelhot M, Gatter M, Barth G (2014) Production of lycopene in the non-carotenoid producing yeast *Yarrowia lipolytica*. Appl Environ Microbiol. doi:10.1128/AEM.03167-13
43. Mehta BJ, Obraztsova IN, Cerda-Olmedo E (2003) Mutants and intersexual heterokaryons of *Blakeslea trispora* for production of β-carotene and lycopene. Appl Environ Microbiol 69:4043–4048
44. Mortensen A (2006) Carotenoids and other pigments as natural colorants. Pure Appl Chem 78:1477–1491
45. Nanou K, Roukas T, Papadakis E (2012) Improved production of carotenes from synthetic medium by *Blakeslea trispora* in a bubble column reactor. Biochem Eng J 67:203–207
46. Naqvi S, Zhu C, Farrel G, Ramessar K, Bassie L, Breitenbach J, Perez Conesa D, Ros G, Sandmann G, Capell T, Christou P (2009) Transgenic multivitamin corn: Biofortification of corn endosperm with three vitamins representing three distinct metabolic pathways. Proc Natl Acad Sci USA 106:7762–7767
47. Nonomura AM (1989) In: Krinsky NI, Mathews-Roth MM, Taylor RF (eds) Industrial biosynthesis of carotenoids. Carotenoids: chemistry and biology, Plenum Press, New York, pp 365–375
48. Ooi CK, Choo YM, Yap SC, Basiron Y, Ong ASH (1994) Recovery of carotenoids from palm oil. J Americ Oil Chem Soc 71:423–426
49. Paine JA, Shipton CA, Chaggar S, Howells RM, Kennedy MJ, Vernon G, Wright SY, Hinchliffe E, Adams JL, Silverstone AL, Drake R (2005) Improving the nutritional value of Golden Rice through increased pro-vitamin A content. Nat Biotechnol 23:482–487

50. Perry A, Rasmussen H, Johnson EJ (2009) Xanthophyll (lutein, zeaxanthin) content in fruits, vegetables and corn and egg products. J Food Compos Anal 22:9–15
51. Piccaglia R, Marotti M, Grandi S (1998) Lutein and lutein ester content in different types of *Tagetes patula* and *T. erecta*. Industr Crops Prod 8:45–51
52. Raisig A, Sandmann G (2001) Functional properties of diapophytoene and related desaturases of C30 and C40 carotenoid biosynthetic pathways. Biochim Biophys Acta 1533:164–170
53. Raja R, Hemaiswarya S, Rengasamy R (2007) Exploitation of *Dunaliella* for β-carotene production. Appl Microbiol Biotechnol 74:517–523
54. Reyes LH, Gomez JM, Kao KC (2014) Improving carotenoids production in yeast via adaptive laboratory evolution. Metabolic Engineering 21:26–33
55. Rodríguez-Sáiz M, de la Fuente JL, Barredo JL (2010) *Xanthophyllomyces dendrorhous* for the industrial production of astaxanthin. Appl Microbiol Biotechnol 88:645–658
56. Römer S, Lübeck J, Kauder F, Steiger S, Adomat C, Sandmann G (2002) Genetic engineering of a zeaxanthin-rich potato by antisense inactivation and co-suppression of carotenoid epoxidation. Metab Eng 4:263–272
57. SanGiovanni JP, Neuringer M (2012) The putative role of lutein and zeaxanthin as protective agents against age-related macular degeneration: Promise of molecular genetics for guiding mechanistic and translational research in the field. Am J Clin Nutr 96:1223S–1233S
58. Sandmann G (2001) Carotenoid biosynthesis and biotechnological application. Arch Biochem Biophys 385:4–12
59. Sandmann G (2001) Genetic manipulation of carotenoid biosynthesis: strategies, problems and achievements. Trend Plant Sci 6:14–17
60. Sandmann G, Albrecht M, Schnurr G, Knörzer P, Böger P (1999) The biotechnological potential and design of novel carotenoids by gene combination in *Escherichia coli*. Trends Biotechnol 17:233–237
61. Sandmann G, Römer S, Fraser PD (2006) Understanding carotenoid metabolism as a necessity for genetic engineering of crop plants. Metabolic Engin 8:291–302
62. Sandmann G, Misawa N (2002) In: Osiewacz HD (ed) Fungal carotenoids. The Mycota X: industrial application. Springer Verlag Berlin, pp 247–262
63. Schaub P, Al-Babili S, Drake R, Beyer P (2005) Why is golden rice golden (yellow) instead of red? Plant Physiol 138:441–450
64. Schmidt I, Schewe H, Gassel S, Jin C, Buckingham J, Hümbelin M, Sandmann G, Schrader J (2010) Biotechnological production of astaxanthin with *Phaffia rhodozyma/Xanthophyllomyces dendrorhous*. Appl Microbiol Biotechnol 89:555–571
65. Shewmaker CK, Sheehy JA, Daley M, Colburn S, Ke DY (1999) Seed-specific overexpression of phytoene synthase: increase in carotenoids and other metabolic effects. Plant J 20:401–412
66. Shimada H, Kondo K, Fraser PD, Miura Y, Saito T, Misawa N (1998) Increased carotenoid production by the food yeast *Candida utilis* through metabolic engineering of the isoprenoid pathway. Appl Environ Microbiol 64:2676–2680
67. Tyczkowski JK, Hamilton PB (1986) Absorption, transport, and deposition in chickens of lutein diester, a carotenoid extracted from Marigold (Tagetes erecta) petals. Poult Sci 65: 1526–1531
68. Verdoes CJ, Sandmann G, Visser H, Diaz M, van Mossel M, van Ooyen AJJ (2003) Metabolic engineering of the carotenoid biosynthetic pathway in the yeast *Xanthophyllomyces dendrorhous* (*Phaffia rhodozyma*). Appl Environm Microbiol 69:3728–3738
69. Verwaal R, Wang J, Meijnen JP, Visser H, Sandmann G, van den Berg JA, van Ooyen AJ (2007) High level production of beta-carotene in *Saccharomyces cerevisiae* by successive transformation with carotenogenic genes from *Xanthophyllomyces dendrorhous*. Appl Environ Microbiol 73:4342–4350
70. Xie W, Liu M, Lv X, Lu W, Gu J, Yu H (2014) Construction of a controllable β-carotene biosynthetic pathway by decentralized assembly strategy in *Saccharomyces cerevisiae*. Biotechnol Bioeng 111:125–133
71. Xiong L, Zhu JK (2003) Regulation of abscisic acid biosynthesis. Plant Physiol 133:29–36

72. Yamano S, Ishii T, Nakagawa M, Ikenaga H, Misawa N (1994) Metabolic engineering for production of beta-carotene and lycopene in Saccharomyces cerevisiae. Biosci Biotechnol Biochem 58:1112–1114
73. Ye X, Al-Babili S, Klöti A, Zhang J, Lucca P, Beyer P, Potrykus I (2000) Engineering the provitamin A (beta-carotene) biosynthetic pathway into (carotenoid-free) rice endosperm. Science 287:303–305
74. Zhu C, Naqvi S, Breitenbach J, Sandmann G, Christou P, Capell T (2008) Combinatorial genetic transformation generates a library of metabolic phenotypes for the carotenoid pathway in maize. Proc Natl Acad Sci USA, 105:18232–18237

Adv Biochem Eng Biotechnol (2015) 148: 469–470
DOI 10.1007/10_2015_310

Published Online: 27 March 2015

Erratum to: Isoprenoid Drugs, Biofuels, and Chemicals—Artemisinin, Farnesene, and Beyond

Kevin W. George, Jorge Alonso-Gutierrez, Jay D. Keasling and Taek Soon Lee

Erratum to: Adv Biochem Eng Biotechnol, DOI: 10.1007/10_2014_288

We missed to include this line in this chapter "K.W. George and J. Alonso-Gutierrez contributed equally to this paper".

The online version of the original chapter can be found under
DOI 10.1007/10_2014_288

K.W. George · J. Alonso-Gutierrez · J.D. Keasling · T.S. Lee (✉)
Joint BioEnergy Institute, 5885 Hollis St. 4th floor, Emeryville, CA 94608, USA
e-mail: tslee@lbl.gov

K.W. George · J. Alonso-Gutierrez · J.D. Keasling · T.S. Lee
Physical Biosciences Division, Lawrence Berkeley National Laboratory, Berkeley, CA, USA

J.D. Keasling
Department of Bioengineering, University of California, Berkeley, CA, USA

J.D. Keasling
Department of Chemical and Biomolecular Engineering, University of California, Berkeley, CA, USA

Adv Biochem Eng Biotechnol (2015) 148: 471–476
DOI 10.1007/978-3-319-20107-8

Index

The manufacturer's authorised representative in the EU is Springer Nature Customer Service Centre GmbH, Europaplatz 3, 69115 Heidelberg, Germany. If you have any concerns regarding our products, please contact ProductSafety@springernature.com

Printed and bound by CPI Group (UK) Ltd, Croydon, CR0 4YY
22/09/2026
02221052-0004